The Sociology of Health, Illness, and Health Care

The Sociology of Health, Illness, and Health Care

A Critical Approach   FOURTH EDITION

Rose Weitz

Arizona State University

THOMSON

WADSWORTH

Australia • Brazil • Canada • Mexico • Singapore • Spain
United Kingdom • United States

The Sociology of Health, Illness,
and Health Care: A Critical Approach, **Fourth Edition**
Rose Weitz

Senior Acqusitions Editor: Robert Jucha
Assistant Editor: Kristin Marrs
Editorial Assistant: Katia Krukowski
Technology Project Manager: Dee Dee Zobian
Marketing Manager: Michelle Williams
Marketing Assistant: Jaren Boland
Marketing Communications Manager: Linda Yip
Project Manager, Editorial Production:
 Christine Sosa
Creative Director: Rob Hugel
Print Buyer: Judy Inouye
Permissions Editor: Roberta Broyer

Production Service: Mona Tiwary,
 International Typesetting and Composition
Photo Researcher: Terri Wright
Copy Editor: Chris Thillen
Illustrator: International Typesetting
 and Composition
Cover Designer: William Stanton
Cover Image: Doable/Getty Images
Cover Printer: R.R. Donnelley/Crawfordsville
Compositor: International Typesetting
 and Composition
Printer: R.R. Donnelley/Crawfordsville

Library of Congress Control Number:
2005937325

ISBN 0-495-17203-0

Thomson Higher Education
10 Davis Drive
Belmont, CA 94002-3098
USA

For more information about our products,
contact us at:
Thomson Learning Academic Resource Center
1-800-423-0563

For permission to use material from this text or
product, submit a request online at
http://www.thomsonrights.com.
Any additional questions about permissions
can be submitted by e-mail to
thomsonrights@thomson.com.

To Miriam and Lara, and in memory of Marty.

ABOUT THE AUTHOR

Rose Weitz received her doctoral degree from Yale University in 1978. Since then, she has carved an exceptional record as both a scholar and a teacher. She is the author of numerous scholarly articles, the book *Life with AIDS* (Rutgers, 1991), and the book *Rapunzel's Daughters: What Women's Hair Tells Us About Women's Lives* (Farrar, Straus and Giroux, 2004). She also is coauthor of *Labor Pains: Modern Midwives and Home Birth* (Yale University Press, 1988) and editor of *The Politics of Women's Bodies: Appearance, Sexuality, and Behavior* (Oxford University Press, 1998). Professor Weitz has won two major teaching awards at Arizona State University as well as the Pacific Sociological Association's Distinguished Contributions to Teaching Award, and she has been a finalist for other teaching awards numerous times.

BRIEF CONTENTS

CONTENTS

3 The Social Distribution of Illness in the United States 55

PART 3   Health Care Systems, Settings, and Technologies   223

8   The U.S. Health Care System and the Need for Reform   225

9   Alternative Health Care Systems   257

During the last quarter century, the sociology of health, illness, and health care has changed dramatically. Begun primarily by sociologists who worked closely with doctors, taking for granted doctors' assumptions about health and health care and primarily asking questions that doctors deemed important, the field has shifted toward asking a very different set of questions. Some of these new questions have challenged doctors' assumptions, whereas others have focused on issues that lie outside most doctors' areas of interest or expertise, such as whether increases in income inequality affect a nation's average life expectancy or how individuals develop meaningful lives despite chronic illness.

I entered graduate school during this shift, drawn by the prospects of studying how health and illness are socially created and defined and how gender, ethnicity, social class, and, more broadly, power affect both the health care system and individual experiences of health and illness. As a result, over the years I have researched such topics as how medical values affect doctors' use of genetic testing, how midwives and doctors have battled for control of childbirth, and how social ideas about AIDS affect the lives of those who have this disease.

Although I had no trouble incorporating the new vision of the sociology of health, illness, and health care into my research, I consistently found myself frustrated by the lack of a textbook that would help me incorporate it into my teaching. Instead, most textbooks still seemed to reflect older ideas about the field and to take for granted medical definitions of the situation. Most basically, the books assumed that doctors define illness according to objective biological criteria and so failed to examine how social forces affect illness definitions. Similarly, most textbooks ignored power relationships rather than investigating the sources, nature, and health consequences of those relationships. For example, the textbooks gave relatively little attention to how doctors gained control over health care, how politicians and corporations in industrialized nations affect the health of people living in developing nations, and how individuals' race, gender, and social class affect the health care they receive. As a result, the available textbooks used sociology primarily to answer questions posed by those working in the health care field, such as what social factors lead to heart disease and why patients might ignore their doctors' orders. These textbooks often seemed to offer a surprisingly unsociological perspective, with their coverage of some topics differing in only minor ways from that found in health education textbooks.

Similarly, the existing textbooks did not press students to question their own assumptions about health and health care. For example, American students often assume that the United States has the world's best health care system, that medical advances explain the modern rise in life expectancy, or that all Americans receive the same quality of health care regardless of their ethnicity, gender, or social class. These were the sort of assumptions I believed a textbook on sociology of health should challenge. In addition, I believed that a textbook could do so successfully only if it encouraged students to engage with the materials, rather than to memorize already-processed information.

My purpose in writing this textbook was to fill these gaps by presenting a critical approach to the sociology of health, illness, and health care. This did not mean presenting research findings in a biased fashion or presenting only research that supports my preexisting assumptions, but it did mean bringing critical skills to bear in evaluating research findings and pulling these findings together into a coherent "story" in each chapter. In addition, I hoped to tell these stories in a manner that would engage students— whether in sociology classes, medical schools, or nursing schools—and stimulate students to learn actively and think independently. These remain the primary goals of this fourth edition. Both goals led me to decide not to try to please all sides or cover all possible topics, because I believe such a strategy leads both to the intellectual homogenization that makes many textbooks seem lifeless, and to the grab-bag approach that makes them hard to follow.

The Critical Approach

The critical approach, as I have defined it, means using the "sociological imagination" to question previously taken-for-granted aspects of social life. For example, most of the available textbooks in the sociology of health, illness, and health care in essence have examined the issue of patients who do not comply with prescribed medical regimens through doctors' eyes, starting from the assumption that patients should do so. More broadly, previous textbooks have highlighted the concept of a sick role—a concept that embodies medical and social assumptions regarding "proper" illnesses and "proper" patients and that downplays all aspects of individuals' lives other than the time they spend as patients.

In contrast, I emphasize recent research that questions all such assumptions. For example, I discuss patient compliance by examining recent research about how patients view medical regimens and compliance, why doctors sometimes have promoted medical regimens and procedures that later proved dangerous (such as prescribing hormone replacement therapy for almost all postmenopausal women), and how doctors' tendency to cut short patients' questions can reduce patient compliance. Similarly, this textbook explains the concept of a sick role but pays more attention to the broader experience of illness—a topic that has generated far more sociological research than has the sick role in the last decade.

Coverage

Although I have tried in this book to present a coherent critical view, I have not sacrificed coverage of topics teachers have come to expect. Consequently, this book covers essentially all the topics—both micro level and macro level—that have become standard over the years, including doctor-patient relationships, the nature of the U.S. health care system, and the social distribution of illness. In addition, I include several topics that usually receive little coverage, including bioethics, mental illness, the medical value system, the experience of illness and disability, and the social sources of illness in both the developing and industrialized nations. As a result, this text includes more materials than most teachers can cover effectively in a semester. To assist those who choose to skip some chapters, each important term is printed in bold and defined the first time it appears in the text and is printed in bold without a definition the first time it appears in each subsequent chapter, alerting students that they can find a definition in the book's glossary.

In addition, reflecting my belief that sociology neither can nor should exist in isolation but must be informed by and in turn inform other related fields, each chapter includes a historical overview. For example, the chapter on health care institutions discusses the political and social forces that led to the development of the modern hospital, the chapter on medicine as a profession discusses how and why the status of medicine grew so dramatically after 1850, and the chapter on the meaning of illness discusses how people throughout history have explained and responded to illness and ill persons. These discussions provide a context to help students understand the current status of, respectively, hospitals, doctors, and ill persons.

Changes in the Fourth Edition
New and Expanded Topics

- Intersex (Chapter 2)
- The impact of sex and gender on health (Chapter 2)
- The impact of obesity (Chapter 2)
- The social causes of obesity (Chapter 2)
- Medical errors as causes of death (Chapter 2)

- Social stress and illness (Chapters 2 and 7)
- Health behavior and health lifestyles (Chapter 2)
- Chronic pain (Chapter 6)
- Health social movements (Chapter 6)
- The international traffic in human organs (Chapter 6)
- Why the United States lacks a national health care system (Chapter 8)
- The pharmaceutical industry (Chapter 8)
- The backlash against managed care (Chapter 8)
- Health technologies (Chapter 10)
- Commercial institutional review boards, or IRBs (Chapter 13)
- Community research advisory boards (Chapter 13)
- The Terri Schiavo case (Chapter 13)

Up-to-Date Coverage of All Topics

Throughout the textbook, I have thoroughly updated not only statistics and discussions of topical issues (like health care reform, the Internet, and managed care) but also all reviews of the theoretical and empirical literature. As a result, most references in this new edition are from 1995 or later, with about one-third from 2000 or later, and the reader can assume that all statistics are the latest available.

Pedagogical Features

Chapter Openings

Unfortunately, many students take courses only to fill a requirement. As a result, the first problem teachers face is that of interesting students in the topic. For this reason, each chapter opens with a vignette taken from a sociological or literary source and chosen to spark students' interest in the topic by demonstrating that the topic has real consequences for real people— that, for example, stigma is not simply an abstract concept but something that can cost ill persons their friends, jobs, and social standing.

Ethical Debates

To teach students that ethical dilemmas pervade health care, each chapter includes a discussion of a relevant ethical debate. The debates are complex enough that students must use critical-thinking skills to assess them; teachers can use these debates for classroom discussions, group exercises, or written assignments.

Key Concepts

To help students understand particularly important and complex topics, such as the difference between the sociological and medical models of illness, or the strengths and weaknesses of the sick role model, I have included "Key Concepts" tables in several chapters.

Chapter Conclusions

Each chapter in this textbook ends not with a summary that reiterates the materials, but with conclusions that discuss the implications of the chapter and point the reader toward new questions and issues. These conclusions should stimulate critical thinking rather than rote memorization and can serve as the basis for class discussions.

Student Aids

Each chapter includes study questions and suggestions for further reading. The book also includes an extensive bibliography and a glossary that defines all important terms.

In addition, and in the hopes that this book will mark the beginning and not the end of students' interest in the field, each chapter includes a description of nonprofit, activist organizations that can provide students with both more information and opportunities for personal involvement. Internet addresses as well as phone numbers and street addresses are provided for all organizations.

Internet Features

Each chapter contains Internet exercises for students and Internet addresses for all nonprofit organizations described in that chapter. Readers of this textbook also have access to the wide variety of tools and resources available at Wadsworth's sociology website (http://www.thomsonedu.com/sociology). This textbook has its own web page at that site, which contains updated web links for all Internet sites discussed in the textbook as well as links to other health websites that might be useful for students and instructors. Finally, when ordering this textbook, professors can request that their students receive free access via the Wadsworth website to InfoTrac® College Edition, an online archive offering full-text versions of hundreds of scholarly articles, many on health-related topics.

"Making a Difference" Boxes

To help students see how sociological knowledge can translate into effective social action, each chapter includes boxes describing the work of nonprofit organizations that are using sociological insights to "make a difference" in health and health care. For the same reason, I have woven throughout the book descriptions of positive changes that have occurred in recent years in health and health care, such as the rise of more humanistic training in medical schools.

Instructor's Manual with Test Bank

For each chapter, the *Instructor's Manual with Test Bank* contains a detailed summary, a set of multiple-choice questions, and a list of relevant narrative and documentary films. In addition, the manual includes several questions for each chapter that require critical-thinking skills to answer and that teachers can use for essay exams, written assignments, in-class discussions, or group projects. To guarantee the quality of the *Instructor's Manual with Test Bank,* I wrote everything in it rather than relying on student assistants. The manual is available for downloading at the Instructor's web page at this book's website (http://www.thomsonedu.com/sociology). Or, you can also contact your local Wadsworth representative.

Critical Thinking

In this textbook, I have aimed not only to present a large body of data in a coherent fashion but also to create an intellectually rigorous textbook that will stimulate students to think critically. I have tried to keep this purpose in mind in writing each chapter. The chapter conclusions, ethical debates, and essay questions all encourage students to use critical thinking.

PowerPoint Lectures

Beginning with this edition, Microsoft PowerPoint lectures for each chapter, including all tables and figures, can be downloaded from this textbook's website. These lectures should prove useful both for new adopters and for past users who would like to incorporate more visual materials into their classrooms.

Acknowledgments

In writing this textbook, I have benefited enormously from the generous assistance of my colleagues. I am very fortunate to work with several exceptional scholars here at Arizona State University—Victor Agadjanian, Verna Keith, Bradford Kirkman-Liff, Jennie Jacobs Kronenfeld, and Deborah Sullivan—who share my interest in health issues and who critiqued chapters for me. I am also exceptionally fortunate to have had the assistance of several research assistants—Melinda Konicke, Christopher Lisowski, Diane Sicotte, Caroleena Von Trapp, Sarah St. John, Ashley Fenzl, Ann Jensby, Leslie Padrnos, Stephanie Mayer, and, especially, Karl Bryant, Lisa Comer, and Amy Weinberg, who worked on the first edition.

Because, of necessity, this textbook covers a wealth of topics that range far broader than my own areas of expertise, I have had to rely heavily on the kindness of strangers in writing it. One of the most rewarding aspects of writing this book has been the pleasure of receiving information, ideas, critiques, and references from individuals I did not previously know and who, subsequently, have become friends as well as colleagues. My thanks to the many new and old friends who gave me the benefit of their expert advice: Emily Abel (University of California–Los Angeles), James Akré (World Health Organization), Ofra Anson (Ben Gurion University of the Negev), Judy Aulette (University of North Carolina, Charlotte), Miriam Axelrod,

James Bachman (Valparaiso University), Paul Basch (Stanford University), Phil Brown (Brown University), Peter Conrad (Brandeis University), Timothy Diamond (California State University–Los Angeles), Luis Durán (Mexican Institute of Social Security), Michael Farrall (Creighton University), Jill Fisher and the late Kitty Felker (Arizona State University), Arthur Frank (University of Alberta), Frederic W. Hafferty (University of Minnesota–Duluth), Harlan Hahn (University of Southern California), Paul Higgins (University of South Carolina), Allan Horwitz (Rutgers University), David J. Hunter (University of Durham), Michael Johnston (University of California–Los Angeles), Stephen J. Kunitz (University of Rochester), Donald W. Light (University of Medicine and Dentistry of New Jersey), Judith Lorber (City University of New York), William Magee (University of Toronto), Judy Mayo, Peggy McDonough (York University), Jack Meyer (Economic & Social Research Institute), Cindy Miller, Jeanine Mount (University of Wisconsin), Marilynn M. Rosenthal (University of Michigan), Beth Rushing (Kent State University),Wendy Simonds (Georgia State University), Clemencia Vargas (Centers for Disease Control and Prevention), Robert Weaver and his students—especially Cheryl Kratzer (Youngstown State University), Daniel Whitaker, David R. Williams (University of Michigan), Irving Kenneth Zola (Brandeis University), and Robert Zussman (University of Massachusetts–Amherst). This book undoubtedly would have been better if I had paid closer attention to their comments. I apologize sincerely if I have left anyone off this list.

This book also has been substantially improved by the suggestions of the reviewers to the various editions: Thomas E. Allen, Jr. (University of South Dakota), Karen Bettez (Boston College), Linda Liska Belgrave (University of Miami), Karen Frederick (St. Anselm College), Stephen Glazier (University of Nebraska), Linda Grant (University of Georgia), Janet Hankin (Wayne State University), Heather Hartley (Portland State University), Alan Henderson (California State University–Long Beach), Frances Hoffman (University of Missouri), Lilly M. Langer (Florida International University), Larry R. Ridener (Pfeiffer University), Gary Tiedman (Oregon State University), Diana Torrez (University of North Texas), Robert Weaver (Youngstown State University), and Diane Zablotsky (University of North Carolina–Charlotte). Finally, I would like to express my appreciation to Serina Beauparlant, Susan Shook, Eve Howard, and Robert Jucha, the Wadsworth editors who made the four editions of this book possible.

The Sociology of Health, Illness, and Health Care

1 Social Factors and Illness

Illness is a fact of life. Everyone experiences illness sooner or later, and everyone eventually must cope with illness among close friends and relatives.

To the ill individual, illness can seem a purely internal and personal experience. But illness is also a social phenomenon, with social roots and social consequences. In this first part, I demonstrate the role social factors play in fostering illness within societies and in determining which groups in a given society will experience which illnesses with which consequences.

Chapter 1 introduces the sociological perspective and illustrates how sociology can help us understand issues related to health, illness, and health care. In the subsequent chapters, I discuss the role social forces play in causing illness and in determining who gets ill. Chapter 2 provides a brief history of illness in the Western world; I describe how patterns of illness have changed over time and assess the relative roles of social factors and medical advances in those changes. I then look at the social sources of illness in the contemporary United States. In Chapter 3, I investigate how four social factors—age, sex and gender, social class, and race or ethnicity—affect the distribution of illness in the United States and explore why some social groups bear a greater burden of illness than others. Finally, in Chapter 4, I describe the very different pattern of illnesses found in poorer countries and how social forces—from the low status of women to the rise of migrant labor—foster illness in these countries.

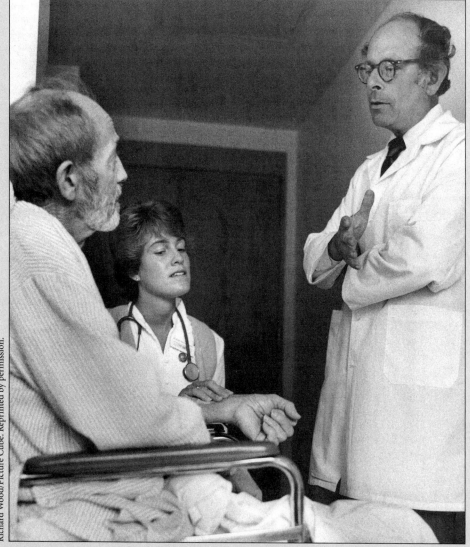

Introduction

In 1996, at the age of 46, my friend Lara learned she had breast cancer. Once her doctor concluded from Lara's mammogram (a form of X-ray) that a lump in her breast seemed cancerous, events followed in quick succession. The next day, a surgeon removed a piece of the suspicious lump for testing. A few days later, Lara learned that the lump was cancerous. That week, she got her affairs in order and signed a "living will" specifying the circumstances in which she would want all treatment stopped and a "medical power of attorney" giving me legal authority to make medical decisions for her if she physically could not do so herself. These two documents, she hoped, would protect her from aggressive medical treatments that might prolong her suffering without improving her quality of life or chances of survival.

Two weeks after the initial tests, her surgeon removed the rest of the lump as well as the lymph nodes under her arm (where breast cancer most often spreads). The surgery went well, but the subsequent laboratory tests showed that the cancer indeed had spread to some of Lara's lymph nodes.

Yet in many ways, Lara was fortunate. Her breast cancer was detected at a relatively early stage, giving her about a 65 percent chance of surviving for at least five years. Although she had no husband or children to turn to, her friends proved uniformly supportive. She received health insurance through her employer and had no fears of losing either her job or her insurance.

Nevertheless, cancer changed Lara's life irrevocably, making it, at times, a nightmare. Having breast cancer shook Lara's faith in her body and changed her sense of her physical self. At the same time, her illness threatened her relationships with others. Despite the supportive responses she received from friends and co-workers, she nevertheless feared they would drift away as her illness continued or that she would chase them away with her all-too-reasonable complaints, worries, and needs.

Although she had far better health insurance than many Americans have, her debts for items not covered by insurance nonetheless mounted. In addition, she had to spend hours fighting her insurance company to obtain relaxation training and expensive but effective antinausea drugs to cope with chemotherapy's side effects. Yet the chemotherapy made her so ill she often found it difficult to function, let alone fight her insurance company. In addition, chemotherapy proved so toxic that it damaged her veins with each painful intravenous treatment. As a result, her doctors suggested inserting a semipermanent plastic tube into her chest wall so they could instead administer the chemotherapy through the tube. Although doing so would have reduced her pain, Lara rejected the suggestion because she felt that, with this sign of her illness physically attached to her body at all times, she would truly become a cancer patient, rather than someone for whom cancer was merely one part of her life.

After a year of surgery, chemotherapy, and radiation, Lara's physical traumas ended, although it took another year before she regained her former energy. Although her future remains uncertain, to date she is free of any signs of cancer.

The Sociology of Health, Illness, and Health Care: An Overview

Lara's story demonstrates the diverse ways illness affects individuals' lives. It also demonstrates the diverse range of topics that sociologists of health, illness, and health care can study. First, sociologists can study how social forces promote health and illness and why some social groups suffer more illness than others do. For example, researchers have explored whether working conditions in U.S. factories help explain why poorer Americans get certain cancers more often than wealthier Americans do. Similarly, sociologists can study how historical changes in patterns of social life can explain changes in patterns of illness. To understand why rates of breast cancer have increased, some researchers have studied the impact of women's changing social roles, and others have studied the impact of political forces that promote increased meat consumption. Second, instead of studying broad patterns of illness, sociologists can study the experiences of those, like Lara, who live with illness on a day-to-day basis—exploring, for example, how illness affects individuals' sense of identity, relationships with family, or ideas about what causes illness. Third, sociologists study how social factors affect health care providers. Some sociologists have analyzed how the status and power of different occupations have shifted over time, and others have investigated how power affects interactions between health care occupations (such as between doctors and nurses). Still others have examined interactions between health care workers and patients, asking, for example, how doctors maintain control in discussions with patients or whether doctors treat male and female patients differently. Finally, sociologists can analyze the health care system as a whole. Sociologists have examined how health care systems have developed, compared the strengths and weaknesses of different systems, and explored how systems can be improved. For example, some have studied how U.S. health insurance companies can make it difficult for people like Lara to get needed care, explored why European countries do better than the United States at providing health care to all who need it, and examined whether European health care policies could work in the United States.

The topics researched by sociologists of health, illness, and health care overlap in many ways with those studied by health psychologists, medical anthropologists, public health workers, and others. What most clearly differentiates sociologists from these other researchers is the **sociological perspective.** The next section describes that perspective.

The Sociological Perspective

Using a sociological perspective means focusing on social patterns rather than on individual behaviors. Whereas a psychologist might help a battered wife develop a greater sense of her own self-worth so she might eventually leave her abusive husband, a sociologist likely would consider therapy a useful but inefficient means of addressing the root causes of wife abuse. Most battered wives, after all, do not have the time, money, or freedom to get help from psychologists. Moreover, even when therapy helps, it takes place only after the women have experienced physical and emotional damage. The sociologist would not deny that individual personalities play a role in wife battering, but instead finds it more useful to explore whether social forces can explain why wife battering is much more common than husband battering, or why battered wives so often remain with abusive husbands. Consequently, whereas the psychologist hopes to enable the individual battered wife eventually to leave her husband, the sociologist hopes to uncover the knowledge needed by legislators, social workers, activists, and others to prevent wife abuse in the first place.

As this example demonstrates, using the sociological perspective means framing problems as *public issues,* rather than simply *personal troubles.* According to C. Wright Mills (1959: 8–9), the sociologist who first drew attention to this dichotomy:

> [*Personal*] *troubles* occur within the character of the individual and within the range of his immediate relations with others; they have to do with his self and with those limited areas of social life of which he is directly and personally aware. Accordingly, the statements and the resolutions of troubles properly lie within the individual as a biographical entity and within the scope of his immediate

milieu. . . . *[Public] issues* have to do with matters that transcend these local environments of the individual and the range of his inner life. They have to do with the organization of many such milieux into the institutions of an historical society as a whole.

For example, whenever a child dies from leukemia, it is a tragedy and a personal trouble for the child's family. If, on the other hand, several children in a neighborhood die of leukemia during the same year, it could suggest a broader public issue such as toxic contamination of the neighborhood water system. A sociologist would be likely to look for such a pattern, and to explore why, for example, polluting industries are more likely to build factories in poor, minority neighborhoods than in affluent, white neighborhoods. The sociological perspective, then, departs radically from the popular American belief that individuals create their own fates and that anyone can succeed if he or she tries hard enough.

The sociological perspective can help us identify critical research questions that might otherwise go unasked. For example, in the book *Forgive and Remember: Managing Medical Failure,* sociologist Charles Bosk (2003: 62–63) described a situation he observed one day on "rounds," the time each day when recently graduated doctors (known as residents) and more senior doctors jointly examine the patients on a service, or ward:

> Dr. Arthur [the senior doctor] was examining the incision [surgical cut] of Mrs. Anders, a young woman who had just received her second mastectomy. After reassuring her that everything was fine, everyone left her room. We walked a bit down the hall and Arthur exploded: "That wound looks like a walking piece of dogshit. We don't close wounds with continuous suture on this service. We worked for hours giving this lady the best possible operation and then you screw it up on the closure. That's not how we close wounds on this service, do you understand? These are the fine points that separate good surgeons from butchers, and that's what you are here to learn. I never want to see another wound closed like that. Never!" Arthur then was silent, he walked a few feet, and then he began speaking again: "I don't give a shit how Dr. Henry [another senior doctor] does it on the Charlie Service or how Dr. Gray does it on Dogface; when you're on my service, you'll do it the way I want."

Dr. Arthur and the residents he supervised undoubtedly viewed this situation as a personal trouble, requiring a personal solution—the residents seeking to appease Dr. Arthur, and Dr. Arthur seeking to intimidate and shame the residents into doing things the way he considered best. Similarly, depending on their viewpoint, most nonsociological observers probably would view this as a story about either careless residents or an autocratic senior doctor. Sociologists, however, would first ask whether such interactions among doctors occur often. If they do, sociologists then would look for the social patterns underlying such interactions, rather than focusing on the personalities of these particular individuals. So, for example, based on his observations in this and other cases, Bosk discovered that cultural expectations within the medical world regarding authority, medical errors, and the importance of personal, surgical experience had enabled Dr. Arthur and the other supervising doctors to humiliate residents publicly and to set policies based more on personal preferences than on scientific data.

Whereas Bosk investigated health issues within hospitals, David Kirp used a sociological perspective to explore health issues in the community. For the book *Learning by Heart: AIDS and Schoolchildren in America's Communities,* Kirp (1989) observed half a dozen communities around the country to determine why they responded in such different ways to the presence of schoolchildren who had acquired immunodeficiency syndrome (AIDS). The following events occurred relatively early in the history of the AIDS epidemic, in Swansea, Massachusetts:

> For the members of the Swansea, Massachusetts, school committee, habituated to brief and sparsely attended bimonthly meetings in the century-old red brick administration building, the evening of September 11, 1985, was an eye-opener. More than seven hundred people, almost all parents, filled the high school auditorium, the biggest meeting place in town.
>
> The people of Swansea are usually polite in their dealings with one another, but these parents were in no mood for good manners. They demanded to know why their superintendent and their school committee had acted differently than every other school official in the entire country. Why had they allowed a thirteen-year-old boy with AIDS—a boy named Mark, known and liked by many of the people, but now fatally tainted in their eyes—to remain in school?
>
> Why, the parents asked, had people they trusted—a school committee they had elected, most of whose members were natives of Swansea, and a superintendent who had been a fixture in their schools for nearly three decades—exposed their children to the bizarre terror of AIDS? (Kirp, 1989: 16–17)

As in the case of Dr. Arthur and his residents, we could view the Swansea furor simply as an isolated event caused, depending on one's viewpoint, by either an unthinking and arrogant school board or uneducated and heartless parents. Probably the school board and the parents saw the problem in these terms and therefore focused, respectively, on calming the parents or overturning the school board's action. By looking at the variety of ways communities responded to the presence of schoolchildren with AIDS, however, Kirp was able to identify a different set of issues—politics, power, and stereotypes—and of causal factors, such as how the media fostered fears and how popular beliefs about the meaning of illness bred bigotry against ill persons.

In sum, the sociological perspective shifts our focus from individuals to social groups and institutions. One effect of this shift is to highlight the role of power. *Power* refers to the ability to get others to do what one wants, whether willingly or unwillingly. Power is what allowed Dr. Arthur to treat his residents so rudely and allowed some school boards to override the wishes of their communities. Because sociologists study groups rather than individuals, the sociological analysis of power focuses on why some social groups have more power than others, how groups use their power, and the consequences of *differential* access to power (i.e., some having more than others), rather than on how specific individuals get or use power. For example, sociologists have examined why doctors as a group proved more successful than nurses did in obtaining the power to control their working conditions and how recent changes in the health care system have limited doctors' power. Similarly, sociologists have explored how *lack* of power exposes poor persons and disadvantaged minorities to conditions that promote ill health, while limiting their access to health care.

A Critical Approach

Although the concept of power underlies the sociological perspective, some sociologists do not emphasize power in their research and writing. Instead, some sociologists essentially take for granted the way power is distributed in our society, examining the current system without questioning why it is this way or how it might be changed. For example, some sociologists have investigated whether lower-class persons are more likely than upper-class persons are to suffer mental illness without first questioning whether definitions of mental illness might reflect an upper-class perspective regarding socially acceptable behaviors or whether the same behaviors might more likely be defined as symptoms of mental illness when performed by lower-class persons.

Those sociologists, on the other hand, who do *not* take for granted existing power relationships and who instead focus on the sources, nature, and consequences of power relationships can be said to use a critical approach. Critical sociologists recognize that, regardless of how power is measured, men typically have more power than do women, adults more power than do children, whites more power than do African Americans, heterosexuals more power than do gays and lesbians, persons with socially acceptable bodies more power than do persons who are disabled, and so on. Critical sociologists who study health, illness, and health care have raised such questions as how this differential access to power affects the likelihood that members of a social group will be exposed to illness-producing conditions or will have access to quality health care.

Critical sociologists also emphasize how social institutions and popular beliefs can support or reflect existing power relationships. For example, many researchers who study the U.S. health care system have looked simply for ways to improve access to care or quality of care within that system, such as offering poor people subsidized health insurance or providing financial incentives to doctors who practice in low-income neighborhoods. Those who use a critical approach have asked instead whether we could provide better care to more people if we changed the basic structure of the system, such as by removing the profit motive from health care to reduce the costs of care for everyone.

Similarly, critical sociologists have drawn attention to how doctors' power and authority enable them to frame our ideas about health, illness, and health care. Most basically, these sociologists have questioned the very terms *health, illness,* and *disability* and have explored whether such terms reflect social values more than they reflect objectively measurable physical characteristics.

In any sociological field, therefore, those who adopt a critical approach will ask quite different research questions than will others. Within the sociology of health, illness, and health care, this approach translates largely to whether sociologists limit their research to questions about social life that doctors consider useful—a strategy referred to as **sociology *in* medicine**— or design their research to answer questions of interest to sociologists in general—a strategy referred to as the **sociology *of* medicine** (Straus, 1957). Research using the latter strategy often challenges both medical views of the world and existing power relationships within health care.

To understand the difference between sociology in medicine and sociology of medicine, consider the sociological literature on patients who do not follow their doctors' advice. Because doctors typically define such patients as problems, over the years many sociologists, accepting medical ideas regarding what questions need asking, have sought to determine how to "bring patients to their senses" and increase their compliance with medical advice. In contrast, sociologists *of* medicine have looked at the issue of compliance through patients' eyes. As a result, they have learned that patients sometimes ignore medical advice not out of stubbornness or foolishness but because their doctors have not explained clearly either how to follow the prescribed regimens or why they should do so. In other circumstances, patients have ignored medical advice because they have concluded rationally that the emotional or financial costs of doing so outweigh the potential medical benefits. Similarly, whereas those practicing sociology *in* medicine have studied various aspects of the experience of *patienthood,* those practicing sociology *of* medicine instead have studied the broader experience of *illness,* which includes but is not limited to the experience of patienthood. The growing emphasis on sociology of medicine and on the critical approach has led to a proliferation of research on the many ways illness affects everyday life and on how ill individuals, their families, and their friends respond to illness.

Chapter Organization

This textbook demonstrates the breadth of topics included in the sociology of health, illness, and health care. The text covers both micro-level issues (those occurring at the level of interactions among individuals and small groups) and macro-level issues (those occurring at the level of the society as a whole). In Part One, I discuss the role social factors play in fostering illness and in determining which social groups experience which illnesses. Chapter 2 describes the major causes of preventable deaths in the United States and how they have changed over time, including both long-standing problems such as cancer and emerging problems such as severe acute respiratory syndrome (SARS) and drug-resistant tuberculosis. This chapter demonstrates how social as well as biological factors affect health and illness. Building on this basis, Chapter 3 describes how age, gender, social class, and race or ethnicity affect which Americans get ill with which illnesses. Finally, Chapter 4 explores the nature and sources of illness in the poorer countries of Asia, Africa, and Latin America.

Part Two analyzes the meaning and experience of illness and disability in the United States. In Chapter 5, I explore what people mean when they label something an illness, as well as how social groups explain both why illness occurs and why illness strikes certain individuals rather than others. This chapter also looks at the social consequences of defining behaviors and conditions as illnesses. With this as a basis, in Chapter 6, I first explore the meaning of disability and then offer a sociological overview of the experience of living with chronic pain, chronic illness or disability, including the experience of seeking care from either medical doctors or alternative health care providers. Chapter 7 provides a parallel assessment of mental illness, describing what people mean when they label something a mental illness, analyzing the relationship between social factors and mental illness, providing a sociological account of the diagnosis and treatment of mental illness, and exploring the experience of living with mental illness.

In Part Three, I move the analysis to a more macro-level perspective. Chapter 8 describes the basic outlines of the U.S. health care system and examines some of the current problems with that system. I begin Chapter 9 by suggesting some basic measures for evaluating health care systems and then use these measures to evaluate the systems found in Canada, Great Britain, the People's Republic of China, and Mexico. I conclude this chapter by asking what useful lessons the United States can take from these other countries and by assessing the prospects for health care reform within the United States. Finally, Chapter 10 examines several health care settings, including hospitals, hospices, nursing homes, and family homes, and provides a social analysis of the technologies used in those settings.

Part Four shifts our focus from the health care system to health care providers. In Chapter 11, we analyze how doctors have achieved both prestige and professional autonomy and the factors now threatening their position. The chapter also describes the process of becoming a doctor, the values embedded in medical culture, and the impact of those values on doctor-patient relationships. Chapter 12 describes the history and social position of various other mainstream and alternative health care occupations, including pharmacists, lay midwives, osteopaths, and Christian Science practitioners. Finally, Chapter 13 presents an overview of bioethics, the study of ethical issues involved in the provision of health care. The chapter discusses how bioethics can inform sociological debate and how sociology can inform bioethical debate. (Reflecting the importance of bioethics to understanding health, illness, and health care, each of the preceding chapters also includes an ethical debate on a topic related to that chapter.)

The glossary at the end of the book defines all essential terms used in the book. The first time a term appears in the book, it is defined and set in bold-face type. In case professors assign the chapters out of sequence, each term also appears in boldface type (without a definition) the first time it appears in any subsequent chapter. In addition, "Key Concepts" tables throughout help explain particularly complex and important topics.

Each chapter ends with suggested readings, review questions, and a section labeled "Getting Involved" that identifies pertinent nonprofit organizations. The suggested readings were selected not only because of the materials they cover but also because they are exceptionally well written and interesting. The study questions are designed to provide an overview of the chapter. Readers who can answer these questions should feel confident that they understand the material. Finally, the "Getting Involved" sections are included both as sources of additional information and as potential means through which readers can become personally involved in working on the issues raised in the chapters. Updated web addresses for these organizations can be found at this textbook's Student Book Companion Site (click Sociology of Health, at http://www.thomsonedu.com/sociology). To show readers how individuals can make a difference, each chapter also includes a boxed discussion on an organization that has fought successfully to prevent illness or improve the lives of those who experience illness or disability.

A Note on Sources

Printed Sources

This book is based primarily on data from three types of printed sources: medical journals, sociological journals and books, and government and United Nations statistics. Before readers can evaluate this book and the conclusions drawn in it, they need to know how to evaluate these sources.

The most influential medical journals in the United States are the *Journal of the American Medical Association* and the *New England Journal of Medicine*. The comparable British journals are the *British Medical Journal* and *Lancet*. These journals are most influential for several reasons. Each has been in existence for several decades, proving its worth through its longevity. Each has a large readership, indicating that doctors take them seriously enough to pay for subscriptions. Each accepts for publication only a small percentage of submitted manuscripts, so these journals publish only the best articles. Finally, each uses peer review, sending every submitted manuscript to two or more reviewers for evaluation before the editors decide whether to publish it.

Much of the medical research presented in this textbook comes from these sources or from the *American Journal of Public Health*, published by the major professional association in public health. Because it is a specialty journal rather than a general medical journal and therefore has a smaller audience, the *American Journal of Public Health* is not as prestigious as the top medical journals. However, the standards for publication in this journal are as high as for the top medical journals, and all health care professionals take seriously anything published in it.

The most influential sociological journals in the United States are the *American Sociological Review,* the *American Journal of Sociology, Social Forces,* and, to a somewhat lesser extent, *Social Problems.* By general agreement, the most important journal in the sociology of health, illness, and health care in the United States is the *Journal of Health and Social Behavior.* Many of the sociological articles cited in this book come from these sources. Sociologists widely respect these journals for the same reasons that doctors widely respect the *New England Journal of Medicine.*

Although all these journals—especially the medical journals—sometimes print articles based on only a few cases, most of the articles cited in this book draw on large samples. As a result, the conclusions presented in these articles are more likely to reflect trends among the population as a whole rather than to reflect individual idiosyncrasies. For the same reason, the most methodologically sophisticated articles use **random samples.** In a random sample, researchers select participants in such a way that each member of a population has an equal chance of being selected (such as drawing names out of a hat, or interviewing every fifth person listed on a class roster). When a sample is randomly selected, researchers can be fairly certain that the selected individuals will statistically represent the population as a whole. In addition, these articles typically use statistical techniques to **control** for the impact of extraneous factors on the observed relationships. For example, researchers studying the relationship between smoking cigarettes and lung cancer can use statistical techniques to control for the impact of smoky work environments. To do so, they would first divide the population under study into those who do or do not work in smoky environments. They could then see if smokers were more likely than nonsmokers were to develop cancer in one environment, both environments, or neither environment.

Finally, this book draws heavily on statistics collected by the U.S. government and by the World Health Organization (WHO), a branch of the United Nations. Because these statistics are collected by nonpartisan bureaucrats whose employment typically continues regardless of shifts in the political climate, rather than by groups with a particular political agenda, they are generally regarded as the most objective data available.

This brief discussion of sources suggests several questions readers should keep in mind while reading this book. First, ask if the data come from a reputable source. Second, ask whether the data were peer reviewed or in some other way checked for quality or potential bias. Third, ask about the size and nature of the study's sample as well as whether the study controlled statistically for possible confounding factors. Fourth, ask what questions the researchers asked in collecting their data and what questions they *should* have asked. For example, countries that define infants who die during the first week after birth as stillborns will appear to have fewer infant deaths than will countries that define these as infant deaths. Finally, ask if the data presented are sufficient to justify the conclusions. If not, ask what additional data are needed to reach a firmer conclusion and how one might obtain that information from reputable sources.

Internet Sources

In addition to using printed sources, I obtained some of the information used in this book through the Internet. The Internet can be an excellent source for current statistics and an efficient way of learning about many topics. However, the vast wealth of materials available via the Internet and the ease with which anyone can post on it make it crucial for users to evaluate these sources critically.

When evaluating materials garnered through the Internet, readers can use the same principles used to evaluate printed materials. Most important, users must determine whether a reputable source provided the information; most of the information used in this book and obtained through the Internet came from either U.S. government sources or the World Health Organization. The source of an Internet page is often apparent in its address. Internet addresses for government agencies usually end with *.gov* and addresses for educational institutions usually end with *.edu.* Nonprofit organizations, like the Sierra Club or the Muscular Dystrophy Association, usually have addresses that end with *.org.* Commercial sites, on the other hand, usually end with *.com;* this ending applies to sites run directly by businesses, such as General Electric (www.ge.com), as well as to sites run by individuals who purchase Internet access from businesses (e.g., addresses ending with *aol.com*). For example, in evaluating information about different treatments for cancer, you should probably give more credence to information obtained from www.healthfinder.gov (a site run by the United States government) or from www.mayohealth.org (run by the nonprofit Mayo Clinic) than to information obtained from a site that ends with *.com* and that might reflect either one individual's views or the views of a business that earns its profits by selling a particular treatment. Be aware, though, that any individual or company can obtain an ".org" address. And keep in mind that websites identified by search engines such as Yahoo and MSN.com are likely to be sponsored by drug companies or others with commercial interests to protect (Green, Kazanjian, and Helmer, 2004).

Unfortunately, Internet sources come and go rapidly, and addresses change constantly, making it difficult to provide a reliable list of useful

---

Box 1.1   *Useful Internet Sources*

Here are several online sources you may find useful when reading this book:

www.healthfinder.gov: Run by the United States Department of Health and Human Services, this site offers a wide range of health information, as well as an extensive set of links to other government and nongovernmental health-related sites.

www.nlm.nih.gov: This site provides access to both published and unpublished materials available at the National Library of Medicine, the largest medical library in the world.

www.mayohealth.org: Run by the Mayo Clinic, this site offers both consumer health information and the opportunity to email questions to physicians.

www.who.int: Run by the World Health Organization, this website provides a vast array of information about health, illness, and health care around the world.

hippo.findlaw.com: An invaluable compendium of information on health law, policy, and regulations. Although run by a for-profit organization, "hippo" is highly regarded in the health field.

scholar.google.com: This branch of the Google® search engine takes viewers only to scholarly journal articles, on health as well as other topics. An excellent starting point for finding reputable information on any topic.

Internet sites. Box 1.1 suggests some currently useful Internet sources for health issues that seem likely to remain stable for the near future. Readers can find updated addresses and additional sources at this textbook's website (located under Sociology of Health, at http://www.thomsonedu.com/sociology).

Suggested Readings

Mills, C. Wright. 1959. *The Sociological Imagination.* New York: Grove Press. The classic statement of the sociological perspective.

Schwalbe, Michael. 2004. *The Sociologically Examined Life: Pieces of the Conversation.* 3rd ed. New York: McGraw-Hill. Another excellent introduction to the sociological perspective.

Review Questions

What is the sociological perspective?

How do the questions sociologists ask differ from the questions asked by psychologists or by health care workers?

What does this textbook mean by a critical approach?

What are some ways a reader can tell if a journal article or Internet website is a reliable data source?

Internet Exercises

1. Try different Internet search strategies to find information about writing a living will. First, try one of the major search engines, such as Excite, Yahoo, or AltaVista, which you can probably access by clicking on the word *Search* or on a "search" icon on the main menu of your Internet browser. What kinds of information (quantity, quality, type) do you find? Then try using Metacrawler (www.metacrawler.com), which searches and combines results from other search engines.

2. If you searched for "living will," you probably found a great deal of irrelevant information. (Your search was probably more productive if you searched for "living wills." Can you figure out why?) To make your search more effective, you'll need to learn how to perform "advanced" or "power" searches. Instructions for doing so, or tips for searching, probably appear somewhere on the web page for your browser. For example, in some browsers, to find web pages on living wills (rather than on every document about living that includes the word *wills*), you must search for "living+wills," whereas in other browsers you would need to search for "living wills." Do your search again, using proper syntax to specify your request. How does this affect your results?

3. Now try the same search, using Medline, the major online archive for medical and other health-related journals. You might be able to access Medline through your college library or its website. Otherwise, you will need to first search for and then connect to the Grateful Med website, the library of a major university, or the National Library of Medicine (a branch of the National Institute of Health). Check your screen, and see if it offers instructions for narrowing your search, power searches, or advanced searches. How does the information you get from Medline differ from the information you found using a web browser?

4. Finally, try looking for articles on living wills in InfoTrac® College Edition, a large online archive of scholarly articles available through Wadsworth Publishing at www.infotrac-college.com/wadsworth. (You have free access to InfoTrac College Edition this semester if your professor ordered it when ordering this textbook.) If you don't find anything after searching for "living wills," try searching for "right to die" or for "advanced directives" (a general term referring to legal documents specifying what types of medical care an individual would want in a given situation).

CHAPTER **2**

The Social Sources of Illness

In 1964, Dr. Margrethe ("Grethe") Rask left her native Denmark to work in Central Africa. For several years, she worked at a primitive hospital in northern Zaire, where, as her former colleagues described (Shilts, 1987: 4–7), basic supplies were woefully lacking:

> You just used needles again and again until they wore out; once gloves had worn through, you risked dipping your hands in your patient's blood because that was what needed to be done. The lack of rudimentary supplies meant that a surgeon's work had risks that doctors in the developed world could not imagine.

In the early 1970s, Rask began working at a major hospital in the capital city of Kinshasa. By Christmas 1976:

> She was thin, losing weight from a mysterious diarrhea. She had been suffering from the vague yet persistent malaise for two years now, since her time in the impoverished northern villages. In 1975, the problem had receded briefly after drug treatments, but for the past year, nothing had seemed to help. The surgeon's weight dropped further, draining and weakening her with each passing day.
>
> Even more alarming was the disarray in the forty-six-year-old woman's lymphatic system, the glands that play the central role in the body's never-ending fight to make itself immune from disease. All of Grethe's lymph glands were swollen and had been for nearly two years. Normally, a lymph node might swell here or there to fight this or that infection, revealing a small lump on the neck, under an arm, or perhaps in the groin. There didn't seem to be any reason for her glands to swell; there was no precise infection anywhere, much less anything that would cause such a universal enlargement of the lymph nodes all over her body. . . .

Suddenly, she could not breathe. Terrified, Grethe flew to Copenhagen, sustained on the flight by bottled oxygen. [Throughout 1977,] the top medical specialists of Denmark had tested and studied the surgeon. None, however, could fathom why the woman should, for no apparent reason, be dying. There was also the curious array of health problems that suddenly appeared. Her mouth became covered with yeast infections. Staph infections spread in her blood. Serum tests showed that something had gone awry in her immune system; her body lacked T-cells, the quarterbacks in the body's defensive line against disease. But biopsies showed she was not suffering from a lymph cancer that might explain not only the T-cell deficiency but her body's apparent inability to stave off infection. The doctors could only gravely tell her that she was suffering from progressive lung disease of unknown cause. And, yes, in answer to her blunt questions, she would die. . . .

On December 12, 1977, Margrethe P. Rask died. She was forty-seven years old.

A scant few years later, the cause of Grethe Rask's death—AIDS—would make headlines around the world. The news of a new, fatal infectious disease stunned both doctors and the public. Yet throughout history, new diseases have appeared and old diseases have disappeared. In this chapter, I provide a brief history of how patterns of disease have shifted over time, from the great epidemics of the past, to the late nineteenth-century decline of infectious diseases, to their modern reemergence. I then describe the current evidence regarding the main sources of premature death in the United States today, including tobacco, alcohol, medical errors, and motor vehicles.

Before we can understand patterns of disease, however, some basic concepts need to be defined.

An Introduction to Epidemiology

The first essential concept that students of health and illness need to understand is disease. To researchers working in health care, **disease** refers to a biological problem within an organism, whereas **illness** refers to the social experience and consequences of having a disease. So, for example, an individual who is infected with the poliomyelitis virus has the disease we call polio. When we refer, however, to subsequent changes in that individual's sense of self and social relationships, we should properly refer to these changes as consequences of the *illness* known as polio, not the *disease*. (I will discuss the meaning of illness in more detail in Chapter 5.)

The study of the distribution of disease within a population is known as **epidemiology.** This chapter and the next focus more specifically on **social epidemiology,** or the distribution of disease within a population according to social factors (such as social class or use of tobacco) rather than biological factors (such as blood pressure or genetics). For example, whereas biologists might investigate whether heart disease is more common among those with high versus low cholesterol levels, social epidemiologists might investigate whether it is more common among smokers versus nonsmokers.

What do we mean when we say that a certain disease is "more common" among one group than another? One way is to look at how many people in each group have the disease. Relying on raw numbers, however, can distort our picture of a population's health. For example, during 2003–2004, more than 500,000 persons in Brazil were infected with the virus that causes AIDS; but in the Bahamas, less than 7,000 persons were infected. On the surface, these numbers suggest that Brazil has a far greater AIDS problem than the Bahamas does. However, Brazil's population is much larger than that of the Bahamas. To take this difference into account, epidemiologists would look at the *rate* rather than the number of AIDS cases in these two countries. **Rate** refers to the proportion of a specified population that experiences a given circumstance. We would use the following formula to calculate the rate of any event (whether disease, disability, birth, or death):

$$\frac{\text{Number of events in a given period}}{\text{Specified population during that period}} \times 10^n$$

Using this formula, we find that the rate of persons known to be infected with the virus that causes AIDS (calculated as the number of infected persons in a country divided by the country's population) was 3,000 per 100,000 adults in the Bahamas but only 700 per 100,000 adults in Brazil (Population Reference Bureau, 2005). This tells us that AIDS affects a greater proportion of the population in the Bahamas than in Brazil and demonstrates the advantage of using rates rather than raw numbers.

Two particularly useful types of rates are incidence and prevalence rates. **Incidence** refers to the number of *new* occurrences of an event (disease, births, deaths, and so on) within a specified population during a specified period. **Prevalence** refers to the *total* number of cases within a specified population at a specified time—both those newly diagnosed and those diagnosed in previous years but still living with the condition under study. So, for example, to calculate the *incidence* rate of lung cancer in the United States this year, we would use the formula:

$$\frac{\text{Number of new cases of lung cancer } \textit{diagnosed} \text{ this year in U.S.}}{\text{Population of U.S. this year.}} \times 100,000$$

To calculate the *prevalence* rate of lung cancer, we would use the formula:

$$\frac{\text{Number of persons } \textit{living} \text{ with lung cancer in U.S. this year}}{\text{Population of U.S. this year.}} \times 100,000$$

In general, incidence better measures the spread of **acute illnesses,** such as chicken pox and cholera, that strike suddenly and disappear quickly—sometimes killing their victims, sometimes causing only a mild illness. Incidence also better measures rapidly spreading diseases such as AIDS. For example, to see how AIDS spread during the first decade after it was identified, we would compare its incidence in 1981 to its incidence in 1991. Prevalence, on the other hand, better measures the frequency of **chronic illnesses.** Chronic illnesses are defined as those illnesses that typically last for many years, such as muscular dystrophy, asthma, and diabetes.

Two final terms often used in epidemiology are *morbidity* and *mortality.* **Morbidity** refers to symptoms, illnesses, and impairments; **mortality** refers to deaths. To assess the overall health of a population, epidemiologists typically calculate the rate of serious morbidity in a population (that is, the proportion suffering from serious illness), the rates of infant mortality and maternal mortality (that is, the proportion of infants and childbearing women who die during or soon after childbirth), and **life expectancy** (the average number of years individuals born in a certain year can expect to live).

But what if we want to compare the health of two populations that differ in some critical way? Most often, this issue arises when one population is younger than another. For example, Arizona's population is younger on average than is North Dakota's, and so we can expect Arizona to have a higher rate of death from drunk driving and a lower rate of death due to heart disease. To deal with this issue, epidemiologists use **age-adjusted rates.** These rates are calculated using standard statistical procedures that eliminate the effect of age differences among populations.

The next section uses epidemiological concepts and data to describe how patterns of disease have changed over time.

A Brief History of Disease

The European Background

The modern history of disease begins during the Middle Ages (approximately A.D. 800 to 1300), as commerce, trade, and cities began to swell (Kiple, 1993). These shifts sparked a devastating series of epidemics. The term **epidemic** refers to any significant increase in the numbers affected by a disease *or* to the first appearance of a new disease. In the fledgling European cities, people lived in close and filthy quarters, along with rats, fleas, and lice—perfect conditions for transmitting infectious diseases such as bubonic plague and small-pox. In addition, because city dwellers usually disposed of their sewage and refuse by tossing them out their windows, typhoid, cholera, and other water-borne diseases that live in human waste flourished. Simultaneously, the growth of long-distance trade helped epidemics spread to Europe from the Middle East, where cities had long existed and many diseases were **endemic** (that is, had established themselves within the population so they maintained a fairly stable prevalence). In addition, religious pilgrimages and crusades to Jerusalem helped spread diseases to Europe.

The resulting epidemics ravaged Europe. Waves of disease, including bubonic plague, leprosy, and smallpox, swept the continent. The worst of these was bubonic plague, popularly known as the "Black Death." Between

1347 and 1351, plague killed at least 25 million people—between 25 percent and 50 percent of Europe's population and as much as two-thirds of the population in some areas (Gottfried, 1983; J. Kelly, 2005).

Although the great **pandemics** (worldwide epidemics) began diminishing during the fifteenth and sixteenth centuries, average life expectancy increased only slightly, for malnutrition continued to threaten health (Kiple, 1993). By the early 1700s, however, life expectancy began to increase. This change cannot be attributed to any developments in health care, for folk healers had nothing new to offer, and medical doctors and surgeons (as will be described in more detail in Chapter 11) harmed at least as often as they helped. For example, former president George Washington died after his doctors, following contemporary medical procedures, "treated" his sore throat by cutting into a vein and draining two quarts of his blood over the course of a day (Kaufman, 1971: 3).

If advances in medicine did not cause the eighteenth-century decline in mortality, what did? Historians commonly trace this decline to a combination of social factors (Kiple, 1993). First, changes in warfare moved battles and soldiers away from cities, protecting citizens from both violence and the diseases that followed in soldiers' wakes. Second, the development of new crops and new lands improved the nutritional status of the population and increased its ability to resist disease. Third, women began to have children less often and at later ages, increasing both women's and children's chances of survival. Fourth, women less often engaged in long hours of strenuous fieldwork, increasing their chances of surviving the physical stresses of childbearing. Infants, too, more often survived because mothers could more easily keep their children with them and breastfeed. (This lifestyle, however, would change soon for those women who became factory workers.)

Disease in the New World

As these changes were occurring in Europe, colonization by Europeans was decimating the native peoples of the New World (Kiple, 1993). The colonizers brought with them about fourteen new diseases—including influenza, measles, smallpox, scarlet fever, yellow fever, cholera, and typhoid—that had evolved in the Old World and for which the Native Americans had no natural immunities. These diseases ravaged the Native American population, in some cases wiping out entire tribes (Crosby, 1986). Conversely, life expectancy *increased* for those who emigrated from Europe to the colonies, for the New World's vast lands and agricultural resources protected them against the malnutrition and overcrowding common in Europe.

The Epidemiological Transition

As industrialization and urbanization increased, mortality rates rose, especially among the urban poor. The main killer was **tuberculosis,** followed by influenza, pneumonia, typhus, and other infectious diseases. By the late nineteenth century, however, deaths from infant mortality, child mortality, and infectious diseases began to decline rapidly. Between 1900 and 1930, life expectancy rose from 47 years to 60 years for whites and rose from 33 years to 48 years for African Americans (U.S. Bureau of the Census, 1975).

As infant mortality declined, families no longer felt obligated to have many children to ensure that one or two would survive long enough to get work and bring income into the household. At the same time, the national economy continued to shift from agriculture to industry, reducing couples' need to have children to work on the family farm. Similarly, employers increasingly offered pensions and other social benefits, so couples had less need to have children to care for them in their old age. Taken together, these trends produced a sharp decline in family size. Consequently, families could devote more resources to each child, further increasing their children's chances of survival.

As infectious diseases declined in importance, chronic and degenerative diseases, which can affect only those who live long enough for symptoms to develop, gained importance. Cancer, heart disease, and stroke became major causes of mortality, while arthritis and diabetes emerged as major sources of morbidity. Increasingly, too, conditions like heart disease, stroke, and hypertension shifted from being primarily diseases of the affluent to being disproportionately diseases of the poor.

The shift from a society characterized by infectious and parasitic diseases and low life expectancy to one characterized by degenerative and chronic diseases and high life expectancy is referred to as the **epidemiological transition** (Omran, 1971). This transition seems to occur around the world once a nation's mean per capita income reaches a threshold level (in 2005 dollars) of about $7,365 (Wilkinson, 1996). As we will see in more detail in the next chapter, some countries have fully made the epidemiological transition but others have not.

Contrary to conventional wisdom, medical interventions such as vaccinations, new drugs, and new surgical techniques played little role in the epidemiological transition, which began more than 200 years ago in Western societies (Leavitt and Numbers, 1985; McKeown, 1979; McKinlay and McKinlay, 1977). In a series of dramatic graphs showing how mortality from several important diseases declined over time, McKinlay and McKinlay (1977) have demonstrated that most of these declines *preceded* the introduction of effective medical interventions (see Figure 2.1). For example, the death rate for tuberculosis declined steadily from greater than 3.5 per 1,000 in 1860 to .34 per 1,000 in 1946. Yet streptomycin, the first effective treatment for tuberculosis, was not introduced until 1947. Only polio and smallpox declined substantially after the introduction of medical interventions. Of these two, only the decline in polio can be confidently attributed to medical intervention, as we cannot separate the possible impact of inoculation on the rate of smallpox from the impact of the myriad other changes that occurred after inoculation was first widely adopted about 200 years ago. Similarly, the introduction of chlorination and filtering to city water

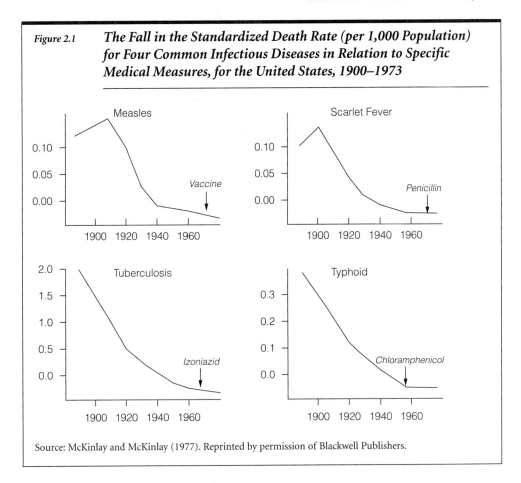

Figure 2.1 **The Fall in the Standardized Death Rate (per 1,000 Population) for Four Common Infectious Diseases in Relation to Specific Medical Measures, for the United States, 1900–1973**

Source: McKinlay and McKinlay (1977). Reprinted by permission of Blackwell Publishers.

systems virtually eliminated waterborne diseases like typhoid fever and dramatically reduced the rates of other infectious diseases, like pneumonia and tuberculosis. Cleaner water systems accounted for almost half of the drop in overall mortality rates and two-thirds of the drop in infant mortality rates between 1900 and 1940 (Cutler and Miller, 2005).

Researchers using generous assessments of the potential impact of modern medical care on life expectancy have concluded that medical care explains no more than one-sixth of the overall increase in life expectancy during the twentieth century (Bunker, Frazier, and Mosteller, 1994). Rather, most of this increase resulted from changes in the social environment (McKinlay and McKinlay, 1977). As nutrition and living conditions improved, so did individuals' ability to resist infection and to survive if they became infected. At the same time, although somewhat less importantly, public health improvements such as the development of clean water supplies and sanitary sewage systems increasingly protected individuals from exposure to disease-causing microbes.

Given the enormous improvements in life expectancy during the twentieth century, it was natural for scientists to assume that life expectancy would continue to rise steadily along with incomes. However, Richard Wilkinson (1996, 2005), using a diverse wealth of data from studies conducted around the world, makes the convincing (and highly influential) argument that increases in average income above about $7,365 (in 2005 dollars) bring only modest increases in life expectancy. Instead, further increases in life expectancy appear to occur not when *absolute incomes* increase but only when the *relative income differential* within a country narrows. In other words, if the gap in income between rich and poor narrows, as it has in Costa Rica, average life expectancy increases. Conversely, if the income gap widens, as happened following the collapse of the former Soviet Union, average life expectancy declines. As a result, life expectancy is greatest within countries like Sweden and Japan, which have experienced the epidemiological transition *and* have the smallest income gap between rich and poor, rather than in countries like the United States, which despite its great wealth has the widest income gap among the industrialized nations.

After weighing all the available evidence, Wilkinson argues that the key to the better health found in societies with small income gaps is the reduction in chronic psychosocial stresses. When income inequality is high, all citizens—regardless of their social class—must worry constantly about maintaining their social status and are susceptible to shame and depression when they cannot do so (Marmot, 2004; Wilkinson, 2005). As a result, individuals live with chronic stress, which makes it more difficult for their immune systems to fight disease. Moreover, to manage these stresses, anxieties, and depression many will turn to tobacco, alcohol, or fatty "comfort" foods, which further endanger their health. In addition, they are far less likely to trust others or to have a sense of belonging to a community and far more likely to be concerned about maintaining face and pride. As a result, they (and those around them) are far more likely to die or be injured by violent crimes; indeed, societies with high income inequality have especially high rates of death linked to accidents, violence, and alcohol. Importantly, Wilkinson notes that when societies reduce income inequality through increasing education, housing, and employment opportunities (as Japan and Korea did following World War II), all members of the society benefit because lower-class persons become both more economically productive and less likely to engage in criminal or violent behaviors.

The New Rise in Infectious Disease

By the second half of the twentieth century, Americans—both health care workers and the public—had come to believe infectious diseases were under control (even though they continued to rage in poorer regions of the world). Partly because of this belief, few paid much attention when on June 5, 1981, the federal government's *Morbidity and Mortality Weekly Report* published a brief article describing a curious syndrome of immune-deficiency disorders in five gay men. Within a few years, however, people around the world would learn to their horror that a deadly new infectious disease, now known as AIDS, had taken root. Since then, other new infectious diseases (such as Ebola virus) have been identified, previously known diseases (such as cholera and streptococcus) have become deadlier, and previously harmless microbes (such as the virus that causes bird flu) have caused important disease outbreaks (Altman, 1994).

The renewed dangers posed by infectious disease partly reflect basic principles of natural selection. Just as natural selection favors animals whose camouflaging coloration hides them from predators so they can survive long enough to reproduce, natural selection favors those germs that can resist drug treatments. As doctors prescribed antibiotics more widely, often under pressure from patients who feel "cheated" if they do not receive a prescription at each visit (Vuckovic and Nichter, 1997), the drugs killed all susceptible variants of disease-causing germs while allowing variants resistant to the drugs to flourish. Similarly, drug-resistant tuberculosis is increasing worldwide, as AIDS and poverty leave individuals both more susceptible to infection and less able to afford consistent, effective treatment. Meanwhile, widespread use of antibiotics in everything from cutting boards to kitty litter, chicken feed, and soaps also encourages the rise of drug-resistant bacteria.

Other forces also promoted the rise in infectious diseases (L. Garrett, 1994). Population growth and the rise of cities had fostered the spread of infectious diseases in Europe centuries ago; these same factors now are causing new epidemics to develop in the rapidly growing cities of Africa, Asia, and Latin America. Meanwhile, older cultural traditions often erode among those who move to these cities, making health-endangering activities like tobacco smoking and sexual experimentation more likely. At the same time, the destruction of ecosystems in these regions, as industrial sites and cities replace forests and farmlands, changes the balance between human, animals, and microbes, encouraging some microbes that previously had infected only animals to begin infecting humans.

All these factors have been heightened by **globalization,** the process through which ideas, resources, people, and trade increasingly operate in a worldwide rather than local framework. The erosion of cultural traditions in Asia, Africa, and Latin America reflects, among other things, the increasingly global spread of Western ideas by tourists, the mass media, businesspeople, and nongovernmental organizations such as the International Monetary Fund (IMF) and the United Nations. Similarly, environmental changes that encourage disease partly stem from actions taken by Western-based industries and corporations, which have found it increasingly easy to operate around the world due to new free trade agreements (such as NAFTA, the North American Free Trade Agreement). Finally, the globalization of business investment and tourism has globalized disease simply by increasing the number of people traveling from one region to another (L. Garrett, 1994). For example, severe acute respiratory syndrome (SARS) is a new, infectious respiratory disease that doctors first identified in China in late 2002. Due to

---

Box 2.1   *The Threat of Bioterrorism*
by Sarah St. John

The terrorist attacks of September 11, 2001, the spread of anthrax through the mail shortly thereafter, and the 2005 bombings in Madrid and London busses and subways have heightened concerns about bioterrorism in the United States. Since World War II, numerous governments (including the United States) have developed stockpiles of dangerous germs—including bubonic plague, typhus, smallpox, and anthrax—as well as technologies for making those germs more toxic and easier to disseminate (T. Brown and Fee, 2001; J. Miller, Engelberg, and Broad, 2001). All of us are endangered by the possibility that a government or terrorist group might use these germs. Such an event could produce massive human casualties, severely reduce food or water supplies if animals or water are infected, and cause great disruption (like the disruption to the postal system after the anthrax mailings).

The United States is particularly vulnerable to bioterrorism for several reasons. First, our status as the world's leading superpower makes us a target for those who envy our economic and political power, fear our cultural influence, or resent our actions. Second, our open society and commitment to individual liberties makes it more difficult to protect against terrorists than would be the case if our country were a dictatorship and we lived under constant governmental surveillance. Third, many Americans lack health insurance, adequate food and housing, and proper immunization against disease, and so are more susceptible to infection and less likely to receive the sort of quick treatment that could stop an epidemic quickly. Fourth, U.S. hospitals lack effective plans for coping with large bioterrorist attacks and lack the vaccinations, treatments, space, equipment, and knowledgeable personnel needed to do so (Daniell, Treser, and Wetter, 2001). Finally, severe cutbacks in the public health system over the last 25 years have weakened both local health departments and the national Centers for Disease Control and Prevention, thus weakening our ability to detect and respond to epidemics (L. Garrett, 2000).

In response to these problems, the U.S. government has dramatically increased its antiterrorism funding. Questions have been raised, however, regarding how much of this money should be devoted to fighting bioterrorism (how great is the risk, and how do we calculate it?) and how such money could best be allocated (to military surveillance? epidemiological surveillance? stockpiling medicines?). However these questions are answered, any monies spent alleviating problems associated with poverty, housing, nutrition, access to health care, and the decaying health care infrastructure will bring benefits whether or not we suffer a serious terrorist attack (Cohen, Gould, and Sidel, 2001).

---

international travel, within less than a year more than 8,000 cases were reported in 29 countries, including the United States (World Health Organization, 2005a). Globalization also can encourage infectious disease through its political effects. Since September 11, 2001, the American public has realized that the U.S. role in world politics can make it a target for terrorists, some of whom may be willing to use infectious diseases as weapons. Box 2.1 discusses this threat.

Table 2.1   *Modes of Transmission for Adults and Adolescent AIDS Cases Diagnosed 2003, United States*

Exposure Category	Percentage of Cases
Men who have sex with men	35
Injecting drug use	16
Men who have sex with men and inject drugs	4
Heterosexual contact	19
Female-to-female sexual contact	0
Unknown*	25
Total	99**

*Typically, mode of transmission is unknown because the case is still under investigation; the individual refused to answer questions or died before being interviewed; or the mode of transmission, while suspected, could not be proved; about half of all cases initially listed as mode of transmission unknown are eventually reclassified.

**Does not add to 100% due to rounding.

Source: Centers for Disease Control and Prevention (2004a).

The Emergence of AIDS

AIDS provides the premiere example of the new rise in infectious disease. Beginning in 1979, a few doctors in New York, San Francisco, and Los Angeles had noticed small outbreaks in young gay men of rare diseases that typically affect only persons whose immune systems have been damaged by disease or chemotherapy. By 1982, the Centers for Disease Control and Prevention (CDC) had officially coined the term **acquired immunodeficiency syndrome (AIDS)** to describe what we now know is the last, deadly stage of infection with **human immunodeficiency virus (HIV).** Because most HIV-infected persons do not in fact have AIDS, this textbook uses the term **HIV disease** rather than AIDS except when reporting statistics based solely on AIDS cases.

HIV disease is spread through sexual intercourse; through sharing unclean intravenous needles; through some still-unknown mechanism from mother to fetus; through blood transfusions or blood products; and, rarely, through breastmilk. The last three modes of transmission are now rare in countries where HIV blood tests, breastmilk substitutes, and drugs for reducing the risk of maternal/fetal transmission are affordable. Studies have demonstrated conclusively that AIDS is not spread through insects, spitting, sneezing, hugging, nonsexual touching, or food preparation (Stine, 2005). Table 2.1 shows the modes of transmission for AIDS cases diagnosed in 2003. In 2005, the number of Americans infected with HIV passed 1 million for the first time.

The rapid spread of HIV disease since 1981 reflects public attitudes as much as biological realities. A handful of behavioral changes could have virtually halted its spread: testing the blood supply for infection, using latex condoms and spermicide with sexual partners, and using clean needles when injecting drugs. Unfortunately, throughout the early years of the epidemic when intervention would have been most effective, the U.S. government (like most other governments) treated HIV disease as a distasteful moral issue rather than as a medical emergency. At critical junctures during the 1980s, federal officials lobbied Congress to restrict funding for HIV research and education (Epstein, 1996). Moreover, the limited funds the government provided early on for HIV education came with many strings attached, such as prohibiting explicit pictures in materials on sexual education, prohibiting language that might offend heterosexuals even in educational materials designed solely for gay men, and—even though substantial proportions of teenagers engage in sexual intercourse—refusing to fund education programs for children and young adults unless the programs taught only abstinence from sex and not how to have sex safely.

Similarly, both federal and local authorities have made it exceedingly difficult for individuals to protect themselves from infection by using intravenous needles safely. By retaining laws making it illegal to purchase or own needles and prosecuting those who distribute needles, the government unwittingly encourages addicts to share needles and thus to spread HIV, hepatitis, and other diseases. At the same time, the government has refused funding to those who would teach drug users how to clean needles. Yet most research suggests that helping drug users to protect themselves reduces the incidence of HIV infection without increasing the rate of drug use (Gostin et al., 1997).

The Modern Disease Profile

Despite the recent reemergence of infectious diseases, however, these diseases still play a relatively small role in U.S. mortality rates. Table 2.2 shows the top ten causes of death in the United States in 2002 and illustrates how these causes have changed since 1900.

As the table demonstrates, whereas the top killers in 1900—influenza, pneumonia, and tuberculosis—were infectious diseases, the top killers currently—heart disease and cancer—are chronic diseases primarily associated with middle-aged and older populations. These diseases now far outpace infectious diseases as causes of death.

But infectious diseases have not disappeared from the list of leading causes of death. Influenza and pneumonia remain significant for the population as a whole, while AIDS remains a leading cause of death among persons ages 25 to 44, with rates especially high among African Americans (National Center for Health Statistics, 2004). The newest drugs for treating HIV disease (the protease inhibitors) do seem to increase life expectancy, but only for those who can tolerate the drugs' side effects, manage the

Table 2.2	*Main Causes of Deaths, 1900 and 2002*		
1900	Rate/100,000	2002	Rate/100,000
Influenza and pneumonia	202	Heart disease	240.8
Tuberculosis	194	Cancer	193.5
Gastritis	143	Cerebrovascular disease	56.2
Disease of the heart	137	Chronic pulmonary disease	43.5
Cerebrovascular diseases	107	Accidents	36.9
Chronic kidney disease	81	Diabetes	25.4
Accidents	72	Influenza and pneumonia	22.6
Cancer	64	Alzheimer's	20.4
Diseases of early infancy	63	Kidney disease	14.2
Diphtheria	40	Septicemia	11.7

Source: M. Greenberg (1987: 5); Kochanek et al. (2005).

required regimen of as many as twenty pills per day taken at strictly regulated times, and afford the cost of about $15,000 per year.

Finally, Table 2.2 shows the continued role social factors play in causing deaths. Accidental deaths mostly stem from motor vehicle accidents (many of them linked to alcohol use), while tobacco use is the main cause of chronic pulmonary disease and a common contributor to heart disease, cancer, and cerebrovascular disease (strokes). Each of these causes of death reflects social behaviors rooted in social conditions. The remainder of this chapter discusses the role social forces play in mortality and morbidity.

The Social Sources of Premature Deaths

In a widely cited article titled "A Case for Refocusing Upstream," sociologist John McKinlay (1994) offers the following oft-told tale as a metaphor for the modern doctor's dilemma:

Sometimes it feels like this. There I am standing by the shore of a swiftly flowing river and I hear the cry of a drowning man. So I jump into the river, put my arms around him, pull him to shore and apply artificial respiration. Just when he begins to breathe, there is another cry for help. So I jump into the river, reach him, pull him to shore, apply artificial respiration, and then just as he begins to breathe, another cry for help. So back in the river again, reaching, pulling, applying, breathing, and then another yell. Again and again, without end, goes the sequence. You know, I am so busy jumping in, pulling them to shore, applying artificial respiration, that I have *no* time to see who the hell is upstream pushing them all in. (McKinlay, 1994: 509–510)

This story illustrates the traditional emphasis within medicine on **tertiary prevention:** strategies designed to minimize physical deterioration and complications among those already ill. Tertiary prevention includes such tactics as providing kidney dialysis to persons whose kidneys no longer function or insulin to those who have diabetes. Doctors much less commonly focus on **secondary prevention:** strategies designed to reduce the prevalence of disease through early detection and prompt intervention. Examples of secondary prevention include screening patients for cervical cancer or glaucoma so these diseases can be detected at still-treatable stages. Those who focus on secondary prevention typically work in public health or in the **primary practice** fields (family practice, pediatrics, or internal medicine). Finally, only a small fraction of doctors, usually in public health—or, less commonly, primary practice—focus "upstream" on **primary prevention:** strategies designed to keep people from becoming ill or disabled, such as discouraging drunk driving, lobbying for stricter highway safety regulations, and promoting vaccination.

Even when doctors and researchers (or, for that matter, the public) have focused on primary prevention, they typically have looked only far enough upstream to see how individual psychological or biological characteristics make some people more susceptible than others to disease or unhealthy behaviors. For example, an increasing number of medical researchers now focus on the genetic roots of disease, such as a possible gene for alcoholism. Similarly, many psychologists focus on understanding the psychodynamic forces that lead individuals to adopt behaviors believed to prevent illness, such as exercising regularly or refraining from smoking. The popular media, meanwhile, usually focuses on how individual "lifestyle choices" such as dieting, smoking, or using a seat belt affect the likelihood of health or illness.

Sociologists agree that biological factors and psychological predispositions affect decisions about whether to adopt healthier behaviors. But sociologists also note that these decisions do not occur in a vacuum. Rather, they occur in particular economic, cultural, and political settings that can make healthy behaviors or health itself either more or less possible. For example, adolescents' decisions regarding whether to drink alcohol are affected significantly by the attitudes of their friends, family, and culture in general. Similarly, the high rates of diabetes found among contemporary Native Americans partially reflect individual patterns of exercise and diet. They also, however, reflect the effects of the reservation system, with its sedentary lifestyle, ready access to fatty and sugary foods, limited access to fresh fruits and vegetables, and limited prospects for employment that make purchasing healthier foods difficult. In both cases, to blame unhealthy behavior patterns on individual choices seems oversimplistic.

Table 2.3   *Estimates of Actual Causes of Premature Death in the United States, 2000*

CAUSE	NUMBER	PERCENTAGE OF ALL DEATHS
Tobacco	435,000	18
Diet/activity patterns[a]	100–400,000	17
Medical errors[b]	>100,000	>13
Alcohol	85,000	4
Bacteria and viruses[c]	75,000	3
Toxic agents	55,000	2
Motor vehicles[d]	43,000	2
Firearms	29,000	1
Sexual behavior	20,000	1
Illicit use of drugs	17,000	1

[a]Estimates vary.

[b]Number of deaths is a rough estimate, because different studies have looked at different locations (in-hospital versus out-of-hospital) and different types of errors (surgical, medical, pharmacological).

[c]Does not include deaths related to HIV, tobacco, alcohol, illicit drugs, or infections caused by nonmicrobial diseases.

[d]Includes motor vehicle accidents linked to drug use, but *not* to alcohol use.

Source: Mokdad et al., 2004; HealthGrades, 2004; Kohn, Corrigan and Donaldson, 1999.

As these examples suggest, truly refocusing upstream requires us to look beyond individual behavior or characteristics to what McKinlay refers to as the **manufacturers of illness:** those groups that promote illness-causing behaviors and social conditions. These groups include alcohol distributors, auto manufacturers that fight against vehicle safety standards, and politicians who vote to subsidize tobacco production.

An article by Ali Mokdad and his colleagues (2004) published in the *Journal of the American Medical Association* provides a useful starting point for refocusing upstream. The article synthesizes the available literature on the major underlying causes of premature deaths (that is, deaths caused neither by old age nor by genetic disease) to identify those causes that we could most readily reduce or eliminate through social or medical interventions.

Mokdad and his colleagues identify nine causes that, they believe, together account for almost half of all premature deaths in the United States. Table 2.3 shows these causes and their prevalences (listed not by disease, but by the factors that cause disease) as well as a tenth cause that other researchers have identified. The next sections look at these ten causes of illness, focusing not on the individual behavior patterns that the authors emphasize in their article, but on the manufacturers of illness that precede these individual behaviors.

Tobacco

As Table 2.3 shows, tobacco causes far more premature deaths in the United States than does any other legal or illegal drug. Whether smoked, chewed, or used as snuff, tobacco can cause an enormous range of disabling and fatal diseases, including heart disease, strokes, emphysema, and numerous cancers (World Health Organization, 1998a). About half of all smokers will die because of their tobacco use, with half of these dying in middle age and losing an average of 22 years from their normal life expectancy. Tobacco use also increases morbidity and mortality among "passive smokers," those who must live and work around smokers (World Health Organization, 1998a). Similarly, both active and passive smoking can cause birth defects and infant mortality. Unfortunately, quitting smoking is difficult, for nicotine (the active ingredient in tobacco) is more addictive than heroin (Weil and Rosen, 1998).

Given nicotine's addictiveness, it is easy to understand why individuals continue smoking once they have started. But why do individuals begin smoking in the first place, especially when many initially find tobacco vile tasting and even nauseating? To answer this question, we need to look at the role of tobacco in American culture and at how tobacco manufacturers have created that role.

Since the 1960s, when research first proved the link between smoking and lung cancer, tobacco manufacturers have labored to convince the public through advertising to associate tobacco with positive attributes and achievements rather than with death and disability. This advertising has especially targeted youths, women, and minorities. According to an article published in the *American Journal of Public Health,*

> Young people are able to name and recognize cigarette ads and can also match cigarette brand name with cigarette slogans. More than half of current adolescent smokers and approximately one quarter of nonsmoking teens own cigarette promotional items and participate in these campaigns. . . . Longitudinal studies of advertising patterns and young people's tobacco use demonstrate a positive association between advertising and teenage smoking. In addition, the vast majority of adolescent smokers prefer the most heavily advertised brands (Schooler, Feighery, and Flora, 1996)

Manufacturers also have targeted their marketing to women by playing on women's desire for equality, excitement, personal fulfillment, and weight loss (a cultural imperative for women in contemporary American culture and a major reason women smoke). This strategy was exemplified by Virginia Slims—the name was not accidental—and its slogan, "You've come a long way, baby." To target minorities, manufacturers advertise heavily in magazines such as *Ebony* and *Jet.* Manufacturers also have gained influence and visibility in minority communities by providing financial sponsorship for charitable and cultural organizations and events and by geographically targeting minority communities with "culturally appropriate" advertising (White, 1988).

Over the last decade, successful legal attacks on tobacco manufacturers and advertisers have begun to erode their ability to attract new customers. For example, tobacco companies can no longer use cartoon characters in advertisements and now must limit their sponsorship of sports and entertainment events. Smoking by both teenagers and adults has declined slowly but steadily since the late 1990s, and the American public is showing growing support for the idea of "smoke-free" areas and a smoke-free culture (Givel and Glantz, 2004; National Center for Health Statistics, 2004: 33).

Diet, Exercise, and Obesity

The second most common cause of premature deaths, according to Mokdad and his colleagues (2004), is a high-fat diet, sedentary lifestyle, and resulting obesity, which the authors argue increases the odds of developing cardiovascular disease, strokes, certain cancers (of the colon, breast, and prostate), and diabetes, among other problems.

The Obesity Myth?

Since 1978, and as Figure 2.2 shows, rates of overweight and obesity in the United States have skyrocketed (Centers for Disease Control and Prevention, 2005). These changes have led the Centers for Disease Control and the Department of Health and Human Services to declare obesity a top national priority.

Without question, life is more difficult for those who are overweight. In contemporary America, overweight persons are not only considered less attractive but also assumed to be less intelligent, less responsible, and even less moral than others. Heavier persons—especially if they are female—are less likely to get dates, job promotions, marriage proposals, and so on. Physically, too, above a certain weight life becomes more difficult. Chairs and clothes don't fit, exercise becomes less fun, and finding the energy for life's daily tasks and pleasures becomes more difficult. Moreover, individuals who are obese, rather than just overweight, are more likely to develop heart conditions, diabetes, sleep difficulties, and other problems that diminish their quality of life, whether or not those problems shorten their life spans (Centers for Disease Control and Prevention, 2005).

Despite all this, however, there is good reason to believe that the dangers of being overweight have been overstated (Campos, 2004; Flegal et al., 2005; Gibbs, 2005). In 2005, the Centers for Disease Control reduced their estimate of yearly mortality due to obesity from 360,000 to 112,000 deaths per year (which still leaves it the second highest cause of premature deaths).

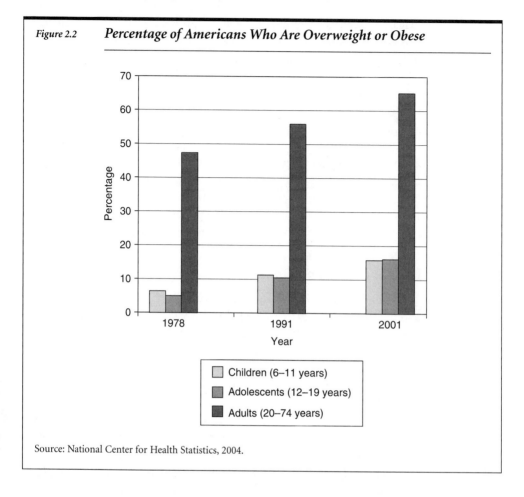

Figure 2.2   ***Percentage of Americans Who Are Overweight or Obese***

Source: National Center for Health Statistics, 2004.

This new estimate came from an article, published by Katherine Flegal and her colleagues (2005) in the *Journal of the American Medical Association*, that combined data from five national **random samples** conducted over a 30 year period by federal government researchers (rather than by researchers funded by the diet industry). Not only did the authors find that previous researchers had substantially overestimated the dangers of obesity, they also found that being overweight actually *reduces* death rates in most cases. The highest death rates overall occurred among those who were obese (e.g., 5'6" and more than 180 pounds), but the second highest death rates occurred among those who were *underweight* (e.g., 5'6" and less than 112 pounds). Moreover, above age 60, and among nonsmokers (regardless of age), underweight persons had the *highest* death rate.

Why have studies disagreed so dramatically about the impact of weight on mortality? Earlier studies were based on narrow populations (such as middle-aged nurses), relied on self-reported weights and heights, **controlled** statistically for few variables, and may not reflect current conditions. In addition, earlier studies compared obese Americans to normal-weight Americans, whereas Flegal and her colleagues explored the full range of weights (including underweight and overweight persons who were not obese.) Finally, the diet industry funded almost all of the earlier research, which may have colored study findings (Campos, 2004).

In addition, some argue that the real issue is not obesity, but physical activity and fitness (Blair and Church, 2004). Reviews of the available research indicate that individuals who are obese but physically fit have half the death rate of individuals who are normal weight but *unfit*. Moreover, research consistently suggests that significant, sustained weight loss is nearly impossible; but that 150 minutes a week of moderate physical activity is sufficient to substantially reduce a person's health risks (Blair and Church, 2004; Kolata, 2004a). Consequently, from a health perspective it seems more important and realistic to encourage Americans to exercise rather than to diet.

Despite these debates about the consequences of being overweight, scientists agree that Americans would be fitter, feel healthier, and perhaps live longer if they increased their activity levels; reduced consumption of fats, sugars, salt, and meat; and increased consumption of fruits, vegetables, and whole grains. Such changes would most benefit poor Americans, who are more likely than others are to eat unbalanced diets, heavy in sugars and fats, because such diets provide energy and satisfy hunger most cheaply (James et al., 1997).

"Supersizing" Americans

Why have Americans gained so much weight over the last generation? To answer this question, we need to look at how biology, economics, and politics combine (Crisler, 2003).

Biologically, humans have a natural craving for sweet and fatty foods and a natural desire to seek food of all sorts for survival. In past generations, when food was scarce, these cravings were healthy. Now, though, most Americans have plentiful access to food and eat more calories than their bodies can use, leading in the long run to overweight or obesity. In addition, the rapid adoption by food manufacturers of high-fructose corn syrup (an inexpensive sweetener) and palm oil (an inexpensive fat) also led to rapid weight gains, for the former is metabolized by the body differently than are other sugars, and the latter is a more saturated fat than even pig lard (Crisler, 2003).

To these changes in *what* Americans eat were added changes in *how much* Americans eat. Longer work weeks, lowered costs for eating out, and the increase in two-earner and single-parent families have led Americans to eat out more than ever before: Whereas in 1970 Americans spent 25 percent of their food budget eating out, by 1996 they spent 40 percent of their food budget eating out (Crisler, 2003: 32). Yet restaurant foods typically include far more fat and sugar than do homemade meals. Moreover, restaurants typically offer a wider variety of foods at any given meal, on appetizer plates, buffet tables, and in packaged "value meals." In addition, since the 1970s, restaurants have increasingly "supersized" portions as a way to increase sales and profits. Unfortunately, few individuals can regulate themselves when presented with large portions of varied foods, and so both these trends increased calorie consumption. As a result, the rise in eating away from home has increased Americans' calorie consumption by an average of 1,400 calories per person per week (Crisler, 2003: 33). Eating at fast-food restaurants is particularly problematic. For example, teenage boys who eat at fast-food restaurants three or more times a week consume 800 more calories *per day* than do those who eschew fast food (French et al., 2001).

Food manufacturers and the fast-food industry have used advertising to reinforce the tendency toward eating a sweet, fatty, high-calorie diet. Because manufacturers earn far less money by selling healthy foods (like fruits, vegetables, and grains) than by selling highly refined products loaded with fat, sugar, and salt (like candy, soft drinks, and convenience foods), they spend more than thirty times as much to advertise the latter foods (Nestlé, 2002: 22). That advertising has grown increasingly insidious, and now pervades every sphere of our society—especially those where children can be found. Soft-drink companies, for example, advertise to children not only on television, in magazines, and on the Internet but also through such tactics as offering toys and clothing with brand logos, placing products in movies, sponsoring school sports teams, providing soda-vending machines to schools, and offering cash bonuses to schools based on how much soda their students purchase. Many schools have felt helpless against this onslaught, because signing deals with soda companies seems the only way they can replace funds lost to budget cuts and to meeting mandates to prepare students for the new, required high-stakes tests. Consumption of sugar-sweetened soft drinks is directly and substantially related to obesity and diabetes among both adults and children (Apovian, 2004).

Meanwhile, as caloric consumption has increased, physical exercise has decreased. School budget cuts and intense political pressure to focus on test scores have led many schools to drop physical education and even recess. Almost half of all U.S. high school students no longer have any physical education classes (Gerberding and Marks, 2004). And at home, few children these days are allowed to spend their afternoons running free or playing non-organized sports. Instead, poor children are admonished to stay indoors to stay safe, and more-affluent children are shepherded from tutors, to classes, to the occasional sports activity. At the same time, physical activity is now an everyday part of life for only a small minority of children or adults, because very few commute by foot or bicycle to work, play, or shopping.

Medical Errors

According to a report released by the federal Institute of Medicine in 2000 (Kohn, Corrigan, and Donaldson, 1999), between 44,000 and 98,000 Americans die each year because of preventable medical errors in hospitals, and at least as many die from errors outside of hospitals. Subsequent research has suggested that these numbers greatly underestimate the problem (HealthGrades, 2004; Leape and Berwick, 2005). Medical errors cause more preventable deaths than do any factors other than tobacco use, and possibly unhealthy diet and activity patterns, and they cost the nation approximately $17 billion per year. Examples include conducting surgery on the wrong patient, giving a patient two drugs that interact dangerously, or misdiagnosing and hence mistreating a patient.

When errors occur, it is natural to focus on identifying which individual is to blame. Yet in most cases errors occur even though health care workers are dedicated, intelligent, and well trained. Consequently, most researchers and public health workers argue that we should focus not on individual malfeasance but rather on problematic systems. For example, many hospitals stock certain drugs only at full strength, even though the drugs are potentially fatal at full strength and must be diluted before use. Stocking these drugs pre-diluted would eliminate this source of death much more effectively than trying to identify every doctor or nurse who might administer the wrong dosage. Similarly, doctors can easily order the wrong drug when very different drugs have very similar names. The consequences can be fatal: A person experiencing epileptic seizures, for example, who receives the antifungal drug Lamisil instead of the antiepileptic Lamictal can die when his seizures continue unabated. In these examples, preventable deaths occur only when human error combines with systems that do not stop and that may even facilitate the error.

Adding to these problems is the lack of a system for identifying when deaths have been caused by medical errors. During the 1950s and 1960s, hospitals routinely autopsied from 50 to 70 percent of all patients who died. Now, due to a combination of economic costs and fear that autopsies will identify errors and result in malpractice claims, hospitals autopsy fewer than 10 percent, thus virtually eliminating one of medicine's most basic tools for identifying and reducing medical error (Lundberg, 2001: 253).

Medical culture, too, makes it difficult to identify and respond to medical errors. Research consistently finds that doctors rarely focus on identifying such errors, even in hospital's routine Mortality and Morbidity Conferences, whose purpose is to help doctors understand why patients have died or worsened unexpectedly (Bosk, 2003; Freidson, 1975; Millman, 1976; Orlander and Fincke, 2003; Pierluissi et al., 2003). First, cases typically are selected for discussion at conferences because they illustrate interesting intellectual puzzles rather than preventable medical errors. Second, medical errors rarely are identified as such. Third, when medical errors *are* identified, discussion rarely focuses on how to avoid such errors in future. Instead, because of professional etiquette, the need to maintain good relations with colleagues, and a medical culture that values individual doctors' right to make their own decisions, most errors are labeled unavoidable or blamed on nonmedical staff or on doctors from other divisions who are not invited to the discussions (Pierluissi et al., 2003).

Despite all these problems, the Institute of Medicine's report seems to have had some significant effects (Leape and Berwick, 2005). Most important, there is now widespread agreement among doctors, insurers, researchers, the public, and the government that medical error is a problem, and there is growing agreement that the problem needs to be dealt with systemically. The federal government has significantly increased its funding for research on patient safety, and hospitals across the country have adopted various systemic changes. For example, Veterans Administration (VA) hospitals now use a computerized record system that gives nurses and doctors access to comprehensive information on their patients. In addition, the record system generates bar-coded strips that are attached to each nurse, patient, and medication. Before administering any medication, a nurse must scan his or her own bar code as well as those of the patient and medication into a computer. The computer then checks that the nurse has the right drug for the right patient, and that the drug will not interact dangerously with any other drug taken by that patient. Since adopting this system, VA hospitals report that medication errors have dropped 70 percent (Leape and Berwick, 2005). In 2005, the government announced that it would make this record system available at no cost to all U.S. doctors who treat patients under **Medicare,** the federally funded insurance program for the elderly and permanently disabled.

Alcohol

Like tobacco, alcohol kills far more people than do all illegal drugs combined. Heavy alcohol use can cause irreversible brain damage, hepatitis, heart disease, cirrhosis of the liver, and cancers of the digestive system, while reducing the body's ability to fight infections such as tuberculosis and pneumonia. In addition, by diminishing individuals' ability to make rational choices, alcohol use contributes to deaths from drownings, fires, violence, and accidents and increases the odds of engaging in unsafe sexual behavior. Finally, withdrawal from alcohol is more dangerous than withdrawal from any other legal or illegal drug and can cause brain damage, heart failure, or stroke. Yet despite the dangers of alcohol, by law the U.S. Office of National Drug Control Policy cannot use any of its funds ($507 million during 2004–2006) to fight problem drinking. Proposals to change this statute, supported by the American Medical Association and the American Public Health Association, have been fiercely and successfully resisted by alcohol manufacturers and distributors (Wren, 1999).

To ensure that the government continues to treat alcohol as a beverage rather than a drug, alcohol manufacturers contribute heavily to political campaigns, giving $11 million to federal candidates alone during the 2004 election cycle (Center for Responsive Politics, 2005). Manufacturers also have worked to define the individual drinker rather than alcohol itself as the problem by promoting the idea that alcoholism is a disease that affects only susceptible individuals; by funding research on biological roots of alcoholism; and, like tobacco manufacturers, by supporting laws that criminalize underage drinking while fighting laws that would criminalize the sale of alcohol to minors (Morgan, 1988; Mosher, 1995).

At the same time, alcohol manufacturers have worked diligently to sell alcohol to the public not as a drug but as a lifestyle. Much of this marketing either directly or indirectly targets youths, despite voluntary industry codes that forbid manufacturers from marketing alcohol to audiences in which a majority are under age 21. During 1997–1998, only four of eight manufacturers studied by the Federal Trade Commission (1999) met even this lenient standard, and manufacturers paid to have their products appear on eight of the fifteen television shows most popular with teenagers. In addition, in recent years alcohol manufacturers have increased sales to youths by developing "alcopops": extra-sweet, fruit-flavored alcoholic beverages like Hard Lemonade, Smirnoff Ice, and Skyy Blue. Advertisements for these and other alcoholic beverages typically associate alcohol with adulthood, sexual adventure, status, freedom, excitement, and pleasure. Meanwhile, alcohol also sells because it offers an effective, if self-destructive, way to dull the emotional pains of daily life and the physical pains of hunger, cold, or abuse.

Bacteria and Viruses

Bacteria and viruses surround us all the time. Yet only rarely do individuals become infected, and even more rarely do these infections lead to deaths. Under what conditions do these deaths occur?

First, individuals will not develop fatal diseases if they are vaccinated against them. Virtually all U.S. children are vaccinated before they begin school, but about one-quarter do not receive all the required vaccinations by the recommended ages (National Center for Health Statistics, 2004: Table 72).

Second, even in the absence of vaccinations, individuals exposed to germs may not become infected unless they already are physically weakened. For example, a significant percentage of all persons admitted to hospitals—a population that obviously is already physically vulnerable—develop infections (some of which are life-threatening) while in the hospitals. Similarly, individuals are far more susceptible to infection if age, malnutrition, poor housing, insufficient clothing, or other difficulties weaken their bodies. This explains why American tourists rarely contract tropical diseases even when traveling in countries where disease is endemic and even when they are neither vaccinated nor taking prophylactic drugs.

Third, the same factors that leave some susceptible to infection help explain why, among those who do become infected with a given disease, some will die but others will experience only minor health problems. Measles, for example, is a minor childhood disease in the United States but a major killer in poorer countries (as Chapter 4 will describe).

Fourth, among those who become ill, death or long-term disability may not occur if individuals have ready access to good health care. For example, doctors can cure most bacterial infections in otherwise healthy individuals, and simply providing intravenous nutrition and fluids can save the lives of many infants suffering from life-threatening diarrhea.

Toxic Agents

Mokdad and his colleagues (2004) trace 2 percent of premature deaths to toxic agents. These agents can be divided into occupational hazards and environmental pollutants. In "light" industries like electronics, workers are often exposed to a wide variety of potentially toxic solvents, such as trichloroethylene (TCE), and in traditional industries such as mining and construction, welders often face substantially increased risks of lung cancer caused by toxic levels of chromium and nickel. Similarly, agricultural workers, as described in the next chapter, often are regularly exposed to dangerous pesticides.

Unlike occupational hazards, environmental pollution poses the greatest dangers to children because of their still-growing bodies and immune systems, the time they spend playing outdoors, and their tendency to play on the ground and put things in their mouths. Many forms of environmental pollution threaten children (U.S. Environmental Protection Agency, 1996). For example, about 900,000 U.S. children under age 6 have elevated levels of lead in their blood from eating old house paint, which can cause retardation, learning disabilities, hearing deficiencies, hyperactivity, and other problems. Each year, 24,000 children are poisoned by eating pesticides; and many more children are exposed to pesticides at lower but still unsafe levels. Similarly, 33 percent of U.S. children now live in areas that do not meet national air quality standards, which partly explains why 4.8 million children have asthma. Finally, 10 million children under the age of 12 live within four miles of a toxic waste dump, thus increasing their risks of cancer and genetic defects (U.S. Environmental Protection Agency, 1996).

In the long run, the greatest environmental health threat may be global warming. During the last quarter century, carbon dioxide and synthetic gases, especially chlorofluorocarbons (CFCs) such as Freon, have mushroomed. According to the Intergovernmental Panel on Climate Change, a joint venture of the World Meteorological Organization and the United Nations Environment Programme (Houghton et al., 1996), these chemical by-products of industrial manufacturing have damaged the ozone level surrounding the planet and caused temperatures to rise around the globe. Debate continues about the consequences of global warming, but many scientists suspect that global warming and the resulting damage to the ozone level will foster genetic mutations, cancers (especially skin cancer), and smog-related health problems such as bronchitis, asthma, and emphysema.

Motor Vehicles

Mokdad and his colleagues (2004) attribute 2 percent of all premature deaths to motor vehicle accidents (including accidents involving drug use but not alcohol use). These deaths are not a necessary by-product of modern life. Rather, they reflect in part a series of decisions regarding the design of automobiles and transportation systems.

Changes in car design can dramatically reduce the chances that an accident will cause death or serious injury. The rate of deaths from motor vehicle accidents has declined substantially since 1966, when Congress established the National Highway Traffic Safety Administration (NHTSA) to regulate motor vehicle design and oversee highway safety programs. NHTSA was founded in response to the public outcry that followed publication of Ralph Nader's book *Unsafe at Any Speed* (1965), which documented how automobile manufacturers for years had ignored evidence of automobile safety hazards that could have been eliminated for a few dollars per car.

Automobile manufacturers have continued to fight against inexpensive improvements that could save thousands of lives yearly, such as strengthening bumpers and side doors to resist impact, covering instrument panels and roof interiors with softer materials to protect against head injuries, and redesigning gas tanks to reduce the likelihood of explosions during crashes. Equally important, legislators and government regulators have continued to exempt vans, multipurpose vehicles, and light trucks—which now account for more than 50 percent of all noncommercial vehicle sales—from passenger car safety regulations, even though most consumers use these vehicles as family cars.

One way, then, to reduce the rate of deaths and disability caused by cars is through simple changes in car design. Another way is to get people out of cars. The most basic reason for the higher rate of motor vehicle accidents in the United States as compared to other Western nations is that U.S. residents drive far more miles per year. Although the size of the United States partially explains this difference, Americans also drive so much because they lack other options. Through a series of local and federal decisions, public transportation in this country has declined significantly since its apex in the 1920s (Yago, 1984). Trains and railroad tracks have decayed while federal dollars have subsidized highway construction and motor vehicle production. Long-distance bus systems run for profit have eliminated money-losing connections to many smaller communities. Meanwhile, cities spend billions for parking facilities, road construction, and road maintenance but offer bus service only to limited locations, during limited hours, on a limited schedule. Consequently, whereas a French citizen can use publicly subsidized trains or buses to go to any town or city in France on any given day and probably at several different times, an American citizen often has no way to go by public transportation from one town to the next. Phoenix, Arizona, for example, is the sixth largest city in the United States but has no passenger-rail service.

Nevertheless, despite these problems, the rate of motor vehicle deaths declined by almost 10 percent from 1990 to 2000. Mokdad and colleagues attribute this decrease to greater enforcement coupled with public education campaigns against drunk driving (Mokdad et al., 2004).

Firearms

According to Mokdad and his colleagues (2004), firearms account for 1 percent of all premature deaths in the United States: 16,586 suicides, 10,801 homicides, 776 accidental deaths, and 270 deaths by police. Death from firearms is very much a U.S. phenomenon. Among young males, the rate of firearm deaths is from 12 to 273 times higher in the United States than in other industrialized nations (Kellerman et al., 1993). No other country has nearly as many privately owned firearms and, not coincidentally, no other country has nearly as many firearm-related homicides; studies have found that having a gun in the home significantly increases the odds of suicide, of homicide, and of unintentional shooting deaths of children (Kellerman et al., 1993).

Those who support firearm ownership typically argue that having a gun protects individuals against attacks by criminals. Yet guns are far more often used against family members than against criminals. Furthermore, *even when a home is forcibly entered or a victim attempts to resist,* owning a gun increases the chances of being killed (Kellerman et al., 1993).

Although interest in gun control rose sharply following the 1999 massacre of twelve students and a teacher at Littleton Colorado's Columbine High School and later similar events, this interest has not translated into widespread legislative changes. Those favoring gun control face heavy financial odds, for the "gun rights" lobby routinely donates about ten times more to federal candidates than does the "gun control" lobby (Center for Responsive Politics, 2005.) Nevertheless, firearm-related violence has decreased since 1993, at least partly because of new restrictions on the sale of guns (Wintemute, 1999). Box 2.2 describes some innovative methods doctors are currently using to curb gun violence.

Sexual Behavior

Mokdad and his colleagues (2004) attribute 1 percent of premature deaths to sexual behavior, primarily via hepatitis B, HIV disease, and cervical cancer. (Although the precise mechanisms causing cervical cancer are unknown, it occurs most often among those who have multiple sexual partners and do not use condoms, diaphragms, or spermicides.) Mokdad and his colleagues also include in this category infant mortality following unplanned and unwanted pregnancies, a situation occurring most commonly among teenagers and poor women.

No "manufacturer of illness" benefits from convincing people to engage in sexual activity without protecting themselves against disease or pregnancy. However, social conditions can encourage such behavior. First, those forced by economic necessity to turn to prostitution to support themselves, whether

---

Box 2.2   *Making a Difference: Physicians for Social Responsibility*

In 1961, a group of doctors concerned about the threat nuclear arms posed to human health and life founded Physicians for Social Responsibility (PSR). With the decline of the Cold War, PSR members have shifted their focus to working, as health professionals, toward ending other forms of violence and encouraging nonviolent means of conflict resolution. Doctors around the country belong to PSR, which has chapters at many medical schools.

In the last few years, PSR members and chapters have begun numerous grassroots efforts aimed at educating both health care professionals and the public about the dangers of handgun violence. The organization does not lobby to ban gun ownership but has supported laws that would keep guns away from children, dangerous individuals, and irresponsible owners. In addition, PSR supports actions designed to stigmatize violent gun use in the same way that drunk driving and tobacco smoking have become increasingly stigmatized in recent years. For example, student members at the University of California–Irvine recently sponsored a "die-in" to raise awareness of gun violence. The die-in was staged in a busy campus location, where many students were gathered for lunch, and took place to the sound of gunfire booming from loudspeakers. Die-in organizers used the event to distribute information and materials on the dangers of gun violence. Similarly, if less vividly, students at the University of Vermont Medical School sponsored a widely publicized program in which medical students gave teddy bears to everyone who turned in a gun. Students used the exchange as a forum for raising public awareness about guns and featured a contest in which children won prizes for drawing posters about the dangers of guns. Finally, the Seattle PSR chapter, together with the Washington State Medical Association, the King County Prosecutor's Office, and the Seattle Police Department, has created a program called "Options, Choices, and Consequences," in which a physician-presenter visits school classes and describes what really happens when someone is shot, highlighting the differences between that reality and what children usually see on television.

---

male or female, often find that they cannot suggest safer sex to clients without either losing business or risking violence. Similarly, those whose intimate relationships are not based on mutual respect and equality sometimes find that suggesting safer sex to their romantic partners results in violence or abandonment (Wingood and DiClemente, 1997). Finally, those who have learned to have little hope for the future—a sentiment particularly common among youths in communities wracked by racism and poverty—sometimes feel they have little to lose by engaging in unsafe sexual activity (Plotnick, 1992).

Other sexually active individuals, however, do fear both sexually transmitted diseases and pregnancy. For these individuals, sexual activity does not need to lead to disease or pregnancy if they have knowledge about safer sexual practices and access to birth control and abortion. Knowledge about safer sex is now widespread, but access to birth control and abortion has declined over the last decade. Cuts in public funding for contraceptive services have limited options for precisely those groups—teenagers and low-income women—most at risk for unplanned pregnancies and infant mortality. Similarly, the federal government will not pay for abortions for women on **Medicaid** (the

---

Box 2.3   *Ethical Debate: Drug Testing in Schools and Workplaces*

Since 1986, the federal government has required all federal job applicants, as well as randomly selected federal employees who hold "safety sensitive" positions, to take urine or blood tests to detect illegal drug use. In addition, many businesses use blood or urine tests to identify job applicants or current employees who use illegal drugs (DeCew, 1994). Many schools require students to test negative for illegal drugs, and sometimes for alcohol and tobacco, before they can participate in extracurricular activities like sports, chess clubs, and language clubs (Steinberg, 1999).

To date, U.S. courts generally have found that use of drug tests by government agencies breaches the Fourth Amendment right to privacy, unless necessary to protect public safety or unless other evidence suggests that a particular individual used drugs. Courts generally have placed no restrictions on private employers' use of drug tests and have permitted schools to require drug tests for extracurricular activities but not for academic courses.

At first glance, the benefits of drug testing seem obvious. Students, employees, and potential employees who know they will be tested may either cease using drugs or never begin doing so, thereby reducing the overall level of drug use in

society. In addition, reducing drug use may reduce rates of both accidents and violence. Moreover, from a strictly financial perspective, reducing drug use may reduce absenteeism, tardiness, and insurance costs, while improving student and worker performance.

But drug testing comes with a price. Those opposed to drug testing argue that testing inherently invades privacy because it involves taking urine or blood from an individual's body. Moreover, the only way to ensure a urine sample comes from a specific individual is to watch that individual urinate—an obvious invasion of Western norms of privacy. In addition, drug testing constitutes an invasion of privacy because it can reveal much more than just illegal drug use. For example, the same tests that identify use of illegal drugs can identify legal use of drugs to control epilepsy, manic depression, or schizophrenia. Individuals identified in this way may experience not only social embarrassment but also discrimination and even loss of employment. Finally, drug testing invades privacy because it measures not only what a person does in school or on the job but also what she does during her free time. An individual who uses drugs only in the evenings or on weekends may test positive for drugs at school

---

government-funded health insurance program for poor persons) unless the woman's life is endangered. Studies have found that between 20 and 37 percent of women who would have had an abortion if Medicaid paid for it instead carry to term (Boonstra and Sonfield, 2000). Meanwhile, cutbacks in government funding for abortions and harassment and violence against abortion providers have reduced the number and geographic distribution of abortion providers. So, too, has the fact that only a small fraction of medical schools teach abortion, which is the only medical procedure that doctors can opt not to learn. As of 2000, thirty-four percent of women ages 15 to 44 lived in counties without any abortion provider (Henshaw and Finer, 2003). Other restrictions, such as requiring waiting periods or parental consent, also cause limited access to abortion, especially for poor and young women. Yet despite these restrictions, abortion remains common: An estimated one-third of all

or work even though the drugs no longer can affect his performance.

In addition, those who oppose drug testing in the workplace also question why, if the purpose of testing is to identify workers whose performance is impaired, we measure drug use rather than job performance. After all, some individuals who use drugs nevertheless will perform adequately while others who do not use drugs will perform poorly. Moreover, most drug-related impairment in the workplace stems from use of alcohol, yet employers usually test only for use of illegal drugs.

Finally, opponents of drug testing argue that the potential benefits of testing are far outweighed by the potential for harm when individuals are falsely labeled as drug users. As many as 40 percent of those identified as drug users by urine tests have not actually used illegal drugs. Urine tests can confuse decongestants with amphetamines, ibuprofen (Advil) with marijuana, cough syrup with morphine, and herbal teas with cocaine. The proportion of false positives is considerably lower when blood rather than urine tests are used, but the latter are more often used because they are cheaper and quicker. Similarly, schools and employers often save money by testing only once, rather than confirming test results with a second, more accurate, test. Conversely, those who use illegal drugs may go undetected if they drink large amounts of water before testing; add small amounts of salt, vinegar, or bleach to their urine sample; or time their drug use so the drugs will have left their bodies before they are tested.

In sum, developing a responsible policy regarding drug testing requires us to find a balance between public safety and protection of individual rights.

Sociological Questions

1. What social views and values about medicine, society, and the body are reflected in this policy? Whose views are these?

2. Which social groups are in conflict over this issue? Whose interests are served by the different sides of this issue?

3. Which of these groups has more power to enforce its view? What kinds of power do they have?

4. What are the intended consequences of this policy? What are the unintended social, economic, political, and health consequences of this policy?

U.S. women will have an abortion at some time during their lives (Allan Guttmacher Institute, 2001). As a result, preserving the safety of abortion services is an important health issue.

Illicit Drugs

The last cause of premature death listed in Table 2.3 is illicit drugs. According to Mokdad and his colleagues (2004), illicit drugs kill users through overdose, suicide, motor vehicle injury, HIV infection, pneumonia, hepatitis, and endocarditis (heart infections); and they kill nonusers by contributing to homicide and birth defects. In addition, illicit drug use can contribute to dangerous behaviors. This chapter's ethical debate discusses drug testing in schools and the workplace, which has emerged in response to these concerns. See Box 2.3.

The two illicit drugs that most often cause mortality and morbidity (although they are not the most commonly used illicit drugs) are heroin and cocaine (including "crack" cocaine). Both heroin and cocaine can cause physical addiction, although cocaine is usually used in quantities too small to do so (Weil and Rosen, 1998). Cocaine provides such great pleasure so briefly, however, that some individuals use it as often as possible, creating the appearance that they are addicted. As a result, both heroin and cocaine can cause people's lives to spin out of control. Although heroin causes no direct damage to the human body, cocaine can cause severe sleep disturbances, which in turn can lead to paranoia and violence (Liska, 1997; Weil and Rosen, 1998). Cocaine also may increase the risk of heart failure or stroke, although evidence for this theory is limited.

In recent years, much public attention has focused on the problems of "crack babies." Infants born to drug users do have higher than average rates of mortality and morbidity. However, a thorough reanalysis, published in the *Journal of the American Medical Association,* of all previous research on the topic strongly suggests that these problems are caused by the mothers' poverty, malnutrition, lower education levels, or tobacco smoking rather than by their illicit drug use (D. Frank et al., 2001). In this reanalysis, the authors found that the higher rates of infant mortality and morbidity among drug users are more apparent than real, because virtually all research used data collected inconsistently after births, and doctors naturally were more likely to collect data on mothers' drug use when babies had problems than when babies were born healthy.

Added to the inherent dangers of illicit drugs are the dangers caused by their illegality. As mentioned earlier, when drug users cannot obtain clean needles legally, they are likely to share needles and thus to increase their risks of HIV disease, hepatitis, and endocarditis. Similarly, users who buy drugs on the street cannot know how powerful the drugs are and thus risk overdose. For example, someone who typically injects heroin that is 30 percent pure can die if he or she accidentally buys heroin that is 60 percent pure and thus doubles his or her normal dosage.

Pneumonia, too, results not from the drugs themselves but from the poverty and disorganized lifestyle that can either lead to drug use or result from trying to obtain steady supplies of illegal drugs at the extraordinarily high prices charged by illegal drug dealers. Similarly, violence among heroin users results not from the drug itself (which makes users more passive) but because users must resort to crime to pay the high prices of illegal drugs. Cocaine, on the other hand, can directly stimulate violent behavior.

Research consistently shows that prevention and treatment programs are both cheaper and more effective than criminal sanctions in reducing the use and social costs of illicit drugs (Amaro, 1999). Unfortunately, about two-thirds of government funding for drug control goes to the criminal justice system, and only one-third goes to prevention and treatment.

Health Behaviors, Social Stress, and Illness

It is no secret that tobacco kills, guns can be deadly, sex without condoms spreads disease, and so on. So why do people continue to engage in behaviors that place their health at risk? And why do some people avoid illness despite apparently poor choices while others die young? The first question takes us to the issue of health behaviors, the second to the impact of social stress.

Health Behaviors and Health Lifestyles

The most commonly used framework for studying compliance is the **health belief model.** This model was developed by Irwin Rosenstock (1966) and extended, most importantly, by Marshall Becker (1974, 1993). Its original purpose was to explain why healthy individuals adopt preventive health behaviors. According to the model, **four factors** affect these decisions (see Key Concepts 2.1): (1) Individuals must believe they are susceptible to a particular health problem; (2) they must believe that problem is serious;

Key Concepts 2.1	*The Health Belief Model*	
PEOPLE ARE MOST LIKELY TO ADOPT HEALTHY BEHAVIORS WHEN THEY:	EXAMPLE: ADOPTING HEALTHY BEHAVIORS LIKELY	EXAMPLE: ADOPTING HEALTHY BEHAVIORS UNLIKELY
Believe they are susceptible	Forty-year-old smoker with chronic bronchitis who believes he is at risk for lung cancer.	Sixteen-year-old boy who believes he is too healthy and strong to contract a sexually transmitted disease.
Believe risk is serious	Believes lung cancer would be painful and fatal, and does not want to leave his young children fatherless.	Believes that sexually transmitted diseases can all be easily treated.
Believe compliance will reduce risk	Believes he can reduce risk by stopping smoking.	Doesn't believe that condoms really prevent sexual diseases.
Have no significant barriers to compliance	Friends and family urge him to quit smoking, and he can save money by so doing.	Enjoys sexual intercourse more without condoms.

(3) they must believe adopting preventive measures will reduce their risks significantly; and (4) they must not perceive any significant barriers to doing so. For example, people are most likely to adopt a low-fat diet if they believe that otherwise, they will face high risks of heart disease, that heart disease will substantially decrease their life expectancy, that a low-fat diet will substantially reduce their risk of heart disease, and that adopting such a diet will not be too costly, inconvenient, or unpleasant. In turn, according to the health belief model, these four factors are affected by demographic variables (such as the individual's gender and age), psychosocial variables (such as personality characteristics and peer group pressures), structural factors (such as access to knowledge about the problem and contact with those who experience the problem), and external cues to action (such as media campaigns about the problem or doctors' advice).

Although this model incorporates the possibility for social factors as well as individual psychological factors to affect health decision making, in practice the model is most often used to identify why individuals make the choices they do. In other words, researchers who use this model tend to emphasize **agency**—individual free will to make choices—over **structure**—social forces that limit the choices individuals truly have available to them (Cockerham, 2005). As a result, such researchers, along with most policymakers, more often promote policies such as educating consumers about the dangers of smoking than policies such as banning smoking in public places. The debate over the relative importance of agency and structure—often referred to as "life choices" versus "life chances"—is at the center of many theoretical discussions within sociology and, even more so, between sociology and other fields such as psychology and medicine.

All human behavior is affected by both agency and structure. No one blindly follows every social rule and expectation. Nor is anyone fully free of socialization, cultural expectations, and social limitations on what options are truly available. Nevertheless, knowing to which social groups an individual belongs allows us to predict the likelihood that he or she will adopt various health behaviors: Lower-class citizens are far more likely than upper-class citizens to smoke, men are far more likely than women to drink heavily, and so on. Consistent patterns such as these led sociologist William Cockerham to propose a new **health lifestyle theory** that acknowledges both agency and structure but emphasizes group rather than individual behaviors. Compared to the health belief model, this new theory offers a more comprehensive analysis of why healthy behaviors are or are not adopted.

Cockerham (2005: 55) defines health lifestyles as "collective patterns of health-related behavior based on [life] *choices* from options available to people according to their life *chances*." (Emphasis mine.) According to this theory (see Key Concepts 2.2), decisions about healthy and unhealthy behavior begin with demographic circumstances, cultural memberships, and living conditions. First, these factors affect individuals' experiences and socialization regarding how to think about healthy and unhealthy behaviors—whether,

Key
Concepts 2.2

Health Lifestyle Theory

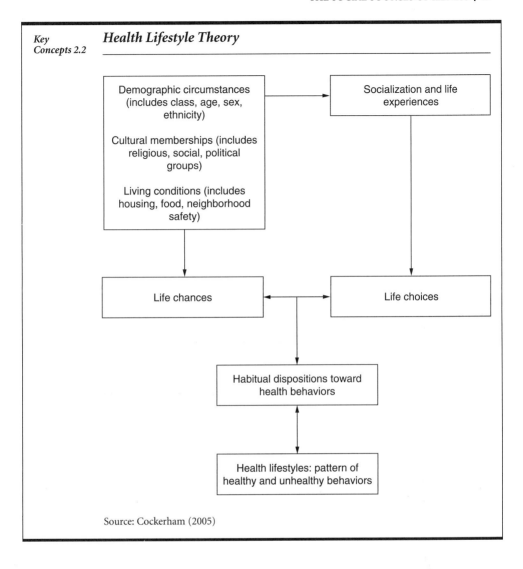

Source: Cockerham (2005)

for example, they grow up learning to consider alcohol a tool of the devil or watching their parents routinely drink for pleasure. These factors also *directly* affect individuals' life chances (such as whether they have the education needed to avoid physically dangerous jobs) and, through their effect on socialization and experiences, *indirectly* affect their life choices (such as the decision to seek dental care). For example, someone who grew up middle class likely learned early to consider dental checkups important and likely has the money to purchase dental care as an adult. In turn, life choices affect life chances, and vice versa. Those who choose to drive safely are more likely to avoid injury and have better chances to get ahead in life, while those who have better chances to get ahead are more likely to try to avoid injury because they are looking forward to the future. As this theory suggests, life choices and life chances come together to create habitual dispositions toward health behaviors—routine, almost instinctual ways of thinking about whether certain behaviors are or are not worth adopting. These dispositions are crucial to the health lifestyles individuals and groups adopt. Finally, Cockerham notes that not only do dispositions affect health lifestyles, but health lifestyles affect dispositions. As people's ways of thinking about behaviors such as smoking change, so do their behaviors; and as their behaviors change, so do their dispositions.

Social Stress

As we saw earlier, social stress helps to explain why life expectancy rises when income inequality rises. But the impact of social stress on illness extends far beyond this one circumstance. In fact, social stress can be considered another underlying cause of illness in general. It is therefore important that we understand both the nature of stress and its impact on health.

The Nature of Social Stress

The term **stress** has three major meanings. First, *stress* refers to situations that make individuals feel anxious and out of balance. (The term **stressor** is also used to refer to any specific source of stress, like buying a new house or losing a job.) Second, *stress* refers to the emotions that result from exposure to such situations. Finally, *stress* refers to the bodily changes that occur in response to these situations and emotions. For example, if we fight with a boyfriend or girlfriend, the fight itself can be referred to as a stress. Both during and after the fight, we are likely to experience the *emotion* of stress, as we wonder what the fight means for our relationship, and the *physical* stress of tensed muscles, rapid heartbeats, and heavy breathing, as our body gathers its resources in preparation for responding to the stress.

Although stress can be either acute or chronic, the latter form has far more serious health consequences. One form of chronic stress that has received considerable attention in sociology is **role strain** (Pearlin, 1989). *Role strain* refers to problems such as unwanted roles, rapidly changing roles, roles that exceed a person's resources and abilities, and conflicting roles (such as lacking the time to be both a good student and a good worker). The second form of chronic stress commonly studied by sociologists is that caused by broader social forces, such as income inequality and racism. In Chapter 3, we will discuss these stresses in more detail.

Stress is a natural, unavoidable, and sometimes beneficial part of life. Thousands of years ago, hunters experienced stress as they anxiously prepared to track wild animals, and farmers experienced stress as they wondered whether their crops would get enough rain. That emotional stress put physical stress on their bodies, but it also kept their minds focused on their tasks. If, for example, a wild animal suddenly attacked, a hunter might survive because the emotional stress resulted in the physical stress response known as the fight-or-flight syndrome. The same quick heartbeat and heavy breathing we experience while fighting with a boyfriend or girlfriend could have saved the life of someone fighting or fleeing from a lion, because these physical changes help our bodies produce additional energy and oxygen and hence respond more quickly and effectively to threats.

The Impact of Social Stress

The fight-or-flight response is highly adaptive for dealing with sudden threats like rampaging lions, speeding cars, and last-minute quizzes. It is far less useful for dealing with chronic stresses like poverty, an ill child, or a racist supervisor. Each time the body responds to a threat, it uses muscles, energy, and other resources. Over the long run, such stresses wear out the body. Chronic stress is especially likely to affect the immune system, leaving the body less able to fight off infection or illness—from herpes to asthma to heart disease (House, 2002; Siegrist, 1996, 2001). It also can lead to mental disorders such as depression (a topic explored in Chapter 7) and to conditions on the border between mental and physical (such as insomnia, migraines, and colitis). In addition, stress can lead individuals to adopt health-threatening behaviors, including smoking tobacco, driving too fast, and participating in unsafe sexual activity.

But everyone does not respond in the same way to stress, whether acute or chronic. The likelihood that stress will affect health depends in part on how individuals *appraise* the stress and how they *cope* with the stress. In turn, both of these responses to stress depend on the *social resources* individuals bring to the situation (Ensel and Lin, 1991; Pearlin and Aneshensel, 1986). For example, flunking an exam is far more stressful for a student who could lose his scholarship as a result than it is for a student who has no such fears. It will also be less stressful if the student copes by quickly seeking out a good tutor, rather than by going out for a drink or blaming his poor score on an incompetent teacher. But the student's ability to respond effectively will be determined in part by his social resources: Has he learned from a young age to turn to alcohol as a coping measure? Do his friends encourage him to continue trying or to drop out? Does he have the funds needed to hire a tutor and the contacts needed to find a good one? The answers to each of these questions will affect whether this acute stress leads to chronic stress and, in the end, to ill health.

Conclusion

Recent years have seen an increasing tendency to blame individuals for their own health problems (a topic discussed further in Chapter 5). Yet as we have seen, patterns of disease stem from social conditions as much as, if not more, than, they stem from individual behaviors or biological characteristics. As Marshall Becker, a sociologist and one of the researchers who has done the most to help elucidate why people engage in health-endangering activities, writes:

> I would argue, first, that health habits are acquired within social groups (i.e., family, peers, the subculture); they are often supported by powerful elements in

the general society (e.g., advertising); and they have proven to be extremely difficult to change. Second, for most people, personal behavior is not the primary determinant of health status and it will not be very effective to intervene at the individual level without concomitant attempts to alter the broader economic, political, cultural, and structural components of society that act to encourage, produce, and support poor health. (1993: 4)

In sum, improving the health of the population will require us to look beyond individual behavior to broader social structural issues—to look, in C. Wright Mills's terms, for public issues rather than personal troubles. Once we do so, we can focus our energies on such problems as restraining the manufacturers of illness and ensuring that public health considerations rather than special interests drive health policy.

Suggested Readings

Campos, Paul. 2004. *The Obesity Myth: Why America's Obsession with Weight Is Hazardous to Your Health*. New York: Gotham Books. A refreshing counterpoint to discussions of the dangers of obesity.

Miller, Judith, Stephen Engelberg, and William Broad. 2001. *Germs: Biological Weapons and America's Secret War*. New York: Simon & Schuster. A truly horrifying account of how terrorists and governments (including that of the United States) have developed biological weapons.

Stine, Gerald J. 2005. *AIDS Update 2005*. San Francisco: Benjamin Cummings. An excellent overview of AIDS in the United States.

Weil, Andrew, and Winifred Rosen. 2004. *From Chocolate to Morphine*. Rev. ed. New York: Houghton Mifflin. An iconoclastic review of both legal and illegal psychoactive drugs, coauthored by a famous medical school professor.

Getting Involved

Handgun Control and Brady Campaign. 1225 I Street NW, Suite 1100, Washington, DC 20005. (202) 898-0792. www.bradycampaign.org. The most influential national organization lobbying for stricter legal limits on handgun ownership.

Planned Parenthood Federation of America. 810 7th Avenue, New York, NY 10019. (212) 541-7800. www.plannedparenthood.org. The nation's foremost organization working for reproductive freedom.

Students Against Destructive Decisions (formerly Students Against Drunk Driving). P.O. Box 800, Marlborough, MA 01752. (877) SADD-INC. www.saddonline.com. Organization created by and for students to educate about the dangers of drunk driving. It has since expanded its mission to educate about depression, suicide, violence, and other dangers that young people face.

Review Questions

What is the difference between morbidity and mortality, incidence and prevalence, and acute and chronic illnesses?

What is the epidemiological transition?

What factors caused the decline in mortality between the nineteenth and early twentieth centuries?

What factors have caused the recent increases in infectious diseases, including tuberculosis and HIV disease?

How is globalization affecting rates of disease?

How have the "manufacturers of illness" increased deaths caused by tobacco? by alcohol? by toxic agents? By diet?

How have social forces and political decisions increased deaths caused by sexual behavior? caused by illicit drugs?

What system-level factors help to explain medical errors? How does medical culture keep doctors from identifying medical errors?

Think of someone you know who smokes or engages in another unhealthy behavior. Use the health belief model to explain what would have to change for him to change his behavior. Then use health lifestyle theory to explain why you yourself do or do not have a generally healthy lifestyle.

Internet Exercises

Find the website for the Center for Science in the Public Interest (CSPI) and see what it has to say about the health issues involved in either alcohol or food manufacturing. How does it define and describe the problem? Then find a website for an alcohol or food manufacturer or manufacturing group and compare its coverage of the issue with CSPI's coverage. Manufacturer groups include the Beer Institute, the Wine Institute, the Distilled Spirits Council, and the National Broiler Council (for poultry-processing manufacturers).

CHAPTER 3

©AP/Wide World Photos

The Social Distribution of Illness in the United States

On August 29, 2005, Hurricane Katrina hit the Gulf Coast of the United States. For more than two weeks afterward, people were trapped and, in some instances, dying in fetid "shelters" and on rooftops without shade, water, or food. Six months later, as I write this, more than a thousand people are known to have died, 3,200 more are still unaccounted for, and most of New Orleans remains uninhabitable.

In the days after the hurricane, the nation was riveted by photos of the dead lying in the streets and in the water, and of refugees who lost everything being scattered around the country or still awaiting evacuation in dangerously squalid conditions. As the photos made clear, these victims were disproportionately very old, very young, chronically ill, or disabled, and they were overwhelmingly African American and poor (as evident in the striking number of refugees interviewed on television who had bad teeth, one of the surest markers that an American grew up poor and unable to afford dental care).

The rich cultural history of New Orleans draws on a legacy of slavery, economic inequality, racial segregation, and racial discrimination. Virtually since the city's founding, poor African Americans were relegated to housing in low-lying areas near the coast, where flooding was most likely. The chances of flooding have only increased with time, as the city and nation's power elite supported straightening the Mississippi River channel and draining the marshes surrounding the city, even though this increased the river's force while depriving the city of its natural protections from both river and sea. Meanwhile, the system of levees that protected the city from flooding was allowed to deteriorate by politicians who believed that taxes should be kept to a minimum and that private enterprise could do a better job than government of providing transportation, housing, emergency aid, and other needs of the citizenry. Moreover, even though it was widely known that the levees would collapse in a major hurricane, the city's emergency disaster plan called only for individuals to evacuate by private car or to go to the city's Superdome. Yet planners knew that at least one-third of New Orleans residents lacked cars and that in a catastrophe the Superdome would lack sufficient water, food, electric generators, medical personnel, and other crucial goods and services.

Once the hurricane struck, tens of thousands of city residents were unable to flee because they lacked cars in a city with minimal public transportation, lacked money for gas (especially at the end of a month, when paychecks have been spent), or lacked a means to pay for hotel rooms if they did leave. Not knowing how severe the storm would be, many lost their lives because they incorrectly guessed that they would be better off chancing homelessness in their own town, among family, friends, and familiar surroundings, rather than guaranteeing they would become homeless elsewhere among strangers. Others lost their lives because they were too old or too young to flee or because, like poor populations everywhere, they were disproportionately likely to be disabled by diabetes, heart disease, and other health problems, leaving them unable to flee and unable to survive without food, water, prescription drugs, and other basic necessities.

When, almost a century ago, the steamship Titanic hit an iceberg and sunk, less than 3 percent of women and children traveling first class—where all the lifeboats were kept—died, compared to almost half in third class (W. Hall, 1986). Similarly, the horrors experienced by New Orleans residents stemmed not only from the hurricane but also from decisions we have made as a society. That this hurricane disproportionately injured, disabled, and killed persons who were poor, minority, and elderly or very young is an all too common pattern. (Likewise, whenever societies are under stress, women and girls may be especially vulnerable to rape, violence, and attendant health risks.) In this chapter, we look at how four social factors—age, sex and gender, social class, and race or ethnicity—combine with biological forces to inequitably distribute illness, disability, and death in the population.

Age

Overview

Not surprisingly, age is the single most important predictor of mortality and morbidity. As noted in Chapter 2, until the twentieth century, deaths during the first year of life were common. Although far less common now, infant mortality remains an important issue because so many years of productive life are lost when an infant dies and because infant mortality so often is caused by preventable social and environmental conditions. However, because infant mortality is so closely linked to social class and race or ethnicity, we discuss infant mortality in more detail later in this chapter.

Once individuals pass the danger zone during and immediately after birth, mortality **rates** drop precipitously. Those rates begin to rise significantly beginning at about age 40 and escalate with age. For those who survive past age 65, **chronic illnesses** (such as cardiovascular disease, diabetes, and arthritis) rather than **acute illnesses** comprise the major health problems, often bringing years of disability in their wake.

Because age and illness are so closely linked, when the average age of a population changes, so does the overall health of that population. Since 1900, the American population has aged steadily, with the population over age 85 growing the fastest.

Although most middle-aged and older persons are relatively healthy, rates of illness, disability, and mortality inevitably rise as the population ages. Similarly, both the total costs for health care and the percentage of health care dollars spent on the elderly—already greatly disproportionate to the size of that population—will increase. At the same time, as young persons become a smaller proportion of the population, the pool of persons who can provide or pay for the care needed by the elderly will shrink. Consequently, it will become more difficult to provide services to all the elderly persons who will need health care or assistance with daily tasks such as shopping or cooking.

These problems are amplified by the **feminization of aging**—the steady rise in the proportion of the population who are female in each older age group, so that women comprise a larger proportion of the elderly than of the young and middle-aged. Because elderly women more often than elderly men are poor and lack a spouse who can or will care for them, and because (as we will see in the next section) women in general experience more illness than men do, the feminization of aging will increase the costs of providing health and social services to the elderly.

Case Study: Prostate Cancer and Aging in Men

Among men, one almost inevitable consequence of aging is cancer of the prostate, a poorly understood bodily organ that produces chemicals believed necessary for reproduction. Most men develop prostate cancer by middle age, and virtually all do so if they live long enough (Kolata, 2005). Members of all racial and ethnic groups can get prostate cancer, but for some still-unknown reason, African Americans are especially susceptible.

Prostate cancer typically grows extremely slowly; most men who have it are killed by something else before the cancer can grow large enough to threaten their health. Because prostate cancer is so common, however, the small percentage of men that do develop this health problem account for about 35,000 deaths per year—slightly fewer than the number of deaths per year caused by breast cancer. Moreover, when prostate cancer does grow, it often leads to excruciatingly painful bone cancer.

Before doctors can treat prostate cancer, they first must identify it. To do so, doctors, since the 1970s, have tested their male patients at periodic intervals for prostate-specific antigen (PSA), a chemical produced by the prostate. If a patient's PSA level has increased significantly, doctors then perform a biopsy—inserting a needle into the prostate to remove a few cells, which they then check for cancer. Unfortunately, PSA tests are highly inaccurate: About 30 percent of those who have cancer are not identified by the test and about two-thirds of those identified by the test as having cancer in fact do not have it. The test brings no benefits to those whose cancers are missed, while those who are falsely identified as having cancer suffer emotional trauma, financial costs, and painful procedures before learning that the test results were incorrect.

If the biopsy suggests cancer, doctors usually perform a prostatectomy (that is, surgical removal of the prostate). The surgery succeeds in removing the cancer in about 80 percent of cases. Even in these cases, however, the risks of surgery can outweigh the benefits. Between 0.5 percent and 2 percent of patients die within a month of surgery, and another 5 percent experience serious and potentially deadly complications (Lu-Yao et al., 1993). In addition, more than 30 percent become impotent and 7 percent develop urinary incontinence, with many more experiencing periodic sexual or urinary problems. Perhaps most important, large studies using **random samples** and **controlling** for other variables have found no significant differences in survival rates between men who do and do not receive prostatectomies, apparently because the short- and long-term dangers of surgery counterbalance the benefits and because untreated prostate cancer rarely causes death (Holmberg et al., 2002; Litwin et al., 1998).

Despite the limitations of current screening techniques and treatments, the American Cancer Society now recommends routine PSA screening for all men beginning at age 50. If this recommendation is followed and the United States implements a nationwide screening and treatment program, the cost of detecting and treating prostate cancer will rise exponentially: Any money saved by treating prostate cancer patients at earlier stages of the disease will be more than counterbalanced by money spent on screening and treating men who probably never would have experienced health problems related to prostate cancer (Mann, 1993).

In sum, at least among older men, the financial, emotional, and physical costs of identifying and treating prostate cancer seem to outweigh the benefits. Consequently, the rapid adoption of these strategies seems "a case study in one of the American medical system's worst shortcomings—its propensity to embrace expensive treatments without considering their long-term social or medical impact" (Mann, 1993: 104). This **technological imperative,** which drives doctors to use all available technology, is discussed in more detail in Chapter 11.

Sex and Gender

Overview

Both sex and gender strongly affect health status. **Sex** refers to the biological categories of male and female, to which we are assigned based on our chromosomal structure, genitalia, hormones, secondary sexual characteristics such as facial hair, and so on; those who have two X chromosomes and a vagina are sexually female, those with one X and one Y chromosome and a penis are sexually male. (Later in this section, we will consider those who do not fit neatly into these categories.) In contrast, **gender** refers to the social categories of masculine and feminine, and the social expectations regarding masculinity and femininity, to which we are assigned based on our behavior, personalities, and so on. Because these categories are social, they vary across time and across culture.

Basic epidemiological data suggest that sex and gender can affect health. For example, before the twentieth century, complications of pregnancy and childbirth often cut short women's lives, and so on average women died younger than did men. These days, however, American women (regardless of race) live longer than men do, as Table 3.1 shows—even though the same set of diseases (including heart disease, cancer, and cerebrovascular disease) eventually kills most people. The *differences* between men and women's life expectancies suggest that sex may directly affect health, while the *changes* in these differences across time suggest that gender affects health: Women now live longer than men not because their biology has changed, but because their social position and access to resources have changed.

But mortality differences tell us only part of the story. If we look only at life expectancies, we might conclude that women are biologically hardier than men. When we look at *morbidity* rates, however, the picture blurs. At each age, men have higher rates of mortality and of fatal diseases, even though women have higher rates of morbidity and of nonfatal disease (Rieker and Bird, 2000). Arthritis, for example, which is the most common chronic,

Table 3.1	*Life Expectancy at Birth and at Age 65, by Race/Ethnicity and Sex*			
	WHITE		AFRICAN AMERICAN	
	MALES	FEMALES	MALES	FEMALES
At birth:	75.0	80.2	68.6	75.5
At age 65:	81.5	84.5	79.4	82.9

Source: U.S. Bureau of the Census (2004).

nonfatal condition among both men and women above age 45, strikes women about 50 percent more often than it does men. In addition, at each age, women experience a 20 to 30 percent greater incidence of *acute* conditions (not including health problems related to their reproductive systems). In sum, women live longer than men but experience more illness and disability, whereas men experience relatively little illness but die more quickly when illness strikes.

How can we explain these paradoxical findings? Some researchers have hypothesized that women's higher rates of illness are more apparent than real—that women do not actually experience more illness than men but simply *label* themselves ill and seek health care more often. Most researchers, however, have concluded based on various measures of health status that the health differences between men and women are real. They trace these health differences to both the biological differences of sex and the socially reinforced differences of gender.

Sex does seem to offer females some biological health benefits (Rieker and Bird, 2000). Perhaps in natural compensation for those females who die from childbearing, in societies where females receive sufficient nourishment, more females than males survive at every stage of life from fetus to old age. Although the exact mechanisms through which this works are unknown, some theorize that estrogen and other "female" hormones (which in fact also occur in males, but in different proportions) somehow protect the heart and other bodily organs and tissues from fatal disease.

Gender also protects women from fatal disease and injury (Rieker and Bird, 2000). Most importantly, because of differences in male and female gender roles, women less often engage in potentially disabling or deadly activities. Men are more likely than women to use legal and illegal drugs, drive dangerously, participate in dangerous sports, or engage in violence. Work, too, more often endangers men, who more often labor in dangerous occupations like agriculture or commercial fishing. Less importantly, gender roles more often bring women than men into routine contact with medical care. Unlike men, who are socialized to downplay physical problems as signs of weakness, women are more comfortable seeking health care when they experience problems. In addition, because they often must obtain health care for children or elderly parents and must seek obstetric or gynecological care for themselves, women are more likely than men to meet with health care providers. As a result, women are more likely to have health problems identified and treated early enough to make a difference.

Sex and gender may also help explain why, despite women's lower rates of mortality, their rates of morbidity are higher than are those for men. Research on this topic, however, is far less conclusive (Barker, 2005). Most commonly, theories suggest that women are more susceptible to nonfatal illnesses because of their hormones (a sex effect) or their relatively high stress levels and low control over their lives (a gender effect).

A Sociology of Intersex

So far, we have been talking about sex as if it were a *binary category*—one with two and only two conditions, male or female. However, up to 2 percent of babies are born with genitalia that appear neither clearly male nor clearly female (Blackless et al., 2000). Such babies are referred to as **intersex**: having characteristics of both sexes. Intersexuality refers to biological sexual characteristics, and it is not the same as homosexuality, which refers to sexual desires and practices.

Intersexuality can be caused by hormonal factors, chromosomal factors, or both. During their first eight weeks of development, the only sex differences among fetuses are their chromosomes (XX among females, XY among males). After that point, the production of male hormones leads some fetuses to develop male genitalia, while the same fetal tissue becomes female genitalia in the absence of these hormonal changes. A slightly different hormonal balance produces fetuses that have both male and female external genitalia (penis, testicles, clitoris, vagina, labia) or internal genitalia (gonads, uterus, fallopian tubes). This can happen for many reasons. For example, some fetuses inherit unusual hormonal patterns or sex chromosome patterns (such as XO or XXY rather than the typical male XY or female XX), and others are affected by hormones or environmental pollutants their mothers are exposed to.

The social response to intersex conditions varies greatly across cultures. Some cultures revile the condition, and expect parents or midwives to kill intersex babies at birth. Other cultures assume that three or more sexes occur naturally in the population and consider intersex to be merely a normal human variation. These cultures typically integrate intersex individuals into normal social life. Still others assign special, valued roles to intersexed individuals. Modern Western culture, however, generally supports hiding intersex, stigmatizing it, or eliminating it in some way.

Beginning in the 1950s, surgery and hormonal manipulation became the standard medical practice for handling intersex children (S. Kessler, 1998). Under the leadership of Dr. John Money of Johns Hopkins University, doctors urged parents to have their intersex children surgically reassigned to be either male or female as early as possible, on the assumption that this would help children develop into the "appropriate" gender. Decisions about which sex to assign reflected doctors' cultural assumptions about gender: Children were assigned to be boys if doctors considered their penises sufficiently large, and were assigned to be girls if their internal organs would allow them to give birth. Boys with penises considered too small had their penises surgically removed and artificial vaginas constructed, even if their hormonal and chromosomal makeup were indisputably male. Girls with clitorises considered unattractively large had their clitorises surgically removed or reduced, even though this meant removing healthy organs and impairing their adult ability to experience sexual pleasure. To assist the children in adopting their assigned sex, parents were instructed to socialize them strictly to their new gender, to hide their history from them, and to place them on a steady (if secret) diet of sexual hormones to change the children's bodies to better match their assigned sex.

Currently surgery is performed on about one to two of every 1,000 babies, with lifelong hormonal injections following (Blackless, 2000). This treatment became the norm because doctors assumed it was the most humane option, although no research was available on its psychological, social, or physical consequences. Since the 1990s, however, this standard medical treatment has come under considerable attack, both from scholars and from activists who themselves experienced sex reassignment as children. Opponents of sex reassignment point out that this treatment is based not on scientific evidence but on gender beliefs: that small penises are "unmanly," that large clitorises are frightening, that children need strict socialization into "appropriate" gender behaviors, and that a vagina need only permit penile penetration, not provide natural lubrication, elasticity, or the possibility of female sexual pleasure (S. Kessler, 1998; Preves, 2003). Moreover, opponents argue, sex reassignment reinforces children's sense of difference, reduces their ability to enjoy sexual pleasure as adults, and depends on webs of deception among children, parents, and doctors that create their own psychological nightmares (S. Kessler, 1998; Preves, 2003).

At this point, there is insufficient evidence to say whether sex assignment more often helps or harms these children. Surgical intervention remains the norm, but doctors increasingly are holding off on surgery at least briefly to allow parents time to consider other options, rather than presenting sex reassignment as the only possibility (M. Navarro, 2004).

Case Study: Woman Battering as a Health Problem

One health issue in which gender plays an especially critical role is woman battering. Although neither health care workers nor the general public typically thinks of battering as a health problem, woman battering is a major cause of injury, disability, and death among American women, as among women worldwide.

The best data currently available on the extent of woman battering come from a national, random survey of 16,000 women and men, conducted during 1995–1996 by researchers cosponsored by the U.S. Centers for Disease Control and Prevention and the U.S. National Institute of Justice (Tjaden and Thoennes, 1998). Half of the surveyed women (51.9 percent) had been physically assaulted during their lives, and 17.6 percent had experienced rape or attempted rape. Three-quarters of those who were raped or assaulted as adults had been attacked by a current or former husband, lover, or date. Women were about twice as likely as men to report that they were seriously injured during an attack, and about one-third of the seriously injured women needed emergency health care. Extrapolating from these data, the researchers estimate that more than a half million women per year seek care at hospital emergency rooms for injuries resulting from assault by an intimate partner. Other studies have concluded that about 35 percent of women patients in hospital emergency rooms go there to seek treatment for injuries caused by battering (Council on Scientific Affairs, 1992; Novello et al., 1992).

That assaults by men should far surpass battering by women should not surprise us. Before 1962, U.S. courts consistently ruled that women could not sue their husbands for violence against them—in essence declaring wife battering a man's legal and even moral right. Even after that date, most police refused to arrest men for wife battering and most courts refused to prosecute, a situation that did not begin to change for more than a decade.

Woman battering continues to exist because it reflects basic cultural and political forces in our society and, indeed, around the world (Dobash and Dobash, 1998). Through religion, schools, families, the media, and so on, women often are taught to consider themselves responsible for making sure that their personal relationships run smoothly. When problems occur in relationships, women are taught to blame themselves, even if their husbands respond to those problems with violence. Moreover, once violence occurs, women's typically inferior economic position can leave them trapped in these relationships. Men, meanwhile, often receive the message—from sources ranging from pornographic magazines to religious teachings that give husbands the responsibility to "discipline" their wives—that violence is an acceptable response to stress and that women are acceptable targets for that violence. Although most men resist these messages, enough men absorb these messages to make woman battering a major social problem.

Battering occurs most often among men who believe that their power within the family is threatened. For example, men are significantly more likely to batter their wives if they are unemployed or in economic trouble, if their wives have higher educational or occupational levels than they do, or if their wives in some way appear to challenge their power (Lips, 1993: 311–314). In addition, battering occurs most often among men who have a high need for power and who support traditional gender roles. Taken together, these data tell us that woman battering is not only an individual response to social stress, but, at a broader and largely unconscious level, a form of **social control** (that is, a way social expectations and power relationships are reinforced—in this case, reinforcing men's power over women and women's inferior position within society). Consequently, as long as gender inequality remains the norm, so will woman battering.

Recognition of battering as a health risk has led various health-related organizations to enter the fight against woman battering. During the last decade, the U.S. Centers for Disease Control and Prevention has begun funding research on the causes, consequences, and prevention of battering, and the U.S. Public Health Service has evaluated and helped develop violence prevention programs, trained health professionals and others in violence prevention, and encouraged health care workers to learn how to identify battered women in emergency rooms. Similarly, the American College of Obstetricians and Gynecologists now requires medical schools to teach how to identify and respond to battered women and publishes materials designed to aid health professionals in doing so.

Social Class

Overview

Social class refers to individuals' position within their society's economic and social hierarchy. Most often it is measured by looking at individuals' education, income, or occupational status, with some researchers using only one of these indicators and some combining two or more. Other researchers have argued for additional measures, with wealth perhaps the most important. For example, imagine two students who work together at Starbucks, earning the same income. Now imagine that one receives a new wardrobe and a trip to Europe from her parents every summer, whereas the other receives only a bus ticket home. These students have the same income, education, and occupation, but differ in social class because they differ in wealth.

In addition to being a characteristic of individuals, social class is also a characteristic of groups, activities, occupations, and geographic areas. Bowling, for example, is most popular in working-class neighborhoods and would be characterized by most as a working-class activity. Most bowling teams are working class, and most team members belong to the working class. Like individuals' social class positions, these structural elements of social class also affect health.

Finally, social class is part of the structure of a society. As we saw in Chapter 2, some societies are more characterized than others are by income inequality—which is largely the same as social class inequality. In such societies, individuals are highly likely to remain in the social class into which they were born, and the difference between the lives of those at the bottom and those at the top of the class structure is very great.

The link between social class and ill health is strong and consistent. For example, the food, shelter, and clothing available to poor Americans 200 years ago differed greatly from that available to poor Americans now, which in turn differs greatly from that available to poor Brazilians these days. Even so, in each place and era, poor persons experience more illness than wealthier persons do. Because of this very strong link between social class and health, some sociologists label social class a "fundamental cause" of illness (Link and Phelan, 1995; Phelan et al., 2004).

The impact of social class on health is obvious: Within the United States as elsewhere, at each age and within each racial or ethnic group, those with higher social class status have lower rates of morbidity and mortality

(Feinstein, 1993; Marmot, 2002, 2004; Marmot and Shipley, 1996; V. Navarro, 1990; D. Williams and Collins, 1995). This relationship holds true for all major and most minor causes of death and illness, and regardless of how researchers measure social class (Wilkinson, 1996, 2005). For example, heart disease occurs three times as often and arthritis twice as often among low-income persons compared to more affluent persons. Moreover, these health differences appear not only when the poorest and the wealthiest are compared but also across the entire income scale, with each group on the social class ladder having better health than the group just below it (Wilkinson, 1996, 2005; Marmot, 2004). Controlling for all known individual risk factors (such as obesity and smoking) only slightly reduces the impact of social class on mortality and morbidity rates (Wilkinson, 1996, 2005).

The relationship between social class and ill health begins at birth, with infant mortality significantly higher among those born to poor women (Nersesian, 1988). Similarly, poor children are more likely than other children to become ill or to die (Federal Interagency Forum on Child and Family Statistics, 1999). Only 65 percent of poor children are described by their parents as having very good or excellent health, compared with 84 percent of other children. Similarly, poor children are almost twice as likely as other children to be physically disabled by chronic health problems.

Journalist Laurie Kaye Abraham (1993), in her book *Mama Might Be Better Off Dead*, vividly describes the overwhelming toll that poverty can take on a family's health. Abraham traces the health history of Jackie Banes and her family, who live in Chicago's predominantly African American North Lawndale neighborhood, where unemployment is the norm and almost half of all residents are on welfare. According to Abraham,

> accompanying this kind of poverty is a shocking level of illness and disability that Jackie and her neighbors merely take for granted. Her husband's kidneys failed before he was thirty; her alcoholic father had a stroke because of uncontrolled high blood pressure at forty-eight; her Aunt Nancy, who helped her grandmother raise her, died from kidney failure complicated by cirrhosis when she was forty-three. Diabetes took her grandmother's legs, and blinded her great-aunt Eldora, who lives down the block. . . .
>
> For the most part, the diseases that Jackie and her family live with are not characterized by sudden outbreaks but long, slow burns. As deadly infectious diseases have largely been eliminated or are easily cured—with the glaring exceptions of AIDS and drug-resistant tuberculosis—chronic diseases have stepped into their wake, accounting for much of the death and disability among both rich and poor. Among affluent whites, however, diabetes, high blood pressure, heart disease, and the like are diseases of *aging*, while among poor blacks, they are more accurately called diseases of *middle-aging*. In poor black neighborhoods on the West Side of Chicago, including North Lawndale, well over half of the population dies before the age of sixty-five, compared to a quarter of the residents of middle-class white Chicago neighborhoods. (Abraham, 1993: 17–18)

Importantly, health is affected more by social class than by race or ethnicity—which, in the United States, is highly correlated with class (Baquet et al., 1991; V. Navarro, 1990; Nersesian, 1988; Otten et al., 1990; D. Williams and Collins, 1995). For example, data from one national random sample found that apparent race differences in mortality rates between Mexican Americans, Asian Americans, and white Americans disappeared once social class was controlled for, and differences between African Americans and white Americans diminished substantially (Rogers et al., 1996). Looking at the same issue from a different angle, another study also using a national random sample found that class differences in mortality and morbidity were almost twice as great as race differences (V. Navarro, 1990). For example, morbidity was 4.6 times more common among those making $14,858 or less per year (in 2005 dollars) compared with those making more than $52,000, but only 1.9 times more common among African Americans than among whites. These numbers suggest that social class is a more powerful predictor of mortality and morbidity than is race or ethnicity. This does not, however, reduce the importance of race or ethnicity, for both contemporary and historical racial discrimination remain at the root of minority poverty. Rather, it suggests that if incomes and social positions of minorities rise, the racial gaps in health status will diminish (Farmer, 1999; D. Williams and Collins, 1995).

The Sources of Class Differences in Health

How can we explain the link between poverty and illness? One possible explanation is that illness causes poverty: As people become disabled or ill, their abilities to earn a living or attract an employed spouse decline, and they fall to a lower social status than that of their parents. This explanation is known as **social drift** theory. Studies that have tracked cohorts of Americans over time, however, have found that social drift explains only a small proportion of the poor ill population (D. Williams and Collins, 1995). Instead, and far more often, poverty causes illness (Marmot, 2002, 2004).

But how does poverty cause illness? Most basically, sociologists argue, those who belong to the lower class experience worse health because, compared to wealthier persons, they are subject to more stress, have less control over that stress, and have less access to health-preserving resources (Link and Phelan, 1995; Phelan et al., 2004). These problems play themselves out in many aspects of everyday life. The most important of these are work conditions, environmental conditions, housing, diet, and access to health care.

First, the work available to poorly educated lower-class persons—when they can find it—can cause ill health or death by exposing workers to physical hazards. A coal miner, for example, is considerably more likely than a mine owner to die from accidental injuries or lung disease caused by coal dust. In addition, lower-status workers typically experience both demanding work conditions and low control over those conditions. For example, factory workers must keep pace with the production line but cannot control either the speed of the line or even when they take bathroom breaks. Numerous studies have found that workers who face high demands with little control over their work conditions are particularly likely to experience stress, resulting in both physical and psychological illness (North et al., 1996; Marmot, 2004; Wilkinson 2005).

Second, environmental conditions can increase rates of morbidity and mortality among poorer populations. Chemical, air, and noise pollution all occur more often in poor neighborhoods than in wealthier neighborhoods both because the cheap rents in neighborhoods blighted by pollution attract poor people and because poor people lack the money, votes, and social influence needed to keep polluting industries, waste dumps, and freeways out of their neighborhoods (Bullard, Warren, and Johnson, 2001; Camacho, 1998). Pollution fosters cancer, leukemia, high blood pressure, and other health problems, as well as emotional stress. Because of this, both poor and middle-class persons who live in poor neighborhoods have higher mortality rates than do persons with similar incomes who live in more-affluent neighborhoods (Haan, Kaplan, and Camacho, 1987).

Third, inadequate, overcrowded, and unsafe housing increases the risk of injuries, infections, and illnesses, including lead poisoning when children eat peeling paint, gas poisoning when families must rely on ovens for heat, and asthma triggered by cockroach droppings, rodent urine, and mold (Reading, 1997). For example, Dr. Arthur Jones, who runs a clinic in Lawndale, told author Laura Abraham of his initial response to a patient with severe cat allergies who nonetheless refused to give away her cat:

> "I really got kind of angry," Dr. Jones remembered, "and then she told me that if she got rid of the cat, there was nothing to protect her kids against rats." Another woman brought her 2-year-old to the clinic with frostbite, so Dr. Jones dispatched his nurse . . . to visit her home. . . . The nurse discovered icicles in the woman's apartment because the landlord had stopped providing heat. (Abraham, 1993: 18)

Fourth, the food poor children eat—or don't eat—affects lifetime risks of illness. Federal researchers estimate that during 2003 almost 17 percent of poor families with children sometimes or often did not have enough food to eat (Nord, Andrews, and Carlson, 2004). Children who live in such circumstances have significantly more colds each year and are significantly more likely to be in poor health, lack sufficient iron, experience chronic headaches or stomachaches, or have a disability. This situation is likely to worsen over the next few years, as more families reach the five-year lifetime limits on welfare and food stamp benefits that were implemented during the 1990s "welfare reform" movement (Hancock, 2002).

The "diet of poverty" also increases health risks among the poor (James et al., 1997). This diet relies heavily on fast foods children can prepare for themselves while their parents work and fatty or sweet foods that satisfy hunger and provide energy inexpensively but offer little nutrition. Such a diet saps children's concentration and intellectual abilities, making it difficult for them to succeed in school and continuing the cycle of poverty.

Poor children also suffer nutritionally because they are less likely than others are to be breast-fed (U.S. Department of Agriculture, 1999). Infants who are not breast-fed are more likely than others to develop infections, diabetes, allergies, and other health problems and to die in infancy or early childhood (Lawrence, 1997; Raisler, Alexander, and O'Campo, 1999). Lower rates of breast-feeding among the poor reflect cultural differences in attitudes, more limited education about how and why to breast-feed, and less control over their daily circumstances. For example, women professors are more likely to have the option of breast-feeding at work than are waitresses or maids.

Fifth, poverty limits individuals' access to health care. In the United States, only the poorest can receive free health care under the **Medicaid** health insurance program (described in more detail in Chapter 8). Even these individuals still can find it difficult to obtain care if they cannot afford time off from work for medical visits, transportation to the doctor, or child care while there. Many more Americans, referred to as the **medically indigent,** earn too much to qualify for Medicaid but too little to purchase either health insurance or health care. Not surprisingly, and as Table 3.2 shows, even if they have health insurance poor children and adults are less likely than others to have a regular source of medical care (National Center for Health Statistics, 2004: 256). Instead, they receive care in hospital clinics or emergency rooms, where quality of care is necessarily lower than in less-rushed and less-crowded settings. Similarly, poor children are less likely to receive all necessary vaccinations by the recommended ages (National Center for Health Statistics, 2004: 251).

Access to health care cannot eliminate class differences in mortality and morbidity—differences that exist even in countries where access to care is universal—because it cannot eliminate the other factors that leave poor people more susceptible to illness in the first place (Marmot, 2002, 2004). For this

Table 3.2	*Percentage of Children Under Age 18 With No Usual Source of Health Care, by Insurance Status and Income, 2001–02*		
INSURANCE STATUS	POOR (%)	NEAR POOR (%)	NOT POOR (%)
Insured	5.5	4.8	2.3
Uninsured	39.7	27.9	19.7

Source: National Center for Health Statistics (2004: 256).

reason, access to health care plays a smaller role in the relationship between poverty and ill health than do the other factors discussed so far (Feinstein, 1993; D. Williams and Collins, 1995). Nevertheless, access to health care can protect against some problems, such as debilitating dental disease preventable through routine cleaning and disabling illnesses preventable through immunization. In addition, access to health care can improve quality of life dramatically through such simple interventions as providing eyeglasses, hearing aids, and comfortable crutches or wheelchairs. Conversely, lack of access can have deadly consequences (as we will see in Chapter 8). One large-scale study found that by the end of a 10 year period, 18.4 percent of those who lacked health insurance had died, compared with only 9.6 percent of those who had insurance (Franks, Clancy, and Gold, 1993). Even when the researchers statistically controlled for sex, age, race, education, preexisting illnesses, or use of tobacco, they still found 25 percent more deaths among uninsured persons than among insured persons.

In all these ways, then, poverty and illness are linked by underlying social conditions. Unfortunately, these social conditions have worsened over the last few decades, and social class differences in morbidity and mortality rates have continued to grow (D. Williams and Collins, 1995).

Case Study: Health Among the Homeless

The impact of social class on health falls heaviest on the homeless. Homelessness has been a major problem for the United States since the early 1980s, when the federal government slashed funds for low-income housing while increasing subsidies for "gentrifying" good-quality older buildings in inner-city neighborhoods (Aday, 2001). Although the latter policy was intended to improve quality of life in these neighborhoods, its unintended consequence was to raise rents. Meanwhile, the value of the minimum wage (adjusted for inflation) declined, and public assistance became harder to get and lower in value. As a result, an American must earn *twice* the mandated minimum wage to afford a modest, two-bedroom apartment (National Low Income Housing Coalition, 2001).

Not surprisingly, given the physical and emotional strains of life on the streets, homeless persons experience a disproportionate share of chronic and acute illnesses, as well as greatly increased mortality rates. Researchers estimate that 35 percent of homeless people in Los Angeles have active tuberculosis, and more than 30 percent have some other chronic health conditions (Cousineau, 1997; Kleinman et al., 1996). Homeless women face additional risks from rape and violence: One study of 53 long-term homeless women found that 15 percent had been raped and 42 percent battered in the preceding year (B. Fisher et al., 1995). Finally, a random survey of residents of New York City homeless shelters found **age-adjusted** death rates for both men and women four times higher than among other New Yorkers, with rates highest for those who had been homeless the longest (Barrow et al., 1999).

Homeless children—a growing population—face a particular set of health risks (Aday, 2001). Studies have found that about 50 percent of children in New York City's homeless shelters have asthma, compared to 25 percent of children in the city's poorest neighborhoods and 6 percent of children overall (Pérez-Peña, 2004). Asthma can threaten children's lives and, by making breathing so difficult, can make it impossible for them to concentrate in school or enjoy any activities outside of school. Yet only 50 percent of New York City's homeless children with severe asthma have been diagnosed by a doctor, and only *10 percent* are receiving medication to treat it. Similarly, a Massachusetts study found that homeless children experience ear infections, diarrhea, fever, and severe asthma more often than other children and are more likely to be in fair or poor health overall (Weinreb et al., 1998).

All the factors explaining high rates of morbidity and mortality among poor persons also apply to homeless persons. However, maintaining health is even more difficult for homeless persons than for other poor persons. For example, because poverty, malnutrition, and cold weaken their bodies, and because they often can find shelter only in crowded dormitories where infections spread easily, homeless persons are more likely than others to develop upper respiratory infections. If they develop an infection, they cannot rest in bed until they recover, because they have no beds to call their own. Similarly, homeless persons often suffer skin problems such as psoriasis, impetigo, scabies, and lice; if left untreated, these conditions can cause deadly infections. Even if homeless persons receive prompt treatment for these skin problems, their living conditions make it impossible for them to keep their linens and clothing clean enough to prevent reinfection. Finally, homeless persons, regardless of age or sex, often can support themselves only through prostitution, which dramatically increases their risks of rape, battering, and sexually transmitted diseases, including **HIV disease.**

Access to health care is also particularly difficult for homeless persons. The struggles necessary to meet basic needs for food, clothing, and shelter can leave individuals with little time, energy, or money for arranging transportation to health care facilities or for purchasing health care or prescription drugs. In addition, both substance abuse and mental illness—which affect more than 40 percent of homeless persons and can either cause or result from homelessness—can make it harder for individuals to recognize they need health care, to seek care promptly when they recognize it is needed, to follow the instructions of health care workers, and to return for needed follow-up visits (Cousineau, 1997).

In the book *Under the Safety Net,* Brickner and his colleagues describe the true costs homeless people pay and the limited benefits they receive when they seek health care:

A homeless man with severe cellulitis [diffuse inflammation under the skin] of the legs, skin breakdown, and bilateral leg ulcers makes his way to the local hospital emergency room. Because he is not a genuine emergency, he waits for five hours. He loses his opportunity for lunch at a soup kitchen. He loses a bed for the night because he wasn't standing in line at the right time. He finally is examined by a physician and given a prescription for antibiotics, told to stay supine [on his back] for a week with his legs elevated and soaked in warm dressings, and given a return appointment for clinic. The realities of his life prohibit him from carrying out any portion of this treatment plan. (Brickner et al., 1990: 10)

In sum, until the underlying conditions causing homelessness are alleviated, health care workers can offer homeless persons only the most temporary of help.

Race and Ethnicity

The concept of "race" is a social construction, with almost no biological basis. For example, a century ago many "white" Americans considered Jews and Irish people as separate and inferior races (Jacobson, 1998). Similarly, contemporary Americans typically label individuals "African American" if they have any known African ancestors, even if most of their ancestors were European. For this reason, from this point on this textbook uses the term *ethnicity*, which suggests cultural rather than biological differences, rather than the less accurate term *race*.

As noted in the previous section, social class affects health more than does ethnicity. Yet ethnicity remains an important and independent factor in predicting health status. In this section we look at health and illness among African Americans (12.8 percent of the U.S. population), Hispanics (13.7 percent), Asian Americans (4.0 percent), and Native Americans (1.0 percent). As we will see, life expectancy is shortest among African Americans and longest among Asian Americans.

Ethnic differences are also apparent in active and inactive life expectancy. **Active life expectancy** is the number of years a person can expect to live in good health and without disabilities; **inactive life expectancy** measures the years a person can expect to live in poor health and with disabilities (Hayward and Heron, 1999). (The two figures added together equal total life expectancy.) As Figures 3.1 and 3.2 illustrate, on average, Asian Americans not only live longer but also have a higher active life expectancy than do members of other ethnic groups. Both total life expectancy and active life expectancy are greater for white non-Hispanics than for Hispanics, and greater for Hispanics than for African Americans. Finally, although Native Americans on average live as long as white non-Hispanics, the former live more years in poor health than any other group. For the remainder of this section, we will explore in more detail some reasons for these ethnic differences in health.

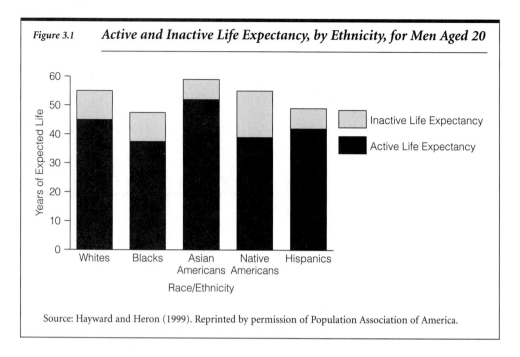

Figure 3.1 **Active and Inactive Life Expectancy, by Ethnicity, for Men Aged 20**

Source: Hayward and Heron (1999). Reprinted by permission of Population Association of America.

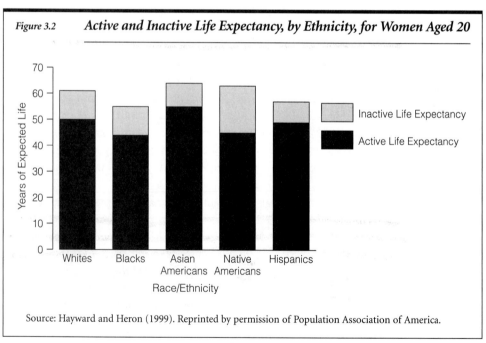

Figure 3.2 **Active and Inactive Life Expectancy, by Ethnicity, for Women Aged 20**

Source: Hayward and Heron (1999). Reprinted by permission of Population Association of America.

African Americans

The impact of ethnicity on health stands out vividly in studies of infant mortality. For all causes of infant deaths, African Americans have higher mortality rates than whites (Anderson, 2001). Moreover, those differences have increased over time: Whereas in 1950 African American infants were

Table 3.3	Infant Mortality Rates per 1,000 Live Births		
COUNTRY	RATE	COUNTRY	RATE
Singapore	2.2	United Kingdom	5.3
Hong Kong	2.4	New Zealand	5.6
Sweden	2.8	**U.S. white, non-Hispanic**	**5.8**
Japan	3.0	Greece	5.9
Finland	3.2	**U.S., all races**	**6.7**
Norway	3.4	Cuba	7.0
Spain	3.7	Hungary	7.3
Czech Republic	3.9	Poland	7.5
France	4.1	Slovakia	7.6
Germany	4.1	Chile	8.3
Denmark	4.4	Puerto Rico	9.6
Switzerland	4.4	Costa Rica	10.0
Austria	4.5	Kuwait	10.0
Australia	4.7	Bulgaria	12.3
Netherlands	4.8	Russia	13.0
Italy	4.8	Uruguay	13.5
Portugal	5.0	Azerbaijan	13.0
Ireland	5.1	**U.S. blacks**	**13.8**
Canada	5.2	Romania	16.7
Israel	5.3	Thailand	20.0
Belgium	5.3	Mexico	25.0

Source: Population Reference Bureau (2004).

1.6 times more likely than white infants to die, by 2004 African American infants were 2.4 times more likely to die (Schoendorf et al., 1992; National Center for Health Statistics, 2004: 131). African Americans have an infant mortality rate considerably higher than that found in such poor countries as Cuba, Poland, and Slovakia and similar to that found in countries like Azerbaijan and Russia (see Table 3.3).

One partial explanation for the high rate of infant mortality among African Americans is their relatively low income, for almost 60 percent of African American children are poor or near poor (National Center for Health Statistics, 2004: 25). To determine whether ethnicity affects infant mortality independent of income, Schoendorf and his colleagues (1992) looked at mortality rates among a national random sample of African American and white infants whose parents were at least 20 years old and college graduates. Even within this relatively well-off sample, and after controlling for age, number of previous births, use of prenatal care, and marital status, African American infants were almost twice as likely to die as white infants, largely because of higher rates of prematurity and low birthweight.

These differences, the authors theorize, reflect a constellation of factors stemming from racism, which although far less common than in the past, remains deeply embedded in American culture (Feagin and Sikes, 1994; D. Williams, 1998). For example, data collected in 2000 by the widely used national, random General Social Survey found that 22 percent of whites believe African Americans are unintelligent, 56 percent believe they prefer to live off welfare, and 58 percent believe they lack the motivation or willpower to pull themselves up out of poverty (General Social Survey, 2002). We can reasonably assume that even more survey respondents held these views, but did not admit it. Because of racism, even middle-class African Americans (like those studied by Schoendorf and his colleagues), who could afford decent housing in neighborhoods free from pollution and violence, sometimes find it impossible to obtain such housing when landlords, realtors, or mortgage bankers flout laws banning housing discrimination (D. Williams, 1998; D. Williams and Jackson, 2005). Other African Americans prefer living in poorer, segregated neighborhoods rather than facing the daily hostility—or, simply, social discomfort—of white neighbors. Consequently, more-affluent African Americans sometimes live in conditions similar to those experienced by poorer African Americans. This hypothesis gains support from studies suggesting (if inconclusively) that African American infant mortality rates are highest among those living in the most segregated cities (LaVeist, 1993; Polednak, 1996).

In addition, the psychosocial stresses of racism can harm health among African Americans (as well as among other minority groups). Several studies have found that as the number of incidents of ethnic discrimination that individuals have experienced increases, their physical and mental health deteriorates (D. Williams et al., 1997; D. Williams, 1998).

The disparities in health status between African Americans and whites do not end in infancy. At each age, and for 13 of the 15 leading causes of death, African Americans have higher death rates and lower life expectancies than whites do. Ethnic differences in life expectancies have declined slowly over the past 30 years, from 7.6 years in 1970 to 4.8 years in 2005 (U.S. Bureau of the Census, 2005: Table 92).

In a much-cited article in the *New England Journal of Medicine,* Colin McCord and Harold P. Freeman (1990) vividly demonstrated these stark differences in life expectancy. The article compared the chances of surviving to old age in Bangladesh, one of the poorest countries in the world, to the chances in Harlem, an overwhelmingly poor, African American, New York City neighborhood. As Figure 3.3 shows, although before age 5 both males

Figure 3.3 *Survival to the Age of 65 in Harlem, Bangladesh, and Among U.S. Whites: 1980*

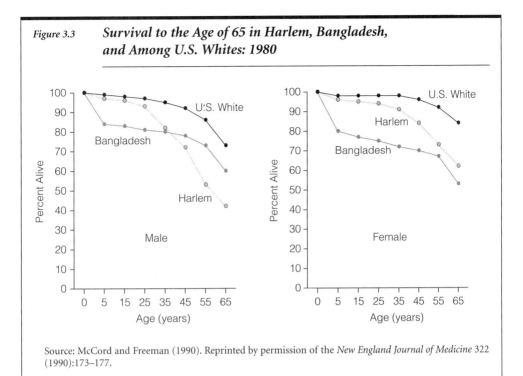

Source: McCord and Freeman (1990). Reprinted by permission of the *New England Journal of Medicine* 322 (1990):173–177.

and females have higher death rates in Bangladesh than in Harlem, after that age the death rate levels off in Bangladesh but rises in Harlem. Consequently, for females, the chances of surviving are lower in Bangladesh than in Harlem, but only because of the differences in the first five years of life. For males, the chances of surviving are lower in Bangladesh only until age 40 and almost solely because of deaths in the first five years of life. Among those who survive to age 5, both males and females have a greater chance of surviving to age 65 in Bangladesh than in Harlem.

Unfortunately, these high death rates among African Americans extend far beyond the borders of Harlem. Table 3.4 shows the age-adjusted death rates for selected causes of death in 2002 (the latest data available as of 2005). This table shows that HIV disease kills ten times more African Americans than white non-Hispanics, and homicide kills seven times more African Americans. Both these causes of death are markers of poverty, hopelessness, and inequality. The table also highlights the disproportionately large role diabetes plays in African American mortality. Diabetes, which is caused by both genetic factors and a diet of poverty, kills African Americans twice as often as it kills whites, mostly by causing kidney disease (Centers for Disease Control and Prevention, 1998).

Yet kidney disease need not kill, if transplants or dialysis can substitute for failing kidneys. However, African Americans are significantly less likely

Table 3.4   ***Age-Adjusted Death Rates per 100,000 for Selected Causes of Death, 2002***

	TOTAL	WHITE, NON-HISPANIC	AFRICAN AMERICAN	NATIVE AMERICAN	HISPANIC	ASIAN AMERICAN OR PACIFIC ISLANDER
All causes	845.3	837.5	1,083.3	677.4	629.3	474.4
Heart disease	240.8	239.2	308.4	157.4	180.5	134.6
Cerebrovascular diseases (strokes)	56.2	54.6	76.3	37.5	41.3	47.7
Cancer	193.5	195.6	238.8	125.4	128.4	113.6
Chronic lower respiratory diseases	43.5	46.9	31.2	30.1	20.6	15.8
Influenza and pneumonia	22.6	22.6	24.0	20.4	19.2	17.5
Chronic liver disease and cirrhosis	9.4	9.0	8.5	22.8	15.4	3.2
Diabetes	25.4	22.2	49.5	43.2	35.6	17.4
HIV disease	4.9	2.1	22.5	2.2	5.8	0.8
Unintentional injuries	36.9	38.0	36.9	53.8	30.7	17.9
Suicide	10.9	12.9	5.3	10.2	5.7	5.4
Homicide	6.1	2.8	21.0	8.4	7.3	2.9

Source: National Center for Health Statistics (2004: Table 29).

than whites to receive transplants or dialysis because standard procedures for selecting patients for these therapies unintentionally discriminate against them (Council on Ethical and Judicial Affairs, 1990; Gaston et al., 1993). Transplant programs generally require near-perfect biological matches between donor and potential recipient before they will perform a transplant, although the difference in survival rates when kidneys are less well matched is small. Because African Americans donate kidneys far less often than whites do, African Americans who need kidneys less often match the available kidneys perfectly and, thus, less often receive transplants. African Americans also receive transplants less often because doctors less often refer them to transplant programs. Even when African Americans are referred to transplant programs, they are more frequently rejected as patients because they lack transportation to care facilities and funds to pay for aftercare, which can costs thousands of dollars per year (Council on Ethical and Judicial Affairs, 1990). This chapter's ethical debate (Box 3.1) looks at the broader problem of allocating scarce health resources.

This pattern recurs among other minority groups and in other areas of health care. A research review conducted by the highly prestigious Institute of Medicine (Nelson, Smedley, and Stith, 2002) found that, after controlling for symptoms and insurance coverage, doctors were more likely to offer whites various life-preserving treatments (including angioplasty, bypass surgery, and the most effective drugs for HIV infection) and more likely to offer minorities various less-desirable procedures (such as leg amputations for diabetes).

Hispanics

Like African Americans, Hispanic Americans experience an unusually high burden of illness—although this is truer for some Hispanic groups than for others. In general, Cubans (3.7 percent of U.S. Hispanics) have fared considerably better than Puerto Ricans (8.6 percent) or Mexican Americans (67.0 percent). Relatively little is known regarding the health status of the newer immigrant groups from Central and South America who comprise 14.3 percent of U.S. Hispanics.

As among African Americans, health problems among Hispanics largely reflect their generally lower social class status (Rogers et al., 1996). Hispanics are two and one-half times more likely than non-Hispanic whites to live in poverty and, except for Cubans, are half as likely to have completed college. In addition, cultural and language barriers as well as social discrimination can make it difficult for Hispanics to take advantage of health care resources even when they can afford them. Partly as a result, Hispanic children are less likely than non-Hispanic white children to receive all necessary vaccinations by age 3 and, regardless of income, are about twice as likely to have no regular source of health care (National Center for Health Statistics, 2004).

For reasons that remain unclear, rates of infant mortality among Hispanics (other than Puerto Ricans) are comparable with those of non-Hispanic white

---

Box 3.1   ***Ethical Debate: Allocating Scarce Health Resources***

You are the chair of a regional organ bank charged with allocating one donated kidney. This kidney will mean the difference between life and death to whoever receives it. Which one of these people would you give it to?

- James Russell, a world-famous pediatrician who is 60 years old, unmarried, and childless

- Julie Brown, a 35 year-old, unmarried mother and sole supporter of four young children, who is a high school dropout and lives on government assistance

- Sally Michaels, a 45 year-old homemaker with children in college, who is married to a lawyer and is active in various local charities

Deciding how to allocate scarce resources has animated public debate since the early 1960s, when kidney dialysis—a treatment that can keep alive those who would otherwise die from kidney failure—first became feasible. Because demand for dialysis far exceeded supply, hospitals had to establish procedures for deciding who would receive treatment and who would not—in essence deciding who would live and who would die.

Since then, demand for dialysis has continued to exceed supply, and so such decisions still must be made. The same dilemma faces all those who must allocate expensive and scarce treatments, for no national policies regulate how to make these decisions.

Probably all observers would agree that medical factors must be considered in allocating scarce resources. For example, it makes little sense to give transplants to someone who is likely to die during or shortly after a transplant operation, such as a patient whose tissue does not adequately match that of the prospective organ donor and whose body is therefore likely to reject the donated organ. In other circumstances, however, the role played by medical factors in these decisions is far less clear. For example, some argue that those who are healthiest should receive highest priority because they are most likely to survive a transplant and to have a good quality of life afterward. Others, however, argue that these individuals can live the longest *without* a transplant and so should have lowest priority.

Although it might seem fairest, relying on medical factors is also problematic because doing so may unintentionally discriminate against minorities and the poor. For example, for various reasons, including generalized mistrust stemming from a history of poor treatment by the medical establishment, African Americans are less likely to donate organs than are whites. As a result, African Americans more often die while waiting for a closely matched donor kidney. Similarly, selecting the healthiest persons first discriminates against poorer persons, who on average are in worse health.

Using other "objective" criteria for selection also can unintentionally discriminate. Individuals are most likely to benefit from a procedure if they have family members who can take care of them while they recover; can afford to pay all necessary costs of receiving care, including costs for drugs, any special diet, and transportation to and from the health care delivery site; have the intellectual and emotional ability to follow the prescribed treatment and follow-up regimen; and have a stable life that allows them to do so.

Yet all these factors encourage the selection of middle- and upper-class persons who share not only social status but also cultural values with those who control access to health care.

But this selection bias is not necessarily a problem. In fact, some consider it perfectly reasonable to use social characteristics overtly in making decisions, and probably most would agree that it makes more sense to allot scarce health resources to a 40-year-old than to an equally healthy 60-year-old because more years of productive life would be lost should the 40-year-old die.

Implicit in such a decision is a notion of social worth—that a younger person is automatically worth more than an older one. Similarly, many would argue that scarce resources should be allocated to those most likely to benefit the community. This generally translates into those who are married, parents of young children, educated, and employed. Such decision rules, of course, reflect the values of the middle- and upper-class persons who sit on hospital selection committees and are likely to work against minorities and the poor.

The difficulties with establishing equitable decision rules have led some to propose mechanisms for eliminating the need to make decisions, such as lotteries. These proposals assume that all persons have equal social worth. Yet most people *do* consider some people more morally worthy than others, and so find such proposals unacceptable.

Another way to avoid making these difficult decisions is to allot scarce resources on a "first come, first served" basis. Such a policy, however, would benefit more-affluent patients because they typically receive accurate diagnoses and learn how to join waiting lists earlier in the course of their disease. Consequently, this system would be inequitable in practice.

Finally, some argue that instead of trying to establish equitable decision rules, we should allocate scarce resources simply based on the ability to pay. Proponents of this view see no reason to treat scarce health resources differently from any other valued resource, like shoes or houses. Opponents argue that doing so is equivalent to declaring the lives of some individuals more valuable than others simply because they are wealthier.

In sum, decisions regarding how to allocate scarce health resources always rely on social and cultural as well as medical factors. Perhaps the best we can hope for is that decision makers will recognize how these factors affect their decisions and use that recognition to work for more equitable policies.

Sociological Questions

1. What social views and values about medicine, society, and the body are reflected in this policy? Whose views are these?

2. Which social groups are in conflict over this issue? Whose interests are served by the different sides of this issue?

3. Which of these groups has more power to enforce its view? What kinds of power do they have?

4. What are the intended consequences of this policy? What are the unintended social, economic, political, and health consequences of this policy?

Table 3.5	*Infant Mortality Rate by Ethnicity, United States*

MOTHERS' ETHNICITY	RATE
African American	13.8
Native American	8.6
White non-Hispanic	5.8
Hispanic origin	5.6
Asian or Pacific Islander	4.8
All mothers	**7.0**

Source: National Center for Health Statistics (2004: 131).

Americans (see Table 3.5). On other measures of health, however, Hispanics fare less well. Life expectancy is lower for Hispanics than for non-Hispanic whites, even though the main causes of death are the same for both groups. Like African Americans, Hispanics are at greater risk than non-Hispanic whites for diabetes and for its more serious complications. Hispanics also die at higher rates from violence and from liver disease (typically linked to heavy alcohol use). Finally, Hispanics are almost twice as likely as whites to die from HIV disease. Conversely, Hispanics have lower death rates from heart disease, cerebrovascular disease, and cancer simply because they are less likely to live long enough to develop these diseases.

Health status is particularly poor among those who are migrant workers (Azevedo and Bogue, 2001; Greenhouse, 2001). Of course, most Hispanics are not migrant workers, but the majority of migrant workers are Hispanic, and most other migrant workers belong to other minority communities. Consequently, issues of minority status and social class are tightly interwoven, and both must be considered in order to understand why these individuals are so vulnerable to health problems.

About half of the 2.5 million migrant laborers working in agricultural fields in the United States are illegal aliens (P. Martin, 2002). The work itself is physically hazardous, with long days of repetitive stooping and bending, heavy lifting, and exposure to toxic pesticides (Gwyther and Jenkins, 1998; Sandhaus, 1998). Access to clean water and sanitary toilets is often limited, and workers are routinely exposed to weather extremes. Living conditions, too, are often poor, with many individuals crowded together in poorly heated or cooled rooms with insufficient water and toilets and low wages that make it difficult to obtain nutritious foods. Yet because so many migrant laborers are illegal aliens, they cannot protest these conditions without risking deportation. Finally, lack of transportation, cultural differences, and communication problems make it difficult for laborers and their families to obtain good health care. As a result, life expectancy is substantially reduced among migrant workers and their families, and chronic health problems, infectious diseases (including tuberculosis, typhoid, and hepatitis), miscarriages, and infant mortality are several times more common than among the rest of the population (Gwyther and Jenkins, 1998; Sandhaus, 1998).

The same pressures that lead undocumented immigrants to take dangerous jobs leave some immigrant groups more vulnerable than others. Recent years have seen a surge in immigration (legal and illegal) from the poorer countries of Central America (such as El Salvador and Guatemala), where living conditions are poorer than in Mexico. Because of the longer distance to the United States and the fact that migrants must cross more than one national border to reach this country, immigration from Central America is more dangerous and expensive than it is from Mexico. As a result, Central Americans are more likely to stay with whatever job they first get in the United States rather than risk attracting the attention of immigration authorities while seeking other work. Consequently, Central Americans are more likely than Mexicans are to stay in low-paying, dangerous occupations. For example, Arizona health data identify roof building as the most dangerous job in construction—14 percent of roofers reported injuries during 2002, and undoubtedly many more were injured without reporting—and indicate that most of those employed in this work are undocumented Central American immigrants (Gonzales, 2005).

Native Americans

As is true with any ethnic group, Native Americans are highly diverse. Native Americans in the United States belong to more than five hundred different tribes, each with a distinct language and culture. Slightly more than half of Native Americans live off reservations, often in large urban areas.

Native American life expectancy has improved substantially since the 1950s. Official statistics now indicate that average life expectancy for Native Americans almost equals that of white Americans. However, these figures are misleading. Because Native Americans who die at hospitals off of reservations are often listed as "white" on their death certificates, federal researchers estimate that death rates for Native Americans are underestimated by 21 percent (National Center for Health Statistics, 2001). In addition, these death rates include both highly assimilated persons with little Native American background living in suburbia and traditional Native Americans living on reservations. In Arizona, for example, where most live on reservation, average life expectancy is 55 (Nichols, 2002).

Even when looking only at national averages, sharp differences between Native and white Americans are apparent in the particular patterns of disease these two groups experience (Kunitz, 1996; U.S. Department of Health and

Human Services, 1990). These differences begin at birth. Although lower than among African Americans and lower than in the past, infant mortality (see Table 3.5) remains considerably higher among Native Americans than among whites (National Center for Health Statistics, 2004: 131).

The differences between whites and Native Americans become clearer when we divide infant mortality into **neonatal infant mortality** (deaths occurring during the first 27 days after birth) and **postneonatal infant mortality** (deaths occurring between 28 days and 11 months after birth). The *neonatal* infant mortality rates are essentially the same among Native Americans and whites—4.2 per 1,000 live births versus 3.8 (National Center for Health Statistics, 2004: 131). However, the *postneonatal* infant mortality rate is almost three times higher among Native Americans as compared to whites—5.4 per 1,000 live births versus 1.9. These figures reflect differences in rates of pneumonia and gastritis. Although less common than in the past, these easily preventable diseases—precipitated by poverty, malnutrition, and poor living conditions and normally controllable through prompt medical attention—still occur more often among Native Americans than among others. Box 3.2 describes the benefits and limitations of the Indian Health Service, the federally funded program charged with providing health care to Native Americans.

For Native Americans who survive past infancy, heavy alcohol use stands out as an especially serious health risk (see Table 3.4). Although alcohol-related deaths among Native Americans have decreased in recent years, liver disease, which is typically linked to alcohol use, remains 2.5 times more common than in the U.S. population as a whole and more common than in any other ethnic group. In addition, Native Americans are significantly more likely than others are to die from unintentional injuries, with alcohol use often contributing to these deaths. Because of these factors leading to early deaths, Native Americans are less likely than white non-Hispanics to die from heart disease, cerebrovascular disease, or cancer.

Native Americans differ from other Americans in their pattern of diseases as well as their pattern of deaths. The rate of respiratory disease is 31 percent higher than in the U.S. population as a whole, partly due to high rates of tobacco use. Native Americans have higher rates of tobacco use than any other ethnic group in the United States, are the only group in which rates have not declined since the 1970s, and are the only group in which women are as likely to smoke as men (*Morbidity and Mortality Weekly Report,* 1998). Native Americans also have mortality rates from infectious diseases twice as high as those found among white Americans, primarily due to inadequate sanitation, lack of access to clean water, and the general physical debilitation associated with poverty. In addition, diabetes affects approximately 9 percent of Native American adults, who are three times more likely to die from it than whites are (Centers for Disease Control and Prevention, 1998; Claiborne, 1999).

Box 3.2   *The Indian Health Service*

Since the 1830s, under the provisions of various treaties, the U.S. government has provided health care to Native Americans (Kunitz, 1996; Dixon and Roubideaux, 2001). Today, more than 1 million Native Americans, living in urban and rural areas both on and off reservations, receive comprehensive health services from the Indian Health Service (IHS).

The IHS offers both "direct" health care and "contract" care. Direct care programs, generally located on Indian reservations, provide access to generalist medical care from internists, family doctors, and pediatricians and are open to all Native Americans. In addition, the IHS contracts with private health care providers to offer specialty care. This contract health program, however, is open only to Native Americans who live either on a reservation or in the contract area affiliated with their tribe. For example, a Navajo who moves to Flagstaff, Arizona, where the IHS contract health program includes Navajos, can obtain care through that program. The same individual could not receive services in Phoenix, where the IHS contract health program does not include Navajos, or in Minneapolis, where the IHS has no contract health program.

Since the 1970s, the IHS increasingly has moved toward local control (Kunitz, 1996; Dixon and Roubideaux, 2001). Tribes now can sign agreements to take over some services offered by the IHS or to provide additional services; about half of all Indian health programs in the country are now run by tribes (Nichols, 2002). Unfortunately, the IHS can afford to spend only $1,920 per capita each year. In contrast, the federal government spends $3,859 per person on Medicaid and $5,600 per person on Medicare, while private insurers spend $4,392 per capita (Nichols, 2002). As a result, only 15 of the 515 IHS health care facilities can provide the kinds of services offered in large hospitals, and funds for these 15 facilities usually run out early in each fiscal year. Similarly, in 1994, the IHS had 90 doctors per 100,000 patients, compared with 229 doctors per 100,000 patients in the United States as a whole (Claiborne, 1999). Because of problems like these, the move toward tribal control of health care has pitted tribes against each other in the fight for limited federal dollars—a battle that has particularly hurt smaller, poorer tribes and tribes located in isolated regions where finding qualified health care providers is difficult and expensive. The need for additional funds to pay for tribal health care costs partly explains why many tribes have aggressively pursued casino gambling in the last two decades (Kunitz, 1996).

Asian Americans

Overall, Asian Americans enjoy far better health than do other American minority groups. The largest Asian American groups (Chinese, Japanese, and Filipino) have life expectancies and infant mortality rates equal or superior to those of white Americans (see Table 3.5, Figures 3.1 and 3.2). As a group, Asian Americans experience the same causes of death as whites but at significantly lower rates.

These statistics, however, tell only part of the story. Since 1975, a substantial portion of Asian immigration has come from the war-torn countries of Southeast Asia. These immigrants typically have far lower income and education levels than those of established Asian Americans. In addition to having the health problems that always accompany poverty, these individuals often suffer from unavoidable dietary changes, culture shock, tropical diseases for which diagnosis and treatment can prove elusive, and the long-lasting traumas of warfare and refugee life.

The limited available data on the health status of Southeast Asians in the United States suggest that they have significantly higher mortality and morbidity rates than those for whites or other Asians (Association of Asian Pacific Community Health Organizations, 1997). For example, only 22.7 percent of Vietnamese Americans report that their health is excellent, as compared with just over 40 percent of Americans who are white non-Hispanic, Japanese, or Asian Indian (Kuo and Porter, 1998). Compared with white Americans, Southeast Asian immigrants are 13 times more likely to have tuberculosis and 25 times more likely to have hepatitis B. Higher rates of hepatitis B mean higher rates of liver cancer. Lung cancer, too, is more common among male Southeast Asian immigrants largely because they are two to three times more likely to smoke than other American men are.

At the same time, Southeast Asians typically have more limited access to health care (Association of Asian Pacific Community Health Organizations, 1997). Rates of health insurance coverage are low, and even those who have insurance sometimes find that linguistic or cultural barriers make it nearly impossible to communicate with health care workers and to obtain quality health care. As a result, Southeast Asians are less likely than are other Americans to use Western health care (although some continue to use traditional Asian healers and therapies).

Writer Anne Fadiman poignantly describes the communication barriers between new immigrants and their doctors, and the problems these barriers create for both groups, in her prize-winning book, *The Spirit Catches You and You Fall Down: A Hmong Child, Her American Doctors, and the Collision of Two Cultures* (1997). Fadiman describes the completely divergent worldviews of American doctors and Hmong patients in Merced, California, where many Hmong refugees from Laos have settled:

> Most Hmong believe that the body contains a finite amount of blood that it is unable to replenish, so repeated blood sampling [for lab tests] . . . may be fatal. When people are unconscious, their souls are at large, so anesthesia may lead to illness or death. If the body is cut or disfigured, or if it loses any of its parts, it will remain in a condition of perpetual imbalance, and the damaged person not only will become frequently ill but may be physically incomplete during the next reincarnation; so surgery is taboo. If people lose their vital organs after death, their souls cannot be reborn into new bodies and may take revenge on living relatives; so autopsies and embalming are also taboo. . . .
>
> Not realizing that when a man named Xiong or Lee or Moua walked into the Family Practice Center with a stomachache he was actually complaining that the entire universe was out of balance, the young doctors of Merced frequently failed

to satisfy their Hmong patients. How could they succeed? . . . They could hardly be expected to "respect" their patients' system of health beliefs (if indeed they ever had the time and the interpreters to find out what it was), since the medical schools they had attended had never informed them that diseases are caused by fugitive souls and cured by [sacrificing] chickens. All of them had spent hundreds of hours dissecting cadavers . . . but none of them had had a single hour of instruction in cross-cultural medicine. To most of them, the Hmong taboos against blood tests, spinal taps, surgery, anesthesia, and autopsies—the basic tools of modern medicine—seemed like self-defeating ignorance. They had no way of knowing that a Hmong might regard these taboos as the sacred guardians of his identity, indeed, quite literally, of his very soul. [Moreover], what the doctors viewed as clinical efficiency the Hmong viewed as frosty arrogance. And no matter what the doctors did, even if it never trespassed on taboo territory, the Hmong, freighted as they were with negative expectations accumulated [during years under military siege and in refugee camps] before they came to America, inevitably interpreted it in the worst possible light. (Fadiman, 1997: 33, 61)

Growing recognition of problems like these has spurred medical schools to incorporate training in working with culturally diverse populations in their programs, as we will consider in more detail in Chapter 11.

Case Study: Environmental Racism

One health issue that cuts across America's minority communities is **environmental racism.** Environmental racism refers to the disproportionate burden of environmental pollution experienced by ethnic minorities. According to Benjamin F. Chavis,

> Environmental racism is racial discrimination in environmental policymaking. It is racial discrimination in the enforcement of regulations and laws. It is racial discrimination in the deliberate targeting of communities of color for toxic waste disposal and the siting of polluting industries. It is racial discrimination in the official sanctioning of the life-threatening presence of poisons and pollutants in communities of color. And, it is racial discrimination in the history of excluding people of color from the mainstream environmental groups, decision-making boards, commissions, and regulatory boards. (1993: 3)

Environmental racism first became a subject for widespread discussion following the 1983 publication of a groundbreaking study by sociologist Robert D. Bullard. Bullard documented how, since the 1920s, the city of Houston had located all of its landfills and 75 percent of its garbage incinerators in African American neighborhoods, even though those neighborhoods constituted only a tiny fraction of the city. After Bullard's study appeared, federal agencies, social activists, and scholars around the country began collecting evidence demonstrating that minority communities bear a

| Table 3.6 | Percentage Living in Polluted Areas, by Ethnicity | | |

TYPE OF POLLUTION	WHITES (%)	AFRICAN AMERICANS (%)	HISPANICS (%)
Particulate matter	15	17	34
Carbon monoxide	34	46	57
Ozone	53	62	71
Sulfur dioxide	7	12	6
Lead	6	9	19

Source: U.S. Environmental Protection Agency (1992).

disproportionate share of the nation's environmental hazards, from Hispanic farmworkers exposed to dangerous pesticides to Navajo communities poisoned by deadly uranium mines and inner-city African Americans plagued by asthma-inducing air pollution (Bullard et al., 2001; Camacho, 1998). The most important of these environmental hazards, because it is so widespread and devastating, is lead—found in polluted air, contaminated soil, and the paints and pipes of older residences. Among children under age 5 who are known to have high levels of lead in their blood, 17 percent are white non-Hispanic, 16 percent are Hispanic, and *60 percent* are African American (Meyer et al., 2003). Compared with whites, minorities are exposed more often to dust and soot, carbon monoxide, ozone, sulfur, and sulfur dioxide, as well as to pesticides, emissions from hazardous waste dumps, and other hazardous substances. Researchers have found that exposure to environmental pollution is more highly correlated with race than with any other factor, including poverty (Bullard, 1993; Stretesky and Hogan, 1998). Table 3.6 provides some examples.

Environmental racism exemplifies the workings of internal colonialism. The term **internal colonialism** highlights the similarities between the treatment of minority groups within a country and of native peoples by foreign colonizers, such as under the former apartheid system (Blauner, 1972). Scholars and activists who write about environmental racism argue that just as colonizers exploit native labor power and lands and keep native peoples economically dependent for the benefit of the colonizing power, so majority groups can exploit internal colonies of minority group members. In the case of environmental racism, racial discrimination enables industrialists, with the tacit approval of government bureaucrats and politicians, to place environmental hazards in these internal colonies without worrying that those communities will have sufficient political power or financial resources to resist. Poverty and lack of other job opportunities can even encourage minority

---

Box 3.3   *Making a Difference: The Center for Health,*
              Environment & Justice

During the 1970s, a series of unexplained deaths from cancer and leukemia plagued children living in Love Canal, New York. Eventually, local community activists traced the children's deaths to a nearby toxic waste site and won federal funding to relocate their families to safer areas. Perhaps more important, the activists' work led to passage of the federal Superfund program to clean up toxic waste sites around the country.

In 1981, some of these activists founded the Center for Health, Environment & Justice (CHEJ) to assist other grassroots groups in similar battles (www.chej.org, accessed August 2005). Since then, CHEJ has served as an invaluable resource. Each day CHEJ workers answer letters and phone calls from individuals and grassroots organizations seeking information about toxic threats. In addition, CHEJ publishes two magazines, *Everyone's Backyard* and *Environmental Health Monthly,* and more than one hundred guidebooks and information packages on issues related to chemical hazards and to environmental justice more broadly. CHEJ also puts interested individuals in touch with appropriate organizations and runs workshops to train environmental activists and to help environmental organizations work more effectively.

Since its founding, CHEJ has had many successes. Working with local activists, CHEJ has helped win legislative approval for laws and regulations establishing state Superfund programs, prohibiting corporations convicted of dumping toxic wastes in one state from setting up business in another, and prohibiting corporations forced to clean up toxic waste in one state from dumping it in another.

---

communities to welcome polluting industries for the jobs they will bring. This does not mean, however, that those who make decisions about where to locate environmental hazards *intend* to discriminate against minorities— certainly those who make these decisions would argue that they decide solely on economic and technical considerations—only that their actions have the *effect* of discriminating.

Currently, dozens of grassroots groups of African Americans, Hispanics, Asian Americans, and Native Americans are working to fight for environmental justice (Sandweiss, 1998), as are numerous national civil rights and environmental organizations; Box 3.3 describes the work of one of these groups. Similarly, the Environmental Protection Agency (EPA) a few years ago began using the Civil Rights Act of 1964, which forbids racial discrimination in any federally funded programs, as grounds for investigating how companies and local governments decide where to locate environmental hazards. The first EPA study found that 90 percent of major industrial polluters in Louisiana were located in predominantly African American areas and resulted in the cancellation of a hazardous waste permit in that state (Sandweiss, 1998).

Conclusion

Far from being purely biological conditions reflecting purely biological factors, health and illness are intimately interwoven with social position. In the United States as elsewhere, those who are poor or are targets of racial discrimination die younger than others do. Sex and gender have more complex health consequences: Women enjoy longer life spans than men do, but they are subject to more illness and disability.

Because social forces as well as biological factors affect health, understanding social trends can help us predict future health trends. For example, as women's social roles have changed, their rates of tobacco use and lung cancer have approached those for men, while their ability to protect themselves from the health consequences of male violence has increased. Similarly, if economic and ethnic inequality either increase or decrease, we are likely to see changes in the health status of currently disadvantaged economic and ethnic groups.

Suggested Readings

Boston Women's Health Book Collective. 2005. *Our Bodies, Ourselves: A New Edition for a New Era.* New York: Touchstone. An excellent overview of women's health issues, emphasizing self-help while discussing the political and social aspects of health and health care.

Eugenides, Jeffrey. 2002. *Middlesex.* New York: Picador. This funny, poignant, engaging, Pulitzer Prize–winning novel recounts the story of "Cal" Stephanides, who is born with a (real) genetic condition that shifts his body from female to male at adolescence.

Marmot, Michael G. 2004. *The Status Syndrome: How Your Social Standing Directly Affects Your Health and Life Expectancy.* London: Bloomsbury. Epidemiologist Michael Marmot, who received a knighthood for his research, explains why at each step on the social status ladder, persons live longer than those even one step below them.

Schneider, Andrew and David McCumber. 2004. *An Air That Kills: How the Asbestos Poisoning of Libby, Montana Uncovered a National Scandal.* New York: Putnam's Sons. Journalists Schneider and McCumber tell how the actions of a multinational mining corporation led to an epidemic of cancer deaths, and how the community fought back.

Getting Involved

Association of Asian Pacific Community Health Organizations. 439 23rd Street, Oakland, CA 94612. (510) 272-9536. www.aapcho.org. Excellent source of information about health and health care among both new and old Asian American communities.

Center for Health, Environment & Justice. PO Box 6806, Falls Church, VA 22040. (703) 237-2249. www.chej.org. Central clearinghouse for the environmental justice movement; assists grassroots organizations located primarily in poor and minority communities.

Habitat for Humanity. 121 Habitat Street, Americus, GA 31709. (912) 924-6935. www.habitat.org. Ecumenical Christian organization that helps poor families build low-cost housing.

National Coalition Against Domestic Violence. PO Box 18749, Denver, CO 80218. (303) 839-1852. www.ncadv.org. A national organization that can refer you to organizations in your region.

National Women's Health Network. 514 10th St. NW, Suite 400, Washington, DC 20004. (202) 628-7814. www.womenshealthnetwork.org. Educational and lobbying group concerned with all issues affecting women's health.

Review Questions

What are the health care consequences of an aging population and the feminization of aging?

Why might sociologists and other observers argue *against* early detection and treatment of prostate cancer?

Why do men have higher mortality rates than women but lower morbidity rates?

What are the sources and consequences of woman battering? Why do some health care workers consider woman battering a serious health problem?

How and why does social class affect people's health?

What are the special health problems of homeless persons? of migrant farmworkers?

How does ethnicity affect health separately from social class? How does social class affect health separately from ethnicity? How can you tell which is the more powerful factor?

How and why do the particular health problems of African Americans, Hispanics, Native Americans, and Asian Americans differ from those of whites?

What is environmental racism?

Internet Exercises

1. Both the United Nation's World Health Organization (www.who.int) and the U.S. National Institutes of Health (www.nih.gov) have websites devoted to health problems associated with aging. Find those sites, and compare the major health problems identified by the World Health Organization with the major problems identified by the National Institute of Health. How do you explain the differences?

2. The U.S. Census Bureau (www.census.gov) provides a wealth of information about the U.S. population. Find out what percentage of Americans now live below the poverty line.

3. To find out how social class affects individuals' perceived health status, first locate the website for the University of California's Survey Documentation and Analysis (SDA) archive. This archive contains data from several national random surveys. Enter the SDA archive; then click on the GSS Cumulative Datafile, 1972–2002, full analysis. Find the "Select an Action" section; then click the button for "Frequencies or Crosstabulations." Next, click on "Start." A form with several blank spaces will appear on your screen. For row variable, type "health." For column variable, type "class." Click on the boxes to the left of Column Percentaging, Statistics, and Question Text. Then click the button "Run the Table." What effect does social class have on people's perceptions of their health status?

Illness in the Developing Nations

For almost 20 years, Paul Farmer, an American doctor and anthropologist, has worked among Haiti's rural poor. One of his patients is Jean Dubuisson, who

> lives in a small village in Haiti's Central Plateau, where he farms a tiny plot of land. He shares a two-room hut with his wife, Marie, and their three surviving children. All his life, recounts Jean, he's "known nothing but trouble." His parents lost their land [when] the Péligre hydroelectric dam [was built and flooded their village]—a loss that plunged their large family into misery. Long before he became ill, Jean and Marie were having a hard time feeding their own children: two of them died before their fifth birthdays, and that was before the cost of living became so intolerable.
>
> And so it was a bad day when, some time in 1990, Jean began coughing. For a couple of weeks, he simply ignored his persistent hack, which was followed by an intermittent fever. There was no clinic or dispensary in his home village, and the costs of going to the closest clinic . . . are prohibitive enough to keep men like Jean shivering on the dirt floors of their huts. But then he began having night sweats. Night sweats are bad under any conditions, but they are particularly burdensome when you have only one sheet and often sleep in your clothes. (Farmer, 1999: 187–188)

Although Jean and Marie both recognized that he needed to seek medical care, doing so was unaffordable. Over the next few months, however, Jean's health continued to decline and his weight to drop. Even more frightening, in December 1990 Jean began to cough up blood, which, given how common tuberculosis is in Haiti, they easily recognized as a symptom of the disease. At that point, Jean agreed to go to a clinic:

> At the clinic, he paid $2 for multivitamins and the following advice: eat well, drink clean water, sleep in an open room and away from others, and

go to a hospital. Jean and Marie recounted this counsel without a hint of sarcasm, but they nonetheless evinced a keen appreciation of its total lack of relevance. In order to follow these instructions, the family would have been forced to sell off its chickens and its pig, and perhaps even what little land they had left. They hesitated, understandably.

Two months later, however, a second, massive episode of [coughing up blood] sent them to a church-affiliated hospital [some distance away, where Jean] was charged $4 per day for his bed; at the time, the per capita income in rural Haiti was about $200 a year. When the hospital's staff wrote prescriptions for him, he was required to pay for each medication before it was administered. Thus . . . he actually received less than half of the medicine prescribed. . . . [Jean] discharged himself from the hospital when the family ran out of money and livestock. (Farmer, 1999: 188–189)

Some months later, Jean learned of a nonprofit clinic Farmer had founded in a nearby village, and sought care there. As Farmer describes,

Jean was cured of his tuberculosis, but this cure, in many respects, came too late. Although he is now free of active disease, his left lung was almost completely destroyed . . . forever compromising his ability to feed his family—a precarious enough enterprise in contemporary Haiti, even for the hardy. (Farmer, 1999: 197)

As Jean and Marie's story suggests, the sources and patterns of illness and health care in poorer countries differ dramatically from those found in more affluent countries. We begin this chapter by comparing some of these differences and then focus on health problems in the poorer countries.

Disease Patterns Around the World

In making international comparisons, politicians, social scientists, medical researchers, and others typically divide the world into two broad groups, the **industrialized nations** and the **developing nations.** Essentially, this division reflects the economic status of the various nations. The industrialized nations are primarily defined by their relatively high gross national income (GNI) per capita compared with developing nations. In addition, the industrialized nations are characterized by diverse economies made up of many different industries, whereas the developing nations have far simpler economies, in some cases still relying heavily on a few agricultural products such as rubber or bananas. Because of these economic differences, the developing nations as a group have higher infant and maternal mortality, lower life expectancies, and a greater burden of infectious and parasitic diseases than do the industrialized nations. Table 4.1 shows life expectancies for various developing and industrialized nations.

Table 4.1	*Life Expectancy at Birth*			
COUNTRY	LIFE EXPECTANCY		COUNTRY	LIFE EXPECTANCY
Japan	82		Philippines	70
France	79		Egypt	68
Singapore	79		Bolivia	63
Costa Rica	79		India	62
United States	77		Haiti	51
Cuba	76		Somalia	47
Mexico	75		Ethiopia	46
Thailand	71		Zimbabwe	41
China	71		Sierra Leone	35

Source: Population Reference Bureau (2004).

Although dividing the globe into industrialized versus developing nations is a useful analytic tool, it is important to remember that development level is a scale, not a dichotomy. So, for example, the most rapidly developing nations like Mexico and Thailand have many complex industries as well as traditional agricultural crops and enjoy infant mortality rates and life expectancies approaching those found in the United States and Europe. In addition, although infectious and parasitic diseases remain more common in the rapidly developing nations than in the industrialized nations, chronic diseases are now the most common sources of mortality in both sets of nations (Murray and Lopez, 1996; World Health Organization, 2005b). In contrast, in thirty-five of the *least developed* nations, life expectancy remains less than 50 years and infectious and parasitic diseases still claim most lives (Population Reference Bureau, 2004). Table 4.2 compares the patterns of disease in developing and industrialized nations.

This division between developing and industrialized nations also should not keep us from recognizing that social conditions and, hence, health patterns vary from community to community and from social group to social group within each nation. Thus, as noted in Chapter 3, conditions in Harlem in some ways resemble those in Bangladesh, whereas conditions in wealthy sections of Bangkok resemble those in wealthy sections of U.S. cities. Within the developing nations, the income gap—and consequently the "health gap"—between rich and poor has increased in the last two decades. These growing gaps in income and health largely stem from "structural adjustment" policies adopted by the International Monetary Fund. These structural

Table 4.2   *Percentage of Deaths by Causes, Industrialized and Developing Nations, 2002*

DISEASE	DEVELOPING NATIONS (%)	INDUSTRIALIZED NATIONS (%)
Infectious and parasitic disease	29	6
Circulatory disease (e.g., heart problems)	28	38
Cancers	11	27
Respiratory disease	7	6
Infant and maternal mortality	6	0.4
All other and unknown	19	22.6
Total	100	100

Source: World Health Organization, 2005.

adjustment policies have required developing nations to cut back social programs such as food subsidies and health care for the poor in exchange for economic aid (Kolko, 1999; Peabody, 1996).

Finally, although the terms *developing* and *industrialized* nations imply linear progression from one status to the other, this is not necessarily the case. For example, economic and health conditions worsened in Eastern Europe following the collapse of the Soviet Union and in parts of Africa due to the AIDS epidemic.

With these caveats, in the remainder of this chapter we will explore the sources and nature of disease in the developing nations. Keep in mind, though, that *diseases respect no national borders.* Because of **globalization,** diseases and disease-causing conditions spread rapidly from developing to industrialized nations and vice versa. Researchers have proven that air pollution from Asia—caused by deforestation, overgrazing, and the use of toxic chemicals in agriculture and manufacturing—is now affecting air quality in the western United States (Polakovic, 2002). This pollution increases risks of heart attacks, respiratory failure, and asthma in both continents. Conversely, most used electronics equipment collected in the United States for recycling is shipped to Asia, where the recycling process poisons water supplies with acids, heavy metals (such as lead), and other toxic products (Markoff, 2002).

Because only an imaginary line divides the United States and Mexico—two countries that, at least along their borders, share the same water, air, and, to a large extent, economies—U.S. citizens need to be especially concerned about health conditions in Mexico. For example, only one-third of the sewage generated by the more than 1 million people living in Juarez,

Table 4.3	***Gross National Income per Capita by Life Expectancy***

GNI PER CAPITA*	LIFE EXPECTANCY
Less developed countries (average per capita GNI=US $3,850)	65 years
More developed countries (average per capita GNI=US $23,690)	76 years

*Figures are given in "international dollars," in which $1 equals the amount of goods and services a person could buy in the United States with one dollar.

Source: Population Reference Bureau (2004).

Mexico, is appropriately treated, and thousands of people living in the neighboring city of El Paso, Texas, lack sanitary septic systems (Schmidt, 2000; Skolnick, 1995). As a result, from both sides of the border human wastes drain into the Rio Grande, which provides water for drinking and for agriculture in these two cities and in downstream communities, including Laredo and Brownsville, Texas. This untreated sewage has made gastrointestinal disease a leading cause of infant mortality in both Juarez and El Paso. Diseases like cholera or hepatitis also could easily take root in these areas and spread into the interiors of both countries. Thus, as this example suggests, those who live in the industrialized nations have a vested interest in understanding health and illness in the developing nations.

Sources of Disease in the Developing Nations

Poverty, Malnutrition, and Disease

The primary cause of low life expectancies in the developing nations is poverty. In Chapter 3 we saw how, in the United States, wealthier people experience less illness and live longer than do poorer people. In the same way, wealthier nations have lower rates of illness and mortality than do poorer nations. As Table 4.3 shows, residents of less developed nations (where the per capita GNI averages $3,850) die an average of 11 years earlier than do residents of more developed nations (where the per capita GNI averages $23,690).

In large part, poverty causes disease and death by causing chronic malnutrition. According to the **World Health Organization (WHO),** malnutrition accounts for 53 percent of deaths before age 5 in the developing nations (World Health Organization, Child and Adolescent Health, 2005).

Malnutrition indirectly causes disease and death by damaging the body's immune system, leaving individuals more susceptible to all forms of illness and contributing to both infant and maternal mortality. In addition, malnutrition directly causes numerous health problems, including brain damage caused by iodine deficiency, blindness caused by vitamin A deficiency, and mental retardation caused by anemia.

The Roots of Chronic Malnutrition

Given the link between malnutrition, illness, and death, the importance of investigating the roots of chronic malnutrition is clear. At first thought, we might easily assume that malnutrition in developing nations that have not yet experienced the **epidemiological transition** results naturally from overpopulation combined with insufficient natural and technological resources. Yet food production has surpassed population growth in most countries, including most of those where hunger is common (Lappé, Collins, and Rosset, 1998). In fact, most of the "hungry" countries export more food than they import, and almost every country has access to sufficient food to feed its entire population.

Nor can malnutrition be blamed on population density (Lappé et al., 1998). The Netherlands, for example, is one of the most densely populated countries in the world, yet chronic malnutrition no longer occurs there. Similarly, Honduras has twice as much cropland per person as Costa Rica, yet malnutrition remains common only in Honduras.

If overpopulation, lack of food, population density, and lack of cropland do not explain chronic malnutrition, what does? The answer lies in the social distribution of food and other resources: *Malnutrition occurs most often in those countries where resources are most concentrated.* In other words, malnutrition occurs not in countries where resources are scarce, but in countries where a few people control many resources while many people have access to very few resources (Dreze and Sen, 1989; Lappé et al., 1998). Similarly, within each country, malnutrition occurs most often among those groups—typically females and the poor—with the least access to resources (Messer, 1997). In essence, then, malnutrition is a disease of powerlessness.

If powerlessness causes malnutrition, then eliminating inequities in power should eliminate malnutrition. Evidence from China and Costa Rica supports this thesis. These two nations—the first essentially communist and the second essentially capitalist—both adopted in past decades socialistic strategies for redistributing resources somewhat more equitably. By giving farmland to formerly landless peasants, extending agricultural assistance to owners of small farms, working to raise the status of women, and so on, they made chronic malnutrition almost unknown within their borders. On the other hand, China has not proved immune to *acute* malnutrition caused by famines. According to Nobel Prize–winning economist Amartya Sen, famines occur only when (1) natural events reduce harvests *and* (2) nondemocratic governments (like that in China) need not fear being voted out of office if they do not meet their citizens' basic needs (Sen, 1999).

The Role of International Aid

Similarly, in democratic developing nations, international aid—both food aid and development projects—has helped improve citizens' standard of living and health status. But in *nondemocratic* nations, aid often has had the opposite effect (World Bank, 1998). Most international food aid comes from the United States, under the 1954 Food for Peace Act, or PL-480. The primary purpose of this law is to protect U.S. economic and military interests (Lappé et al., 1998). By sending U.S. farm surpluses overseas as food aid, agricultural producers can maintain prices for their goods at home while opening new markets to U.S. agricultural commodities. In addition, because the United States sells food aid on credit rather than giving it away, food aid helps offset U.S. trade deficits. Food aid also helps protect U.S. military interests by bolstering the governments of nations with strategic military importance for the United States. This explains why U.S. food aid primarily goes not to the hungriest countries, but to countries where the U.S. has military interests, such as Egypt, Israel, El Salvador, Pakistan, and Turkey.

Once food aid reaches the developing nations, its distribution can unintentionally reinforce inequities in access to resources and thus malnutrition (Lappé et al., 1998). Food aid goes directly to foreign governments, which can distribute it as they choose. In countries run by democratic governments committed to social equality, aid is likely to benefit those who need it most. Unfortunately, many developing nations are run by small, economically powerful elites, who sometimes instead sell on the open market any food their governments receive and pocket the profits, thus accentuating social inequities.

Because the hungriest people cannot afford to buy food aid sold in the marketplace, food aid does not improve their nutritional status. Rather, food aid *contributes* to the malnutrition of the landless tenants, sharecroppers, and day laborers who form the overwhelming bulk of those suffering from malnutrition (Lappé et al., 1998). When the United States sells its surplus agricultural commodities in the developing nations, the prices of those commodities in those nations plummet. As a result, owners of small farms may no longer be able to earn a living and must sell their land to larger landowners who can take advantage of economies of scale. Thus land ownership and power become more concentrated, as do the inequities that underlie malnutrition and illness.

Like international food aid, internationally sponsored development projects have had mixed impacts on malnutrition and on health in general (World Bank, 1998). According to the politically conservative World Bank, carefully designed projects, sensitive to local conditions and culture and located in countries with democratic governments, open trade, social safety nets, and conservative economic policies can reduce malnutrition and its root causes. In Pakistan, for example, school enrollment of girls soared in 1995 when local communities received development money to open new schools on the condition that they increase the enrollment rate for girls (World Bank, 1998). In the long run, this approach should increase the status of women, which, as we will see, is directly linked to malnutrition, infant mortality, and maternal mortality.

On the other hand, although projects like the Péligre Dam in Haiti, the Akosombo Dam in Ghana, and the Aswan Dam in Egypt have brought electricity to urban elites and industrial sites run by multinational corporations, they are flooding and destroying agricultural fields and rural villages and bringing plagues of waterborne diseases to rural dwellers (Basch, 1999: 280–281; Farmer, 1999). Agricultural development projects have been particularly likely to contribute to malnutrition among women and children (Lappé et al., 1998). These projects often start from the assumption, based on Western ideas about the family and the economy, that raising cash crops will benefit families more than raising food crops will and that men rather than women should be responsible for agricultural efforts. However, cultural traditions in many developing nations hold women responsible for growing food and feeding the family (Lappé et al., 1998). When development projects encourage men to grow cash crops, the men sometimes take over land women had used to grow food and, because men consider feeding the family a woman's responsibility, use their profits not to purchase food but, rather, to purchase high-status goods for themselves such as tobacco or Western clothes. As a result, malnutrition increases among women and children.

Infectious and Parasitic Diseases

One result of malnutrition and, more broadly, of poverty is a high rate of infectious and parasitic disease. As Table 4.2 shows, although such diseases have declined rapidly in recent years in the developing nations, they still account for far more deaths there than in the industrialized nations.

As in Europe and the United States before the twentieth century, the high rates of infectious and parasitic diseases reflect the dismal circumstances in which many people live. In addition to malnutrition, overcrowding promotes the spread of airborne diseases like tuberculosis, while contamination of the water supply with sewage spreads waterborne diseases such as cholera and intestinal infections. Similarly, poor housing and lack of clean water for bathing result in frequent contact with disease-spreading rats, fleas, and lice.

The infectious and parasitic diseases that cause the most deaths in the developing nations are **HIV disease**, tuberculosis, diarrheal diseases, and malaria. In addition, measles is a major cause of death for children. In the next sections we consider these diseases in more detail.

HIV Disease

HIV disease now kills more persons in the developing world than does any other infectious or parasitic disease. More than 90 percent of HIV-infected persons in the world live in developing nations (World Health

Organization, 2002). HIV infection is now endemic in parts of the Caribbean and in much of Africa, and it is spreading especially rapidly in Asia (particularly in India and Southeast Asia).

From the beginning of the epidemic, heterosexual intercourse has been the major mode of HIV transmission in the developing nations. Consequently, women account for half or more of all cases in these nations. Because many of these countries lack the funds needed to test blood for HIV, transmission via blood transfusions remains common. Similarly, in part because they lack the funds to supply infected women with the drug azidothymidine (AZT), which can prevent transmission from mother to fetus, such transmission remains common. (However, the recent development of new, inexpensive drug regimens to prevent maternal-fetal transmission could make it less common in future.) Infection is most common in urban areas but is spreading rapidly in the countryside, especially in areas where war has disrupted families and increased both consensual and nonconsensual sexual intercourse between soldiers and villagers. (Box 4.1 discusses in more detail how war and militarism affect health.) In the hardest-hit countries (most located in sub-Saharan Africa), more than 25 percent of adults are infected, while life expectancies have dropped below 40 years (UNAIDS/ WHO, 2004).

As stunning as these numbers might appear, they understate the impact of HIV disease. Unlike most illnesses, HIV disease most commonly strikes at midlife, normally the most economically productive years. In the hardest-hit countries, agricultural production is declining steeply, causing food shortages. Moreover, HIV disease disproportionately has affected the most educated segments of the population in the developing nations; in parts of Uganda and Malawi, almost one-third of all teachers have the disease (Schemo, 2002). Consequently, HIV disease has crippled both schools and the economy in numerous countries. The resulting increase in unemployment and poverty is sending ripples of illness and death throughout these countries. In addition, HIV disease typically strikes during the child-rearing years. This situation has produced a rise in child deaths, for whenever mothers die, their children are also more likely to die, especially if they have no surviving relatives to care for them (UNAIDS/WHO, 2004). Map 4.1 shows the worldwide distribution of children who have lost one or both parents to HIV disease.

Several theories have been proposed to explain why HIV disease has hit Africa so hard. Two theories that have gained the most supporters are the cultural theory used by demographer John Caldwell and his colleagues and the economic and political theory, known as "world systems theory," used by sociologist Charles Hunt (see Key Concepts 4.1).

Caldwell and his colleagues (1989, 1991, 1992) have argued that because soils are poor throughout much of Africa, farming there always has been highly labor intensive, and farming families have needed to have many children to help them in the fields. Consequently, a cultural system developed

| Box 4.1 | *War and Health* |
| | by Lisa Comer |

In addition to poverty, malnutrition, germs, and parasites, another important, although often overlooked, source of death and disease in the developing nations is war (Geiger and Cook-Deegan, 1993; Toole and Waldman, 1993). War is an ongoing fact of life in many developing nations, and high rates of civilian deaths from bombs and guns are the norm rather than the exception. For example, a report published by the prestigious British medical journal, *Lancet,* estimated that the risk of violent death among Iraqi civilians increased 58-fold after the invasion by U.S.-led coalition forces, and that 100,000 civilians, mostly women and children, were killed by military forces in the first 18 months after the invasion (Roberts et al., 2004). In addition to the risk of death from military violence, civilians in war-torn lands also risk death or injury from forced labor, chemical warfare, torture, mutilation, rape, and malnutrition (when forced to leave their farms and herds). For example, it is estimated that for every civilian killed by combatants in the Congo during 2005, sixty-two—most of them women and children—were killed by malnutrition, disease, or other war-related causes (Lacey, 2005). These physical traumas typically are magnified by the psychological traumas of losing one's family, community, and, frequently, dignity; the events of September 11 brought home some of these issues to U.S. citizens.

Human rights violations committed during wars often lead to a rise in illness among civilians. Forcing refugees into overcrowded, unsanitary relocation camps frequently results in epidemics of communicable diseases, which often go untreated because health care workers and medical facilities—insufficient in the best of times—are targeted for destruction by the military (Geiger and Cook-Deegan, 1993; Toole and Waldman, 1993). For the same reason, wars often disrupt public health services, including

that valued fertility over chastity or monogamy and valued ties between parents and children more than ties between spouses. As a result, individuals tended to have relatively high numbers of sexual partners over their lifetimes. In past centuries, Africans typically obtained these sexual partners within small social and geographic circles. Since the rise of European colonization, however, and the attendant growth of towns, bars, transportation networks, and a commercial sex industry, both the size and geographic spread of Africans' social circles have broadened, causing dramatic increases in average numbers of sexual partners as well as the geographic diversity of those partners. As a result, Africans are particularly likely to be exposed to sexually transmitted diseases, including HIV.

Whereas Caldwell and his colleagues give primary emphasis to cultural factors in explaining the devastating rates of HIV disease in Africa, Charles Hunt (1989, 1996) emphasizes the impact of economic and political conditions. Hunt's argument is based on **world systems theory,** which divides the world's nations into **core nations, peripheral nations,** and a few **semiperipheral nations** (Chase-Dunn, 1989; Wallerstein, 1974). The core nations,

vaccination programs for children. Moreover, an estimated 300,000 children as young as age 8 in forty nations are serving (usually involuntarily) as soldiers (Crossette, 2001). These children are exposed to all the horrors and dangers of warfare and to increased risks of malnutrition, disease, landmine injuries, sexual abuse, and substance abuse, while losing opportunities for education and normal family life that might protect their mental and physical health as adults.

Given the profound impact of war on public health, the medical community can and sometimes does play a critical role in documenting and preventing war crimes and related human rights violations. Health care workers' documentation of these horrors is especially important, because politicians are more likely to believe testimony about war crimes received from health care workers as compared to that received from other civilians (Geiger and Cook-Deegan, 1993; Swiss and Giller, 1993).

Consequently, health care workers can help to awaken public awareness of war crimes and human rights violations. By so doing, they can speed health care and other assistance to war survivors and bring war criminals to justice.

Over the years, individual health care workers and nonprofit groups, such as Physicians for Human Rights and *Medécins Sans Frontières* (Doctors Without Borders)—which won the Nobel Peace Prize in 1999—have eased the burdens of war victims substantially. In addition, beginning in the early 1990s, the American medical community moved toward officially asserting a commitment to war survivors. For example, the *Journal of the American Medical Association* in 1993 devoted part of an issue to this topic. If this pattern continues, doctors may play a growing role in documenting, treating, and perhaps even preventing this significant source of death and disease in the developing nations.

such as France and the United States, are in effect an upper class of nations—enjoying highly diversified, industrialized economies that provide a high standard of living for most citizens. Conversely, the peripheral nations form a lower class of nations, where modernization and industrialization have developed slowly if at all, and the standard of living is low for all but a small elite.

World systems theory argues that the core nations have achieved and maintained their present economic position by exploiting the resources of the peripheral nations. This is done through the work of multinational corporations based in the core nations. Rather than establishing industries in peripheral nations that would help those nations modernize their economies, multinational corporations instead have established industries that extract raw goods (such as rubber, minerals, or specialized food crops). Profits from the sale of those raw goods and from the finished goods made from those raw goods are brought back to the core nations. Lacking their own modern industries, peripheral nations must buy most manufactured goods and, sometimes, basic foods from the core nations. In this way, the

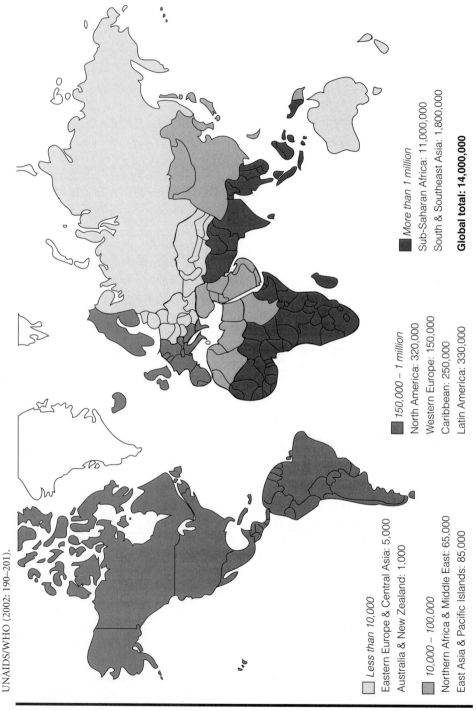

UNAIDS/WHO (2002: 190–201).

More than 1 million
Sub-Saharan Africa: 11,000,000
South & Southeast Asia: 1,800,000

Global total: 14,000,000

150,000 – 1 million
North America: 320,000
Western Europe: 150,000
Caribbean: 250,000
Latin America: 330,000

Less than 10,000
Eastern Europe & Central Asia: 5,000
Australia & New Zealand: 1,000

10,000 – 100,000
Northern Africa & Middle East: 65,000
East Asia & Pacific Islands: 85,000

Map 4.1   ***Children Under Age 15 and Currently Living Who Have Lost One
or Both Parents to AIDS***

		ECONOMIC AND POLITICAL:
TYPE OF THEORY	CULTURAL	"WORLD SYSTEMS THEORY"
Theorist	John Caldwell	Charles Hunt
Central dynamic	Labor-intensive farming requires large families for success.	Core nations exploit peripheral nations for their own profit.
Central effect	Fertility is valued more than chastity, monogamy, or fidelity.	Multinational corporations from core nations create only low-wage jobs in peripheral nations, extracting raw materials in a few, centralized locations.
Social consequence	People have many sexual partners.	Men must leave their farms to seek paid work in extractive industries. But they don't earn enough money to support their families, and women can't grow enough crops on their own to survive. Men turn to prostitutes while away from home, and women become prostitutes to survive.
Health consequence	HIV spreads.	HIV spreads.

Key Concepts 4.1 — *Understanding the Spread of HIV in Africa*

core nations maintain a favorable trade balance with peripheral nations, force the peripheral nations to rely for their economic well-being on inherently unstable markets for raw materials, and perpetuate the underclass position of the peripheral nations.

Applying this theory to HIV in Africa, Hunt (1989, 1996) argues that the African nations remain largely under the economic control of corporations based in the former colonial powers. To increase their profits, those corporations have concentrated industries in a few sites, rather than distributing manufacturing, mining, and corporate agriculture around the continent. Attracted by the prospects of cash income and faced with little means of earning a living in their home villages, native men leave the countryside to seek employment at these sites, often living apart from their wives and families for weeks, months, or even years at a time. These conditions foster the use of prostitutes and, in turn, the spread of sexually transmitted diseases, including HIV disease. Once workers become ill, their employers fire them and send them back to their villages, where they spread infection still further.

Meanwhile, health conditions also deteriorate among women and children left in rural villages. The loss of men's labor makes it more difficult for women to grow sufficient crops to feed themselves and their children. As a result, women typically adopt agricultural practices and crops that require less labor, even though these changes deplete the soil and provide less nutrition. Those left in rural villages grow progressively malnourished and susceptible to disease. Faced with these conditions, women's only option is to seek employment in cities, where many find prostitution the only available job. This completes the cycle through which multinational corporations indirectly encourage HIV infection among both men and women, in rural and urban areas.

Support for this theory comes from data suggesting that HIV was most common and appeared earliest in areas where migrant laborers worked, was next most common in the rural areas from which migrant laborers were recruited, and was least common in areas without links to migrant labor (Hunt, 1989). Other studies similarly have found that economic and structural factors better explain the explosive spread of HIV in Africa than do cultural factors (Simmons, Farmer, and Schoepf, 1996). At this point, however, the poor quality of data on HIV rates in Africa makes it difficult to test any theory with confidence. Moreover, neither the cultural theory used by Caldwell and his colleagues nor the materialist theories used by Hunt and others can account fully for the geographic distribution of HIV infection in Africa (Hunt, 1996). Thus, neither theory can be considered fully supported.

Tuberculosis

Each year, tuberculosis infects about 9 million people and kills about 2 million (World Health Organization, 2002). The disease is most common in Asia, followed by Africa and the Middle East. Tuberculosis is particularly devastating because, like HIV, it typically hits people during their prime work years, and so sharply curtails family incomes.

As described in Chapter 2, the incidence of **tuberculosis** is increasing around the world for two reasons. First, developing nations cannot afford to treat the new, drug-resistant strains of the disease. Second, the rise of HIV infection, which makes individuals more susceptible to other infections, has led to soaring rates of tuberculosis: During the last decade, in the countries hardest hit by HIV, the number of cases of tuberculosis has doubled or tripled (World Health Organization, 2002).

Diarrheal Diseases

In industrialized nations, diarrhea is generally a source of passing discomfort. In developing nations, diarrheal diseases can be fatal, especially among children under age 2. WHO estimates that diarrheal diseases kill more than

2 million children yearly, accounting for 27 percent of all child deaths (World Health Organization, Child and Adolescent Health, 2005). Diarrhea is a symptom, not a disease, and can result from infection with any of several bacteria, viruses, or parasites. Diarrhea kills by causing dehydration and electrolytic imbalance. It also leads to malnutrition, because affected children not only eat less but also absorb fewer nutrients from the foods they do eat. In turn, malnutrition leaves children susceptible to other fatal illnesses. Conversely, other illnesses can leave children susceptible to both diarrheal diseases and malnutrition.

Diarrheal diseases (including dysentery, cholera, and infection with *E. coli*) occur when individuals ingest contaminated water or foods. The likelihood of severe diarrhea is greatest when families lack refrigerators, sanitary toilets, sufficient fuel to cook foods thoroughly, or safe water for cooking and cleaning. Using government reports from 2002, WHO estimates that about 1 billion people lack access to "improved" water supplies, and even more lack access to truly safe water (WHO/UNICEF, 2004). These figures undoubtedly overestimate access, because governments may report to WHO that citizens of their countries have access to clean water even if the only water source is a single, sporadically working faucet, a mile or more away, and shared by many families. The *number* of persons without safe water is greatest in Asia, whereas the *percentage* of those without safe water is highest in sub-Saharan Africa.

Survival rates for children with diarrheal diseases in developing nations have improved rapidly in recent years. Before the 1960s, those suffering from diarrheal diseases could be treated only by using expensive intravenous fluids, thus making treatment unfeasible for many in the developing nations. Since then, however, scientists have demonstrated that a simple and inexpensive solution of dried salts and water is just as effective, and the World Health Organization has actively and successfully promoted this "oral rehydration therapy."

Malaria

Each year, about 300 million people (mostly in tropical Africa) become infected with malaria, and more than a million die from the resulting anemia, general debility, or brain infections (World Health Organization, 2002). In addition, many of those who survive will experience disabilities from the intermittent chills, fevers, and sweats that malaria brings.

Malaria poses the greatest threat to pregnant women, infants, and young children. Among pregnant women, malaria increases the risks of miscarriage, anemia, and premature labor, each of which increases the risk of potentially fatal hemorrhaging. Infants born to malaria-infected women typically have lower than average birthweights and, hence, higher chances of death or disability. Malaria is often fatal among young children, whose immune systems have not yet developed sufficiently to fight infection. About 90 percent of those who die from malaria are under age 5, and malaria accounts for 12.7 percent of deaths among children under age 5 (World Health Organization, 2002; World Health Organization, Child and Adolescent Health, 2005).

Malaria is caused by protozoan parasites belonging to the genus *Plasmodium*. Malaria is transmitted only by *Anopheles* mosquitoes and, consequently, exists only where those mosquitoes live. (*Anopheles* mosquitoes and malaria used to exist throughout the United States and appear to be making a comeback; in 1999, for the first time in decades, malaria was diagnosed in a U.S. resident who had neither lived nor traveled in another country.) The disease cycle begins when a mosquito bites an infected individual and ingests the parasite from the individual's blood. The parasite reproduces in the mosquito's stomach and then migrates to the mosquito's salivary glands. The next time the mosquito bites someone, it transmits the parasite to that person.

Because of this transmission cycle, eliminating *Anopheles* mosquitoes will eliminate malaria. Since the 1940s, antimalaria campaigns have depended heavily on using pesticides to kill mosquitoes. Although these campaigns initially work well, over time pesticide-resistant mosquitoes evolve and the pesticides lose their potency. As a result, nations must constantly search for new and more toxic pesticides, each of which can endanger birds, fish, and insects that benefit humans. Because of these problems, some recent campaigns have instead focused on encouraging the use of insect repellents, mosquito netting, and screens to prevent infection. These campaigns also have focused on encouraging the use of drugs, such as chloroquine and mefloquine, which can both prevent and treat malaria. Unfortunately, because these drugs can cause debilitating side effects and cost more than many residents of developing nations can afford, infected individuals often stop taking the drugs before they are cured. This continual undertreatment of malaria, like the undertreatment of tuberculosis, has encouraged the evolution of drug-resistant strains of the disease around the globe. Consequently, although malaria has been eliminated in some regions, the situation in the rest of the world has worsened during the past decade.

Measles

To persons living in the industrialized nations, where measles is considered a minor childhood illness, it might seem odd to see measles listed as a major cause of death. Yet measles kills 6.3 percent of children under age 5 in the developing nations (World Health Organization, Child and Adolescent Health, 2005). These deaths occur when children, already weakened by malnutrition and poor living conditions, become further weakened by measles. Their bodies' ability to fight disease diminishes, leaving them susceptible to potentially deadly pneumonia, respiratory infections, and diarrhea. Unlike tuberculosis and malaria, however, rates of measles have declined almost by half since 1990, following a worldwide WHO measles vaccination campaign.

Immunization rates have remained unchanged in Africa, however, due to ongoing and severe economic problems on that continent.

Unfortunately, even if vaccination becomes more widespread and rates of measles continue to decline, the overall health of children in the developing nations will not improve unless social conditions also improve. As long as conditions in the developing nations continue to foster diseases of all kinds, children who do not die from measles are still likely to die young from other diseases; at least one study has found that reducing a country's death rate from measles has no effect on its rate of childhood mortality (Turshen, 1989). Only when the basic inequities in living conditions that underlie death and disease are substantially reduced will more children survive.

Infant Mortality

Like infectious and parasitic diseases, infant mortality is far more common in the developing nations than in the industrialized nations. As of 2004, the average infant mortality rate in the developing nations was 62 per 1,000 live births—nine times higher than the rate in the industrialized nations (Population Reference Bureau, 2004). These averages, however, hide the great range in infant mortality rates within the developing nations. (See Table 4.4.)

The most common causes of infant mortality in the developing nations are malnutrition and infections (particularly respiratory infections and diarrheal diseases). Because we examined these factors earlier in this chapter,

Table 4.4	*Infant Mortality per 1,000 Live Births*		
COUNTRY	INFANT MORTALITY RATE	COUNTRY	INFANT MORTALITY RATE
Somalia	207	Philippines	29
Afghanistan	165	Mexico	25
Ethiopia	105	Thailand	20
Haiti	80	Costa Rica	10
India	70	United States	7
Zimbabwe	65	Cuba	7
Bolivia	54	France	4
Egypt	38	Japan	3
Brazil	33	Singapore	2
China	32		

Source: Population Reference Bureau (2004).

the focus here is on two other important sources of infant mortality: women's status and infant formula manufacturers.

The Role of Women's Status

The low status of women plays a critical role in infant mortality in developing nations. Infant mortality occurs most often among babies with low birthweights. In the industrialized nations, low birthweight typically occurs when babies are born prematurely. In the developing nations, low birthweight typically occurs among babies born at full term to mothers who have malaria, are underfed, routinely perform heavy labor, or suffer from anemia, which affects more than 50 percent of pregnant women in developing nations (World Health Organization, 1998b).

These conditions reflect women's typically low status. Throughout the developing nations, girls and women often spend long hours in heavy labor and, in many nations, receive less nutrition than do boys and men (Messer, 1997). In addition, girls are less likely than boys to be immunized against disease, to receive health care when ill, and to receive health care promptly (Messer, 1997). Girls are thus more likely to become ill and less likely to survive their illnesses. Consequently, women often enter their childbearing years already ill and malnourished—a situation that worsens as pregnancies further stress their bodies and drain their energy.

Similarly, infant mortality is highest among infants born to very young or very old mothers and to infants born less than 18 months after a sibling. This situation occurs most commonly in cultures that expect women to marry at young ages and that judge women's worth by the number of sons they produce. In part, these cultural values reflect the economic realities of agricultural life: In agricultural societies, children produce more economic resources than they consume, so a family with many children is more likely to succeed than a family with few children. Further, in the absence of any formal provisions for social security, individuals can guarantee their security in old age only by having sons. (Having daughters usually does not help, because daughters in most cultures are expected to take care of their husbands' parents rather than their own.)

Nevertheless, even in these societies many women would like to limit their fertility. This desire is so great that throughout the world, women often choose illegal abortion over childbearing: 44 percent of all abortions performed worldwide (and 54 percent in developing nations) are illegal (Henshaw, Singh, and Haas, 1999). In fact, statistics from Romania, where abortion was outlawed between 1966 and 1989, suggest that making abortion illegal has almost no long-term impact on either the abortion rate or the birth rate—although it dramatically increases the number of women who die or become infertile following unsafe abortions (World Health Organization, Division of Reproductive Health, 1998b). Meanwhile, the slums of Bombay and Rio de Janeiro, like the orphanages of Romania after 1966, are filled with abandoned children whose families could not support them. Similarly, in parts of Asia, infanticide of girl babies continues to occur among families that want babies only if those babies are male, and abortions now often occur when women learn through genetic testing that they are carrying a female fetus (Banister, 1999; Lawn, Cousens, and Zupan, 2005). This chapter's ethical debate (Box 4.2) discusses some of the moral quandaries posed by using abortion for sex selection.

In sum, research suggests that if women's social status were higher, they would enter their childbearing years with healthier bodies, wait longer before having babies, wait longer between babies, and have fewer babies in total, with each of these factors lowering the infant mortality rate. For all these reasons, many researchers and public health workers have suggested that the most effective way to reduce infant mortality is to improve the status of women, thereby increasing their power to make decisions for themselves. This explains at least partly why infant mortality is so much lower in Costa Rica, China, and Zimbabwe than in other countries at similar levels of development. Box 4.3 describes the actions of one nonprofit agency that is working to improve the health of developing nations by improving women's status.

The Role of Infant Formula Manufacturers

A final cause of infant mortality in the developing nations is the use of infant formula and other foods instead of or in addition to breast-feeding. In Chapter 3 we noted the basic biological benefits of breast-feeding for both infants and mothers. The benefits are even greater in the developing nations, where babies who are fed alternatives to breastmilk (whether infant formula, juice, water, or any other substances) are twenty-five times more likely than breast-fed babies are to die from infections (*Lancet,* 1990). The World Health Organization (1993) estimates that about 1.5 million babies die unnecessarily each year because they are not breast-fed.

In the developing nations, several factors contribute to the especially high rates of death and disease among infants who are not breast-fed. First, in addition to the inherent limitations of substitutes for breastmilk, the process of bottle-feeding itself can expose infants to tremendous risks. Infant formula is typically sold as a powder that must be mixed with water and then transferred to a bottle before it can be used. In most developing nations, this water contains dangerous infectious organisms. Those organisms can be killed if the water, bottle, and nipple are boiled. However, families do not necessarily understand how or why they should do so. Moreover, throughout the developing nations, many women and children already spend hours each day getting water and firewood and lack the time and energy to get the extra supplies needed for sterilization.

Second, other foods cost far more than breastmilk (which is not actually free, because it reduces mothers' nutritional stores and can prevent their return to paid employment). To conserve money, families often stretch infant formulas by diluting them with water. Babies fed diluted formula in essence starve to death even while filling their stomachs.

Box 4.2   ***Ethical Debate: The Ethics of Sex Preselection***

Zhang Zhiquan and his wife Mei live in a rural village in the People's Republic of China. Growing up in rural China, they learned early that couples needed sons to prosper and to care for them in their old age. They also learned that sons were essential for passing on the family name, that wives who produced no sons deserved mockery and abuse, and that girls were so useless that in the past many rural families did not even bother to name them. When Mei became pregnant, therefore, they had to decide what they would do if the baby were female. In the past, should they have felt unable or unwilling to raise a daughter, their only options would have been to kill the baby or give it up for adoption—choices that some families still make. Now, however, they had one additional option: having a health care worker identify the fetus's sex through ultrasound or amniocentesis and perform an abortion if the fetus were female.

Half a world away, the same issues of sex preselection and selective abortion arise, although in a different form:

Sharon and James Black live in Denver, Colorado, with their two young daughters. Because they both believe that children need a parent home at the end of the school day, Sharon works only part-time as a secretary, while James works two jobs so they can make ends meet. Sharon has just learned she is pregnant again. Although they had only planned on having two children, James always wanted a son with whom he can share his interests in sports and automobiles. Having another child, however, will further strain their finances and make it difficult for Sharon to return to full-time work for several more years. Consequently, continuing the pregnancy does not seem worthwhile unless they know the fetus is male.

Is sex preselection ethically justified in these cases? Although the circumstances differ enormously, for both families the birth of a daughter would bring substantial economic hardship. For both families, too, a daughter would enter life unwanted and already having failed to meet her parents' expectations. In addition, for the Chinese family and possibly (although to a lesser extent) the American family, the birth of another daughter might lower the wife's status and strain the marriage. Given these circumstances, wouldn't it be best for all concerned if the families use the available medical technology to test their fetuses' sex and to abort them if they are female?

For hundreds of thousands of couples in Asia and a growing number in the West, the answer, resoundingly, is yes. In China and India, 117 boys are born for every 100 females overall,

---

Finally, by altering the hormonal levels in a woman's body, breast-feeding serves as a moderately effective contraceptive. Breast-feeding thus helps women to space out pregnancies and gives each baby a better chance for survival. For all these reasons, WHO (2001) recommends that children throughout the world, in both industrialized and developing nations, receive only breastmilk during the first six months of life and a combination of breastmilk and other foods until at least age 2.

with an even more skewed sex ratio in rural areas (Eckholm, 2002; Lawn, Cousens, and Zupan, 2005). The same forces are at work in the industrialized nations, although not as strongly: In one study, 47 percent of surveyed geneticists and genetic counselors in these nations had received requests from couples desiring fetal sex selection (D. Wertz and Fletcher, 1998). Twenty-nine percent of the respondents reported that they would test fetal sex for a couple with four daughters who intend to abort if their fetus is female, and another 20 percent would refer the couple to someone they knew would do so.

Those who support prenatal sex selection argue that selective abortion causes little harm, whereas the birth of unwanted girls financially strains families, leaves mothers open to ridicule or even physical abuse, and results in child neglect, abuse, or abandonment. Those who oppose sex preselection argue that it does more harm than good because it reinforces the low status of females. Although in rare circumstances families use medical technologies to ensure that their babies are female (such as families with a history of hemophilia, a disease that affects males but not females), almost always sex preselection means selecting males. In the United States, both women and men prefer boys as their first child and prefer two boys and a girl to two girls and a boy; families are most likely to have three children if their first two are female (B. Rothman, 1986).

When families select male fetuses over female fetuses, they proclaim male babies preferable. Moreover, when health care workers help families to select male babies, the workers in essence validate this preference. Finally, when health care workers assist in sex preselection—whether helping families to select males or females—they reinforce the idea that males and females are inherently different. After all, if male and female personalities, interests, and aptitudes were more similar than different, why would families need to choose one over the other?

In sum, to assess the ethics of sex preselection we need to weigh the potential benefits and costs for families and for society as a whole.

Sociological Questions

1. What social views and values about medicine, society, and the body are reflected?

2. Which social groups are in conflict over this issue? Whose interests are served by the different sides of this issue?

3. Which of these groups has more power to enforce its view? What kinds of power do they have?

4. What are the intended consequences of this policy? What are the unintended social, economic, political, and health consequences of this policy?

Given all the benefits of breast-feeding, why don't more women in developing nations breast-feed? Part of the answer lies in traditional cultural beliefs, such as the conviction that children require certain traditional foods for health, or that it is unsafe to have sex with breast-feeding women (Dettwyler, 1995). Part of the answer lies in practical economic and social issues, such as the difficulty of meshing breast-feeding with paid work. And part of the answer lies with multinational food corporations (most of them

---

Box 4.3   *Making a Difference: Freedom from Hunger*

Freedom from Hunger (FFH) began in the 1940s as a traditional food aid program, providing food relief to the hungry in the developing nations of Africa, Asia, and Latin America. By the 1980s, however, the organization had concluded that the only way to reduce hunger in the long run was to help poor women in the developing nations to become economically self-sufficient. As a result, in 1989 FFH committed all its resources to providing **microcredit** to women in developing nations through its Credit with Education program. Microcredit refers to the practice of awarding very small, short-term loans (typically between $10 and $300 for 4 to 6 months) to poor women who have no meaningful assets or other access to affordable cash credit. FFH distributes these loans through community-based credit associations that it establishes, made up of 20 to 30 women living in the same town. The association is then responsible for allocating credit to individual women and collecting debt

payments from them. More than 100,000 women have participated in the Credit with Education program. To date, FFH's credit associations have had exceptional success in repaying their loans, so that the system has become largely self-sustaining.

Although the amounts given in these loans may seem too small to be meaningful, they can transform women's lives. Women who receive microcredit loans no longer have to purchase supplies or raw materials from local vendors on credit at usurious rates and, instead, can start investing in their own businesses, such as raising chickens or making clothing. In addition, FFH links microcredit to health and nutrition education, using its credit associations to provide basic information about such topics as breast-feeding and treating infant diarrhea. FFH provides women and, in the long run, their children and families, with information on how to improve their health as well as the resources necessary to do so.

---

based in the industrialized nations), which have convinced women in the developing nations that infant formula is superior to breastmilk.

To create a market in the developing nations, corporations have provided free or subsidized formula to patients in maternity hospitals (Gerber, 1990; *Lancet,* 1990). If these women use the formula instead of breast-feeding while in the hospital, they may find it physiologically impossible to switch to breast-feeding later. Corporations also have mounted massive advertising campaigns throughout the developing nations to convince women that bottle-feeding produces healthier babies and even lightens babies' skin— a status symbol in many developing nations. One particularly pernicious strategy is to dress saleswomen as nurses and send them to villages and maternity hospitals to encourage women to bottle-feed.

During the 1970s, recognition of bottle-feeding's role in infant mortality led to the rise of an international, consumer-led campaign, based in the United States and Europe, against the multinational corporations that produce infant formula (Gerber, 1990). The campaign focused especially on Nestlé, the most aggressive marketer of infant formula in the developing

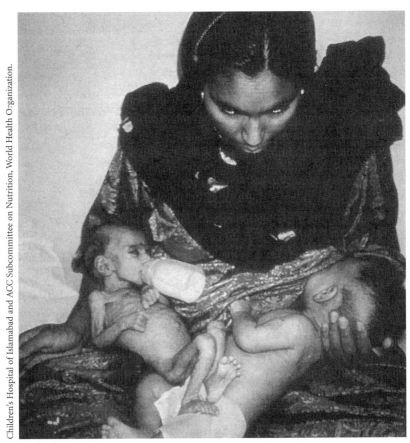

Children's Hospital of Islamabad and ACC Subcommittee on Nutrition, World Health Organization.

This woman knew breastmilk was healthier; but fearing she would not have enough breastmilk, she breast-fed only her son and bottle-fed his twin sister.

nations. The campaign's main tools were an international awareness campaign and a consumer boycott of infant formula and other products made by Nestlé.

In 1981, and partly in response to this campaign, the WHO Assembly adopted an International Code of Marketing of Breastmilk Substitutes, aimed at sharply limiting the promotion and sale of formula in the developing nations. (The sole nation to vote against the code was the United States, which finally ended its opposition in 1996.) Among its provisions, the code calls for manufacturers to refrain from advertising infant formula, providing free samples to mothers, promoting infant formula through health care facilities, hiring nurses or women dressed as nurses to promote infant formula, providing gifts or personal samples to health care workers, and providing free or low-cost supplies to hospitals.

By 1984, all the major formula producers had agreed to accept the WHO code, bringing an end to the boycott. Within the developing nations, however, the mistaken notion that bottle-feeding was more "modern" and healthier already had taken root. Moreover, it soon became obvious that the manufacturers had reneged on their promise to abide by the code. To call attention to these code violations, the International Baby Food Action Network began a new boycott in 1988. Partly due to this consumer pressure, billboards and other advertisements for infant formula have become less common, and health care workers in developing nations now more often actively support women's efforts to breast-feed. Manufacturers continue to break the code, however, although they now focus more on encouraging mothers to stop breast-feeding early rather than encouraging mothers never to begin (Wise, 1998). One study conducted through **random sampling** in four developing nations found that 10 percent of mothers with children under six months old and 25 percent of health care facilities had received free samples of bottle-feeding supplies from manufacturers, in direct violation of the WHO code (Taylor, 1998). This survey probably underestimates the problem because it studied only some of the developing nations known for reasonably good compliance with the code (A. Costello and Sachdev, 1998). As of 2005, activist organizations continue to report ongoing code violations around the world (International Baby Food Action Network, 2005).

Maternal Mortality

Although maternal mortality is now rare in the industrialized nations, in the developing nations it remains the primary cause of death among women of reproductive age. For example, the lifetime risk of dying from childbirth complications is one in 1,400 in Europe, one in 65 in Asia, and one in 16 in Africa (World Health Organization, 1998b).

How can we account for the tremendous toll maternal mortality takes in the developing nations? Patricia Smyke, writing for the United Nations, explains:

> If you ask, "Why do these women die?" the technical response is: "The main causes of maternal death are hemorrhage, sepsis (infection), toxemia, obstructed labor and the complications of abortion." But looking beneath those immediate causes, one must ask why they occurred or why they were fatal. The answer to that is: lack of prenatal care; lack of trained personnel, equipment, blood or transport at the moment the obstetrical emergency arose, or earlier, when it might have been foreseen and avoided; lack of family planning to help women avoid unwanted pregnancies, too many or too closely spaced births, or giving birth when they were too young or too old; preexisting conditions like malaria, anemia, fatigue and malnutrition that predispose to obstetrical complications; problems arising from female circumcision. From that list of intermediary causes one must go deeper still to identify the cultural and socioeconomic factors that put young girls, almost from birth, on this road to maternal death: . . . low status of women and discrimination against them; poverty; lack of education; local customs; and government policies that give low priority to the needs of women. (Smyke, 1991: 61–62)

Like infant mortality, maternal mortality occurs most often among women who suffer from malnutrition or illness (most commonly, malaria). Hemorrhage more often occurs during abortion or childbirth in women who develop anemia because of malaria or inadequate diets. Maternal mortality is also most common among women who give birth before age 20 or after age 35. In Bangladesh, for example, where half of all women marry by age 15, maternal mortality is five times higher among those ages 10 to 14 than among those ages 20 to 24 (Basch, 1999: 208). Mortality also rises with each birth after the third. Finally, maternal mortality is more common among women who give birth in unsanitary conditions and among those who have been circumcised; Box 4.4 provides further details on this dangerous practice.

Another cause of maternal mortality in the developing nations—accounting for 13 percent of deaths—is unsafe abortion (World Health Organization, Reproductive Health and Research Department, 2004). Abortion is a technically simple procedure, far safer than childbirth when performed by trained professionals working in sterile conditions with proper tools. However, most developing nations have criminalized or legally restricted abortion because of cultural traditions, religious beliefs, a desire by political elites to increase population, or financial and political pressures from the United States— which since 1973 has withheld family planning funding from any agencies that offer abortions. In other countries, abortion is legal, but many women cannot afford to obtain abortions from a trained health care worker. Consequently, almost 20 million women yearly—most of whom are married with several children—receive unsafe abortions. Unsafe abortion accounts for about 13 percent of maternal mortality in the developing nations, most commonly because of infections caused by unsterile instruments, hemorrhage when those instruments pierce the uterus, or poisoning when women try to abort themselves by swallowing toxic chemicals (World Health Organization, Reproductive Health and Research Department, 2004). Unsafe abortion can also cause illness or permanent disability: Hospitals in the developing nations spend as much as 50 percent of their resources on treating the aftereffects of unsafe abortion.

Respiratory Diseases

Finally, respiratory diseases, such as emphysema, are also major killers in the developing nations, as in the industrialized nations. As with all disease in the developing nations, poverty and malnutrition increase individual susceptibility to illness. In addition, long periods spent cooking over open fires in closed rooms expose millions of women to cancer-causing toxins; the effects are equivalent to smoking several packs of cigarettes daily. Meanwhile, those who live in cities like Caracas, Mexico City, or Calcutta risk their health daily because of pollution from automobiles and industries. Unfortunately, in some developing nations, government officials lack the political or economic power to control polluting industries; in other such nations, officials are

Box 4.4   *Female Circumcision*

According to the World Health Organization, between 100 and 140 million girls and women across Africa as well as in Malaysia, Indonesia, Yemen, and elsewhere have experienced the ordeal of female circumcision, and about 2 million additional girls are circumcised each year (World Health Organization, 2000a). Female circumcision is a brutal and sometimes fatal procedure, in no way analogous to male circumcision. In clitoridectomy, the first and least common of the three types of female circumcision, either the tip of the clitoris or the skin over the clitoris is cut off. In excision, which comprises about 80 percent of cases, the entire clitoris and labia minora are removed but the vulva is left untouched. In infibulation, which comprises about 15 percent of cases, the clitoris, labia minora, and parts of the labia majora are removed and the sides of the vulva are stitched together, leaving only a small opening for urine and menstrual fluid to escape. Most commonly, a midwife or other lay healer performs the circumcision using a razor blade, knife, or piece of broken glass.

Those who support circumcision believe it makes women more docile and reduces their sex drives, making them better wives and less likely to disgrace their families by engaging in premarital or extramarital sexual relationships. In addition, supporters of circumcision believe that circumcised women are cleaner, healthier, more fertile, and prettier. In countries where circumcision is the norm, these beliefs leave uncircumcised women with few marriage prospects and pressure parents to have their daughters circumcised even if the parents disapprove of the practice.

Circumcision substantially impairs the health of young girls and women. Given the unsanitary conditions in which it is usually performed, the operation can cause life-threatening shock, hemorrhage, infections, or tetanus. Those who survive often experience pain during intercourse and chronic urinary, vaginal, or pelvic infections, sometimes resulting in infertility. If they do become pregnant, scar tissue and the narrowed vaginal opening can make it difficult for a baby to emerge, causing women to die from hemorrhage and babies to die from brain damage. These health problems have convinced some doctors and nurses to perform circumcisions to protect girls who would otherwise be circumcised under more dangerous conditions.

To date, most nations where circumcision occurs officially oppose female genital mutilation, and Senegal and Egypt have outlawed the practice. However, these actions have had little impact on its prevalence (World Health Organization, 1997). Western opposition has proven similarly ineffective, because it is difficult if not impossible for Westerners to condemn circumcision without appearing to condemn the cultures in which it is embedded. Thus, the most effective opponents of female circumcision are those who come from within these cultures. With this in mind, feminists and health care workers native to these cultures have formed alliances aimed at stopping this practice, such as the Inter-African Committee on Traditional Practices Affecting the Health of Women and Children.

---

unwilling to do so because they benefit economically from these industries. Equally important, officials in developing nations sometimes believe that pollution and the attendant morbidity and mortality are short-term costs they must pay to industrialize and to improve their nation's health in the long run.

To these factors must be added the growing role of tobacco, which, in the developing nations as in the industrialized nations, is a major cause of chronic obstructive pulmonary disease. In addition, tobacco serves as a catalyst that increases the risks of other diseases (World Health Organization, 1998a). For example, compared with nonsmokers, smokers who have parasitic bladder infections are more likely to get bladder cancer, and smokers who work in uranium mines are more likely to develop leukemia. In addition, tobacco use promotes disease by taking a large bite out of small incomes. Smokers spend as much as 15 percent of family income in Brazil and as much as 10 percent in India on tobacco; in Egypt wives name their husband's smoking as the main reason their children go hungry (Nichter and Cartwright, 1991). WHO (1998a) estimates that by 2020, tobacco use will cause 11 percent of all deaths in developing nations (and 18 percent of deaths in industrialized nations).

Tobacco use has grown steadily in the developing nations since 1964, when the U.S. Surgeon General declared tobacco a cause of lung cancer and sales of cigarettes plummeted in North America. To maintain their profits, tobacco manufacturers (most of which are based in the United States) turned to the developing nations for new markets (Hammond, 1998; Nichter and Cartwright, 1991). Manufacturers now devote enormous sums to advertising tobacco in those nations. In countries where direct advertising of tobacco on television or radio is restricted, manufacturers instead sponsor cultural and athletic events, especially those oriented toward youths. For example, the Chinese national soccer league is now named the "Marlboro Professional Soccer League." Today, most tobacco users live in the developing nations.

Conclusion

In this chapter, we have seen how poverty and inequality—rather than overpopulation, tropical environments, lack of natural resources, or other biological factors—underlie the high rates of illness and death found in the developing nations. Consequently, reducing poverty and inequality in the developing nations should raise them to the health levels found in the industrialized nations. Conversely, the situation in the former Soviet Union demonstrates how an industrialized nation can slide toward health levels lower than those found in some developing nations (Feshbach, 1999; Feshbach and Friendly, 1992).

With the political and economic upheaval of the last 15 years, poverty has spread across the former Soviet Union and living conditions have deteriorated. The decline in income in these countries during the early 1990s exceeded that in the United States during the Great Depression and seems to have become permanent (Little, 1998). Increasingly across this vast territory, people live in inadequately heated, overcrowded, and ramshackle housing. Almost three-fourths of the water supply is polluted, with one-fourth completely untreated. At the same time, the growing realization that the government can no longer guarantee citizens a minimum standard of living has demoralized people, encouraging many to find solace in drugs. Partly as a result, more than three times as many Russians die each year of acute alcohol poisoning as die from all sorts of poisoning combined in the United States (Wines, 1999).

To these problems must be added those caused by environmental degradation. In past decades, the Soviet Union expanded its economic base as rapidly as possible, with little regard for the human or environmental toll. The Soviet government rarely established and almost never enforced regulations designed to protect the environment from industrial pollution. As a result, industries wreaked far greater environmental havoc in the Soviet Union than in other industrialized nations, polluting farmlands and waterways beyond repair and leaving radioactivity, lead, and other dangerous toxins behind. Similarly, the emphasis on increasing agricultural yields as quickly as possible led to overplowing, which has caused perhaps permanent soil erosion, and to overuse of herbicides, chemical fertilizers, and pesticides, which have poisoned the water, the land, and food crops.

This environmental damage and downturn in living conditions is now taking its toll in human lives. As *New York Times* reporter Michael Specter summarized:

> There is almost no current demographic fact about Russia that would fail to shock: Per capita alcohol consumption is the highest in the world, nearly double the danger level drawn by the World Health Organization; a wider gap has developed in life expectancy between men (59) and women (73) than in any other country; the mortality rate of 15.1 deaths per 1,000 people puts Russia ahead of only Afghanistan and Cambodia among the countries of Europe, Asia and America (the rate for the United States is 8.8); the death rate among working age Russians today is higher than a century ago. (1997: A1)

Although government officials claim that infant mortality is now 16 per 1,000, informed observers believe that it is far higher. Compared with less-poisoned nearby regions, infant mortality is twice as high in agricultural areas where pesticides were used heavily. Meanwhile, **incidence** rates for numerous infectious diseases have increased. For example, in 1998 the former Soviet Union experienced the first large diphtheria epidemic in an industrialized nation in 30 years (Vitek and Wharton, 1998), and tuberculosis—which has a mortality rate thirty-four times higher in Russia than in the United States—is quickly becoming a more common cause of death than cancer and heart disease combined (Feshbach, 1999). In addition, the collapse of the social structure and economy has contributed to a proliferation of sexually transmitted diseases, with rates of both syphilis and AIDS skyrocketing. For all these reasons, life expectancy for males has fallen to only 58 years, compared to 75 in the United States (Population Reference Bureau, 2005).

In sum, no natural progression leads countries toward an increasingly healthy citizenry. Rather, as the political and economic fortunes of a country shift, and as the natural environment improves or declines, so too will the health of its population. Only by continued commitment to eliminating poverty and inequality and to protecting the environment can a nation guarantee that it will keep whatever health gains it has achieved.

Suggested Readings

Farmer, Paul. 1999. *Infections and Inequalities: The Modern Plagues.* Berkeley: University of California Press. A brilliant analysis of the link between disease and social inequality, written by a physician-anthropologist who for many years has divided his time between a clinic in inner-city Boston and one in rural Haiti.

Kidder, Tracy. 2004. *Mountains Beyond Mountains: The Quest of Dr. Paul Farmer, a Man Who Would Cure the World.* New York: Random House. An inspiring book about Paul Farmer, a Harvard professor and medical doctor who spends half of each year working in inner-city Boston and the other half in desperately poor, rural Haiti.

Lappé, Frances Moore, Joseph Collins, and Peter Rosset. 1998. *World Hunger: Twelve Myths.* 2nd ed. New York: Grove Press. Excellent summary of the issues. Lappé is one of the most important figures in this field.

Getting Involved

Amnesty International USA. 322 8th Avenue, New York, NY 10001. (212) 807–8400. www.amnesty-usa.org. Powerful international organization working to end torture and the death penalty and to obtain fair trials and freedom for persons jailed solely because of their beliefs, color, sex, ethnicity, language, or religion.

Freedom from Hunger. 1644 DaVinci Court, Davis, CA 95617. (800) 708–2555. www.freefromhunger.org. Provides small loans to women in developing nations to enable them to become economically self-sufficient and, in the long run, to reduce the chances that they or their families will experience hunger.

International Baby Food Action Network. 10 Trinity Square, Toronto M5G IBI, Ontario PO Box 781, Canada. (+1) 416-595-9819. www.ibfan.org. Educational, lobbying, and activist organization concerned with the sale of infant formula in both the developing and industrialized nations.

The Institute for Food and Development Policy. 398 60th Street, Oakland, CA 94618. (510) 654–4400. www.foodfirst.org/index.html. Popularly known as Food First, this nonprofit organization was founded in 1975 by Frances Moore Lappé and Joseph Collins to promote awareness of the social causes of hunger and poverty around the world.

Review Questions

How do social conditions limit the effectiveness of modern medicine in developing nations?

How do social factors contribute to illness in developing nations?

How do international politics and multinational corporations contribute to illness in developing nations?

How do the role and status of women contribute to illness in developing nations?

Internet Exercises

1. One way to identify the range of opinions on a given topic is to browse listservs or electronic bulletin boards. Listservs and bulletin boards are online discussion groups in which any eligible individual can post a question or an answer to someone else's question. (Some discussion groups are open to everyone, but some are open only to certain groups of individuals, such as members of an organization.) For example, there are a wide variety of opinions regarding female genital mutilation, and regarding what, if anything, westerners should do about it. Go to groups.google.com, and search for posts on female genital mutilation. A large list of posts will appear on your computer screen. Note that these posts come from a wide variety of discussion groups. (The name of the discussion group appears on the last line summarizing each post.) Identify and summarize three different views. How does the nature of the different discussion groups affect the nature of the questions posed and answers given?

2. Obtain current information from the nonprofit Population Reference Bureau's website (www.prb.org) regarding life expectancy at birth by country. Compare that information with the information contained in your textbook. Are there any countries in which life expectancy has changed markedly since this textbook was printed? If so, what might explain those changes?

2 The Meaning and Experience of Illness

Our commonsense understandings of the world tell us that illness is a purely biological condition, definable by objectively measured biological traits. As we will see in Part Two, however, definitions of illness vary considerably over time and space and across social groups. In Chapter 5, we explore the social meanings of illness and consider how ideas about the nature and causes of illness have changed historically, from biblical explanations that attributed illness to punishment for sin to modern New Age explanations that attribute illness to lack of self-love. We also examine how defining something as an illness can act as a form of social control.

Whereas Chapter 5 discusses the meaning of illness in the abstract, Chapter 6 looks at the consequences of chronic illness and disability for individuals. Beginning with a discussion of how Western society historically has treated those who have chronic illnesses and disabilities, we then consider the modern experience of illness, from responding to initial symptoms to searching for mainstream or alternative therapies to coming to terms with a changed body and self-image.

In Chapter 7, we examine parallel questions regarding mental illness. That chapter explores what people mean when we say something is a mental illness. Then we look at how and why mental illness is distributed among social groups; how Western society historically has treated persons with mental illnesses; and how individuals experience mental illness, from initial symptoms, to treatment, to social status following treatment.

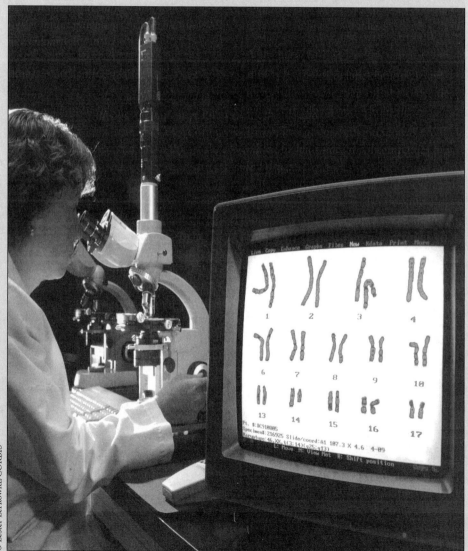

The Social Meanings of Illness

All Marco Oriti has ever wanted, ever imagined, is to be taller. At his fifth birthday party at a McDonald's in Los Angeles, he became sullen and withdrawn because he had not suddenly grown as big as his friends who were already five: in his simple child's calculus, age equaled height, and Marco had awakened that morning still small. In the six years since then, he has grown, but slowly, achingly, unlike other children. "Everybody at school calls me shrimp and stuff like that," he says.

"They think they're so rad. I feel like a loser. I feel like I'm nothing." At age 11, Marco stands 4 feet 1 inch—4 inches below average—and weighs 49 pounds. And he dreams, as all aggrieved kids do, of a sudden, miraculous turnaround: "One day I want to, like, surprise them. Just come in and be taller than them."

Marco, a serious student and standout soccer player, more than imagines redress. Every night but Sunday, after a dinner he seldom has any appetite for, his mother injects him with a hormone known to stimulate bone growth. The drug, a synthetic form of naturally occurring human growth hormone (HGH) produced by the pituitary, has been credited with adding up to 18 inches to the predicted adult height of children who produce insufficient quantities of the hormone on their own—pituitary dwarfs. But there is no clinical proof that it works for children like Marco, with no such deficiency. Marco's rate of growth has improved since he began taking the drug, but his doctor has no way of knowing if his adult height will be affected. Without HGH, Marco's predicted height was 5 feet 4 inches, about the same as the Nobel Prize–winning economist Milton Friedman and . . . Masters golf champion, Ian Woosnam, and an inch taller than the basketball guard Muggsy Bogues of the Charlotte Hornets. Marco has been taking the *shots for six years, at a cost to his family and their insurance company of more than $15,000 a year [$21,000 in 2005 dollars]. . . .*

A Cleveland Browns cap splays Marco Oriti's ears and shadows his sparrowish face. Like many boys his age, Marco imagines himself someday in the NFL. He also says he'd like to be a jockey—making a painful incongruity that mirrors the wild uncertainty over his eventual size. But he is unequivocal about his shots, which his mother rotates nightly between his thighs and upper arms. "I hate them," he says.

He hates being short far more. Concord, the small Northern California city where the Oriti family now lives, is a high-achievement community where competition begins early. So Luisa Oriti and her husband, Anthony, a bank vice president, rationalize the harshness of his treatment. "You want to give your child that edge no matter what," she says, "I think you'd do just about anything." (Werth, 1991)

Does Marco have an illness? According to his doctors, who have recommended that he take an extremely expensive, essentially experimental, and potentially dangerous drug, it would seem that he does. To most people, however, Marco simply seems short.

In the first part of this chapter, we look at the medical and sociological models of illness—two opposing ways of thinking about what illness means. Then we will explore how the public in general thinks about illness, and some of the consequences of these views. In the second part of this chapter, we consider how medicine can act as an institution of social control, highlighting the process through which behaviors or conditions become defined as illnesses and the consequences of these definitions.

Models of Illness

The Medical and Sociological Models of Illness

What do we mean when we say something is an illness? As Marco's story suggests, the answer is far from obvious. Most Americans are fairly confident that someone who has a cold or cancer is ill. But what about the many post-menopausal women whose bones have become brittle with age, and the many older men who have bald spots, enlarged prostates, and urinary problems? Or the many young boys who have trouble learning, drink excessively, or enjoy fighting? Depending on who you ask, these conditions may be defined as normal human variations, as illnesses, or as evidence of bad character. As these questions suggest, defining what is and is not an illness is far from a simple task. In this section we explore the **medical model of illness:** what doctors typically mean when they say something is an illness. This medical model is not accepted in its entirety by all doctors—those in public health, pediatrics, and family practice are especially likely to question it—and is not rejected by all sociologists, but it is the dominant conception of illness in the medical world. The **sociological model of illness** summarizes critical sociologists' retort to the medical model of illness. This sociological model reflects sociologists' view of how the world currently operates, not how it ideally should operate. Key Concepts 5.1 compares these two models, using as an example female sexual dysfunction (FSD), a recently developed and still contentious diagnosis.

The medical model of illness begins with the assumption that illness is an *objective* label given to anything that deviates from normal biological functioning (Mishler, 1981). Most doctors, if asked, would explain that polio is caused by a virus that disrupts the normal functioning of the neurological system, that menopause is a "hormone deficiency disease" that, among other things, impairs the body's normal ability to regenerate bone, and that men develop urinary problems when their prostates grow excessively large and unnaturally compress the urinary tract. Doctors might further explain that, because of scientific progress, all educated doctors can now recognize these problems as illnesses, even though they were not considered as such in earlier eras.

In contrast, the sociological model of illness begins with the statement that illness (as the term is actually used) is a *subjective* label, which reflects personal and social ideas about what is normal as much as scientific reasoning (Weitz, 1991). Sociologists point out that ideas about normality differ widely across both individuals and social groups. A height of 4 feet 6 inches would be normal for a Pygmy man but not for an American man. Drinking three glasses of wine a day is normal for Italian women but could lead to a diagnosis of alcoholism in American medical circles. In defining normality, therefore, we need to look not only at individual bodies but also at the broader social context. Moreover, even within a given group, "normality" is a range and not an absolute. The median height of American men, for example, is 5 feet 9 inches, but most people would consider someone several inches taller or shorter than that as still normal. Similarly, individual Italians routinely and without social difficulties drink more or less alcohol than the average Italian. Yet medical authorities routinely make decisions about what is normal and what is illness based not on absolute, objective markers of health and illness but on arbitrary, statistical cutoff points—deciding, for example, that anyone in the fifth percentile for height or the fiftieth percentile for cholesterol level is ill. Culture, too, plays a role: Whereas the American Society of Plastic and Reconstructive Surgeons recommends breast enlargement for small breasts, which it considers a disease ("micromastia") and believes "results in feelings of inadequacy, lack of self-confidence, distortion of body image and a total lack of well-being due to a lack of self-perceived femininity" (1989: 4–5), in Brazil large breasts are denigrated as a sign of African heritage and breast *reduction* is the most popular cosmetic surgery (Gilman, 1999).

Because the medical model assumes illness is an objective, scientifically determined category, it also assumes there is no *moral* element in labeling a condition or behavior as an illness. Sociologists, on the other hand, argue that illness is inherently a moral category, for deciding what is illness always means

---

Medical and Sociological Models of Illness

MEDICAL MODEL	SOCIOLOGICAL MODEL
Illness is an *objective* label: All educated people agree on what is normal and what is illness.	Illness is a *subjective* category: Educated people sometimes disagree on what should be labeled illness.
Example: Female sexual dysfunction (FSD) is a biological disease characterized by lack of sexual responsiveness.	*Example: Female sexual dysfunction (FSD) is a* label *given to women who are distressed by their lack of sexual responsiveness with their* current *sexual partner.*
Illness is *nonmoral*: Conditions and behaviors are labeled *illness* scientifically, without moral considerations or consequences.	Illness is a *moral* category: Conditions and behaviors are labeled *illness* when they are considered bad (deviant).
Example: Labeling FSD an illness and labeling individuals as having FSD are neutral biological statements that do not reflect moral judgments of the condition or individual.	*Example: We label sexual nonresponsiveness an illness because we find it repugnant, and we typically look down on those who have FSD.*
Illness is an *apolitical* label.	Illness is a *political* label: Some groups have more power than others to decide what is an illness and who is ill.
Example: FSD was first identified by doctors through scientific research.	*Example: The concept of FSD was promoted by pharmaceutical companies to sell drugs.*

---

deciding what is good or bad. When, for example, doctors label menopause a "hormonal deficiency disease," they label it an undesirable deviation from normal. In contrast, many women consider menopause both normal and desirable and enjoy the freedom from fear of pregnancy that menopause brings (E. Martin, 1987). In the same manner, when we define cancer, polio, or diabetes as illnesses, we judge the bodily changes these conditions produce to be both abnormal and undesirable, rather than simply normal variations in functioning, abilities, and life expectancies. (Conversely, when we define a condition as healthy, we judge it normal and desirable.)

Similarly, when we label an individual as ill, we also suggest that there is something undesirable about that *person*. By definition, an ill person is one whose actions, ability, or appearance do not meet social **norms,** or expectations

MEDICAL MODEL	SOCIOLOGICAL MODEL
Illness is a *concrete, unchanging reality* that all informed observers agree on. *Example: If Victorian doctors had been more educated, they would have realized that FSD was a disease.*	Illness is a *social construction*: Each cultural group, at each point in time, assigns the label *illness* to whatever it considers both biological and problematic. *Example: Victorians considered women bad—and ill—if they experienced sexual pleasure. Modern society considers women deviant—and ill—if they are sexually unresponsive.*
Each illness has specific, universally recognizable, features, so diagnosis is objective and consistent across doctors and populations. *Example: All women who lack sexual responsiveness share biological markers (such as low testosterone) and will be diagnosed with FSD, regardless of their doctors or their social characteristics.*	Illness is neither specific nor universally recognized, so diagnosis is subjective and culturally bound. *Example: White women are more likely than others to be diagnosed with FSD, and doctors in cultures that consider female sexuality shameful do not consider FSD an illness.*
Each illness is caused by unique biological forces. *Example: Women can become sexually unresponsive when their hormonal balance is awry.*	Illness is caused by a combination of social, psychological, and biological causes. *Example: Women become sexually unresponsive when their cultures deny female sexuality or their partners lack sexual skills, among other reasons.*

within a given culture regarding proper behavior or appearance. Such a person will typically be considered less whole and less socially worthy than those deemed healthy. Illness, then, like virginity or laziness, is a **moral status**: a social condition that we believe indicates the goodness or badness, worthiness or unworthiness, of a person.

From a sociological standpoint, illness is not only a moral status but (like crime or sin) a form of **deviance** (Parsons, 1951). To sociologists, labeling something deviant does not necessarily mean that it is immoral. Rather, deviance refers to behaviors or conditions that socially powerful persons within a given culture *perceive*, whether accurately or inaccurately, as immoral or as violating social norms. We can tell whether behavior violates norms (and, therefore, whether it is deviant) by seeing if it results in

negative social sanctions. This term refers to any punishment, from ridicule to execution. (Conversely, **positive social sanctions** refers to rewards, ranging from token gifts to knighthood.) These social sanctions are enforced by **social control agents** including parents, police, teachers, peers, and doctors. Later in this chapter we will look at some of the negative social sanctions imposed against those who are ill.

For the same reasons that the medical model does not recognize the *moral* aspects of illness labeling, it does not recognize the *political* aspects of that process. Although some doctors at some times are deeply immersed in these political processes—arguing, for example, that insurance companies should cover treatment for newly labeled conditions such as fibromyalgia or multiple chemical sensitivity—they rarely consider the ways that politics underlie the illness-labeling process in general. In contrast, sociologists point out that any time a condition or behavior is labeled as an illness, some groups will benefit more than others, and some groups will have more power than others to enforce the definitions that benefit them. As a result, there are often open political struggles over illness definitions (a topic we will return to later in this chapter). For example, vermiculite miners and their families who were constantly exposed to asbestos dust and who now have strikingly high rates of cancer have fought with insurance companies and doctors, in clinics, hospitals, and the courts, to have "asbestosis" labeled an illness; meanwhile, the mining companies and the doctors they employed have argued that there is no such disease and that the high rates of health problems in mining communities are merely coincidences (A. Schneider and McCumber, 2004).

In sum, from the sociological perspective, illness is a **social construction,** something that exists in the world not as an objective condition but *because we have defined it as existing.* This does not mean that the virus causing measles does not exist, or that it does not cause a fever and rash. It does mean, though, that when we talk about measles as an illness, we have organized our ideas about that virus, fever, and rash in only one of the many possible ways. In another place or time, people might conceptualize those same conditions as manifestations of witchcraft, as a healthy response to the presence of microbes, or as some other illness altogether. To sociologists, then, *illness,* like *crime* or *sin,* refers to biological, psychological, or social conditions subjectively defined as undesirable by those within a given culture who have the power to enforce such definitions.

In contrast, and as we have seen, the medical model of illness assumes that illness is an objective category. Based on this assumption, the medical model of health care assumes that each illness has specific features, universally recognizable in all populations by all trained doctors, that differentiate it both from other illnesses and from health (Dubos, 1961; Mishler, 1981). The medical model thus assumes that diagnosis is an objective, scientific process.

Sociologists, on the other hand, argue that diagnosis is a subjective process. The subjective nature of diagnosis expresses itself in three ways. First, patients with the same symptoms may receive different diagnoses depending on various social factors. Women who seek medical care for chronic pain, for example, are more likely to receive psychiatric diagnoses than are men who report the same symptoms. Similarly, African Americans (whether male or female) are more likely than whites are to have their chest pain diagnosed as indigestion rather than as heart disease (Hoffman and Tarzian, 2001; Nelson, Smedley, and Stith, 2002). Second, patients with the same underlying illness may experience different symptoms, resulting in different diagnoses. For example, the polio virus typically causes paralysis in adults but only flu-like symptoms in very young children, who often go undiagnosed. Third, different cultures identify a different range of symptoms and categorize those symptoms into different illnesses. For example, U.S. doctors assign the label of attention deficit disorder (ADD) to children who in Europe would be considered lazy troublemakers. And French doctors often attribute headaches to liver problems, whereas U.S. doctors seek psychiatric or neurological explanations (Payer, 1996). In practice, the American medical model of illness assumes that illnesses manifest themselves in other cultures in the same way as in American culture and, by extension, that American doctors can readily transfer their knowledge of illness to the treatment and prevention of illness elsewhere.

Finally, the medical model of illness assumes that each illness has not only unique symptoms but also a unique **etiology,** or cause (Mishler, 1981). Modern medicine assumes, for example, that **tuberculosis,** polio, **HIV disease,** and so on, are each caused by a unique microorganism. Similarly, doctors continue to search for limited and unique causes of heart disease and cancer, such as high-cholesterol diets and exposure to asbestos. Yet even though illness-causing microorganisms exist everywhere and environmental health dangers are common, relatively few people become ill as a result. By the same token, although cholesterol levels and heart disease are strongly correlated among middle-aged men, many men eat high-cholesterol diets without developing heart disease, and others eat low-cholesterol diets but die of heart disease anyway. The doctrine of unique etiology discourages medical researchers from asking why individuals respond in such different ways to the same health risks and encourages researchers to search for **magic bullets**—a term first used by Paul Ehrlich, discoverer of the first effective treatment for syphilis, in referring to drugs that almost miraculously prevent or cure illness by attacking one specific etiological factor. Box 5.1 describes the work of Doctors Without Borders, an organization that offers an inspiring example of doctors and other health care workers who take a truly broad view of the causes and treatment of illness.

Popular Explanations for Illness

Although medicine as an institution certainly affects how the general public thinks about illness, it does not fully control popular beliefs about illness. Consequently, we also need to look at those popular beliefs. As we will see,

---

Box 5.1   *Making a Difference: Doctors Without Borders*

Doctors Without Borders/Médecins Sans Frontières (MSF) is an independent humanitarian organization, founded in 1971, that assists people around the globe whose health has been damaged by disasters, war, or political violence. After an enormous tsunami killed more than 200,000 Indonesians in December 2004, for example, MSF sent doctors, nurses, and other health care workers to treat those who were injured by debris carried by the tsunami, infected by diarrheal diseases spread when sewage systems washed away, or overwhelmed psychologically when loved ones died. Once these "first aid" needs were met, MSF members began working on the broader infrastructure needed to protect the health of the tsunami survivors: organizing vaccination campaigns against tetanus and measles (which had started spreading following the tsunami), food distribution programs (so that malnutrition in the wake of the tsunami would not lead to further mortality), sanitation programs (to prevent disease transmission through unsafe water supplies), and home- and boat-building programs (so people had shelter and a means of earning a living once again).

As this example suggests, MSF's model of illness and how to treat it goes far beyond treating specific symptoms of specific diseases. MSF not only attempts to treat the underlying causes of disease but also includes in its mission the responsibility to publicly bear witness to the problems it sees. Because of its impeccable nonpartisan reputation—taking no sides in any conflict other than on behalf of the people it assists—the doctors and other workers of MSF speak with great moral authority. On its website (www.doctorswithoutborders.org) and in frequent news releases, articles, opinion columns, testimony given at the United Nations General Assembly, and the like, MSF speaks out about illness as well as the social causes of illness. MSF has spoken publicly about how attitudes toward women underlie the use of rape as a military tactic, how international economic dynamics contribute to the short and brutal lives of street-children in developing nations, how pharmaceutical company policies have made treatment for AIDS and other diseases unaffordable in the developing world, how governments use violence to subdue their own populations, and so on. The doctors and other workers of MSF exemplify a broad-based, sociological understanding of illness and health care.

---

because people consider illness undesirable and because it can strike anyone at any time, they most often react with fear and confusion. To relieve their anxiety and make the world seem less capricious and frightening, they typically seek explanations for why illness occurs and why it strikes some rather than others. Most often, those explanations define illness as a deserved punishment and blame individuals for their own illnesses (Brandt and Rozin, 1997; Weitz, 1991). Such explanations provide psychological reassurance by reinforcing people's belief in a "just world," in which punishment falls only on the guilty (Meyerowitz, Williams, and Gessner, 1987).

According to George Foster (1976), all traditional, prescientific theories of illness causation around the world divide into only two, somewhat overlapping, categories: personalistic and naturalistic. **Personalistic theories,** the more common type (Murdock, 1980), hold that illness occurs when a god, witch, spirit, or other supernatural power lashes out at an individual, either deservedly or maliciously. **Naturalistic theories** assert that illness occurs when heat, cold, wind, damp, or other natural forces upset the body's equilibrium. Both personalistic and naturalistic theories blame ill persons for causing their illness, whether by displeasing supernatural beings or by exposing themselves to harmful natural elements. And both define ill persons as less morally worthy than others, whether as sinners or as fools.

Personalistic theories have played an especially important role in the Western world, which in the past often equated illness with divine punishment for sin (Murdock, 1980: 42–52). For example, both the Jewish and Christian Bibles describe leprosy as punishment for an individual's sin. Biblical explanations for leprosy, coupled perhaps with some awareness that leprosy was contagious, led Western societies for centuries to isolate affected individuals. Throughout the Middle Ages and until the Reformation, Christian society required anyone diagnosed with leprosy to participate in a special mass for the dead, known as the lepers' mass. Following the mass, a priest would shovel dirt on the individual's feet to symbolize his or her civil and religious death. From then on, the individual was legally prohibited from entering public gathering places, washing in springs or streams, drinking from another's cup, wearing anything other than the special "leper's dress," touching anything before buying it, talking to anyone without first moving downwind, and so on (Richards, 1977: 123–124). This social banishment continued even after death: Like those who committed suicide or other mortal sins, persons with leprosy could not be buried in church graveyards.

By the early nineteenth century, prescientific ideas about illness had begun to erode as the idea grew, especially among the elite, that scientific principles controlled the natural order. According to the new scientific thinking, illness occurred when biological forces combined with personal susceptibility. Doctors (still lacking a concept of germs) argued that illness occurred when persons whose constitutions were naturally weak or had been weakened by unhealthy behaviors came in contact with dangerous **miasma,** or air "corrupted" by foul odors and fumes. According to this theory, therefore, individuals became ill because of unhealthy rather than immoral behavior.

As the history of cholera shows, however, these new ideas still allowed the healthy to blame the ill for their illnesses. Cholera first appeared in the Western world in about 1830, killing its victims suddenly and horrifyingly, through overwhelming dehydration brought on by uncontrollable diarrhea and vomiting. Cholera is caused by waterborne bacteria, generally transmitted when human wastes contaminate food or drinking water. Because of the link to sanitation, cholera most often strikes poor persons who lack clean water and are weakened by insufficient food, clothing, or shelter.

To explain why cholera had struck, and why it struck the poor especially hard, early nineteenth-century doctors asserted that cholera could attack only individuals who had weakened their bodies through improper living

(Risse, 1988; Rosenberg, 1987). According to this theory, the poor caused their own illnesses, first by lacking the initiative required to escape poverty and then by choosing to eat an unhealthy diet, live in dirty conditions, or drink too much alcohol. Thus, for example, the New York City Medical Council could conclude in 1832 that "the disease in the city is confined to the imprudent, the intemperate, and to those who injure themselves by taking improper medicines" (Risse, 1988: 45). Conversely, doctors (and their wealthy patrons) assumed that wealthy persons would become ill only through gluttony, greed, or "innocently" inhaling some particularly noxious air.

Using this theory, doctors, foreshadowing what would happen with HIV disease, divided patients into the "guilty" (the overwhelming majority), the "innocent," and the "suspect," and hospitals provided or refused care accordingly (Risse, 1988; Rosenberg, 1987). This theory of illness allowed the upper classes to adopt the new, scientific explanations for illness while retaining older, moralistic assumptions about ill people and avoiding any sense of responsibility for aiding the poor or the ill. In sum, instead of believing that immorality directly *caused* illness, people now believed that immorality left one *susceptible* to illness.

Despite the tremendous growth in medical knowledge about illness during the last century, popular explanations for illness have remained remarkably stable. Theories connecting illness to sin continue to appear, as do theories that conceptualize illness as a direct consequence of poorly chosen and hence irresponsible (although not necessarily sinful) behavior (Brandt and Rozin, 1997; Zola, 1972). For example, although most Americans know that viruses cause influenza and the common cold, most continue to hold essentially naturalistic theories regarding these illnesses— warning their children to eat warm foods, wear hats and gloves, and cover up against the rain to avoid infection.

Similarly, the mass media, public health authorities, and the general public now often blame illness on individual lifestyles (Brandt and Rozin, 1997; Tesh, 1988). Magazines regularly print articles such as "Beat Your Risk Factors" (Libov, 1999) and "Ten Easy Ways to Boost Your Immunity" (Strote, 2002), exhorting individuals to protect or restore their health through diet, exercise, stress reduction, and the like. Simultaneously, the U.S. government— even while continuing to subsidize the tobacco and beef industries—spends millions on education campaigns encouraging the public to stop smoking and to eat healthier diets.

Another popular ideology ties illness not to individual actions but to individual personalities (Sontag, 1978). For example, a newspaper account of comedian Gilda Radner's death from ovarian cancer quoted her "therapist" explaining how

> Gilda always had this wonderful will to live. Yet she also exhibited the same pre-conditioning virtually all [cancer patients] have. Fear. Hopelessness. Negativity. What . . . Gilda came to appreciate [in her therapy], is that a positive outlook can

improve the quality of life—up to and including the immune system. (Kahn, 1989)

Similarly, the media continue to warn that individuals with aggressive and competitive type A personalities are at risk for heart problems (Siegman and Dembroski, 1989), despite considerable scientific evidence refuting this link (Aronowitz, 1998).

In its most extreme form, this sort of theorizing has led some to claim that illness occurs not because individuals ignore their bodies or have illness-producing personalities but because they *choose* to become ill. The most influential statement of this theory appears in the best-selling book *Love, Medicine and Miracles* by surgeon Bernie Siegel (1990). Siegel postulates that people become ill because they "need" their illness—to escape a stressful work situation, receive sympathy from their spouses, punish themselves for misdeeds, and so on—and because they do not love themselves enough to take care of their emotional needs. Consequently, Siegel advises ill persons that they will find lasting cures only when they truly desire a healthy, long life.

Theories such as Siegel's draw on research suggesting that stress, personality, and lifestyle can increase personal susceptibility to illness. Such factors may indeed affect the distribution of illness in society. Yet by focusing on these factors as the primary source of illness, these theories encourage the healthy to devalue and reject those who are ill and promote depression and lowered self-esteem among those who blame themselves for their illnesses.

In addition, by emphasizing how individuals cause their own illnesses, these theories encourage policymakers to ignore how social and environmental factors can foster illness (Crawford, 1979; Tesh, 1988; Waitzkin, 1981; Zola, 1972). For example, magazines that emphasize how individuals make themselves ill rarely discuss how factors largely beyond individual control (such as poverty, malnutrition, pollution, or unsafe conditions in our houses, cars, or workplaces) can produce ill health. Nor do these magazines discuss how social factors (including the advertisements for alcohol and cigarettes in some of these same magazines) can pressure individuals to adopt unhealthy lifestyles—how unemployed teenagers with poor job prospects sometimes smoke cigarettes to demonstrate their adulthood, how young mothers who lack assistance with child care probably also lack time for the recommended three sessions per week of aerobic exercise, or how workers sometimes suffer injuries because of unsafe equipment rather than because of personal carelessness. As Barbara Katz Rothman notes,

> Think of the anti-smoking, anti-drinking "behave yourself" campaigns aimed increasingly at pregnant women. What are the causes [as identified in these campaigns] of prematurity, fetal defects, damaged newborns—flawed products? Bad mothers, of course—inept workers. One New York City subway ad series shows two newborn footprints, one from a full-term and one from a premature infant. The ads read, "Guess which baby's mother smoked while pregnant?" Another asks,

"Guess which baby's mother drank while pregnant?" And yet another: "Guess which baby's mother didn't get prenatal care?" I look in vain for the ad that says "Guess which baby's mother tried to get by on welfare?"; "Guess which baby's mother had to live on the streets?"; or "Guess which baby's mother was beaten by her husband?" (1989: 21)

In sum, whether or not they are accurate, theories of illness that focus on individual responsibility reinforce existing social arrangements and help us justify our tendency to reject, mistreat, or simply ignore those who suffer illness.

Medicine as Social Control

Creating Illness: Medicalization

The process through which a condition or behavior becomes defined as a medical problem requiring a medical solution is known as **medicalization** (Conrad and Schneider, 1992; Conrad, 2005). For example, as social conditions have changed, activities formerly considered sin or crime, such as masturbation, homosexual activity, or heavy drinking, have become defined as illnesses. The same has happened to various natural conditions and processes, such as uncircumcised penises, limited sexual desire, aging, pregnancy, and menopause (e.g., E. Armstrong, 2000; Barker, 1998; Figert, 1996; Rosenfeld and Faircloth, 2005). The term *medicalization* also refers to the process through which the definition of an illness is *broadened*. For example, when the World Health Organization (WHO) in 1999 lowered the blood sugar level required for diagnosis with diabetes, the number of persons eligible for this diagnosis increased in some populations by as much as 30 percent (Shaw, de Courten, Boyko, and Zimmet, 1999).

For medicalization to occur, one or more organized social groups must have both a vested interest in it and sufficient power to convince others (including doctors, the public, and insurance companies) to accept their new definition of the situation. Not surprisingly, doctors often play a major role in medicalization, for medicalization can increase their power, the scope of their practices, and their incomes. For example, during the first half of the twentieth century, improvements in the standard of living coupled with the adoption of numerous public health measures substantially reduced the number of seriously ill children. As a result, the market for pediatricians declined, and their focus shifted from treating serious illnesses to treating minor childhood illnesses and offering well-baby care. Pediatrics thus became less well-paid, interesting, and prestigious. To increase their market while obtaining more satisfying and prestigious work, some pediatricians have expanded their practices to include children whose behavior concerns their parents or teachers and who are now defined as having medical conditions such as attention deficit disorder or antisocial personality

disorder (Halpern, 1990). Doctors have played similar roles in medicalizing premenstrual syndrome (Figert, 1996), drinking during pregnancy
(E. Armstrong, 1998), impotence (Loe, 2004; Tiefer, 1994), and numerous
other conditions.

In other instances, however, doctors have proved indifferent or even
opposed to medicalization. For example, although some doctors believe that
woman battering is a medical problem and that doctors should accept
responsibility for identifying it and intervening when it occurs, others believe
that women provoke their own battering, that doctors can do little to help,
or that woman battering is best dealt with by the police rather than by doctors (Kurz, 1987). As a result, many doctors oppose medicalizing woman battering and prefer to treat women's injuries without delving into their causes.

In circumstances such as these, pressure for medicalization can instead
come from consumers and consumer groups (Conrad, 2005). Alcoholics
Anonymous, for example, has fought to medicalize alcoholism partly to
reduce the stigma of that condition. Other consumer groups similarly have
argued for medicalization in the hope that medical control will be more
humanitarian than legal control, in such areas as compulsive gambling,
erratic and violent behavior, and homosexuality. In addition, individuals
sometimes press for medicalization as a way of gaining validation for their
experiences and stimulating research on treatments and cures (Barker, 2005;
Ziporyn, 1992). For example, much of the pressure to define premenstrual
syndrome, chronic fatigue syndrome, and fibromyalgia as illnesses has come
from persons who believe they suffer from these syndromes.

The third major force behind medicalization is the pharmaceutical industry (Conrad, 2005). This industry has a vested economic interest in medicalization whenever it can provide a drug as treatment. The medicalization of
shortness exemplifies this process (Conrad, 2005; S. M. Rothman and D. J.
Rothman, 2003; Werth, 1991). In 1985, the pharmaceutical company
Genentech patented a genetically engineered and mass-produced form of
human growth hormone (HGH). At that time, the available data suggested
that HGH could increase final height in children whose pituitary glands did
not naturally produce enough HGH, but not in children without pituitary
defects. Moreover, it was known that HGH could promote a drastic loss of
body fat and increase in muscle, with unknown consequences in growing
children. Nevertheless, Genentech and, subsequently, Eli Lilly Pharmaceuticals
(which patented a slightly different synthetic hormone) embarked on a major
campaign to sell HGH. Together, they underwrote two-thirds of the budget of
the Human Growth Foundation, a nonprofit advocacy group that works to
increase public awareness of the problems experienced by short children.
With the pharmaceutical companies' help, the foundation began broadcast-
ing news of HGH across the nation at health fairs, shopping malls, and the
like. The pharmaceutical companies also began spending millions of dollars
annually to underwrite medical research supporting HGH, to advertise the
drug to doctors, and to sponsor in-school screening programs that first

identified the shortest 3 percent of students and then informed the students' parents that their children needed medical treatment.

By 1999, about 30,000 children—20 percent of whom have no disease other than shortness—were being treated with HGH in the United States (B. Greenberg, 1999). As of 2004, treatment costs about $20,000 a year, and most children are treated for three to six years (Conrad and Potter, 2004). According to the only long-term study (partially funded by Genentech) of the drug's effectiveness on children with normal pituitary glands, these children can expect to add about two inches to their adult height (Hintz et al., 1999). Because of HGH's limited effectiveness and potential for long-term health problems (such as tumors and diabetes) and because identifying short children as "diseased" and treating them with daily injections over several years can lead to social stigma and lowered self-esteem, the American Academy of Pediatrics recommends against its use in short but otherwise healthy children, even though the Food and Drug Administration (FDA) has approved its use in this population. Meanwhile, doctors increasingly are prescribing estrogen—also a potentially dangerous drug—to *stunt* the growth of girls who are expected to exceed six feet in height. In addition, increasing numbers of doctors are prescribing HGH to older men as an "antiaging" drug, even though research strongly suggests that the drug offers significant risks but no benefits to this population (Conrad and Potter, 2004). Genotropin, the best-selling HGH drug, earned $475 million in 2003 (S. M. Rothman and D. J. Rothman, 2003).

The final major force in battles over medicalization is managed care organizations (MCOs). MCOs (which are discussed in detail in Chapter 8) are health insurance providers that restrain costs (and, ideally, improve quality of care) by monitoring closely which health services are given by which health care providers to which patients. Unlike pharmaceutical companies, MCOs either support or oppose medicalization, depending on which will best protect their interests (Conrad, 2005). For example, in the past MCOs typically rejected requests for gastric bypass surgeries to help obese patients lose weight, implicitly arguing that obesity was a personal and not a medical issue. More recently, MCOs have started approving these surgeries in the belief that they will reduce the long-term complications of obesity and thus reduce overall costs for MCOs.

Case Study: Working Together to Medicalize Hyperkinesis

Neither doctors, nor consumer groups, nor pharmaceutical companies have enough influence to medicalize a condition on their own. Successful medicalization depends on the interwoven interests and activities of these three groups and sometimes others. The history of hyperkinesis illustrates this process.

As originally defined, hyperkinesis lacked any definitive biological markers and instead referred to children above age 5 who were overactive, impulsive, and easily distracted but who had no brain damage (Diller, 1998). Since the late 1930s, doctors have known that amphetamines (including methamphetamine or "speed") can reduce distraction in children and adults, regardless of their mental health or illness. In addition, even though biologically amphetamines are stimulants, they cause an intense focus that can make users appear less active. These characteristics made amphetamines a natural choice for treating hyperkinesis. However, because amphetamines are highly addictive and have dangerous side effects, physicians avoided prescribing them.

In the absence of a viable treatment, physicians rarely made the diagnosis of hyperkinesis. This situation changed in the 1960s, when the amphetamine Ritalin (methylphenidate) appeared on the market (Conrad and Schneider, 1992). Ritalin has fewer short-term side effects than other amphetamines have and, in the short term, improves the ability to concentrate, reduces the tendency to act impulsively, and increases willingness to accept discipline. Yet Ritalin is far from a panacea. Chemically, it acts much like cocaine (Vastag, 2001). Its immediate side effects can include addiction, loss of appetite, sleep deprivation, headache, and stomachache. Its long-term side effects are unknown, and its long-term benefits seem minor at best: The little available research suggests that it does not improve users' chances of graduating high school, holding a job, refraining from illicit drugs, or avoiding trouble with the law (Diller, 1998).

Following the development of Ritalin, pharmaceutical companies embarked on a huge campaign to "sell" hyperkinesis to doctors. According to Peter Conrad and Joseph Schneider:

> After the middle 1960s it is nearly impossible to read a medical journal or the free "throw-away" magazines [mailed by pharmaceutical companies to doctors] without seeing some elaborate advertising for either Ritalin or Dexedrine [another amphetamine]. These advertisements explain the utility of treating hyperkinesis . . . and urge the physician to diagnose and treat hyperkinetic children. The advertisements may run from one to six pages. They often advise physicians that "the hyperkinetic syndrome" exists as "a distinct medical entity" and that the "syndrome is readily diagnosed through patient histories and psychometric testing" and "has been classified by an expert panel" of the Department of Health, Education and Welfare as MBD [minimal brain dysfunction]. These same pharmaceutical firms also supply sophisticated packets of "diagnostic and treatment" information on hyperkinesis to physicians, pay for professional conferences on the subject, and support research in the identification and treatment of hyperkinesis. (1992: 159–160)

Pediatricians proved a ready audience for this marketing campaign, which promised a way to boost their flagging income and prestige. This market further increased in the late 1980s, when the diagnosis of hyperkinesis was replaced by "attention deficit disorder" (ADD). Unlike hyperkinesis, the definition of ADD sets no age limits and includes girls who daydream as well as boys who express their boredom or dissatisfaction through physical activity.

Like pediatricians, many teachers readily adopted the concept of ADD, if for different reasons (Diller, 1998). Faced with cuts in staffing and larger classes at the same time that school boards began placing an increased emphasis on testing and competition at earlier and earlier ages, teachers can hardly be blamed for looking with favor on drugs that make their students more manageable. In addition, diagnosing a student with ADD shifts blame for poor student performance from teacher to student. Not surprisingly, the suggestion to place a child on Ritalin now often comes initially from a teacher (Diller, 1998).

Pharmaceutical companies also promoted Ritalin directly to the public, spending $610 million on direct-to-consumer advertisements in 1996, up from $44 million in 1990 (Diller, 1998: 139). Like teachers, parents often are relieved to find an explanation other than poor parenting for their child's behavioral or educational problems. In addition, like those who argue that alcoholism or compulsive gambling is a disease, these parents hope to remove blame from their children, reduce the chances of legal sanctions against their children, and stimulate research on treatment. Finally, parents also seek diagnoses of ADD to help them obtain educational assistance for their children under federal antidiscrimination statutes (Diller, 1998). These statutes set aside funds for individualized educational services for students who suffer disabilities (including ADD), while making it extremely difficult for schools to discipline children for any problem behaviors that could be considered part of their disability. Thus, many parents find that having their child diagnosed with ADD increases the child's educational opportunities while reducing the chances that the child will be suspended or expelled. For this reason, children are much more likely to be diagnosed with ADD if they are wealthy and white than if they are poor or nonwhite. Similarly, adults with ADD can legally request accommodations in the workplace, such as quiet space or extra time to finish tasks, as long as their disability does not substantially interfere with their job performance. To get these accommodations, increasing numbers of adults now seek an ADD diagnosis for themselves (Diller, 1998).

Taken together, these factors produced an astounding increase in the number of persons diagnosed with ADD, from about 150,000 U.S. children in 1970 to almost 5 million in 1998 (Diller, 1998: 2, 27). Almost 14 percent of boys who visit an American doctor's office now leave with a prescription for Ritalin or a related drug, and use of Ritalin is growing rapidly in preschools (National Center for Health Statistics, 2004: 63).

The Consequences of Medicalization

In some circumstances, medicalization can be a boon, leading to social awareness of a problem, sympathy toward its sufferers, and development of beneficial therapies. Persons with epilepsy, for example, lead far happier and more productive lives now that drugs usually can control their seizures and few people view epilepsy as a sign of demonic possession. But defining a condition as an illness does not necessarily improve the social status of those who have that condition. Those who use alcohol excessively, for example, continue to experience social rejection even when alcoholism is labeled a disease. Moreover, medicalization also can lead to new problems, known by sociologists as **unintended negative consequences** (Conrad and Schneider, 1992; Zola, 1972).

First, once a situation becomes medicalized, doctors become the only experts considered appropriate for diagnosing the problem and for defining appropriate responses to it. As a result, the power of doctors increases while the power of other social authorities (including judges, the police, religious leaders, legislators, and teachers) diminishes. For example, now that troublesome behavior by children is increasingly diagnosed as ADD, parents, teachers, and the children themselves have lost credibility when they disagree with this diagnosis. Similarly, doctors are now given considerable authority to answer questions such as who should receive abortions or organ transplants, how society should respond to drug use, and whether severely disabled infants should receive experimental surgeries, while the authority of the church and family members to answer these questions has diminished.

As this suggests, medicalization significantly expands the range of life experiences under medical control. For example, the existence of "fetal alcohol syndrome"—a constellation of birth defects including mental retardation believed caused by alcohol use during pregnancy—was widely accepted by American doctors based on extremely limited data, collected in a handful of studies that used neither random samples nor statistical controls (E. Armstrong, 1998). Moreover, these studies suggested that the problem was rare, even among severe alcoholics. Nonetheless, doctors have campaigned to forbid restaurants and bars from serving alcohol to pregnant women; to require liquor manufacturers, restaurants, and bars to post warning labels and signs warning of the dangers of drinking during pregnancy; and for legal codes that declare drinking during pregnancy a form of child abuse.

Second, once a condition is medicalized, medical treatment may become the only logical response to it. For example, if woman battering is considered a medical condition, then doctors need to treat women and the men who batter them. However, if woman battering is considered a social problem stemming from male power and female subordination, then it makes more sense to arrest the men, assist the women in developing financial and emotional independence, and work for broader structural changes that will improve all women's status and options.

Third, when doctors define situations in medical terms, they reduce the chances that these situations will be understood in *political* terms. For example, China, Pakistan, and other countries have removed political dissidents from the public eye by committing them to mental hospitals. By so doing, these governments discredited and silenced individuals who might otherwise have offered powerful dissenting voices. In other words, medicalization allowed these governments to **depoliticize** the situation—to define it as a medical rather than a political problem.

Fourth, and as the example of China and Pakistan illustrated, medicalization can justify not only voluntary but also involuntary treatment. Yet treatment does not always help and sometimes can harm. For example, beginning in the 1980s, U.S. courts have forced women to submit to cesarean deliveries, in which babies are surgically removed from their mothers' uteruses rather than delivered naturally through the vagina (Daniels, 1993). In these cases, doctors argued successfully that childbirth is a dangerous medical condition, not a natural process, and that therefore mothers lack the expertise to decide whether cesarean deliveries are in their and their babies' best interests. Yet doctors' judgment is not infallible. In six of the first fifteen cases in which doctors sought court orders to force cesarean deliveries, the mothers in the end delivered healthy babies vaginally (Kolder, Gallagher, and Parsons, 1987); the remaining nine women were forced to have cesareans, so we cannot know whether they might have safely delivered vaginally. Moreover, as of 2005, 29 percent of American women are having cesarean deliveries, even though the WHO recommends a rate of only 10 to 15 percent (Hamilton, Martin, and Sutton, 2004; World Health Organization, 1985: 437), suggesting that U.S. doctors are far too ready to perform this potentially life-threatening surgery. This chapter's ethical debate (Box 5.2) explores the issues involved in forced obstetrical interventions, and the broader issue of "fetal rights."

The Rise of Demedicalization

The dangers of medicalization have fostered a countermovement of **demedicalization** (R. Fox, 1977). A quick look at medical textbooks from the late 1800s reveals many "diseases" that no longer exist. For example, nineteenth-century medical textbooks often included several pages on the health risks of masturbation. One popular textbook from the late nineteenth century asserted that masturbation caused "extreme emaciation, sallow or blotched skin, sunken eyes, . . . general weakness, dullness, weak back, stupidity, laziness, . . . wandering and illy defined pains," as well as infertility, impotence, consumption, epilepsy, heart disease, blindness, paralysis, and insanity (Kellogg, 1880: 365). Today, however, medical textbooks describe masturbation as a normal part of human sexuality.

Like medicalization, demedicalization often begins with lobbying by consumer groups. For example, medical ideology now defines childbirth as an inherently dangerous process, requiring intensive technological, medical assistance. Since the 1940s, however, growing numbers of American women have attempted to redefine childbirth as a generally safe, simple, and natural process and have promoted alternatives ranging from natural childbirth classes, to hospital birthing centers, to home births assisted only by midwives (Sullivan and Weitz, 1988). Similarly, and as described in Chapter 7, gay and lesbian activists have at least partially succeeded in redefining homosexuality from a pathological condition to a normal human variation. More broadly, in recent years, books, magazines, television shows, and popular organizations devoted to teaching people to care for their own health rather than relying on medical care

---

Box 5.2   *Ethical Debate: Medical Social Control and Fetal Rights*

In 1985, Pamela Rae Stewart became pregnant. Her doctor, knowing her history of drug use, warned her to stop using amphetamines. Later, when problems developed during her pregnancy, he advised her to stay off her feet, avoid sexual intercourse, and seek medical attention if she began to bleed heavily.

On November 23, 1985, Stewart gave birth to a severely brain-damaged baby. On the day her child was born, according to police reports, Stewart took amphetamines and had intercourse with her husband. She subsequently began bleeding but did not go to the hospital for several hours. Six weeks later, the baby died, and the District Attorney filed criminal charges against Stewart for child neglect.

Since 1990, about 300 pregnant women—most of them drug users—have been arrested or involuntarily hospitalized to force them to follow medical advice (K. Johnson, 2004). Ironically, pregnant drug users are most likely to face criminal sanctions if they are poor or minorities, even though such women are least likely to have access to substance abuse treatment (Chasnoff, Landress, and Barrett, 1990). Less commonly, doctors and the courts have forced women to have cesarean sections in the belief that these operations were in the babies' best interests. A 1987 study identified the first twenty one cases nationally in which doctors sought court orders to force obstetrical interventions, and found that the doctors succeeded in 86 percent of these cases (Kolder et al., 1987). In these successful suits, 81 percent of the women were African American or Hispanic, 44 percent were unmarried, 24 percent were not fluent in English, and all were poor.

These actions reflect a growing tendency among doctors, lawyers, and the general public to view mother and fetus as separate beings, with separable and sometimes conflicting rights, and to see the fetus rather than the mother as obstetricians' primary patient (B. Rothman, 1989; Daniels, 1993). This tendency reflects both technological and political changes. The growth of technologies like ultrasound, electronic fetal monitoring, and fetal surgery, which allow doctors to view and act on the fetus, have made fetuses seem more like independent beings than ever before (Casper, 1998). And the antiabortion movement has convinced many Americans to think of fetuses as children or "almost children," even though less than one-quarter of Americans believe abortion should be illegal in all circumstances (PollingReport.com, 2005).

The state has a legal obligation to protect children from parents who abuse or otherwise endanger them. Similarly, both ethical and legal guidelines require doctors who learn of child abuse to report it to the state. Should doctors and the state have a similar obligation to protect the fetus even if it means superseding parents' wishes?

Those who argue in favor of medical intervention find it illogical to protect children from bodily harm *after* birth but to deny them protection that might ensure their health *before* birth. Children born prematurely, addicted to drugs, or with birth defects because their mothers did not follow medical advice may suffer short, painful lives or may survive with mental or physical disabilities. In addition, these children cost hospitals and taxpayers vast sums every year. Those costs alone, one could argue, give the medical and legal systems the right to intervene when women endanger their fetuses.

(continued)

Box 5.2   *Ethical Debate (continued)*

Others, however, have raised several objections to placing **fetal rights** above mothers' wishes. First, these individuals question whether doctors necessarily know better than mothers what is in their fetuses' best interest. During the 1950s, for example, doctors routinely X-rayed women's abdomens to check fetal growth; this technique is now known to lead to miscarriages and cancer (B. Rothman, 1989). At any rate, almost all well-structured research studies have found that mothers' drug use during pregnancy causes little if any long-term harm to their children (E. Armstrong, 1998; Koren et al., 1989; Pollitt, 1990; Singer et al., 2002). This information has had relatively little impact on public or medical attitudes, partly because of cultural bias against illicit drugs and partly because of the bias in publishing (including medical journals) toward "breaking news." As a result, well-designed research studies suggesting that illicit drugs do not affect fetuses are regarded as uninteresting and go unpublished more often than do poorly designed studies suggesting that drugs do matter (Koren and Klein, 1991; Koren et al., 1989).

In addition, opponents argue, arresting or forcibly hospitalizing pregnant drug users may encourage other such women to avoid health care altogether, further endangering their fetuses. Moreover, forcibly withdrawing pregnant women from the drugs their bodies have become accustomed to can endanger the fetus more than does steady drug use (Pollitt, 1990).

Opponents of forced intervention further argue that doctors cannot make better decisions than mothers do, because they cannot understand fully the circumstances in which mothers make those decisions. For example, many women continue to use drugs during pregnancy only because they cannot obtain access to treatment programs, which usually have long waiting periods and often will not accept pregnant women. In addition, to enter a treatment program, women almost always have to leave their existing children with relatives or in foster care; for example, Arizona currently has an estimated 5,000 drug-addicted parents but only one treatment facility, with a total of ten beds, that allows parents to keep their children with them (Bland, 1999). Yet leaving children with relatives or in foster care may place children at greater risk than having a drug-using mother, given that women often begin drug use because of problems in their family and that foster care sometimes results in physical, sexual, or mental abuse.

Opponents of forced intervention also argue that the benefits of intervention do not

---

have proliferated. For example, in the early 1970s, the Boston Women's Health Book Collective published a 35-cent mimeographed booklet on women's health. From this, they have built a virtual publishing empire that has sold to consumers worldwide millions of books (including the best-selling *Our Bodies, Ourselves*) on the topics of childhood, adolescence, aging, and women's health.

Social Control and the Human Genome Project

The potential for medicine to act as a form of social control may soon grow through the work of the internationally funded Human Genome Project. The project's goal is to map the locations of all human genes and to determine the role each gene plays in health and illness.

justify the costs to women's civil liberties. Once we decide that women must put their fetuses' welfare above their own, where do we draw the line? Given that tobacco poses a far greater threat to fetuses than does any illicit drug, do we prosecute or hospitalize women who continue to smoke during pregnancy? What about women who continue to eat junk food rather than eating healthy meals? Or women who work two jobs and get insufficient rest? Already, some employers have used the language of fetal rights to bar women (but not men) from work involving toxic chemicals (Nelkin and Tancredi, 1989).

Finally, the effect of fetal rights on women's rights leads to questions regarding the true purposes of the fetal rights movement. Although we require parents to guard their children's health and welfare, we do not require them to donate kidneys, bone marrow, or even blood for their children's sake. Why, then, should we require women—and only women—to protect their fetuses? After all, fathers' use of tobacco, alcohol, and other drugs may damage sperm and therefore fetuses, but no court yet has charged a man for fetal abuse. Similarly, working in toxic environments damages sperm as well as ova and fetuses, yet no employers have tried to "pro-

tect" men from holding such jobs. And during Pamela Stewart's pregnancy, her husband not only used amphetamines and had sexual intercourse with her but also beat her periodically. Yet no district attorney arrested him for wife abuse or fetal abuse. These facts have led some to conclude that the true, if perhaps unconscious, motive behind the rhetoric of fetal rights is not to protect fetuses or children but to restrict women's lives—especially the lives of those women who are most different from and hence considered most suspect by those who make laws and policy.

Sociological Questions

1. What social views and values about medicine, society, and the body are reflected in this policy? Whose views are these?

2. Which social groups are in conflict over this issue? Whose interests are served by the different sides of this issue?

3. Which of these groups has more power to enforce its view? What kinds of power do they have?

4. What are the intended consequences of this policy? What are the unintended social, economic, political, and health consequences of this policy?

Genes affect health in two ways: by causing "true" genetic diseases and by increasing individuals' predisposition to develop disease. True genetic diseases, such as hemophilia, are caused directly by specific genes. Such diseases are relatively uncommon and typically become apparent at birth or early in life. Some can be treated, but none can be cured. As researchers learn which genes cause these diseases and develop tests to determine the presence of those genes, they can offer individuals the opportunity to learn whether they, their children, or (for pregnant women) their fetuses carry the gene. Individuals who learn they have a genetic defect may choose to avoid becoming pregnant; to abort any fetuses that also carry the defect; or to continue a pregnancy to term, knowing that the fetus carries the defect and hoping that this foreknowledge will better prepare them for the birth of an ill or disabled child. Finally, individuals who know they have a genetic defect but who want to have a child that is biologically theirs can have fetuses created through in vitro fertilization (in which eggs removed from the woman's body are mixed with the man's sperm in the laboratory). They can then have their doctors test the resulting fetuses for genetic defects and implant any nondefective fetuses in the woman's uterus. This strategy is rare because the physical costs to the woman and the financial and psychological costs to the couple are extremely high, and the odds of success are low.

In other cases, genes do not directly cause disease but can increase the likelihood of disease developing. For example, no single gene causes Alzheimer's disease, breast cancer, heart disease, or diabetes. These diseases occur more often in some families than others, however, which suggests that the diseases may occur only in those who have some genetic predisposition. In these cases, if doctors can learn which genes correlate with the disease and develop ways of identifying which individuals have those genes, doctors might find it easier to convince at-risk individuals to take potentially health-preserving actions. For example, women who learn that they have the BRCA-1 gene, which correlates with an increased risk of breast cancer, might choose to adopt a low-fat diet or to have their breasts removed before any cancer appears.

The Human Genome Project brings with it tremendous potential for both good and harm. Those who learn they are at increased risk can adopt healthier behaviors, and those who learn that they are *not* at risk can gain peace of mind. Testing could even benefit those who learn that they will develop a genetic disorder, for some will prefer certainty to the anxieties of uncertainty.

Yet the potential harm this knowledge can cause is also great. First, although some might cope well with the knowledge that they or their children will develop an unpreventable genetic disease later in life, others will be overwhelmed by this knowledge. It is hard, for example, to imagine how it can help individuals to learn at age 21 that by their forties they will develop Huntington's disease, a devastating neurological disorder that invariably causes progressive insanity, total disability, and death.

Second, as the knowledge and technologies developed by the Human Genome Project increase and become part of everyday medicine, the use of genetic testing will undoubtedly spread rapidly; already individuals can order genetic tests for themselves on the Internet. Genetic *counseling*, on the other hand, will probably spread more slowly because it is considerably more expensive to provide. In the future more people, especially those who are poor or live far from medical centers, thus are likely to receive complicated, confusing, and potentially devastating information from genetic tests without receiving the counseling necessary to help them understand and cope with this information.

Third, individuals identified through genetic testing as having an illness or being at high risk for illness may experience discrimination and stigma as a result. Individuals have been refused jobs, health insurance, or life insurance because they are carriers of a genetic disease, have a genetic defect although they are still asymptomatic, or are suspected of having or carrying a genetic disease (Billings et al., 1992; Natowicz, Alper, and Alper, 1992). The **Americans with Disabilities Act (ADA)** (described in Chapter 6) outlaws employment discrimination based on illness, disability, or genetic characteristics, but it is legally unclear whether the ADA applies to discrimination in other areas of life (Gostin, Feldblum, and Webber, 1999). Most states have outlawed genetic discrimination in health insurance and in the workplace, and federal legislators are debating similar national legislation, but such laws can help only those who know about them, have evidence of discrimination, and can afford legal assistance (National Genome Research Institute, 2005).

Fourth, genetic tests can tell whether an individual carries the gene for a disease, but not how soon or how severely he or she will be affected. For example, although doctors can tell if a fetus has Down syndrome, they cannot tell if the fetus will become a child who could be self-supporting or a child who could neither walk nor talk. Increasingly, too, tests are identifying genetic anomalies whose effects, if any, are unknown. As a result, couples often must decide whether to abort a genetically abnormal fetus with little idea what their child's life might be like.

Fifth, except for true genetic diseases, genetic tests can only suggest the *probability* that a fetus, child, or adult will develop an illness. For example, prospective parents might learn that their fetus has a 60 percent chance of developing breast cancer as an adult. No one can offer any logical rules for making decisions based on such probabilities. Parents in these circumstances will face far more complex decisions than will parents who know their child would have a genetic disease. Moreover, genetic testing cannot tell the former group of parents any more than the latter regarding when or how severely the illness will affect their children.

Finally, the Human Genome Project raises the potential for genetic controls far beyond anything now available. Relatively few persons oppose programs to prevent the birth of children with Tay-Sachs disease, which causes initially healthy children to deteriorate totally—both mentally and physically—and to die between the ages of 3 and 5. Yet many geneticists hope in the future to expand vastly the number of conditions for which genetic tests are run. Already many fetuses are aborted simply because they are female, as described in Chapter 4 (Banister, 1999; Wertz and Fletcher, 1998). Would the world really be a better place if we could abort fetuses because they would be mentally slow or predisposed toward fatness?

The potential impact of the Human Genome Project is magnified by the treatment it has received in the news media. Like illness, news is a social construction, for news media first decide which stories are newsworthy and then decide how those stories will be told. Research conducted by sociologist Peter Conrad (1997) suggests that the media consistently overplay the impact of genes in presenting news stories. Conrad looked at all coverage of genetics in five major newspapers (including the *Los Angeles Times* and the

Wall Street Journal) and three news magazines (*Time, Newsweek,* and *U.S. News and World Report*) between 1965 and 1995 and found that the media routinely gave prominent coverage to the discovery of a supposed link between a gene and a condition or illness, but either ignored later disconfirmations of the link or relegated them to back pages. For example, all eight news outlets gave prominent and optimistic coverage to a 1990 article published in the *Journal of the American Medical Association* that reported a link between a specific gene and alcoholism. Yet none of the magazines and only a few of the newspapers covered an article, published eight months later in the same journal, refuting the findings of the first article. Moreover, all news stories on the second article were relegated to the back of newspapers, and all suggested that new evidence of genetic links would surely be found soon.

These findings led Conrad to conclude that the news media has adopted a **genetic paradigm,** a way of looking at the world that emphasizes genetic causes. This paradigm

> has considerable appeal. It promises primary causes, located on a basic level of biological reality. Genes are often depicted as an essence, what one is really made of . . . We now can be tempted by the lure of specificity, associating specific genes and particular problems. Identifying specific genes seems so much neater than complex, messy, epidemiological and social analyses. This specificity feeds hopes for genetic "magic bullets" to alleviate human problems. (Conrad, 1997: 142)

Social Control and the Sick Role

Until now, we have looked at how medicine functions as an institution of social control by defining individuals either as sick or as biologically defective. Medicine also can work as an institution of social control by pressuring individuals to *abandon* sickness, a process first recognized by Talcott Parsons (1951).

Parsons was one of the first and most influential sociologists to recognize that illness is deviance. From his perspective, when people are ill, they cannot perform the social tasks normally expected of them. Workers stay home, homemakers tell their children to make their own meals, students ask to be excused from exams. Because of this, either consciously or unconsciously, people can use illness to evade their social responsibilities. To Parsons, therefore, illness threatened social stability.

Parsons also recognized, however, that allowing some illness can *increase* social stability. Imagine a world in which no one could ever "call in sick." Over time, production levels would fall as individuals, denied needed recuperation time, succumbed to physical ailments. Morale, too, would fall while resentment would rise among those forced to perform their social duties day after day without relief. Illness, then, acts as a kind of pressure valve for society—something we recognize when we speak of taking time off work for "mental health days."

From Parsons's perspective, then, the important question was how did society control illness so that it would increase rather than decrease social stability? The author's emphasis on social stability reflected his belief in the broad social perspective known as **functionalism.** Underlying functionalism is an image of society as a smoothly working, integrated whole, much like the biological concept of the human body as a homeostatic environment. In this model, social order is maintained because individuals learn to accept society's norms and because society's needs and individuals' needs match closely, making rebellion unnecessary. Within this model, deviance— including illness—is usually considered **dysfunctional** because it threatens to undermine social stability.

Defining the Sick Role

Parsons's interest in how society manages to allow illness while minimizing its impact led him to develop the concept of the **sick role.** The sick role refers to social expectations regarding how society should view sick people and how sick people should behave. According to Parsons, the sick role as it currently exists in Western society has four parts. First, the sick person is considered to have a legitimate reason for not fulfilling his or her normal social role. For this reason, we allow people to take time off from work when sick rather than firing them for malingering. Second, sickness is considered beyond individual control, something for which the individual is not held responsible. This is why, according to Parsons, we bring chicken soup to people who have colds rather than jailing them for stupidly exposing themselves to germs. Third, the sick person must recognize that sickness is undesirable and work to get well. So, for example, we sympathize with people who obviously hate being ill and strive to get well and question the motives of those who seem to revel in the attention their illness brings. Finally, the sick person should seek and follow medical advice. Typically, we expect sick people to follow their doctors' recommendations regarding drugs and surgery, and we question the wisdom of those who do not.

Parsons's analysis of the sick role moved the study of illness forward by highlighting the social dimensions of illness, including identifying illness as deviance and doctors as agents of social control. It remains important partly because it was the first truly sociological theory of illness. Parsons's research also has proved important because it stimulated later research on interactions between ill people and others. In turn, however, that research has illuminated the analytical weaknesses of the sick role model.

Critiquing the Sick Role Model

Many recent sociological writings on illness—including this textbook— have adopted a **conflict perspective** rather than a functionalist perspective. Whereas functionalists envision society as a harmonious whole held together largely by socialization, mutual consent, and mutual interests, those who hold a conflict perspective argue that society is held together

Key Concepts 5.2	Strengths and Weaknesses of the Sick Role Model	
ELEMENTS OF THE SICK ROLE	MODEL FITS WELL:	MODEL POORLY FITS:
Legitimate reason for not fulfilling obligations	Appendicitis, cancer	Undiagnosed chronic fatigue
Individual not held responsible	Measles, hemophilia	AIDS, lung cancer
Should strive to get well	Tuberculosis, broken leg	Diabetes, epilepsy
Should seek medical help	Strep throat, syphilis	Alzheimer's, cold

largely by power and coercion, as dominant groups impose their will on others. Consequently, whereas functionalists view deviance as a dysfunctional element to be controlled, conflict theorists view deviance as a necessary force for social change and as the conscious or unconscious expression of individuals who refuse to conform to an oppressive society. Conflict theorists therefore have stressed the need to study social control agents as well as, if not more than, the need to study deviants.

The conflict perspective has helped sociologists to identify the strengths and weaknesses in each of the four elements of the sick role model (see Key Concepts 5.2). That model declares that sick persons are not held responsible for their illnesses. Yet, as we saw earlier in this chapter, and as Eliot Freidson (1970a), the most influential critic of Parsons, has noted, society often *does* hold individuals responsible for their illnesses. In addition, ill persons are not necessarily considered to have a legitimate reason for abstaining from their normal social tasks. Certainly no one expects persons with end-stage cancer to continue working, but what about people with arthritis or those labeled malingerers or hypochondriacs because they cannot obtain a diagnosis after months of pain, increasing disability, and visits to doctors (Ziporyn, 1992)? Parsons's model also fails to recognize that the social legitimacy of adopting the sick role depends on the socially perceived seriousness of the illness, which in turn depends not only on biological factors but also on the social setting; a nonunionized factory worker, for example, is less likely than a salaried worker with good health benefits to take time off when sick.

Other aspects of the sick role model are equally problematic. The assumption that individuals will attempt to get well fails to recognize that much illness is **chronic** and by definition not likely to improve. Similarly, the assumption that sick people will seek and follow medical advice ignores the many people who lack access to medical care. In addition, it ignores the many persons, especially those with chronic rather than **acute** conditions, who have found mainstream health care of limited benefit and who therefore rely mostly on their own experience and knowledge and that of other non-medical people. Finally, the concept of a sick role ignores how gender, ethnicity, age, and social class affect the response to illness and to ill people. For example, women are both more likely than men are to seek medical care when they feel ill and less likely to have their symptoms taken seriously by doctors (Council on Ethical and Judicial Affairs, 1991; Steingart, 1991).

In sum, the sick role model is based on a series of assumptions about both the nature of society and the nature of illness. In addition, the sick role model confuses the experience of *patienthood* with the experience of *illness* (Conrad, 1987). The sick role model focuses on the interaction between the ill person and the mainstream health care system. Yet interactions with the medical world form only a small part of the experience of living with illness or disability, as the next chapter will show. For these among other reasons, research on the sick role has declined precipitously; whereas *Sociological Abstracts* listed 71 articles on the sick role between 1970 and 1979, it listed only 7 articles between 1990 and 1999, even though overall far more academic articles were published during the 1990s than during the 1970s.

Conclusion

The language of illness and disease permeates our everyday lives. We routinely talk about living in a "sick" society or about the "disease" of violence infecting our world, offhandedly labeling anyone who behaves in a way we don't understand or don't condone as "sick."

This metaphoric use of language reveals the true nature of illness: behaviors, conditions, or situations that powerful groups find disturbing and believe stem from internal biological or psychological roots. In other times or places, the same behaviors, conditions, or situations might have been ignored, condemned as sin, or labeled crime. In other words, illness is both a social construction and a moral status.

In many instances, using the language of medicine and placing control in the hands of doctors offers a more humanistic option than the alternatives. Yet, as this chapter has demonstrated, medical social control also carries a price. The same surgical skills and technology for cesarean sections that have saved the lives of so many women and children now endanger the lives of those who have cesarean sections unnecessarily. At the same time, forcing cesarean sections on women potentially threatens women's legal and social status. Similarly, the development of tools for genetic testing has saved many individuals from the anguish of rearing children doomed to die young and painfully, but has cost others their jobs or health insurance.

In the same way, then, that automobiles have increased our personal mobility in exchange for higher rates of accidental death and disability, adopting the language of illness and increasing medical social control bring both benefits and costs. These benefits and costs will need to be weighed carefully as medicine's technological abilities grow.

Suggested Readings

Barker, Kristin K. 2005. *The Fibromyalgia Story: Medical Authority and Women's Worlds of Pain.* Philadelphia: Temple University Press. In this sensitive and remarkably evenhanded book, Barker analyzes why fibromyalgia emerged as a diagnosis, and why it has proven so controversial.

Conrad, Peter, and Joseph W. Schneider. 1992. *Deviance and Medicalization: From Badness to Sickness.* Philadelphia: Temple University Press. Presents a theoretical framework for understanding medicalization, as well as several case studies of this process.

Rothman, Barbara Katz. 1998. *Genetic Maps and Human Imaginations: The Limits of Science in Understanding Who We Are.* New York: Norton. A fascinating exploration of the sources and consequences of the genetic paradigm.

Getting Involved

ACT UP. 332 Bleecker St., Suite G5, New York, NY 10014. (212) 966–4873. www.actupny.org. Seeks to increase public awareness and government involvement in the fight against AIDS through rallies and demonstrations.

Council for Responsible Genetics. 5 Upland Road, Suite 3, Cambridge, MA 02140. (617) 868–0870. www.gene-watch.org. Works to educate the public about the social implications of genetic technologies and to advocate socially responsible use and development of those technologies.

Review Questions

What does it mean to say that illness is a social construction and a moral status?

How have explanations for illness changed over time, and how have explanations for illness blamed ill people for their illnesses?

What is the medical model of illness, and what are some of the problems with that model?

What is medicalization, why does it occur, and what are some of its consequences?

How might the Human Genome Project act as social control?

What is the sick role model, and what are some of the problems with that model?

Internet Exercises

1. Although medical sociologists, health psychologists, and doctors are all interested in issues related to illness, their specific interests vary greatly. Using your library or the web, obtain access to the major online indexes in these three fields: *Medline, Sociological Abstracts,* and *PsycInfo.* Search each

database for information on *susto* and on medicalization. How does coverage of these issues differ across fields? To what extent does coverage overlap? What does this tell you about these three fields?

2. Using your library or the web, obtain access to *Periodical Abstracts, the Readers Guide to Periodical Literature,* or another index of popular magazine articles. Look for articles on premenstrual syndrome (PMS) published in the last five years. Copy the results of your search onto a diskette, or download it to your hard drive. Based on the titles and abstracts of the articles, sort the articles into those that assume PMS is an objectively defined illness, those that question the nature or existence of PMS, and those whose position is unclear. What does this tell you about the medicalization of PMS?

CHAPTER **6**

The Experience of Disability, Chronic Pain, and Chronic Illness

Nancy Mairs is a writer, teacher, social activist, mother, and wife who has multiple sclerosis (MS). She writes:

I am a cripple. I choose this word to name me. . . . People—crippled or not—wince at the word "crippled," as they do not at "handicapped" or "disabled." Perhaps I want them to wince. I want them to see me as a tough customer, one to whom the fates/gods/viruses have not been kind, but who can face the brutal truth of her existence squarely. As a cripple, I swagger. . . .

I haven't always been crippled. . . . When I was 28 I started to trip and drop things. What at first seemed my natural clumsiness soon became too pronounced to shrug off. I consulted a neurologist, who told me that I had a brain tumor. A battery of tests, increasingly disagreeable, revealed no tumor. About a year and a half later I developed a blurred spot in one eye. I had, at last, the [symptoms] . . . requisite for a diagnosis: multiple sclerosis. I have never been sorry for the doctor's initial misdiagnosis, however. For almost a week, until the negative results of the tests were in, I thought that I was going to die right away. Every day for the past nearly ten years, then, has been a kind of gift. I accept all gifts.

Multiple sclerosis is a chronic degenerative disease of the central nervous system. . . . During its course, which is unpredictable and uncontrollable, one may lose vision, hearing, speech, the ability to walk, control of bladder and/or bowels, strength in any or all extremities, sensitivity to touch, vibration, and/or pain, potency, coordination of movements—the list of possibilities is lengthy and, yes, horrifying. One may also lose one's sense of humor. That's the easiest to lose and the hardest to survive without. . . .

I don't like having MS. I hate it. My life holds realities—harsh ones, some of them—that no right-minded human being ought to accept without grumbling. One of them is fatigue. I know of no one with MS who does not complain of bone-weariness. . . . As a result, I spend a lot of time in extremis and, impatient with limitation, I tend to ignore my fatigue until my body breaks down in some way and forces rest. Then I miss picnics, dinner parties, poetry readings, the brief visits of old friends from out of town. . . My life often seems a series of small failures to do as I ought. . . .

[Over time], I [have] learned that one never finishes adjusting to MS. I don't know now why I thought one would. One does not, after all, finish adjusting to life, and MS is simply a fact of my life—not my favorite fact, of course—but as ordinary as my nose and my tropical fish and my yellow Mazda station wagon. It may at any time get worse, but no amount of worry or anticipation can prepare me for a new loss. My life is a lesson in losses. I learn one at a time. (1986: 9–12, 19)

Nancy Mairs's story illustrates some of the central tasks faced by those who live with **chronic illness**, chronic pain, or disability—searching for an accurate diagnosis, coming to terms with a body that does not meet social expectations for behavior or appearance, nurturing social relationships despite a contrary body, and constructing a viable and life-sustaining sense of self. In this chapter, we look at these and other issues in the lives of people who have chronic illnesses, chronic pain, or disabilities. We also consider the social context in which these individuals live and see how that context can affect individuals' lives at least as much as the bodily changes Mairs describes.

This chapter begins with an exploration of the meaning and history of disability. We then examine the extent and social distribution of disability in the United States. After that, we look at chronic pain, which falls on the borders between disability and illness, and then consider the experience of living with these conditions.

Understanding Disability

Defining Disability

As explained in Chapter 5, the meaning of the term *illness* is far from obvious. The same is true for the term **disability.** Competing definitions of disability reflect competing stances in an essentially political struggle. The **World Health Organization (WHO)** definition is probably the most widely used. WHO defines disability in terms of impairments: "disturbances in body structures or processes which are present at birth or result from later injury or disease . . . [and which cause] loss or abnormality of psychological, physiological, or anatomical structure or function" (1980: 47). WHO defines disability as "any restriction or lack (resulting from an impairment) of ability to perform an activity in the manner or within the range considered normal for a human being." Disability, then, includes some but not all persons who have chronic illnesses (the majority of those with disabilities) as well as, for example, persons who are born deaf, become paralyzed in an auto accident, or experience chronic pain that limits their ability to function.

As many disability activists and social scientists have noted, this definition reflects a **medical model,** which locates impairments—and thus disabilities—solely within the individual mind or body. At first glance, such a definition seems perfectly reasonable. After all, isn't a disability something that an individual has, a defect in his or her body? According to many people with disabilities, the answer is no. Instead, they argue, their disabilities primarily stem not from their physical differences but from the way others respond to those differences and from the choices others have made in constructing the social and physical environment. For example, a man whose energy waxes and wanes unpredictably during the day might be able to work forty hours per week on a flexible schedule but not within a rigid 9-to-5 schedule. Similarly, a woman who uses a wheelchair might find it impossible to work in an office where furniture can fit only persons who walk and are of average height, but she might have no problems in an office with more adaptable furniture. Disability activists argue that making an office accessible to wheelchair users does not mean providing special benefits for the disabled, but rather compensating for the unacknowledged benefits that existing arrangements offer those who walk, such as chairs to sit in, stools for reaching high shelves, and carpeted floors that make walking easier but wheeling more difficult.

This approach reflects a **sociological model of disability** in its emphasis on social forces and public issues rather than on individual physical variations and troubles. In the rest of this chapter, the term *disability* refers to restrictions or lack of ability to perform activities resulting largely or solely from either (1) social responses to bodies that fail to meet social expectations or (2) assumptions about the body reflected in the social or physical environment.

These two models of disability—the medical model and the more sociological model used by disability activists—have strikingly different implications. As Paul Higgins (1992: 31) notes, "To individualize disability [as the medical model does] is to preserve our present practices and policies that produce disability. If disability is an internal flaw to be borne by those 'afflicted,' then we do not question much the world we make for ourselves. Our actions that produce disability go unchallenged because they are not even noticed." Individualizing disability, therefore, exemplifies the broader process of **blaming the victim,** through which individuals (in this case, people with disabilities) are blamed for causing the problems from which they suffer (Ryan, 1976); an example is the common belief that women would not be battered if they did not provoke their husbands in some way (Dobash and Dobash, 1998). In contrast, the sociological model of disability challenges us to look at the problem of disability from a very different perspective. If we conclude that the problem resides primarily in social attitudes and in the social and built environment, then we can solve the problem most efficiently by changing attitudes and environments, rather than by "rehabilitating" people with disabilities.

People with Disabilities as a Minority Group

Once we start thinking of disability as primarily based on social attitudes and built environments rather than on individual deficiencies, strong parallels emerge between people with disabilities and members of minority groups (Hahn, 1985). A **minority group** is defined as any group that, because of its cultural or physical characteristics, is considered inferior and subjected to differential and unequal treatment and that therefore develops a sense of itself as the object of collective discrimination (Wirth, 1985). Few would argue with the assertion that we differentiate disabled persons from others on the basis of physical characteristics. But can we also argue, as the definition of a minority group requires, that people with disabilities are considered inferior and subject to differential and unequal treatment?

Unfortunately, yes. Even a cursory look at the lives of people with disabilities reveals widespread prejudice and discrimination. **Prejudice** refers to unwarranted suspicion, dislike of, or disdain toward individuals because they belong to a particular group, whether defined by ethnicity, religion, or some other characteristic. Prejudice toward disabled persons is obvious in the fact that, throughout history, most societies have defined those who are disabled as somehow physically or even morally inferior and have considered disabilities a sign that either the individual or his or her parents behaved sinfully or foolishly (Albrecht, 1992).

Prejudice typically expresses itself through **stereotypes,** or oversimplistic ideas about members of a given group. Nondisabled people typically stereotype those who are disabled as either menacing and untrustworthy or as childlike—asexual, dependent, mentally incompetent, the passive "victims" of their fate, and suitable objects for pity (Zola, 1985). These attitudes permeate the health care world as well as the general public. In one study, for example, researchers divided a large sample of health care students and practitioners into two groups and showed each group a videotape of a job interview. Both videotapes used the same actors and scripts, but in one the actor playing the job applicant walked, and in the other he used a wheelchair. Those who saw the videotape with the "disabled" applicant rated the applicant significantly more cruel, selfish, incompetent, weak, dependent, and mentally unstable than did those who saw the same actor portraying a nondisabled applicant (Gething, 1992).

Stereotypes about people with disabilities are so strongly held that obvious evidence regarding the falsity of those stereotypes scarcely affects social attitudes. For example, attorney Marylou Breslin, executive director of the Berkeley-based Disability Rights Education and Defense Fund and a wheelchair user, tells of waiting at the airport for a flight in her dressed-for-success businesswoman's outfit, sipping from a cup of coffee. "A woman walked by, also wearing a business suit, and plunked a quarter into the plastic cup Breslin held in her hand. The coin sent the coffee flying, staining Breslin's blouse, and the well-meaning woman, embarrassed, hurried on" (J. Shapiro, 1993: 19).

Stereotypes about people with disabilities are reflected and perhaps reinforced in the popular media, which often portray disabled individuals as pitiful, maladjusted, or evil (Higgins, 1992: 80–97; Safran, 1998). In book and film characters from Captain Hook in *Peter Pan* to Freddie Krueger in *Nightmare on Elm Street* and the Penguin in Batman comics and films, the media have equated physical deformity with moral deformity. Moreover, when the media do not portray persons with disabilities as horrifying, they often portray them as pitiful—whether depicting Tiny Tim in Charles Dickens's classic novel, *A Christmas Carol,* or Maggie Fitzgerald in *Million Dollar Baby,* for whom death was preferable to life. Although contemporary media sometimes do present more positive images, such as stories about people with disabilities who have "heroically" compensated for their physical disabilities, who have chosen to live "saintly" lives, or whose innocence can help the rest of us learn to live better lives (*Riding the Bus with My Sister,* for instance), these stories, too, typically ignore the social nature of disabilities and instead offer simplistic stories about individual character. Exceptions to these rules—films such as *The Station Agent, Murderball,* and *Children of a Lesser God*—remain rare, although they have become far more common in the last 20 years.

All too often, these prejudices against persons with disabilities result in **discrimination,** or unequal treatment grounded in prejudice. As recently as the first decades of the twentieth century, American laws forbade those with epilepsy, leprosy, Down syndrome, and other conditions from marrying and mandated their institutionalization or sterilization (J. Schneider and Conrad, 1983: 32–33; J. Shapiro, 1993: 197). During the 1930s and 1940s, doctors working for the government of Nazi Germany murdered about 100,000 disabled children and adults as *Lebensunwertes Leben*—"life unworthy of life" (Lifton, 1986). Partly due to discrimination, 68 percent of working-age disabled Americans are unemployed, even though two-thirds of these individuals say they could and would work if given the opportunity (National Organization on Disability, 2001).

To fit the definition of a minority group, a group must not only experience prejudice and discrimination but also consider themselves objects of collective discrimination. This is the weakest link in defining disabled people as a minority group (Higgins, 1992: 39–44). Unlike members of other minority groups, disabled individuals are rarely born to disabled parents. As a result, they might have little contact with, let alone sense of connection to, other people with disabilities. Moreover, fewer than 15 percent of people with disabilities are born disabled (J. Shapiro, 1993: 7). Therefore, most establish their sense of individual and group identity before they become disabled, and not all will change their sense of identity following disability. In addition, those who develop a sense of community with others who share their disability do not necessarily feel a connection to persons with other disabilities; deaf people, for example, might identify with others who are deaf, but not with those who have arthritis. Nevertheless, the sense of belonging to a broader group is surprisingly strong. In a national survey conducted in 2001, 47 percent of disabled persons reported feeling a sense of community with other disabled persons (National Organization on Disability, 2001). Thus disabled Americans increasingly have come to believe that they deserve not charity—as exemplified by the Muscular Dystrophy telethon, with its implications of inferiority and pity—but the same rights as other citizens to live, work, study, and play in the community.

These rights have been reinforced by the federal Education for All Handicapped Children Act, which requires school districts to educate all children regardless of disability in the least restrictive environment feasible, and the **Americans with Disabilities Act (ADA),** which outlaws discrimination and requires accessibility in employment, public services, and public accommodations (including restaurants, hotels, and stores). Box 6.1 describes the work of Disability Rights Advocates, an organization that fights to enforce these legal rights.

To explore how the ADA has affected the work environment, sociologists Sharon Harlan and Pamela Robert (1998) interviewed a nonrandom but diverse sample of disabled, nonmanagerial civil service workers in one state. One-third of their subjects (32 percent) had never requested an accommodation, either because they were not familiar with the procedures or because they assumed that doing so would call attention to their disabilities and threaten their jobs rather than result in meaningful accommodations. Instead, they tried to compensate for their disabilities by working longer hours, working even when sick, refusing promotions that would leave them with more difficult work conditions, and so on.

Of those workers who had requested accommodations, 69 percent had been granted those requests or were still awaiting their resolution. Requests were most often granted for men, for whites, and for persons in higher-status jobs. Employers were more likely to grant requests for changes in the physical environment, such as providing adaptable furniture or disabled parking, than for changes in the social environment, such as offering flexible work schedules or personal assistance. These findings led the authors to conclude that employers will offer accommodations only if those accommodations do not threaten the authority structure of the workplace by suggesting that workers should be granted more flexibility or autonomy.

When employers refuse requests for accommodation, workers have the option of bringing lawsuits or filing complaints with the federal Equal Employment Opportunity Commission (EEOC). Of the 107,000 workers whose complaints were resolved in the first five years after the ADA went into

---

Box 6.1   *Making a Difference: Disability Rights Advocates*

Disability Rights Advocates (DRA) is a non-profit law firm that uses individual lawsuits, class action lawsuits, and the threat of lawsuits to fight for the rights of persons with disabilities (www.dralegal.org). Its staff consists of a few paid lawyers (some of whom have disabilities) and numerous volunteer lawyers and law students. In recognition of its excellent work, DRA has received ongoing funding from various foundations and associations, including the Kaiser Family Foundation, the San Francisco Foundation, and various Bay Area bar associations.

Although initially DRA worked solely on California cases, it grew rapidly into a national organization. In addition, since 1995, when DRA received a grant from the private, nonprofit Soros Foundation and matching funds from the U.S. State Department, it has run an advocacy program for disabled persons in Hungary, as well as leadership training programs open to disabled persons from across Eastern Europe.

Some of DRA's successes include advocating for greater access to California public schools, including educating disabled students and their parents about their rights and helping school administrators remove barriers; settling a statewide lawsuit against Denny's restaurants, as a result of which all California Denny's are being made accessible; and initiating the first lawsuit in the country against a city building department for failure to enforce laws requiring the removal of architectural barriers in public and private buildings. The lawsuit was dropped when the city pledged to hire more building inspectors, train its inspectors in disability access laws, and hire a consulting firm to handle the backlog of complaints from disabled persons against the city. In addition, in the last few years legal pressure was used to convince Greyhound Bus Lines to provide accessible rest stops and assist riders with disabilities, to convince a major national car rental company to make vehicles with hand controls reasonably available, to convince a hospital to provide sign language interpreters for patients and their relatives who are deaf, to convince several hotels to increase the number of accessible guest rooms and to remove physical barriers, and to convince a major supermarket chain to begin providing assistance to disabled shoppers.

Finally, to support those who proactively work to advance the rights of disabled Americans, rather than doing so only when threatened with lawsuits, each year DRA gives out its ADA Eagle Awards. For example, awards have been given to Nordstrom, Inc., for its commitment to making its stores accessible to persons with disabilities and its use of models with disabilities; to Marriott International and Noah's Bagels for their efforts to hire and accommodate employees with disabilities; and to NBC's "Dateline" show for its coverage of housing and employment discrimination against people with disabilities.

---

effect, only 11.4 percent both won their cases and received benefits as a result (Equal Employment Opportunity Commission, 1999). Similarly, during approximately the same time period, workers won only 8 percent of cases that went to trial (American Bar Association, 1998). Unfortunately, the impact of the ADA has been limited because most courts have narrowly defined who qualifies for its protection (Gostin, Feldblum, and Webber, 1999).

For example, courts have ruled that the ADA does not apply to individuals whose diabetes is controlled by insulin or whose spinal cord injuries keep them from working rigid hours but do not otherwise interfere with their work.

The Social Distribution of Disability

According to U.S. government researchers, approximately 12 percent of non-institutionalized persons living in the United States have a disability, defined by these researchers as a chronic health condition that makes it difficult to perform one or more activities generally considered appropriate for persons of a given age—play or study for children, work for adults, or basic activities needed to maintain an independent life (shopping, dressing, bathing, and so on) for the elderly (National Center for Health Statistics, 2004).

The proportion of the population living with disabilities has grown significantly over time (Kaye et al., 1996). Only a few decades ago most paraplegics, babies born prematurely or with serious birth defects, persons with serious head or spinal injuries, and soldiers with major wounds died quickly. Now most live, although often with serious disabilities. During the Iraq war, for example, new body armor that protects soldiers' torsos plus advances in military medical care have resulted in far fewer deaths but far more survivors with brain damage or multiple amputations (Glasser, 2005). In addition, average survival times for various common chronic conditions, such as hypertension and cardiovascular disease, have increased. Finally, as the proportion of the population over age 65 has increased—and in the absence of meaningful attempts to remove the social and physical barriers that can prevent individuals from living independent lives—so has the proportion living with disabilities. As Table 6.1 shows, the percentage of Americans with activity limitations (i.e., unable to perform some basic life activity such as shopping or dressing oneself) increases as age increases, for longer lives translate into more years in which to have accidents or develop degenerative diseases. Even among persons above age 75, however, more than half report no disabilities.

As the table also shows, poorer persons are more likely than wealthier persons to report activity limitations. Not surprisingly, ethnicity also affects rates of disabilities, largely because of its relationship to poverty. By their early thirties, about 12 percent of Native American men have been unable to work or are limited in the work they can do because of illness or injury for at least six months (Hayward and Heron, 1999). In contrast, disability does not become equally common among African American men until their late thirties. Even more startling, white and Hispanic men do not reach this rate of disability until their early fifties, and Asian American men do not reach it until their early sixties. Similar patterns emerge when white, Hispanic, African American, Asian American, and Native American women are compared. However, largely because women live longer than men do, most disabled persons are women.

Table 6.1	*Percentage of Americans With Chronic Activity Limitation, 2002*

	% WITH LIMITATION
AGE	
Under 18 years	7.1
18–44 years	6.3
45–54 years	13.7
55–64 years	21.1
65–74 years	25.2
75 years and over	45.1
ETHNICITY	
Hispanic	10.7
White non-Hispanic	12.4
Black non-Hispanic	15.0
INCOME	
Poor	22.9
Near Poor	17.5
Not Poor	9.5

Source: National Center for Health Statistics (2004: 214).

Understanding Chronic Pain

Chronic pain, which affects about 50 million Americans (Bradshaw, Nakamura, and Chapman, 2005), falls on the border between disability and chronic illness. Chronic pain is a symptom, not an illness in itself. Sometimes it can be attributed to an injury or an illness, such as arthritis or cancer, but in other cases no specific cause can be identified; doctors often lack explanations for chronic headaches or back pain, the two most common types of chronic pain. Finally, chronic pain may be attributed to conditions whose existence and diagnosis remains contested within the medical world; examples are irritable bowel syndrome, fibromyalgia, and chronic fatigue syndrome.

Although the causes of chronic pain are often unclear, its consequences are obvious. Chronic pain is the most common underlying reason for disability among working-age adults (American Pain Society, 2000). In addition to its physical toll (which includes sleep deprivation and exhaustion), chronic pain damages social relationships; increases depression, anxiety, and the risk of suicide; and costs the nation $61 billion yearly in reduced productivity alone (W. Stewart et al., 2003).

Treating chronic pain is notoriously difficult, and there is no medical consensus on how to do so (American Pain Society, 2000). Few truly new treatments are available: Most pain medications derive from either morphine or aspirin, both of which were first commercially produced in the 1800s. Although morphine-related drugs such as OxyContin are often the safest and most effective treatments for chronic pain (American Academy of Pain Medicine and American Pain Society, 1996), American doctors are reluctant to use them, both because doctors share popular American beliefs about opiates and addiction and because they fear arrest under strict U.S. drug trafficking laws. Meanwhile, the most popular aspirin-related drugs (including Celebrex and Vioxx) have recently been found both less effective and more dangerous than initially claimed (Abramson, 2004). To make matters worse, few American doctors are specially trained in pain management.

Obtaining appropriate treatment is particularly unlikely for minorities, poorer persons, children, the elderly, and women (Hoffman and Tarzian, 2001). Women—representing the majority of those living with chronic pain—are significantly more likely than men to encounter doctors who ascribe their pain to psychiatric causes and prescribe sedatives or psychotherapy rather than effective pain medications (Barker, 2005; Hoffman and Tarzian, 2001; Werner and Malterud, 2003). To avoid this fate, women with chronic pain must tread a fine line, striving to appear neither too sick nor too well and neither too assertive nor too passive, in order to receive proper treatment and avoid being labeled hysterical or pushy, malingerers or whiners (Werner and Malterud, 2003). Not surprisingly, women consumers have been at the forefront of the movements to medicalize chronic fatigue syndrome, multiple chemical sensitivity, fibromyalgia, and other similar conditions (Barker, 2005).

Living with Disability and Chronic Illness

Living with disability or chronic illness, whether or not it results in chronic pain, is a long-term process that includes responding to initial symptoms, injuries, or diagnoses; making sense of one's situation; and continually reconceptualizing one's future. In this section, we examine this process and explore how illness, pain, and disability affect individuals' lives, relationships with others, and sense of self.

Initial Symptoms and Diagnosis

Becoming a chronically ill or disabled person begins with recognizing that something about the body is troubling. This recognition does not always come easily. Health problems often build gradually, allowing individuals and their families slowly and almost unconsciously to adapt to them and to minimize their importance (Bury, 1982; Charmaz, 1991: 24–28; J. Schneider and Conrad, 1983; D. Stewart and Sullivan, 1982). In addition, the signs of illness and disability often do not differ greatly from normal bodily variations. A child who doesn't walk by 12 months might have a disability or might simply be a slow developer. Similarly, children with epilepsy, for example, can for many years experience "strange feelings," "headaches," "spaciness," "blackouts," and "dizzy spells" before they or their families recognize these as signs of epilepsy. As one man recalled:

> I'd always had the tendency to roll my eyes back in my head . . . to kind of fade out for a while. But I thought that was nothing, but . . . I guess they call them petit mal [epileptic seizures]? I'd lose consciousness for a while. I wasn't really conscious of it and [the only] time anybody would notice it was when the family was all together at the dinner table and I, I'd be like daydreaming for a while and then I'd roll my eyes back and they'd go, "Stop that!" and I'd go "Stop what?," y'know, I didn't know what I was doin'. (J. Schneider and Conrad, 1983: 57–58)

Social scientists refer to this process of defining, interpreting, and otherwise responding to symptoms and deciding what actions to take as **illness behavior** (Mechanic, 1995). A review article by anthropologists Vuckovic and Nichter (1997), summarizing 20 years of research studies, concluded that U.S. residents treat between 70 and 95 percent of all illness episodes without a doctor's assistance. Individuals typically begin by medicating themselves or those under their care with nonprescription medications recommended by friends, families, store clerks, or pharmacists or, more rarely, with prescription medicines left over from previous illnesses.

Research results are mixed regarding whether gender or ethnicity affects use of self-medication, but age clearly has an impact: Persons over age 65 are considerably more likely than others are to self-medicate, with the majority of older persons using one or more nonprescription drugs regularly (Vuckovic and Nichter, 1997). Social class does not affect the *use* of self-medication, but does affect the *reasons* for doing so: Affluent persons are more likely to self-medicate to save time, whereas poorer persons are more likely to do so to save money. For all Americans, however,

> cultural demands to be productive and practical contingencies related to job/household responsibilities make time off for illness a luxury few Americans can afford. As popular commercials for cold and flu remedies remind mothers, construction workers, and teachers, there is simply no time to be ill. Pressures of the clock inherent in modern life often prohibit taking time for the extra sleep necessary to care for a cold or for the relaxation required to relieve a "stress" headache. In the past, individuals who were ill might "tough it out," waiting for symptoms to subside. Today, Americans can avoid delays by taking products "strong enough to tackle even the toughest cold." Medicines obviate the need to devote time and energy to healing activities, or to the "down time" necessitated by ill health. (Vuckovic and Nichter, 1997: 1289)

Key Concepts 6.1	Some Factors Predicting Illness Behavior	
	INDIVIDUALS ARE *LIKELY* TO DEFINE THEMSELVES AS ILL AND SEEK MEDICAL CARE WHEN:	INDIVIDUALS ARE *UNLIKELY* TO DEFINE THEMSELVES AS ILL AND SEEK MEDICAL CARE WHEN:
	Symptoms appear frequently or persistently (e.g., coughing blood once per day for a week).	Symptoms appear infrequently (e.g., coughing blood every few months).
	Symptoms are very visible (e.g., rash on face).	Symptoms are not very visible (rash on lower back).
	Symptoms are severe enough to disrupt normal activities (e.g., epileptic convulsions).	Symptoms are mild (annoying but tolerable headaches).
	Illness is only likely explanation for physical problems.	Alternative explanations are available (e.g., recent stresses may explain headaches).
	They have ready access to health care (e.g., good health insurance).	They have poor access to health care (e.g., no health insurance).
	They have a positive attitude to health care providers (e.g., trust doctors' abilities and motives).	They have a negative attitude to health care providers (e.g., distrust doctors' abilities and motives).

When and whether individuals seek formal diagnosis for acute or chronic medical problems depends on a variety of factors. According to the illness behavior model developed by David Mechanic (1995) and summarized in Key Concepts 6.1, the likelihood of seeking medical care depends, first, on the presence of alternative explanations for symptoms and the frequency, visibility, and severity of those symptoms (including most importantly how much they interfere with usual daily activities). In turn, how individuals interpret these factors depends on the social context; symptoms that seem serious to a middle-class professional who generally enjoys good health might seem quite minor to a homeless or elderly person who expects a certain amount of bodily discomfort. Social networks of friends and relatives also play a large role in determining how individuals will interpret and respond to symptoms because those networks can reinforce either a medical or a nonmedical interpretation of the problem and of how to treat it (Pescosolido, 1992). Finally, access to care and attitude toward health care providers also affect how quickly individuals seek care; those who can afford care only from public clinics and whose experience of clinics has taught them to expect long waits and rude treatment often put off seeking care for some time.

Eventually, however, if symptoms persist—and especially if they progress—individuals and their families are likely to reach a point where they cannot avoid recognizing that something is seriously wrong. As their previous interpretations of their symptoms crumble, individuals find themselves in an intolerable situation, torn by uncertainty regarding the changes in their bodies and their lives. Once they reach this point, the incentive grows to seek diagnosis and treatment because any diagnosis can become preferable to uncertainty.

Seeking a diagnosis, however, does not necessarily mean receiving one. Although some problems are relatively easy to diagnose—a 45-year-old white man who complains to his doctor of pains in the left side of his chest will probably quickly find himself getting tested for a heart attack—others are far less obvious. Persons with multiple sclerosis, for example, often find that doctors initially dismiss their symptoms as psychosomatic or trivial (Register, 1987; D. Stewart and Sullivan, 1982). In addition, the same symptoms may more rapidly produce a diagnosis for some than for others. For example, and as mentioned earlier, doctors more often ascribe women's than men's complaints to emotional problems rather than to physical illness (Council on Ethical and Judicial Affairs, 1991; Steingart, 1991).

Initially, both women and men can find these alternative diagnoses comforting and welcome—after all, it is far easier to hear that you are suffering from stress than that you have a serious illness. When symptoms persist, however, individuals find themselves torn by ambiguity and uncertainty, suffering anxiety about their failing health but receiving little sympathy or help from relatives and colleagues (Bury, 1982; J. Schneider and Conrad, 1983; D. Stewart and Sullivan, 1982; Waddell, 1982). As a result, eventually most people seek more accurate diagnoses. Some go from doctor to doctor, seeking a more believable diagnosis; others research their symptoms, diagnose themselves, and then press their doctors to confirm their self-diagnoses through testing. In the end, even those diagnosed with life-threatening conditions typically conclude that this certainty is preferable to continued uncertainty.

Responding to Illness or Injury

Once newly diagnosed or newly disabled individuals learn the nature of their conditions, responses vary widely. Some individuals with **HIV disease,** for example, find it easiest to cope by immediately considering their diagnosis a "death sentence," thus eliminating any uncertainty from their minds (Weitz, 1991). Others initially assume they can "beat" their illness, refusing to take seriously any dire predictions about their future. Still others cope by accepting their diagnoses intellectually but denying them emotionally. For example, one young man told how, two months after learning he had AIDS, he thought that he had picked up someone else's medical file when he noticed that his file read, "Caution: Patient has AIDS" (Weitz, 1991). Similarly, following traumatic injuries, some individuals refuse to participate in rehabilitation because they consider their situation hopeless; others refuse because they consider their injuries temporary.

Faced by the uncertainties and loss of control that accompany chronic illness and disability, individuals must reconstruct their images of their futures. Two basic strategies are available to these individuals, as to all who confront uncertainty—**avoidance** and **vigilance** (Janis and Mann, 1977; Weitz, 1989). Some cope by avoiding knowledge about their conditions so they can maintain previous images of their futures and ward off depression. Others cope by seeking knowledge vigilantly so that they can feel prepared to respond appropriately to any changes in their bodies. Both strategies reduce uncertainty and give individuals ways of understanding and, thus, responding to their health problems.

Although learning the nature of one's condition answers some questions, it raises new questions about why this has happened. Those who experience serious illness or injury therefore must reconceptualize not only their futures but also their pasts. Only by doing so can individuals make their situations comprehensible and, consequently, tolerable.

This search for explanations is often a painful one, set as it is in the context of a culture that continues at least partially to believe that individuals deserve their illnesses and disabilities. Nevertheless, some individuals do manage to avoid allocating blame to themselves. For example, one gay man with HIV disease stated in an interview: "Nobody deserves it [HIV disease]. I have friends that say 'Well, hey, if we weren't gay, we wouldn't get this disease.' That's bullshit. I mean, I don't want to hear that from anybody. Because no germ has mercy on anybody, no matter who they are—gay, straight, babies, adults" (Weitz, 1991: 68).

Other individuals, however, readily conclude—whether accurately or not—that they caused their own health problems by acting in ways that either contravened "divine laws" or put them at risk (such as smoking tobacco, having multiple sexual partners, or driving fast). As another man with HIV disease stated, "I should have helped people more, or not have yelled at somebody, or been better to my dad even though we have never gotten along. . . . Maybe if I had tried to get along better with him, maybe this wouldn't be happening" (Weitz, 1991: 68). Increasingly, too, individuals conclude that they caused their health problems through their psychological conflicts. As described in Chapter 5, this theory has gained considerable public exposure through the writings of Bernie Siegel and others, who have theorized that individuals become ill because they "need" their illnesses.

Interruptions, Intrusions, and Immersions

According to sociologist Kathy Charmaz, who interviewed more than one hundred chronically ill people, illness can be experienced as an interruption, an intrusion, or something in which an individual is immersed (Charmaz, 1991). Although Charmaz's research addressed only chronic illness, similar patterns undoubtedly apply to at least some individuals with disabilities, especially those that worsen over time.

When illness or disability is an **interruption,** it remains only a small and temporary part of a person's life (Charmaz, 1991: 11–40). Viewing it as an interruption means regarding it essentially as an acute problem—something to be dealt with at the moment, but not something that will have a significant long-term impact. This strategy can work as long as episodes of illness are minor or rare, or the disability is a mild one. For example, because of unexpected physical problems, someone with multiple sclerosis may need to change plans for a given day but not necessarily for the next week.

If the illness or disability progresses, however, it can become an **intrusion,** demanding time, accommodation, and attention and requiring that a person "live day to day" (Charmaz, 1991: 41–72). For example:

> I just take each day as it comes and I don't worry about tomorrow. I know that when I'm feeling good I should try to do as much as I can without overdoing, because sometimes I won't be able to do that. (Register, 1987: 190)

If the illness or disability progresses still further, people can find themselves immersed in their bodily problems (Charmaz, 1991: 73–104). Upon reaching this stage of **immersion,** they must structure their lives around the demands of their bodies rather than structuring the demands of their bodies around their lives. Social relationships often wither, and people often withdraw into themselves. Dealing with the body and illness can take most of a person's day and require the assistance of others. One woman, for example, told Charmaz that her kidney dialysis

> just about takes up the day. . . . I'm supposed to be on at 12:30, but sometimes don't get on until 1:00, then I'm dialyzed for four and a half hours and then it takes approximately half an hour to be taken off the machine and to have it clot. So quite often it's 6:00 or 6:30 before I ever leave there. So the day is shot. (1991: 83)

This chapter's ethical debate (Box 6.2), on the international trade in human organs, discusses one of the extreme solutions some individuals adopt to avoid such overwhelming illness.

Managing Health Care and Treatment Regimens

Persons who live with chronic illness and disability can turn to both conventional and alternative health care for help. And increasingly, they use the Internet to help them in these decisions.

Using Conventional Health Care

Living with chronic illness or disability often means living a life bound by health care regimens. However, in the same manner that, following injury or diagnosis with a chronic illness, some individuals seek and some avoid knowledge, some will strictly follow prescribed regimens of diet, exercise, or medication and others will not. Researchers traditionally have framed this

---

Box 6.2   **Ethical Debate: The Sale of Human Organs**

One of the most extreme situations an individual can face is the failure of a major organ, be it heart, lung, kidney, or liver. Such situations are death sentences unless the organ can be replaced either with a mechanical substitute or with a donated human organ. But mechanical replacements can severely restrict individuals' lives by tethering them to machines, and human organs can be difficult or even impossible to obtain legally; 85,000 Americans were on waiting lists for organs as of 2004, with an average wait of five years (during which time many on the list will die). As a result, a multimillion-dollar international market in human organs has emerged (Rohter, 2004).

Most commonly, the organs sold through this market are kidneys, although livers, lungs, corneas, and other organs also are sold. Because (almost) every human is born with two kidneys, and only one is needed to live, an individual can sell one kidney and still hope to live a normal and healthy life.

Selling an organ carries great risks, but can seem worth it if an individual is poor enough. In Brazil, for example, a person can earn $80 per month working at minimum wage—if work is available—or can sell his or her kidney for $3,000. Such sales are illegal in many countries, but those laws are rarely enforced.

To some observers, the trade in human organs is a natural and reasonable market response, in which supply (organs for sale) develops to fill an obvious need (organs required). These observers see no difference between selling organs and selling any other valued commodity, be it drugs, cars, or food. Similarly, they argue, people should have at least as much right to buy an organ that will save their life as they have to buy a television or a face-lift, and as much right to risk their health by selling an organ as they have to risk their life by selling their labor in dangerous occupations (Cherry, 2005).

Other observers, however, compare the trade in human *organs* to the trade in *humans,*

---

issue as a matter of compliance—whether individuals do as instructed by health care workers.

The most commonly used framework for studying compliance is the **health belief model.** As we saw in Chapter 2, this model was developed to explain why healthy individuals adopt preventive health behaviors. The same model is also used to understand why people who have acute or chronic health problems comply with medical advice regarding treatment (see Key Concepts 6.2). The model suggests that individuals will be most likely to comply if they believe they are susceptible to a health problem that could have serious consequences, believe compliance will help, and perceive no significant barriers to compliance. For example, people who have diabetes will be most likely to comply with their prescribed diet if they believe that they face substantial risks of blindness due to diabetes-induced glaucoma, that blindness would substantially decrease their quality of life, that the prescribed diet would substantially reduce their risk of blindness, and that the diet is neither too costly nor too inconvenient.

The health belief model is a useful but limited one for understanding compliance with medical treatment because it largely reflects the medical model of

and consider selling organs no more ethical than selling slaves. They argue that no one truly sells their bodily organs freely, but rather does so because they are coerced by poverty. They also argue that whenever a highly profitable commodity is for sale, and that sale is unregulated by laws, unscrupulous individuals will find ways to profit from the sale and vulnerable individuals will be exploited—whether they are buyers or sellers. Individuals who purchase black-market organs have no guarantees that the donor was healthy or that the organ will be a good match for them; and those who sell organs have no guarantee that the surgery will be conducted safely, that it will not harm their health, and that they will receive needed health care afterward. A study conducted in the Indian state of Tamil Nadu found that virtually all who (illegally) sell their kidneys did so to pay crippling debts. Yet because most (86 percent) were in worse health in the years following surgery, their average family incomes declined by one-third, even though average income in the state increased during the same period (Goyal et al., 2002). Despite these problems, though, the trade in organs is likely to continue so long as demand continues to outstrip supply.

Sociological Questions

1. What social views and values about medicine, society, and the body are reflected in this policy? Whose views are these?

2. Which social groups are in conflict over this issue? Whose interests are served by the different sides of this issue?

3. Which of these groups has more power to enforce its view? What kinds of power do they have?

4. What are the intended consequences of this policy? What are the unintended social, economic, political, and health consequences of this policy?

illness and disability. First, the health belief model assumes that noncompliance with medical recommendations stems primarily from psychological processes internal to the patient. Although this is sometimes true, in other cases patients do not comply because health care workers did not sufficiently explain either the mechanics of the treatment regimen or the benefits of following it (Conrad, 1985). Patients also might not comply because they lack the money, time, or other resources needed to do so.

Second, the health belief model implicitly assumes that compliance is always good (that is, that health care workers always know better than patients what patients should do). Yet, although health care workers often can help their patients considerably, this is not always the case, especially with chronic conditions (and it is one reason the **sick role** model does not fit chronic illnesses well). Bodies rarely respond precisely as medical textbooks predict. Nor can those textbooks determine whether an individual will consider a given treatment worth the impact it has on his or her quality of life. For example, persons with bipolar disorder (manic depression) often resist taking medications because the medications leave them feeling sedated and deprive them of the sometimes pleasurable highs of mania.

Key Concepts 6.2	**The Health Belief Model and Medical Compliance**		
	PEOPLE ARE MOST LIKELY TO COMPLY WITH MEDICAL ADVICE WHEN THEY:	EXAMPLE: COMPLIANCE LIKELY	EXAMPLE: COMPLIANCE UNLIKELY
	Believe they are susceptible.	Fifty-year-old man with hypertension who believes he is at risk for a heart attack	Fifteen-year-old boy diagnosed with epilepsy who has had only minor problems. Does not believe he is at risk for convulsions.
	Believe risk is serious.	Believes that heart attack could be fatal	Believes that convulsions would not be physically dangerous
	Believe compliance will reduce risk.	Believes he can reduce risk through taking medication regularly	Believes he doesn't really have a problem, so doesn't see how medication could help
	Have no significant barriers to compliance.	Medication is affordable and has no serious or highly unpleasant side effects.	Medication makes the boy feel drowsy, dull, and set apart from his peers.

Moreover, for numerous chronic conditions, the only available treatments are disruptive to normal routines, experimental, only marginally effective, unpleasant, or potentially dangerous. As a result, many people who at first diligently follow prescribed regimens eventually abandon them and lose some of their faith in mainstream health care (Conrad, 1985). Meanwhile, health care providers who do not understand why their patients did not respond to treatment as expected will often blame the problem on patient noncompliance, further eroding relationships between patients and providers and leading to future noncompliance.

As people's faith in mainstream medicine declines, some begin experimenting with their treatment regimens, learning through trial and error what works best for them not only physically but also socially, psychologically, and economically (Conrad, 1985). Others begin using **alternative** or **complementary therapies** (defined broadly as treatments not widely integrated into medical training or practice in the United States).

Using Alternative Therapies

Interest in alternative therapies has grown rapidly in the United States, both among healthy persons interested in avoiding illness and among those with chronic or acute illnesses. The most widely cited data on use of alternative therapies comes from three national, random surveys of English-speaking U.S. residents, conducted by a Harvard-based research team in 1990, 1997, and 2002 (Eisenberg et al., 1998; Tindle et al., 2005). The researchers looked at use of fifteen alternative therapies, including chiropractic, acupuncture, megavitamins, "folk" remedies, and biofeedback. Thirty-five percent of respondents reported using at least one alternative therapy in 2002, with more than 40 percent of these individuals using more than one therapy.

Users of alternative therapies are disproportionately likely to be female, upper income, below age 65, college educated, white, and suffering from chronic health problems (Astin, 1998; Kessler et al., 2001; Tindle et al., 2005). Currently, the most commonly used therapies are relaxation techniques and herbal medicine (used by 19 percent and 14 percent of Americans, respectively). Chiropractic, massage, and yoga are next most popular (used by 5 to 7 percent of Americans).

Most who use alternative therapies do so because conventional treatments have not helped them (most commonly, for dealing with chronic pain). Individuals typically use alternative therapies to complement rather than to replace mainstream medicine: Whereas 32 percent of those who sought help from a medical doctor also used an alternative therapy, virtually all—96 percent—of those who visited an alternative therapist also visited a medical doctor (Eisenberg et al., 1998). However, more than 60 percent of those who use alternative therapies do not tell their doctors that they have done so (Tindle et al., 2005). Moreover, 95 percent of those who use herbal medicine—the category that grew most rapidly between 1997 and 2002—choose their herbs without advice from a practitioner of any sort.

The popularity of alternative therapies rests on belief—or at least hope—in the efficacy of these treatments. These beliefs are supported both by personal experience and by recommendations from friends and acquaintances who believe alternative therapies have helped them. In some of these cases the therapies no doubt did help, either because of the biological effects of the therapies or because consumers' belief in the therapy helped the body to heal itself, as happens in about 30 percent of all persons treated with **placebos** (drugs known to have no biological effect). In other cases, individuals attribute cures to alternative therapies when actually the problem went away on its own, as happens with 70 to 80 percent of health problems (Lundberg, 2001: 123). Finally, people sometimes convince themselves that the therapies helped them even though their health did not actually improve.

Use of alternative therapies also rests on the belief that "natural" treatments are unlikely to do harm. This can be a dangerous assumption. For example, the Chinese herb, ma huang, helps dieters but can cause heart attacks and strokes. Kava kava tea can reduce anxiety but also can cause liver damage, and gingko biloba both stimulates circulation and increases bleeding during surgery (McNeil, 2002). Moreover, whereas the federal Food and Drug Administration is responsible for regulating the safety, potency, and effectiveness of prescription drugs, no governmental agency regulates herbal remedies or supplements. Current law does not permit manufacturers to claim that alternative herbs and supplements are cures, but does allow them to claim that their products *might* help.

To convince people to try alternative therapies and to believe in their efficacy and safety, manufacturers and retailers now spend millions yearly on promotion. For example, GNC, which sells nutritional supplements and other alternative and natural products, contracted with the Rite Aid drugstore chain to open outlets in 1,500 Rite Aid stores and to jointly run an Internet website where consumers can learn about and purchase their products. The two companies agreed to spend $30 million during the first year to market the stores and website (Janoff, 1999).

Other, less obvious, means are also now used to promote alternative therapies. Mainstream supermarkets routinely devote large sections in prime locations to "wellness products" and alternative therapies, and newsstands are filled with magazines devoted to informing consumers of the reputed health benefits of various alternative therapies and laced with advertisements for those products. Mainstream media, too, regularly run articles and advertisements promoting alternative therapies; a review for this textbook of articles on chiropractors indexed in the *Reader's Guide to Periodical Literature* during 1998 found 64 that described the potential benefits of chiropractic treatment, but only 4 that adequately described its risks.

The huge amounts corporations spend promoting alternative therapies are justified by the even larger amounts of money consumers spend on such services and products. Eisenberg and his colleagues (1998) conservatively estimated that Americans spent $21.2 billion in 1997 on alternative practitioners; $8.9 billion for herbal therapies and megavitamins; and $7.7 billion on books, classes, and equipment related to alternative therapies, for a total of $44.5 billion in 2005 dollars.

A fascinating study by Matthew Schneirov and Jonathan David Geczik (1996) suggests that neither marketing campaigns nor the potential health benefits of alternative therapies can fully explain the appeal of these therapies. Instead, the authors suggest, alternative healing appeals to individuals as a **new social movement,** a term first coined by German sociologist Jürgen Habermas (1981). Habermas argued that whereas older social movements arose out of discontent with material social conditions such as poverty, the new social movements stem from discontent with modern society's emphasis on science and rationality and its devaluing of the **lifeworld** of everyday human interaction, identity, and needs. Because new social movements focus on the lifeworld, they are less concerned with political strategies for social change and more concerned with creating ways of living that reflect their values. Thus new social movements depend less on formal organizations and more on "submerged networks" (Melucci, 1995) in which like-minded individuals can trade resources and obtain social support for adopting nonnormative ways of life. Although more recent writers tend to argue that movements cannot be neatly dichotomized into "new" versus "old,"

Habermas's insight regarding the importance of the lifeworld to social movement growth is nonetheless an important one.

Using Habermas's model, Schneirov and Geczik argue that the rise of alternative healing reflects dissatisfaction with the lack of match between doctors' concerns and patients' concerns: Whereas doctors typically are concerned with solving the puzzle of diagnosis and identifying a specific body part that requires treatment, patients are primarily concerned with the impact of illness on their lives (Mechanic, 1995). This mismatch can leave patients feeling like depersonalized objects and deeply dissatisfied with the care they receive, even if it is technically competent. In contrast, Schneirov and Geczik argue, alternative healing offers patients the opportunity to work as collaborators with health care providers and the promise to look holistically at the sources of their health problems and the consequences of any treatments.

Using interviews, ethnographic observations, and focus groups, Schneirov and Geczik uncovered two slightly overlapping submerged networks linked to alternative healing in the Pittsburgh area: one made up of working-class conservative Christians, the other of college-educated followers of spiritual, Eastern, or New Age philosophies. The researchers conclude that

> at the core of alternative health is a commitment to an ecological conception of the body, in which biochemical processes, emotional states, beliefs, lifestyle practices (especially nutrition), and spiritual phenomena are thought to be interconnected. Beyond this emphasis on holism is also a commitment to low-tech care; individualized treatment regimes (treating the person not the symptom), in which the patient's intuitions and perceptions of his or her illness are an important part of diagnosis and treatment; an emphasis on the self-healing capacities of the body; a commitment to something more than the absence of disease—to "wellness" or some positive conception of health; a desire to narrow the power imbalances between practitioner and patient; and finally an effort to critically appropriate healing traditions that lie outside of Western allopathic medicine. (Schneirov and Geczik, 1996: 630–631)

Most members joined these networks when confronted by a chronic illness and dissatisfied with the treatment they received from mainstream health care providers, and most of the rest joined while going through some other sort of life crisis. Network members were united by several beliefs: that modern medicine focuses too much on treating symptoms through surgery and medication rather than on preventing illness through lifestyle changes, that government regulation of health care endangers both personal freedom and health, that individuals should take responsibility for their own health, and that doing so means adopting stringent behavior regimens, such as restrictive diets and regular use of laxatives. Through these shared beliefs, users of alternative healing constructed not only a philosophy of health care but also a shared sense of identity and community. Thus, Schneirov and Geczik conclude, "the alternative health movement may be seen as part of a larger wave of discontent with the bureaucratic-administrative state, its reliance on expert systems, and the way it coordinates people's health care practices 'behind their backs'—without their knowledge and participation" (1996: 642).

Seeking Information on the Internet

Whether individuals rely primarily on mainstream or alternative therapies, many seek information about their conditions on their own, rather than relying solely on information provided by health care professionals. In the last few years, public access to information has exploded due to the exponential growth of Internet use. A national random survey conducted in December 2004 found that 51 percent of all Americans have used the Internet to seek health information, and 35 percent did so during the month preceding the poll (*Harris Poll,* 2004a).

Unfortunately, there are no controls on the quality of materials posted on the Internet, and its vast size makes it impossible to police for fraudulent information, such as claims that herbs can cure cancer or AIDS. Moreover, more often than not, popular websites such as Yahoo.com and MSN.com take readers seeking health-related information to websites run by individuals or corporations that have vested economic interests in selling certain drugs or treatments (Green, Kazanjian, and Helmer, 2004). Partly in response to concerns about misleading websites, the U.S. Department of Health and Human Services now runs its own website (www.healthfinder .gov) to link consumers to reliable online sources of health information.

Despite limitations in most people's ability to effectively search the Internet or evaluate the information they find there, the Internet has proven enormously beneficial to those living with chronic health problems. The Internet has allowed individuals to find others who share their troubles and to find information far beyond what they otherwise could access. As a result, those who use the Internet are now better able to negotiate with health care providers regarding appropriate treatment and to navigate the daily difficulties of living with illness or disability.

Dealing with Service Agencies

For those who experience disabilities, whether or not they are chronically ill, dealing with social service agencies can become a major part of life. Unfortunately, and despite the best intentions of many social service providers, the philosophies and structures of those agencies create systems that sometimes harm more than help those they serve (Albrecht, 1992; Higgins, 1992: 151–187).

Typically, social service agencies adopt a medical model of disability, focusing on how individuals can compensate for their individual deficiencies rather than on how social arrangements handicap them (Phillips, 1985). This approach has several **unintended negative consequences.** First, to accept someone as a client, agencies must define him or her as disabled. As a result, workers spend much of their time certifying individuals as disabled—identifying internal individual problems rather than looking for individual strengths. Through this process, individuals learn to think of themselves as disabled. According to Paul Higgins (1992: 132), "When service agencies evaluate, place, categorize, transfer, educate, rehabilitate, and so much more, the agencies are informing people who they are and who they are becoming." At the same time, because agencies receive funding based on how many clients they serve, agencies sometimes unintentionally encourage individuals to remain dependent on their services.

Second, because agencies use a medical model that defines people with disabilities as inherently flawed, agencies typically define "progress" as making those with disabilities as much like the nondisabled as possible (Albrecht, 1992; Higgins, 1992). Therefore rehabilitation workers might, for example, encourage someone to use a false leg even though the individual could move more quickly and less painfully on crutches or in a wheelchair. Box 6.3 describes how this philosophy has affected the education of deaf persons.

Third, the medical model encourages agencies to adopt a hierarchical pattern of care. This pattern of care is based on the premise that social service providers understand clients' needs, desires, problems, and strengths better than the clients themselves do and that social service providers are thus better equipped than clients to make decisions regarding clients' lives. Like other health care professionals, those who work in service agencies "evaluate, plan, treat, monitor, revise, discharge, and in other ways manage people. Disabled people (and their families) are expected to do what they are told" (Albrecht, 1992: 178). Thus, unwittingly, agencies encourage dependency.

Social Security, the major governmental program for persons with disabilities, further encourages dependency by economically penalizing those who obtain paid employment. Persons with disabilities who accept paid employment risk losing their government benefits, including both financial assistance and health care. Yet the costs of living with a disability are high; for example, as of 2005, modifying a van for a wheelchair-using driver costs anywhere from $10,000 to $27,000. Thus, unless individuals can get well-paid professional jobs with full health benefits, they may find employment unaffordable (Burns, Batavia, and DeJong, 1993).

Illness, Disabilities, and Social Relationships

For better or worse, chronic illness and disability alter relationships not only with health care providers and service agencies but also with friends, relatives, and colleagues. Illness and disability can strengthen social relationships, as families pull together to face health problems, old wounds are healed or put aside, and individuals realize how much they mean to each other. Illness and disability, however, can also strain relationships. Friends and family might help each other willingly during acute illnesses or the first few months of a chronic illness or traumatic injury, but they might become more loath to do so over time. This is especially true for male friends and family, who less often

---

Box 6.3   ***American Sign Language and the Education of Deaf Children***

American Sign Language (ASL) is the native language of the U.S. deaf community. (English Sign Language is quite different.) Until recently, nonsigners considered ASL little more than a crude collection of gestures. In fact, however, ASL is a fully functioning language with a coherent and unique grammatical structure that allows people to communicate complex ideas as quickly and fluently as any spoken language (Klima and Bellugi, 1979). The history of ASL and its place in the education of deaf children demonstrates the disabling impact of prejudice (Lane, 1992; Neisser, 1983; J. Shapiro, 1993).

Before the nineteenth century, no national American sign language existed, although deaf individuals, scattered around the country, typically developed their own "home signs." European schools had begun teaching deaf children, but American educators considered them incapable of learning. In 1813, Thomas Gallaudet, a Congregationalist minister, distressed by the isolation of a neighbor's deaf child, decided to travel to Europe to observe deaf education there.

In France, Gallaudet for the first time saw sign language used to teach deaf children. He became convinced that deaf children could learn if taught in a language they could understand. Gallaudet returned to the United States accompanied by Laurent Clerc, a deaf teacher who communicated via French Sign Language.

Once back in the United States, Gallaudet and Clerc opened a school in Hartford, Connecticut. Most of the teachers were deaf, and all could sign fluently in the new language—American Sign Language—that developed naturally out of the combination of French Sign Language and American home signs. The school boasted impressive results as deaf children, taught to communicate in ASL, learned a wide variety of academic skills, including reading and writing English.

This "golden age" of ASL was brief, however. In 1867, the Clarke School for the Deaf was established to promote "oralism," the philosophy that deaf children would learn to speak English and lip-read only if forbidden to use ASL. In 1880, the International Congress of Educators of the Deaf—a Congress that included only one deaf educator—voted to make oralism the sole method for teaching

---

than women are socialized to be caregivers (Cancian and Oliker, 2000; Fine and Asch, 1988b). Moreover, the growing burden of gratitude can make those who have chronic illnesses or disabilities reluctant to ask for needed help. Problems are especially acute among elderly persons, who have outlived their close relatives and friends and thus must rely on more distant social connections. For all these reasons, relationships may wither.

Relationships also suffer if individuals no longer can participate in previous activities. How do you maintain a relationship with a tennis partner once you no longer can hold a racket? How do you maintain a relationship with a friend when architectural and transportation barriers keep you from going to movies or restaurants? And how do you maintain a relationship with a spouse or lover when your sexual abilities and interests have changed dramatically—or when your partner no longer finds you sexually attractive?

deaf children. This decision remained in force for more than a century.

The decision to adopt oralism reflected the times (Neisser, 1983). With immigration rising in the United States, many Americans feared (as some do now, more than 100 years later), that "inferior" languages would soon replace English. The movement to eliminate ASL from the classroom paralleled the movement to ban these other languages. ASL seemed especially foreign and even sinister because its reliance on gestures made it seem less like English and more like the stigmatized languages of low-status Jewish and Italian immigrants. In addition, ASL seemed heretical to many because it seemed to refute the popular belief that God had separated humans from animals through speech.

Following the adoption of oralism, schools removed deaf teachers from the classrooms and in some cases began punishing children caught using ASL. Yet, except for the very small proportion of deaf children and adolescents who lose their hearing after learning English—and even for many of those—communicating in English usually remained an empty promise. Students would now spend hours each day practicing lip-reading and forming sounds they could not hear. Despite this, by the time they graduated, the vast majority could lip-read only a small fraction of spoken English and spoke English so poorly even their teachers could not understand them (Lane, 1992: 129). Moreover, the hours devoted to studying oral skills left little time for scholarly subjects, which, at any rate, were taught in incomprehensible spoken English. It was as if U.S. public schools taught children mathematics in Japanese! As a result, most deaf people remained functionally illiterate (Lane, 1992: 130). Since the 1970s, the ban on manual communication in the classroom has eased. In its place, though, most educators have adopted not ASL but artificially developed systems that substitute signs for words within grammatically English sentences. Whether deaf students are best taught in English-based systems or in ASL remains a highly contentious subject among educators and the deaf community, while the average reading level of 18- to 19-year-old deaf students remains no better than that of 9- to 10-year-old hearing students (Paul, 1998: 23).

Declines in financial standing also strain relationships. An individual might, for example, have the physical ability to go to a movie with a friend but lack the price of admission. Women and minorities are especially hard hit because they typically earned lower wages and had more erratic work histories before becoming ill or disabled, and so qualify for lower Social Security benefits, if any (DeJong, Batavia, and Griss, 1989). At the same time, the stress caused by financial pressures can damage relationships with children, lovers, and spouses.

Managing Stigma

Illness and disability affect not only relationships with friends and family but also less intimate relationships. Most basically—and despite the predictions of the sick role model—living with illness or disability means living with **stigma.** Stigma refers to the social disgrace of having a deeply discrediting attribute, whether a criminal record, a gay lifestyle, or a socially unacceptable illness. The term *stigma* does not imply that a condition *is* immoral or bad, only that it is commonly viewed that way.

Some illnesses, especially acute illnesses such as influenza or streptococcal infections, produce relatively little stigma; but others, such as leprosy or HIV disease, are so stigmatized that they can affect even relationships with health care providers. Individuals whose illnesses carry a heavy burden of stigma can manage that stigma in various ways. First, individuals can attempt to **pass,** or to hide their illnesses or disabilities from others (Charmaz, 1991: 68–70, 110–119; Goffman, 1963: 73–91; J. Schneider and Conrad, 1983; Weitz, 1991: 128–132). For example, an elderly man who bumps into furniture because of failing eyesight might try to convince others that he is merely clumsy, and one who sometimes does not respond to questions because of hearing problems might try to convince others that he is merely absentminded. Similarly, those who have chronic illnesses can choose to go out only on days when their symptoms are least noticeable.

Although passing offers some protection against rejection, it carries a high price. Fear of disclosure means constant anxiety. Relationships with friends and families suffer when disabled or ill individuals lie about their conditions. In addition, those individuals forfeit the emotional or practical support they might receive if others understood their situations. Individuals also risk losing jobs or flunking courses when they cannot explain their reduced productivity and increased absences.

Those who cannot tolerate the stresses of passing can instead adopt a strategy of **covering**—no longer hiding their condition but instead trying to deflect attention from it (Goffman, 1963: 102–104). A woman with a visible leg brace can wear eye-catching jewelry, and persons with mobility limitations can arrive early to social gatherings to accustom themselves to the setting, identify potential physical hazards, and find accessible seats. Similarly, elderly persons who no longer see well enough to drive at night can schedule their social activities during daylight hours.

Conversely, those who have invisible disabilities sometimes find advantages in **disclosing** their disability to elicit sympathy or aid (Charmaz, 1991: 119–133). For example, a woman might choose to wear a leg brace or tell co-workers about her arthritis in order to avoid being labeled lazy when she cannot do certain tasks.

Other people deal with the potential for stigma through a process of **deviance disavowal,** that is, convincing others that they are the same as "normal" people (Davis, 1961). These individuals do not try to pass or cover their deviance, but instead try to prove that their illnesses or disabilities make them no different from others. Such "supercrips"—in the slang of disabled activists—often appear in the pages of popular magazines: the quadriplegic who paints holding a brush between her teeth, the blind man who is a champion skier, the participants in Special Olympics, and so on.

Each of these strategies can ease ill or disabled people's lives in an intolerant society. None, however, challenges the basis of that intolerance. Those who pass or cover in no way threaten the prejudices of those who would reject them. Even those who attempt to disavow their deviance do not challenge social prejudices regarding disabilities as much as proclaim they are not like others who have disabilities.

In contrast, other people take the more radical step of rejecting their rejecters and **challenging** the stigma of illness and disability. These individuals reject the social norms that denigrate them and refuse to adopt the accommodative strategies of passing, covering, or disavowing deviance. Instead, they argue that their deviations from bodily norms should not limit their civil rights or social status. Rather than accepting the stigma of illness and disability, these individuals attempt instead to label those who discriminate against them as foolish or immoral (Weitz, 1991: 132–133). They disclose their illness or disability not to elicit sympathy or aid but to affirm their dignity and pride in the lives they have made for themselves despite—or perhaps because of—the ways their bodies differ from social expectations. For example, a woman born without a hand who, after a year of wearing a hot, uncomfortable, and functionally useless artificial hand, decided to switch to a metal hook told an interviewer about her habit of looking at herself when passing store windows:

> I never failed to get a reaction from people, so I always looked too. What the hell are they looking at? I looked and I saw a woman with a *surprisingly* short arm! But when I got the [cosmetic] hand, I looked and I thought, oh my God, that's what I would have looked like [if I had been born with a hand]! And I saw this person that I would have been. But maybe I would have been an asshole just like all the rest of them [the nondisabled]. . . . And [now] when I see the hook, I say, boy, what a *bad* broad. And that's the look I like the best. (Phillips, 1990: 855)

This quote illustrates how individuals can construct an alternative view of both themselves and "normals"—in this case, redefining the self as feisty, independent, and rebellious and defining "normals" as voyeuristic "assholes."

Similar sentiments help explain the 1998 student rebellion at Gallaudet University (Lane, 1992: 186–191). Although Gallaudet is the only American college or university devoted to serving deaf and hearing-impaired students, all its presidents before 1988 had been hearing. That year, when the college's board of trustees (80 percent of whom were hearing) once again chose as president a hearing person who could not communicate in sign language, the students, along with many faculty, staff, and others, rose in protest. To these students, there was nothing wrong with deaf people, only with those who considered them inferior; many referred to themselves as "Deaf" rather than "deaf" to signify that they are linked by a minority culture rather than by a physical deficit. The protesters' anger grew after the board's chairperson reportedly told a group of students that the university needed a hearing president because "deaf people are incapable of functioning in a hearing world" (Lane, 1992: 188); the chairperson's later disclaimer that her remark had been

mistranslated into sign language only highlighted the incongruity of allowing hearing people who knew no sign language to run Gallaudet. The students' protests and the groundswell of support they received from alumni,
staff, faculty, and the general public led to the resignation of the chairperson, the appointment of a new board with a majority of deaf and hard-of-
hearing people, and the appointment of a deaf president who knew sign
language.

Health Social Movements

Like the Gallaudet students, others who live with or are at risk of illness or
disability increasingly have turned to collective political action to address
their grievances. Like other social movements, **health social movements** are
collective (rather than individual) efforts to change something about the
world that movement members believe is wrong (P. Brown et al., 2004).

Health social movements have a variety of goals. Many are organized
around obtaining equal access to health care. For example, both doctors and
consumers have fought to loosen health insurers' restrictions on what they
will cover or for the adoption of a national health insurance system that will
provide coverage to all citizens. Other health social movements are primarily concerned with meeting the needs (including access to health care) of a
particular group. For example, the women's health movement, through
organizations such as the National Women's Health Network, has fought to
obtain for women the same access to heart disease treatments that men
have, to halt the unnecessary use of cesarean sections and hysterectomies,
and to increase the number of women physicians. Finally, a growing
number of health social movements have as their goal challenging medical
understandings of diseases based at least in part on their personal experiences with illness or disability. For example, few doctors believe in the existence of "multiple chemical sensitivity" (MCS), which is theorized to make
some individuals ill whenever they contact any of the many chemicals that
are part of everyday modern life. Persons who believe they have this condition have organized to lobby for medical recognition of their condition and
to sue insurance companies that refuse to cover their treatment.

The rise of health social movements reflects a variety of factors (P. Brown
et al., 2004). The civil rights, women's rights, and gay rights movements of
the 1950s through the 1970s set the stage for a broader discussion of rights
and a broader acceptance of political action across American culture.
Health social movements are partly a product of this changed cultural climate. In addition, the same cultural forces that increased use of alternative
health care and the same technological changes that increased Internet
usage have fostered health social movements, by reinforcing the idea that
individuals have the right and the obligation to challenge medical authority. Individuals are probably most likely to participate in health social movements when they come to believe that medical authorities have failed to

protect them from diseases, to identify their diseases, or to treat their diseases appropriately. For example, the environmental breast cancer movement was organized primarily by women diagnosed with breast cancer who questioned why medical research has focused almost exclusively on early diagnosis and treatment of breast cancer rather than on its prevention. As this example suggests, people who live with illness and disability are not simply victims of their fate, but may actively work to better their situation and those of others like them.

The Body and the Self

Regardless of a person's political stance toward his or her condition, all disabilities and chronic illnesses challenge the self (Brooks and Matson, 1987; Bury, 1982; Charmaz, 1991; Corbin and Strauss, 1987; Fine and Asch, 1988a: 10–11; J. Schneider and Conrad, 1983). Those whose bodies differ in some critical way from the norm must develop a self-concept in the context of a culture that interprets bodily differences as signs of moral as well as physical inferiority. The resulting stigma leads such individuals to feel set apart from others (Conrad, 1987; Kutner, 1987; Weitz, 1991).

Illness and disability threaten self-concept in various ways. People who become physically deformed or less attractive often find it difficult to maintain their self-images, as do those who lose their financial standing or their social roles as worker, student, spouse, or parent (Brooks and Matson, 1987; Weitz, 1991: 97). In addition, the need to rely on others for assistance can shake individuals' images of themselves as competent adults.

Disability and illness create different problems for women than for men. American society expects men to be emotionally, physically, and financially independent, and the threat to self-esteem when men cannot meet these expectations can be great. Conversely, American society expects women (except for African American women) to be dependent, so disability typically does not threaten women's self-esteem as much as it threatens men's self-esteem. For African American women, however, and for all other American women who cherish their independence, illness or disabilities can hamper the struggle to obtain that independence, because prejudice and discrimination based on illness and disability compound prejudice and discrimination based on gender.

The sexual changes accompanying disability and illness also affect women and men differently. Social norms for both persons with and without disabilities expect men to be sexually active but regard women's sexual desires with suspicion. Following disability, men can lose esteem in both their eyes and those of their partners if they no longer can perform as before. Women, on the other hand, often find that others assume they have no sexual feelings at all once they no longer meet social norms of sexual attractiveness. This denial of women's sexuality narrows women's lives and diminishes their self-images (Lonsdale, 1990).

To cope with these threats to the self, individuals sometimes attempt intellectually to separate their essential selves from their recalcitrant bodies. Cheri Register (1987: 33), a writer who has a rare, chronic disease, describes

> a need that many of us feel to visualize the illness as smaller than ourselves. Rather than letting the illness overtake our identities, we try to find some confined space within ourselves or our lives to contain it, and then draw boundaries around it: "*Here* is the illness. I will only let it make *this* much difference."

This strategy succeeds best when symptoms follow a predictable course and the problem affects only one part of the body.

Yet the impact of disability and illness on the self is not solely negative. Indeed, disability and illness can bring improved self-esteem and quality of life. Over time, individuals may learn to devalue physical appearances, derive self-esteem from other sources, and focus on the present rather than on an intangible future (Weitz, 1991: 136–140). They may learn to set priorities in their lives so that, often to a greater extent than before, they accomplish their most important goals rather than wasting precious energy on trivial concerns (Charmaz, 1991: 134–166). Finally, they may come to define their condition simply as part of who they are, with good points and bad points, and to recognize that much of their personalities and accomplishments exist not *despite* their physical condition, but *because* of it (Higgins, 1992: 141). As Barbara Rosenblum, a sociologist and artist who died of breast cancer at age 44, wrote:

> I am a very different person now: more open, much more honest, and more self-knowing. . . . I turned it [cancer] into a possibility of opening up to myself, for discovering, and for exploring new areas.
>
> I've realized that I want to list the ways in which cancer can do that. You can get courage to take larger risks than you ever have before. I mean, you're already sick, so what can happen to you? You can have much more courage in saying things and in living than you ever had before. . . .
>
> And you can do things you've always wanted to do. Cancer, by giving you the sense of your own mortality, can entice you into doing those things you have been postponing. . . .
>
> You have this sense of urgency. And you can turn this urgency—you can harness this energy that propels you—so that you go ahead and do these things and discover new parts of yourself. All the things you ever wanted to do, all the dreams you had. And the dreams that you couldn't even dream, because you didn't allow yourself. . . .
>
> Cancer has put me in touch with that. And then also, it has taught me to enjoy the tenderness and the preciousness of every moment. Moments are very important because there may not be any after that—or you may throw up. Cancer exquisitely places you in the moment.
>
> I have become very human to myself in a way that I would never have imagined. I've become a bigger person, a fuller person. This to me is one of the

greatest lessons: just being human. Having cancer doesn't mean that you lose yourself at all. For me it meant that I discovered myself. (Butler and Rosenblum, 1991: 160–161)

Conclusion

Given the progressive aging of the American population and the increasing ability of medical technology to keep alive ill and disabled individuals, many more of us can expect eventually to live with illness, chronic pain, and disability—whether our own, our parents', or our children's. Consequently, understanding what it means to live with these conditions has never been more important.

As both social constructions and social statuses, illness and disability affect all aspects of life. In addition to forcing those who are ill or disabled to interact with health care providers and to manage health care regimens, illness and disability affect relationships with family and friends, work and educational performance and opportunities, and, perhaps most important, one's sense of self and relationship with one's own body. Living with illness and disability also requires people to come to terms—or to refuse to come to terms—with uncomfortable questions and harsh realities regarding their past, present, and future.

Illness and disability can confer social disadvantages similar to those experienced by members of traditionally recognized minority groups. Yet the impact of illness and disability is not always negative, for illness and disability at times can provide individuals with the basis for increased self-esteem and enjoyment of life. Moreover, like other minorities, those who live with illness and disability have in recent years moved from pleas for tolerance to demands for rights. Those demands have produced significant changes in American architecture, education, transportation, and so on, and have laid the groundwork for the changes still needed.

Suggested Readings

Hockenberry, John. 1995. *Moving Violations: War Zones, Wheelchairs, and Declarations of Independence.* New York: Hyperion. A vivid and honest memoir by radio and television correspondent John Hockenberry, who has been a paraplegic since age 19.

Kamen, Paula. 2005. *All In My Head: An Epic Quest to Cure an Unrelenting, Totally Unreasonable, and Only Slightly Enlightening Headache.* Cambridge, MA: Da Capo. Kamen's wry but horrifying story of her decade-long headache illuminates the problem of chronic pain, the stigma attached to it, and the steps needed to improve the situation.

Getting Involved

Disability Rights Education and Defense Fund. 2212 6th Street, Berkeley, CA 94710. (510) 644-2555. www.dredf.org. Activist group promoting independent living and civil rights for persons with disabilities.

Review Questions

How do the medical and sociological models of disability differ?

Are disabled people a minority group?

What is the Americans with Disabilities Act?

How common is disability, and which social groups have the highest rates of disability?

What difficulties do individuals face in responding to initial symptoms of illness or disability, obtaining diagnoses, and coming to terms with their diagnoses?

What is illness behavior?

How can illness serve as an interruption, an intrusion, or something in which a person is immersed?

Why do individuals sometimes ignore medical advice?

Why do individuals use alternative health care?

How can illness or disability affect social relationships and self-image?

How can individuals manage the stigma of illness or disability?

What is a health social movement, and why have they become more common?

Internet Exercises

1. Find the Pharmaceutical Research and Manufacturers of America website. Read and critique their essay on direct-to-consumer advertising. What are the problems with the surveys they cite? What issues are they glossing over?

2. Find three sites devoted to disability rights. (Hint: Each site will probably have links to other sites.) Browse the sites. In what ways are the problems identified by these sites similar to or different from the problems identified in this chapter?

3. Find a website that sells human growth hormone (HGH) direct to the public. (Hint: Search for "purchase HGH.") Critique the website: What kinds of information is the website highlighting? What kinds of necessary information about the drug is either not available on the website, hard to find, or hard to read? What techniques is the website using to convince the viewer to purchase the drug (for example, suggesting that the drug is more "natural" than other available drugs, or recommended by medical "experts")?

The Sociology of Mental Illness

At the age of 18, Susan Kaysen was committed to a private mental hospital, where she spent the next two years. In her book *Girl, Interrupted,* she describes her experience in making the transition from mental hospital to the outside world:

The hospital had an address, 115 Mill Street. This was to provide some cover if one of us were well enough to apply for a job while still incarcerated. It gave about as much protection as 1600 Pennsylvania Avenue would have.

"Let's see, nineteen years old, living at 1600 Pennsylvania Avenue— Hey! That's the White House!" This was the sort of look we got from prospective employers, except not pleased.

In Massachusetts, 115 Mill Street is a famous address. Applying for a job, leasing an apartment, getting a driver's license: All problematic. The driver's license application even asked, Have you ever been hospitalized for mental illness? Oh, no, I just loved Belmont so much I decided to move to 115 Mill Street.

"You're living at One Fifteen Mill Street?" asked a small basement-colored person who ran a sewing-notions shop in Harvard Square, where I was trying to get a job.

"Uh-hunh."

"And how long have you been living there?"

"Oh, a while." I gestured at the past with one hand.

"And I guess you haven't been working for a while?" He leaned back, enjoying himself.

"No," I said. "I've been thinking things over."

I didn't get the job.

As I left the shop my glance met his, and he gave me a look of such terrible intimacy that I cringed. I know what you are, said his look. (Kaysen, 1993: 123–124)

As Susan Kaysen's story suggests, mental illness is a social as well as a psychiatric condition, and mental hospitalization has social as well as psychiatric consequences. We begin this chapter by considering the extent and distribution of mental illness. We then examine contrasts between the medical model of mental illness, which views mental illness as an objective reality (if subjectively experienced), with the sociological model, which views mental illness as largely a social construction. Finally, we look at the history of treatment and the experience of mental illness.

The Epidemiology of Mental Illness

The importance of understanding mental illness becomes clearer once we realize how many people are affected. The following section discusses research on the extent, distribution, and causes of mental illness.

The Extent of Mental Illness

Since the 1920s, social scientists have tried to ascertain the extent of mental illness. These researchers essentially have adopted medical definitions of mental illness (which, as we will see later in this chapter, are problematic). However, whereas doctors and other clinicians have focused on how biological or psychological factors can foster mental illness, social scientists have focused on how *social* factors can do so.

Over the years, researchers using a variety of methods have reached two consistent conclusions regarding the extent of mental illness. First, all societies, from simple to complex, include some individuals who behave in ways considered unacceptable and incomprehensible (Horwitz, 1982: 85–103). Second, symptoms of mental disorder are fairly common. According to the National Comorbidity Survey Replication (NCS-R), the largest national survey on the topic based on a **random sample** (R. Kessler et al., 2005a), during the course of a year approximately 31 percent of working-age adults experience a diagnosable mental illness, with 20 percent experiencing a moderate or severe disorder. The most common illnesses are major depression and problems with alcohol use, reported by 17 percent and 13 percent respectively. These estimates, however, are probably high, because they are based on reports of symptoms, not medical diagnoses of illnesses (Horwitz, 2002). Survey researchers can't know, for example, if someone has lost weight because of depression or because they are getting ready for a wrestling match.

Social Stress and the Distribution of Mental Illness

So far we have seen how common mental illness is across the population. But mental illness does not burden all social groups equally. In this section we look at how ethnicity, gender, and social class affect rates of mental illness.

Why do some social groups experience more mental illness than others do? For many sociologists, the answer lies in their different levels of exposure to social stress.

In the past, sociologists interested in the link between mental illness and stress largely focused on the **acute** stresses of **life events,** such as divorce, losing a job, or a death in the family. Researchers looked not only at the sheer number of life events individuals experienced but also at the *meaning* life events have for people and the *resources* individuals have for dealing with those life events. For example, an unplanned pregnancy means something quite different to an unmarried college student from a poor family than it does to a married, middle-class housewife. Similarly, some individuals have resources that can reduce the stresses of life events (such as money, social support networks, and psychological coping skills), whereas others lack such resources (Ensel and Lin, 1991; Pearlin and Aneshensel, 1986). For example, a person whose marriage fails but who has enough income to maintain his or her current lifestyle, close friends to provide companionship and social support, and good stress management skills will probably experience less stress than will someone whose economic standing following divorce plummets, who has few friends, and who responds to stress by drinking.

As we saw in Chapter 2, recent research finds that **chronic** stress is more important than acute stress for predicting poor *physical* health. Similarly, researchers have shown that acute stresses like life events often mask the more powerful impact that chronic stresses have on mental illness, as well (Turner and Avison, 2003). As is true for research on physical illness, one important line of research in this field explores how mental illness can result from the chronic stresses of **role strain** (Pearlin, 1989). Role strain refers to problems such as unwanted roles, rapidly changing roles, roles that exceed a person's resources and abilities, and conflicting roles (such as lacking the time to be both a successful college student and a good parent). Currently, however, the main focus of research in this field looks at how exposure to chronic social stress may explain ethnic, gender, and social class differences in rates of mental illness.

The Impact of Ethnicity: Social Class or Discrimination?

Researchers have uncovered few significant ethnic differences in rates of schizophrenia or other major mental illnesses. Compared to non-Hispanic whites, African Americans seem less likely to develop anxiety or mood disorders but more likely to report psychological distress, which overlaps with but is not the same as diagnosable mental illness (R. Kessler et al., 2005a). The former remains unexplained, but the latter is not surprising, because exposure to chronic stress is significantly higher among African Americans than among whites (Turner and Avison, 2003). African Americans report higher levels of distress than white Americans do at all income levels, although these differences taper off as income rises. Researchers theorize that psychological distress among African Americans results from the chronic daily stresses of living with racism and declines at upper income levels because those with higher incomes can better shield themselves from at least some of the effects of racism (R. Kessler and Neighbors, 1986).

Little recent research is available on psychological distress among other U.S. minority groups, and studies are divided as to whether Hispanics experience more or less distress than non-Hispanic whites or African Americans do (Rogler, 1991). However, Hispanics are less likely to develop anxiety disorders, mood disorders, or substance abuse problems (R. Kessler et al., 2005a). The largest study available (as of 2005) on Mexican Americans found that new immigrants' rate of mental disorders initially is half that of U.S.-born Mexicans, but after immigrants live in the United States for 13 years or more, the two rates converge (Vega et al., 1998).

The researchers hypothesize that the Mexican culture's strong emphasis on extended families protects immigrants from mental illness by offering social support and thus reducing chronic stress among persons who are single, childless, less educated, or employed in low-prestige jobs. As Mexicans integrate into American culture, they lose these protections.

The Impact of Gender: Socialization Effects

The impact of gender on mental illness is at least as complex as the impact of ethnicity. Gender has no consistent effect on the rate of schizophrenia or other major psychiatric illnesses. However, men consistently display higher rates of substance abuse problems and personality disorders (conditions characterized by chronic, maladaptive personality traits, such as compulsive gambling or antisocial tendencies), whereas women consistently display higher rates of anxiety disorders and of depression (R. Kessler et al., 2005a).

These differences in mental illness parallel differences in gender roles. Consistently, men display higher rates of disorders linked to violence, such as paranoid schizophrenia and antisocial personality disorder. As a result, some researchers hypothesize that these forms of mental illness occur when men become "oversocialized" to their gender roles. The symptoms of antisocial personality disorder (listed in Box 7.1), for example, essentially parallel expectations within lower-class communities for male behavior. Within these communities, men who meet these expectations are typically considered dangerous but not mentally ill, because their behavior is comprehensible. Although they might be labeled criminal, they are unlikely to be labeled mentally ill unless they somehow come to the attention of doctors from outside their communities.

Similarly, many sociologists hypothesize that depression results when traditional female roles cause chronic stress by reducing women's control over their lives (Horwitz, 2002: 173–179). Research has found that rates of depression are considerably higher among those women with the least control over their lives: nonworking women and married mothers. By the same

---

Box 7.1   *Diagnostic Criteria for Antisocial Personality Disorder*

A. There is a pervasive pattern of disregard for the rights of others since age 15, as indicated by three (or more) of the following:

  (1) failure to conform to social norms with respect to lawful behavior . . .

  (2) deceitfulness . . .

  (3) impulsivity or failure to plan ahead

  (4) irritability and aggressiveness, as indicated by repeated physical fights or assaults . . .

  (5) reckless disregard for the safety of self or others

  (6) consistent irresponsibility such as repeated failure to honor financial obligations

  (7) lack of remorse at having hurt, mistreated, or stolen from another

B. The individual is at least age 18 years.

C. Before age 15, a history of three or more of the following:

  (1) often bullied, threatened, or intimidated others

  (2) often initiated physical fights

  (3) used a weapon that could cause serious physical harm . . .

  (4) was physically cruel to other people

  (5) was physically cruel to animals

  (6) stole while confronting a victim . . .

  (7) forced someone into sexual activity

  (8) deliberately engaged in fire-setting

  (9) deliberately destroyed others' property . . .

  (10) was often truant from school, beginning before age 13 years

  (11) often lied

  (12) stole without confronting a victim . . .

  (13) often stayed out at night despite parental prohibitions, beginning before age 13 years

  (14) ran away from home overnight at least twice . . .

Source: *Diagnostic and Statistical Manual of Mental Disorders-IV-TR* (Arlington, VA: American Psychiatric Association, 2000), pp. 98–99, 706.

---

token, depression is especially common among men who have less power than their wives do, have little control over their work, or lose their jobs.

The Impact of Social Class: Social Stress or Social Drift?

Of all the demographic variables researchers have investigated, social class shows the strongest and most consistent impact on mental illness. As social class increases, the rate of both diagnosable mental illness and psychological distress decreases (Eaton and Muntaner, 1999; R. Kessler et al., 1994). But does lower social class status cause mental illness, or does mental illness cause lower social class? In other words, do the social stresses associated with lower-class life lead to greater mental disorder, or do those who suffer from mental disorder drift downward into the lower social classes? These two theories are referred to, respectively, as **social stress** versus **social drift.**

    Researchers interested in social class have focused primarily on schizophrenia, the disease that shows the most consistent relationship to social class; studies have found that schizophrenia and related disorders occur two to five times more often among those who have not graduated from college compared with those who have. Those who favor the social drift argument have shown that, at first admission to a mental hospital, schizophrenic patients hold jobs lower in social class than one would expect given their family backgrounds. This suggests that mental problems caused these individuals to drift downward in social class (Eaton and Muntaner, 1999).

Those who favor the (more commonly held) social stress theory, on the other hand, argue that instead of looking at the jobs schizophrenic patients held at first *admission* to a mental hospital, we should instead look at their first *jobs*. When researchers do this, they find no difference in educational attainment or in prestige levels of first jobs between schizophrenic patients and comparable others in their communities (Link, Dohrenwend, and Skodol, 1986). Therefore, these researchers argue, whatever causes downward social drift occurs *after* a person completes his or her education and obtains a first job but *before* first admission to a mental hospital. They further note that compared with the general public, a higher proportion of schizophrenic patients have worked in unusually noisy, hazardous, hot, cold, smoky, or humid environments, leading researchers to conclude that the chronic social stress of these working conditions precipitated mental disorder in vulnerable individuals. Similarly, other researchers have found that mental health problems increase among workers laid off because of plant closings, again suggesting that the chronic stresses of unemployment and lower-class status lead to mental disorder, rather than mental disorder leading to lower-class status (R. Kessler, House, and Turner, 1987). These findings are bolstered by research showing that chronic stress is significantly higher among lower-class persons and is a strong predictor of depression (Turner and Avison, 2003).

Defining Mental Illness

As with **disability** and physical illness, doctors and sociologists typically have very different ways of thinking about mental illness. In this section, we look at the contrasts between the medical model of mental illness and the sociological model. Neither of these models is absolute, however, for both sociologists and doctors often blend elements from each in their work. Nevertheless, the contrast between these two "ideal types" provides a useful framework for understanding the broad differences between the two fields.

The Medical Model of Mental Illness

To doctors and most other clinicians in the field, mental illness is an illness essentially like any other. To understand what this means, it helps to understand the history of medical treatment for syphilis, the disease that first demonstrated the power of medicine to control mental illness and that in many ways established the frame through which doctors would understand all mental illnesses.

Since the fifteenth century, doctors had recognized syphilis as a discrete disease. Because of its mild initial symptoms, however, only in the late nineteenth century did doctors realize the full damage syphilis can inflict on the nervous system, including blindness, deformity, insanity, and death. Unfortunately, doctors could do little to help those with syphilis. The best available treatment consisted, essentially, of poisoning patients with arsenic and other heavy metals in the hopes that these poisons would kill whatever had caused the disease before they killed the patients.

In 1905, scientists first identified the bacterium *Treponema pallidum* as the cause of syphilis. Five years later, Paul Ehrlich discovered the drug salvarsan as a cure for syphilis. Salvarsan, an arsenic derivative, was the first drug that successfully targeted a specific microorganism. As such, it opened the modern era of medical therapeutics. Doctors now could cure completely those who sought early treatment for syphilis, whereas people who put off treatment risked irreversible neurological damage and a horrible death.

The history of salvarsan and syphilis provided ideological support for a **medical model of mental illness.** This medical model is composed of four assumptions about the nature of mental illness. These are (Scheff, 1984):

1. Objectively measurable conditions define mental illness, in the same way that the presence of a specific bacterium defines syphilis.

2. Mental illness stems largely or solely from something within individual psychology or biology, even if researchers (like those who studied syphilis before 1905) have not yet identified its sources.

3. Mental illness, like syphilis, will worsen if left untreated, but may diminish or disappear if treated promptly by a medical authority.

4. Treating mental illness, like treating syphilis, rarely harms patients, and so it is safer to treat someone who might really be healthy than to refrain from treating someone who might really be ill.

The Sociological Model of Mental Illness

The sociological model of mental illness questions each of these assumptions (see Key Concepts 7.1). Perhaps most important, sociologists argue that definitions of mental illness, like the definitions of physical illness and disability discussed in Chapters 5 and 6, reflect subjective social judgments more than objective scientific measurements of biological problems.

What do we mean when we say someone is mentally ill? Why do we diagnose as mentally ill people as disparate as a teenager who uses drugs, a woman who hears voices, and a man who tries to kill himself? According to sociologist Allan Horwitz (1982), behavior becomes labeled mental illness when persons in positions of power consider that behavior both unacceptable and inherently incomprehensible. In contrast, we tend to define behavior as crime when we consider it unacceptable but comprehensible; we do not approve of theft, but we understand greed as a motive. (The judgment of not guilty by reason of insanity falls on the border between crime and mental illness.) Similarly, we might not understand why physicists do what

Key Concepts 7.1	*Models of Mental Illness*	
	THE MEDICAL MODEL	THE SOCIOLOGICAL MODEL
	Mental illness is defined by objectively measurable conditions.	Mental illness is defined through subjective social judgments.
	Mental illness stems largely or solely from something within individual psychology or biology.	Mental illness reflects a particular social setting as well as individual behavior or biology.
	Mental illness will worsen if left untreated but may improve or disappear if treated promptly by a medical authority.	Persons labeled mentally ill may experience improvement regardless of treatment, and treatment may not help.
	Medical treatment of mental illness can help but never harm.	Medical treatment for mental illness sometimes can harm patients.

they do, but we assume that those with appropriate training find their behavior comprehensible.

According to Peggy Thoits (1985), behavior leads to the label of mental illness when it contravenes **cognitive norms, performance norms,** or **feeling norms.** Someone who thinks he is Napoleon, for example, breaks cognitive norms (that is, norms regarding how a person should think), whereas someone who can't hold a job breaks norms regarding proper role performance. Thoits argues that the last category—breaking feeling norms—accounts for most behavior labeled mental illness. Feeling norms refer to socially defined expectations regarding the "range, intensity, and duration of feelings that are appropriate to given situations" and regarding how people should express those feelings (Thoits, 1985: 224). For example, laughing is highly inappropriate at a Methodist funeral but perfectly acceptable at an Irish wake, and feeling sad that your pet cat died is considered reasonable for a few days but unreasonable if it lasts for a year.

Different social groups consider different behaviors comprehensible and acceptable. The friends of a drug-using teenager, for example, might consider drug use a reasonable way to reduce stress or have fun. Their views, however, have little impact on public definitions of drug use. Similarly, members of one church might consider a woman who reports talking to Jesus a saint, whereas members of another church consider her mentally ill. The woman's fate will depend on how much power these opposing groups have over her life. The definition of mental illness, then, reflects not only socially accepted ideas regarding behavior but also the relative power of those who hold opposing ideas.

Researchers who use this sociological definition of mental illness do not mean to imply that emotional distress does not exist or that people do not feel real pain when they cannot meet social expectations for thought, behavior, or emotions. Nor do these researchers mean to imply that biology has no effect on behavior or thought. They do, however, question the purpose and consequences of using medical language to describe such problems and question why we label certain behaviors and individuals but not others.

Not all sociologists raise these questions, however. Many, especially those working in health care settings and in **epidemiology,** employ a **sociology *in* medicine** approach and use essentially medical definitions of mental illness in their research and writing. Nevertheless, sociologists are united in assuming that mental illness, like physical illness and disability, stems at least partially from social life rather than solely from individual psychology or biology. For example, beginning in the 1960s the number of young women diagnosed with eating disorders such as anorexia and bulimia skyrocketed (Brumberg, 1997). Those who use a medical model trace these disorders to biological defects such as endocrine or biochemical imbalances or to psychological factors such as poor adjustment to normal life changes, a need for personal perfection, poor relationships with parents, and adolescent identity crises (see, for example, T. Costello and J. Costello, 1992: 151–152). In contrast, those who use a sociological model of mental illness argue that eating disorders have mushroomed partly because of the increased cultural pressures on women to be slim (Brumberg, 1997). Thus, sociologists shift the focus from individual biology and psychology to the social context.

The Problem of Diagnosis

The sociological model of mental illness gains credibility when we look at research on the problems with psychiatric diagnosis. These problems became a political embarrassment for psychiatrists (medical doctors who specialize in treating mental illness) following a well-publicized experiment by psychologist David Rosenhan (1973). Rosenhan and seven of his assistants had presented themselves to twelve mental hospitals and complained of hearing voices, but otherwise had acted normally. The hospitals diagnosed all eight "pseudopatients" as mentally ill and admitted them for treatment. Once admitted, all behaved normally, leading 30 percent of the other patients to identify them as frauds. None of the staff, however, noticed anything unusual about these pseudopatients. It took an average of 19 days for them to win their release, with their symptoms declared "in remission."

When these results were published, psychiatrists objected vociferously that the results were some sort of fluke. In response, Rosenhan agreed to send pseudopatients to another hospital and challenged the staff at that hospital to identify the pseudopatients. During the three months of the experiment, the staff identified 42 percent of their new patients as pseudopatients, even though Rosenhan really had not sent any!

These two experiments vividly demonstrate the subjective nature of psychiatric diagnosis and its susceptibility to social expectations. Within the context of a mental hospital, staff members quite reasonably assume patients are ill and interpret everything patients do accordingly. When, for example, one bored pseudopatient began taking notes, a worker officially recorded this "note-taking behavior" as a symptom. Conversely, when staff members expected to find pseudopatients, they interpreted similar behaviors as signs of mental health.

The problems with diagnosis are particularly acute when therapist and patient do not share the same culture. With the rise in immigration to the United States over the last generation, doctors increasingly must diagnose and treat patients whose symptoms do not appear in Western textbooks (Goleman, 1995). For example, whereas Americans sometimes fear that their bodies will embarrass *them,* Japanese people sometimes experience disabling fears (known as "taijin kyofusho") that their bodies will embarrass *others.* Malaysian men may be stricken by "koro," the sudden and intense fear that their penises and testicles will recede into their bodies and kill them, and Latin Americans by "boufee delirante," characterized by sudden outbursts of excited, confused, violent, or agitated behavior. In response to growing concerns about cross-cultural misunderstandings, the American Psychiatric Association (APA) in 1995 adopted new guidelines that recommend psychiatrists consider cultural and ethnic factors in their work and require psychiatric training programs to cover cross-cultural issues.

The Politics of Diagnosis

To reduce the problems with diagnosis, psychiatrists over the years have attempted to refine the definitions of illnesses in the *Diagnostic and Statistical Manual of Mental Disorders (DSM).* Since the APA first published the *DSM* in 1952, virtually all psychiatrists have relied on this manual for assigning diagnoses to patients. So, too, do most other clinicians, because insurers usually require a *DSM* diagnosis before they will reimburse clinicians for treating a patient. *DSM* and the subsequent *DSM-II,* published in 1968, instructed clinicians to reach diagnoses based on the clinicians' inferences about such intrapsychic processes as defenses, repression, and transference. Because clinicians cannot measure these processes, the same behavior often elicited quite different diagnoses from different clinicians (Helzer et al., 1977).

Partly because of these problems, the APA in 1974 announced its decision to revise *DSM-II* (Spitzer, Williams, and Skodol, 1980). Ironically, although the resulting *DSM-III,* published in 1980, was designed to quiet questions about the ambiguities of psychiatric diagnosis, it instead illuminated those ambiguities because its writing became an overtly political battle, involving active lobbying by both professional and lay groups (Kirk, 1992). This battle revealed wide differences among clinicians regarding what behaviors signified mental illness, what caused those behaviors, who should treat them, and how they should be treated.

These differences already had surfaced during earlier and openly contentious battles regarding homosexuality (Conrad and Schneider, 1992).

DSM-I and *DSM-II* had listed homosexual behavior and desires as conclusive evidence of mental illness. By the early 1970s, however, gay rights activists had begun challenging this definition of the situation, arguing instead that homosexuality was a natural human variation. Active lobbying by gay activists and sympathetic professionals led the APA to hold a referendum in 1974, in which its members voted to drop homosexuality from *DSM-II*. This decision was based as much on political and moral considerations as on new scientific evidence.

The battle over the meaning of homosexuality began again with the writing of *DSM-III*. In the end a compromise was reached, declaring only "ego-dystonic" homosexuality a mental illness. Ego-dystonic homosexuality referred to individuals whose homosexuality caused them emotional pain and who had proved unsuccessful in changing their sexual orientations. This compromise did not end differences over treatment, for those who considered homosexuality merely an alternative sexual orientation treated ego-dystonic homosexuality by helping individuals become comfortable with their sexuality, whereas those who considered homosexuality pathological treated it by trying to change individuals' sexual orientation. DSM-IV, published in 1994, was the first edition that included neither the diagnosis of egodystonic homosexuality nor its symptoms under another name.

Debate over other diagnoses revealed equally divergent views on causation and treatment (see, for example, Scott, 1990). Clinicians trained in Freudian psychiatry (described later in this chapter) traced the roots of mental illness to unresolved childhood sexual conflicts and favored treating it with intensive psychoanalysis. Other clinicians traced mental illness to problematic interpersonal relationships, inappropriate social learning, or biological defects and favored treating it with, respectively, psychotherapy, behavioral conditioning, or drug therapies.

To encourage support for *DSM-III* and to avoid open political battles among psychiatrists, its authors decided to stress symptomatology and avoid discussing either causation or treatment (Kirk, 1992). In addition, to increase the odds that clinicians would use *DSM-III*, the authors described the various diagnoses based not on available research but, rather, on the consensus among practicing psychiatrists. These two strategies, they hoped, would produce a widely used and highly reliable document. **Reliability** refers to the likelihood that different people who use the same measure will reach the same conclusions—in this case, that different clinicians, seeing the same patient, would reach the same diagnosis. Yet even this modest goal was not achieved, for studies continue to find high rates of disagreement over diagnosis (Kirk, 1992; Mirowsky and Ross, 1989). Moreover, reliability in the absence of validity is not particularly useful. **Validity** refers to the likelihood that a given measure accurately reflects what those who use the measure believe it reflects—in this case, that persons identified by *DSM-III* as having a certain illness actually have that illness. As Phil Brown (1990: 393) notes, "anyone can achieve interrater reliability by teaching all people the 'wrong' material, and getting them to all agree on it.... The witch trials [of earlier centuries] showed a much higher degree of interrater reliability than any DSM category, yet we would not impute any validity to those social diagnoses."

Finally, even if the diagnostic categories used by clinicians are reliable and valid, clinicians will not necessarily apply them in an objective fashion. Research suggests that ethnicity and gender of both patient and clinician affect diagnosis. For example, Marti Loring and Brian Powell (1988) asked 290 randomly selected psychiatrists to diagnose two cases based on a brief description. Both cases had experienced hallucinations and extreme anxiety, had symptoms severe enough to damage their family lives, and had proved unable to keep a job. Both also met the *DSM-III* definition of undifferentiated schizophrenic disorder with a dependent personality disorder, a serious psychiatric illness with roots in childhood or adolescence.

The case descriptions the psychiatrists received were identical except for the descriptions of the cases' sex and ethnicity. When sex and ethnicity either were not given or matched those of the psychiatrist, the psychiatrists' diagnoses matched those of the researchers. In the other situations, however, bias seemed to affect the diagnoses. Male psychiatrists proved more likely to diagnose the female cases as having either depression or histrionic personality disorder, a diagnosis given to individuals with a long-standing tendency to express emotions intensely, act charmingly and seductively, feel helpless and therefore act dependent, and engage in romantic fantasies. Both depression and histrionic personality disorder fit stereotypical notions of female psychology and are diagnosed more often in women. In addition, white clinicians and, to a lesser extent, African American clinicians, more often diagnosed African Americans as paranoid schizophrenics. Paranoid schizophrenia is characterized by violence and is considered extremely difficult to treat, and so is considerably more serious than the researchers' diagnosis.

Only nineteen psychiatrists could not reach a diagnosis based on the information they had received. Of these, almost two-thirds (63 percent) had not received information about sex or ethnicity, further suggesting that psychiatrists base their diagnoses at least in part on social stereotypes of gender and ethnicity rather than on symptoms.

Despite all these problems, *DSM-III* and *DSM-IV* gained great support among clinicians because they served a variety of political needs (Horwitz, 2002). By stressing (even if inaccurately) the "objective" nature of diagnosis, clinicians were able to gain respect in the medical world, access to reimbursement from insurance companies, and funding from agencies that sponsor research. By assigning discrete diagnoses to all the different client groups and combinations of symptoms treated by different types of clinicians, they could gain widespread acceptance of the system from both clinicians and clients; *DSM-IV* contains almost 400 different diagnoses. Finally, a system that emphasized diagnosis and symptoms rather than underlying causes of illness both stemmed from and was reinforced by the increasing reliance on psychotropic medications as the main treatment for mental illness.

A History of Treatment

The history of treatment for mental illness further reveals the role social values play in medical responses to problematic behavior. In this section we trace the treatment of mental illness from the prescientific era to the present.

Before the Scientific Era

Although the concept of mental illness is relatively new, all societies throughout history have had individuals whose behavior set them apart as unacceptably and incomprehensibly different. However, premodern societies more often could find informal ways of coping with such individuals (Horwitz, 1982). First, premodern societies could offer acceptable, low-level roles to those whose thought patterns and behaviors differed from the norm. Second, because work roles rarely required individuals to function in highly structured and regimented ways, many troubled individuals could perform at marginally acceptable levels. Third, in premodern societies, work occurred within the context of the family, whether at home or in fields or forests. As a result, families could watch over those whose emotional or cognitive problems interfered with their abilities to care for themselves. These three factors enabled families to **normalize** mental illness by explaining away problematic behavior as mere eccentricity. As a result, unless individuals behaved violently or caused problems for civil authorities, their families and communities could deal with them informally.

In some cases, however, individuals behaved too unacceptably or incomprehensibly for their communities to normalize. In these cases, and as is true with all illnesses (as described in Chapter 5), communities needed to find explanations to help them understand why such problems struck some people and not others. Such explanations helped to make the world seem more predictable and safe by convincing the community that such bad things would never happen to "good people" like themselves.

Until the modern scientific age, societies typically viewed disturbing behavior as a punishment for sin or for violating a taboo; a sign that the afflicted individual was a witch; or a result of evildoing by devils, spirits, or witches. Therefore they assigned treatment to religious authorities—whether shamans, witch doctors, or priests—who relied on prayer, exorcism, spells, and treatments such as bloodletting or trepanning (drilling a hole in the skull to let "bad spirits" out). Religious control of socially disturbing behavior reached a spectacular climax with the witchcraft trials of the fifteenth to seventeenth centuries, during which religious authorities brutally killed at least 100,000 people, including some we would now label mentally ill (Barstow, 1994).

As a capitalist economy began to develop, both religious control and informal **social control** began to decline (Horwitz, 1982; Scull, 1977). Under capitalism, work moved from home and farm to workshops and factories, making it more difficult for families to care informally for problematic relatives. In addition, a capitalist economy could less readily absorb those whose productivity could not be scheduled and regimented. At the same time, widespread migration from the countryside to cities weakened families and other social support systems, as did migration from Europe to the United States in subsequent centuries. Meanwhile, other changes in society weakened religious systems of social control.

These changes fostered a need for new, formal institutions to address mental illness. By the end of the eighteenth century, however, only a few hospitals devoted to treating the mentally ill existed, along with a few private "madhouses" run by doctors for profit. Instead, most of those we would now label mentally ill were housed with the poor, the disabled, and the criminal in the newly opened network of public **almshouses,** or poorhouses.

Conditions in both almshouses and madhouses were generally miserable, but they were especially bad for those considered mentally ill. Doctors and the public typically considered that persons with mental illness were incurable and essentially animals. As a result, institutions treated the mentally ill like animals—chaining them for years to basement walls or cells, often without clothing or proper food, and beating them if they caused problems.

The Rise and Decline of Moral Treatment

By the late eighteenth century, however, attitudes toward persons with mental illness began to moderate (Scull, 1989: 96–117). In place of punishment and warehousing, reformers proposed **moral treatment:** teaching individuals to live in society by showing them kindness, giving them opportunities to work and play, and in general treating mental illness more as a moral than a medical issue. The stunning successes that resulted convinced the public that mental illness was curable. The first American hospital designed to provide moral treatment, the Friends' (or Quakers') Asylum, was founded in 1817.

Despite this strong beginning, moral treatment in the end could not compete with medical models of mental illness (Scull, 1989: 137–161). Because those who promoted moral treatment continued to use the language of medicine, talking of illnesses and cures, medical doctors could argue successfully that only they should control this field. In addition, because moral treatment required only kindness and sensitivity, which theoretically any professionals could offer, no professional group could claim greater expertise than that of doctors. As a result, by 1840, doctors largely had gained control over the field of mental illness both in the United States and Europe.

As care gradually shifted from laypersons to doctors, custodial care began to replace moral treatment. This shift reflected that communities

Benjamin Rush, the "Father of American psychiatry," invented this device to treat mental illness through removing distractions from the patient.

were more interested in controlling problematic individuals than in treatment, especially when those individuals were poor, nonwhite, or immigrants. It also reflected the growing belief that illness was genetic and untreatable.

By the 1870s, moral treatment had been abandoned. Yet the number of mental hospitals continued to grow exponentially (D. Rothman, 1971). Historians refer to this change, and the similar but earlier developments in Europe, as the **Great Confinement.**

The rise of institutions reflected the need to respond to public **deviance.** The Great Confinement drew energy from the well-meaning efforts of reformers—most notably, Dorothea Dix—to close down the brutal and anarchic almshouses and to provide facilities specifically designed to care for the mentally ill, instead of warehousing them with criminals, disabled persons, and the poor (Sutton, 1991). Because no agreed-upon definitions of mental illness existed, however, families and communities found it relatively easy to move troublesome relatives into the newly established mental hospitals. Indeed, a substantial proportion of those found in these new hospitals suffered primarily from old age and poverty coupled with a lack of relatives who could or would care for them (Sutton, 1991). So, except for those wealthy enough to obtain care in small, private mental hospitals, most of those labeled mentally ill continued to find themselves housed with others whom society had rejected. The only difference was that instead of residing in institutions filled with a varied group of deviants, they now lived in large institutions officially devoted to the "care" of the mentally ill.

Freud and Psychoanalysis

By the beginning of the twentieth century, then, doctors controlled the mental illness field. Yet medicine was torn by internal divisions. From the nineteenth century to the present, although doctors overwhelmingly traced mental illness to sources internal to individuals, some emphasized the emotional roots of mental illness while others emphasized physical causes.

This split grew wider with the rise of Freudian psychiatry. According to Sigmund Freud, a Viennese doctor, to become a mentally healthy adult one had to respond successfully to a series of early childhood developmental issues. Each issue occurred at a specific stage, with each stage linked to biological changes in the body and invested with sexual meanings. For example, during the oral stage, infants and toddlers derived their greatest satisfaction from sucking a breast or bottle. Those who did not learn how to signal and fulfill those needs, Freud concluded, would later develop traits such as dependency and narcissism.

The phallic stage (between about ages 3 and 6) plays an especially important role in Freud's model because that is when the **superego**—that portion of the personality that has internalized social ideas about right and wrong—is hypothesized to develop. During the phallic stage, according to Freud, children start noticing and responding to their genitalia. They begin experiencing sexual attraction toward the opposite-sex parent and viewing their same-sex parent as a rival. When boys first learn that girls do not have penises, however, they naturally (according to Freud) conclude that girls have been castrated by their fathers as punishment for some wrongdoing. Fearing the same fate, boys abandon their attraction to their similarly castrated mothers and identify with their fathers, whose love they try to obtain by adopting their fathers' values. Through this process, boys develop a strong superego.

But what of girls, who lack penises? According to Freud, once they realize they lack penises, girls immediately recognize their inferiority (1925 [1971] 241–260:). They descend into jealousy and narcissism, which they can relieve only partially and only by marrying and having baby boys who vicariously give them penises of their own. Thus, girls can never develop strong superegos because they lack the fear necessary for their development.

Freud based this theory on his interpretations of the lives and dreams of his upper-middle-class patients; no scientific data underpin this theory. Looking back at this theory from the present, it is hard to comprehend how anyone could have believed in such notions as three-year-olds lusting after their parents or girls naturally feeling jealous of boys' penises (rather than feeling jealous of the social power maleness confers). Yet Freudianism's long-standing popularity should not surprise us. Freudianism both reflected and supported contemporary cultural notions holding that men's anatomy, intellect, and moral capabilities naturally surpassed women's, that women lacked the necessary maturity and selflessness to hold positions of authority in society, and that women were destined to become wives and mothers. These notions have not been totally abandoned; although no longer widely used in its pure form and rarely used by modern psychiatrists, Freud's conception of human nature and of mental illness continues to permeate American culture and vocabulary and to affect ideas about both normal and abnormal psychology.

For those who accepted Freud's theory, the only way to cure mental illness was to help patients resolve their developmental crises. To do so, Freud and his followers relied on psychoanalysis, a time-consuming and expensive form of psychotherapy geared to patients without major mental illnesses. In psychoanalysis, patients recounted their dreams and told a largely silent therapist whatever came to mind for the purpose of recovering hidden early memories and understanding their unconscious motivations.

Because psychoanalysis was so costly, most mental patients during the first half of the 1900s instead received far cheaper physical interventions (Valenstein, 1986). Insulin therapy became immediately popular from its inception in 1933, followed by electroconvulsive (shock) therapy in 1938. These therapies caused comas or seizures, which psychiatrists believed improved mental functioning. Neither therapy had received scientific testing before becoming popular, nor did later studies find evidence of their effectiveness. Similarly, lobotomies—operations that permanently destroy part of the brain—became popular during the 1940s and 1950s. An estimated 50,000 Americans received lobotomies, and the procedure's originator, Dr. Egas Moniz, received the Nobel Prize in Medicine in 1949. Yet the only proven effects of lobotomies are diminished memory, intelligence, creativity, and emotional capacity (Valenstein, 1986). At any rate, therapy of any sort occupied only a minuscule proportion of patients' time in mental hospitals. Instead, patients spent their days locked in crowded wards with little other than radio or, later, television to ease their boredom.

The Antipsychiatry Critique

By the middle of the twentieth century, mental hospitals had become a huge and largely unsuccessful system (Mechanic, 1989). Patients with mental illnesses occupied half of all hospital beds in the United States. Virtually all (98 percent) were kept in public mental hospitals; insurance rarely covered mental health care, so private hospitals had no interest in the field. At their peak in 1955, public mental hospitals held 558,000 patients, most of them involuntarily confined, for an average of eight years.

Beginning in the 1960s, many voices would challenge this system. Civil rights, antiwar, and feminist movements all brought issues of individual rights to the forefront and stimulated a broader questioning of authority and social arrangements. These ideas contributed to a growing critique of mental health treatment by sociologists, psychologists, and even some psychiatrists such as R. D. Laing (1967) and Thomas Szasz (1970, 1974).

One of the most powerful critiques of large mental institutions appeared in a classic study by sociologist Erving Goffman (1961). Goffman's work fell within the tradition of **symbolic interactionism** theory. According to this theory, individual identity develops through an ongoing process in which individuals see themselves through the eyes of others and learn through social interactions to adopt the values of their community and to measure themselves against those values. In this way, a **self-fulfilling prophecy** is created, through which individuals become what they are already believed to be. So, for example, children who constantly hear that they are too stupid to succeed in school might conclude that it is senseless to attend classes or study. They then fail in school, thus fulfilling the prophecies about them.

Goffman used symbolic interactionism theory to analyze mental hospitals and the experiences of mental patients. He pointed out that mental hospitals, like the military, prisons, and monasteries, were **total institutions**—institutions where a large number of individuals lead highly regimented lives segregated from the outside world. Goffman argued that these institutions necessarily produced **mortification** of the self. Mortification refers to a process through which a person's self-image is damaged and is replaced by a personality adapted to institutional life.

Several aspects of institutional life foster mortification. Persons confined to mental hospitals lose the supports that usually give people a sense of self. Cut off from work and family, these individuals' only available role is that of patient. That role, meanwhile, is a **master status**—a status considered so central that it overwhelms all other aspects of individual identity. Within the mental hospital, a patient is viewed solely as a patient—not as a mother or father, husband or wife, worker or student, radical or conservative. According to Goffman's observations, and as in Rosenhan's (1973) experiment, all behavior becomes interpreted through the lens of illness. In addition, because each staff member must manage many patients, staff members necessarily deal with patients en masse. In these circumstances, patients typically lose the right to choose what to wear, when to awaken or sleep, when and what to eat, and so on. Moreover, all these activities occur in the company of many others. Individuals thus not only experience a sense of powerlessness but also can lose a sense of their identity—their desires, needs, personalities—in the mass of others. As a result, patients experience **depersonalization**—a feeling that they no longer are fully human, or no longer are considered fully human by others. At the same time, the hierarchical nature of mental hospitals reinforces the distinctions between inmate and staff and constantly reminds both parties of the gulf between them. Consequently, patients can avoid punishment and eventually win release only by stifling their individuality and accepting the institution's beliefs and rules.

| Table 7.1 | Average Daily Census of Adult Mental Patients, by Type of Organization, 1969–1988 | | | | | |

TYPE OF FACILITY	1969	1975	1979	1983	1986	1988
Inpatient, public mental hospitals	414,800	225,500	167,300	136,500	128,200	119,400
Inpatient, private psychiatric hospitals or wards	29,400	34,800	37,000	50,800	57,800	65,600
Outpatient facilities	6,240	10,989	11,026	20,970	19,670	19,673

Source: Manderscheid and Sonnenschein (1992: 26).

These forces producing mortification are so strong that even Rosenhan's pseudopatients—knowing themselves sane and hospitalized only briefly—experienced depersonalization.

Implicit in Goffman's work is the idea that mental hospitals may be one of the worst environments for treating mental problems. Later research supports this conclusion. A review of ten **controlled** studies on alternatives to hospitalization, including halfway houses, day care, and supervised group apartment living, found that all could boast equal or better results than those of traditional hospitalization, as measured by subsequent employment, reintegration into the community, life satisfaction, and extent of symptomatology (Kiesler and Sibulkin, 1987).

Deinstitutionalization

By the time the anti-psychiatry critique appeared, the Great Confinement already had begun to wane. Beginning in 1955, the number of mental hospital inmates declined steadily, as treatment shifted from **inpatient** care (in hospitals) to **outpatient** care (see Table 7.1). This process of moving mental health care away from large institutions, known as **deinstitutionalization,** gained further support during the 1970s, as mental patients successfully fought in the courts against involuntary treatment, against hospitals that provided custodial care rather than therapy, and for the right to treatment in the "least restrictive setting" appropriate for their care.

Explaining Deinstitutionalization

Those who adopt a medical model of illness typically assume that deinstitutionalization resulted from the introduction, beginning in 1954, of drugs known as phenothiazines. These drugs, such as chlorpromazine (Thorazine), significantly reduce severe symptoms such as hallucinations in many patients. To these drugs would later be added antidepressants and antianxiety drugs such as diazepam (Valium). Yet the number of patients in public mental hospitals did not fall rapidly until more than a decade after these drugs were introduced.

Although phenothiazines did facilitate deinstitutionalization by making mental patients compliant enough for communities to tolerate their release, financial changes more fully explain this shift (Mechanic and Rochefort, 1990). Increasingly during the 1960s and 1970s, private insurers covered the costs of mental health care, making the treatment of mental illness profitable for private hospitals. As a result, these hospitals began aggressively developing psychiatric facilities and admitting patients who in the past would have gone to large public hospitals (Mechanic, 1999; Mechanic and Rochefort, 1990). Table 7.1 shows the growth in private facilities for treating mental illness. General hospitals also sought psychiatric patients as a means of filling beds emptied during the 1950s and 1960s by the overbuilding of general hospitals and during the 1980s by pressures from insurers to control costs by releasing patients quickly (P. Brown, 1985: 116–117; Gray, 1991).

Changes in public benefit programs played an even more important role in fostering deinstitutionalization. With the establishment in 1965 of the federal health care programs **Medicare** and **Medicaid, nursing homes** realized they could now receive federal funds for caring for chronically mentally ill persons and began aggressively seeking this market. States happily supported this shift, because public mental hospitals were largely funded through state tax dollars but Medicare and Medicaid were largely paid for by the federal government. During the same years, Social Security increased the monthly benefits it paid to persons with chronic mental illness, making it possible for mental hospitals to release patients who previously would have been unable to support themselves. Three-quarters of the reduction in the total number of mental hospital patients occurred after these changes in Medicare, Medicaid, and Social Security, suggesting that these changes were the most important factor behind deinstitutionalization.

Finally, deinstitutionalization also stemmed from the rise of **individualism**—a set of "sociocultural beliefs and practices that encourage and legitimate the autonomy, equality and dignity of individuals" (Horwitz and Mullis, 1998: 122). In past generations, individuals' identities depended on their places within family or community. Because families and communities were far more important social units than were individuals, laws typically upheld the right of these groups over any rights of the individual. Thus, for example, until about 1900, parents had near-absolute rights to discipline their children without interference from the law. Similarly, most psychiatric inpatients were committed by their families, and most requests by families to commit individuals were honored (Horwitz and Mullis, 1998).

During the last few decades, however, this "moral sovereignty" of the family has weakened; families are no longer assumed to know what is best for their members, and family ties of all sorts have weakened. In its stead, individualism has become dominant. Although families still are the most common source of requests for commitment, they now must demonstrate that commitment is in the best interest of the individual. Similarly, mental hospitals now must demonstrate that an individual needs continued treatment rather than the individual having to demonstrate that he or she does not.

Perhaps more important, as family ties have weakened, increasingly families simply abandon their more problematic members, rather than either caring for them or arranging for them to be cared for by others. At the same time, now that laws increasingly protect the right of individuals to dress and behave in unusual ways, communities no longer police unusual public behavior so closely. For both these reasons, the rise of individualism has resulted in fewer commitments to mental hospitals.

The Consequences of Deinstitutionalization

Following deinstitutionalization, persons with mental illness no longer found themselves locked for years in the often brutal conditions of large mental institutions. Yet the promise that deinstitutionalization would herald a new era in which individuals would receive appropriate therapy in the community, avoiding the **stigma,** degradation, and mortification of mental hospitalization, has been met only partially. Unfortunately, individuals who were released from hospitals to the community found few services available to help them with their continuing problems. The situation worsened further beginning in the 1980s, when the federal government began cutting funding for Medicaid and Medicare, the federal health care programs that are supposed to help disabled and poor Americans. As a result, many chronically mentally ill persons could no longer afford treatment. According to the NCS-R, less than half (40 percent) of those with serious mental illnesses currently receive even minimally adequate treatment (R. Kessler et al., 2005b).

During the same years that government funding for the nation's health care system declined, funding for the criminal justice system dramatically *increased* (Butterfield, 1999). As a result, public mental hospitals now find that the best way to pay their bills is to accept for treatment persons sent to them by the criminal justice system: mentally ill prison inmates, people found innocent by reason of insanity, and violent offenders who under new "sex predator" laws can be involuntarily confined even after finishing their prison sentences. For example, at California's Napa State Hospital, almost 75 percent of patients during 1999 came from the criminal justice system (Kligman, 1999).

Simultaneously with these changes, the federal government also reduced funding for low-income housing. As a result, many mentally ill persons who cannot afford treatment also cannot find housing. Consequently, many persons with chronic mental illness now cycle between homelessness, brief jail stays when they prove too troublesome for local authorities who lack other alternatives, and acute episodes in public mental hospitals; a report released by the U.S. Department of Justice in 1999 estimated that 16 percent of jail and prison inmates have a mental illness (Butterfield, 1999). Despite these severe gaps in our mental health system, however, observers generally agree that deinstitutionalization improved the quality of life for most seriously mentally ill persons, whether they live in nursing homes, board and care homes (residential facilities that provide solely assistance in daily living), with relatives, or on their own (Grob, 1997; Horwitz, 1999).

The Remedicalization of Mental Illness

The last 20 years have seen an increasing **remedicalization** of mental illness (P. Brown, 1990). Psychiatrists have developed new techniques for diagnosis and treatment and new theories of illness etiology that link mental illness to individual abnormalities in biochemistry, neuroendocrine functioning, brain structure, or genetic structure and downplay the effects of social factors.

The data for this "biological revolution" consist primarily of simple correlations between biological abnormalities and some serious mental disorders (P. Brown, 1990); no studies have uncovered significant biological differences between those who have minor mental disorders and those who do not. None of this research adequately sorts out other factors that might account for these correlations (such as differences in nutrition or in the use of various drugs) or determines whether either the mental disorders or treatment for them might have caused, rather than resulted from, biological abnormalities.

Despite these weaknesses in the biological model of mental illness, most psychiatrists have adopted it. As a result, psychiatrists now present a more united front in their struggles for control against other mental health occupations such as psychology and social work. In addition, they have increased their political power relative to these other occupations because, having declared mental illness a biological problem, they now can argue that only persons trained in medicine can properly diagnose and treat it (P. Brown, 1990).

Reflecting this medical model, doctors now rely primarily on psychoactive drugs not only to treat mental illness but also to diagnose it. In a process first brought to public attention by psychiatrist Peter Kramer (1993) in his popular book, *Listening to Prozac,* doctors now "listen to drugs," assuming that the reaction to a drug tells us something basic about an individual's mental state. So if Prozac (fluoxetine hydrochloride) or another selective serotonin reuptake inhibitor (SSRI), which increases levels of the neurotransmitter serotonin in the brain, somehow makes an individual feel less depressed, then physicians conclude that lack of serotonin must have caused the depression. Yet as Kramer points out, pneumonia is not caused by a lack of antibiotics nor headaches by a lack of aspirin, but both drugs make ill people feel better. Similarly, doctors increasingly decide whether a patient is clinically depressed based not on whether that patient meets standard criteria for that diagnosis but on whether the patient responds favorably to SSRIs. Yet most people feel better when they take a mood-enhancing drug, whether it is Prozac or cocaine. As a result, during 2001 Americans spent more than $11 billion on SSRIs (Sills, 2002).

Most of the drugs now used to treat mental illness fall into one of three main categories: antipsychotics, mood stabilizers, and antidepressants. Psychiatrists use antipsychotic drugs, such as Clozaril and Risperdal, to help control severe symptoms in persons with major mental illnesses such as schizophrenia. These drugs are considerably less likely than are older drugs such as Haldol and Thorazine to produce loss of alertness and a condition known as "tardive dyskinesia" (uncontrollable, severe, and sometimes permanent muscular spasms). To control anxiety, obsessions, compulsions, and the severe mood swings of bipolar disorder, doctors commonly use mood stabilizers such as Tegretol and Depakote. Finally, psychiatrists use antidepressants to alleviate depression. Unlike previous generations of antidepressants, SSRIs and other new drugs have fewer side effects and cannot be taken to commit suicide, although they are no more effective than the older drugs and increase the odds that individuals will in fact commit suicide. Because drug companies proved successful at both marketing the benefits of SSRIs and downplaying their problems, the use of SSRIs has exploded (Abramson, 2004), primarily among persons who suffer only from minor depression.

The Rise of Managed Care

Beginning in the 1990s and in response to consumer pressure, insurance coverage for mental illness became considerably more common. Still, most insurers offer less coverage for mental illness (especially chronic illness) than for physical illness (R. Frank and McGuire, 1998; Mechanic, 1999: 128–132). Increasingly, too, that coverage is offered through **managed care organizations (MCOs)**. Managed care is described more fully in Chapter 8, but essentially refers to any system that controls health care spending by closely monitoring where patients receive health care, what sorts of providers patients use, what treatments they receive, and with what consequences.

It is too soon to fully assess the impact of managed care on either the cost or quality of care. However, early research suggests that managed care may be able to reduce the costs of mental health treatment, at least for less severe illnesses, by encouraging shorter rather than longer inpatient stays, outpatient rather than inpatient care, conservative rather than aggressive interventions, and use of lower-level clinicians (such as social workers) rather than psychologists or psychiatrists (Mechanic, 1995; Mechanic, 1999: 160–162). According to David Mechanic, probably the most influential sociologist in the area of mental health care, it also may be able to improve the quality of care:

> By reducing inpatient admissions and length of stay, managed care programs potentially make available considerable resources for substitute services and other types of care. Managed care provides incentives to seek closer integration

between inpatient and outpatient and primary and specialized services to achieve cost-effective substitutions.

Managed care also offers the potential to bring . . . science-based mental health care into the mental health system more quickly than traditional programs. . . . Many individual practitioners resist practice guidelines and scientific findings, preferring their own clinical experience, but managed care can put systems in place to measure performance and to enforce adherence to established standards. (1997: 45–46)

But managed care also carries risks. The emphasis on cost containment inherent in managed care has affected who offers mental health services, for how long, and of what type (Scheid, 2001). MCOs encourage the use of clinicians who charge less per hour, preferring those with master's degrees to those with doctorates and preferring those with doctorates to those with medical degrees. To further restrain costs, MCOs press clinicians to restrict care to short-term treatment of immediate problems, rather than longer-term treatment of underlying problems. As a result, therapists increasingly prescribe medications, even if they believe "talking therapies" would be more useful. This shift probably makes sense for most patients who are dealing with mild "problems in living" but is problematic for those with more severe mental problems (Luhrmann, 2000).

Managed care also has affected how mental disorders are diagnosed. One way managed care controls costs is by determining in advance, based on outcome studies of past patients, how much and what type of care patients with specific diagnoses should receive. For this system to work, clinicians must assign a diagnosis to each patient. This in turn reinforces the medical model of mental illness and the idea that every person who seeks mental health services has a specific, diagnosable mental illness.

At the same time, to contain costs, MCOs are trying to curtail the breadth of the diagnostic system (Horwitz, 2002). Because each successive edition of *DSM* has included more diagnoses than its predecessor has, with each edition more individuals have become eligible for mental health care. For this reason, MCOs often oppose new diagnoses or any loosening of the criteria for existing diagnoses. For example, some MCOs deny treatment to individuals who have fewer than five symptoms on a depression checklist, even if individuals' listed symptoms are severe and even if they have other, unlisted symptoms. Box 7.2 describes the National Alliance on Mental Illness, which, among other things, fights for better access to care.

For all these reasons, it remains unclear whether the benefits of managed care will outweigh the disadvantages.

The Experience of Mental Illness

The previous sections described the nature, causes, distribution, and history of mental illness. Next, we look at the experience of mental illness.

Box 7.2   *Making a Difference: The National Alliance for the Mentally Ill*

The National Alliance for the Mentally Ill (NAMI) is a nonprofit, national organization that aids individuals with severe mental illnesses, their friends, and their families. NAMI has several primary missions.

First, NAMI works to increase insurance coverage for mental illness and access to the best treatments. To this end, NAMI's volunteers and staff engage in political advocacy, using NAMI-funded research reports that document both the social costs of untreated mental illness and the ways communities suffer when mental illness is not adequately treated. Second, NAMI works with communities to develop appropriate housing options for persons with severe mental illness and works with employers to develop appropriate jobs. Third, to gain public support for better treatment, housing, and job opportunities, NAMI focuses on fighting the stigma of severe mental illness.

In its fight against stigma, NAMI has used a variety of tactics. For example, each month, NAMI members nationwide are asked to report instances in which national media (television, radio, Internet, etc.) portrayed persons with mental illnesses accurately or inaccurately, demeaningly or sympathetically. These reports are then sent along with the names and addresses of those responsible for these portrayals to the almost 20,000 NAMI members who have volunteered to participate in its StigmaBusters E-mail Alert. Participants are asked to send letters of complaint or commendation, as appropriate, to the responsible parties. State and local NAMI chapters have similar structures to deal with their local media. The flood of email and letters generated by these alerts has helped to reduce ignorance and prejudice and foster more accurate images of mental illness in the mass media.

Becoming a Mental Patient

As already noted, in any given year 31 percent of working-age adults experience a diagnosable mental illness, but only 40 percent of these receive even basic treatment (R. Kessler et al., 2005a, 2005b). Ironically, as the stigma among the middle class against seeking counseling for minor problems has diminished and insurance has increased, levels of treatment have increased among basically well-functioning individuals who experience situational stress, sadness, or lowered self-esteem (R. Kessler et al., 2005b). Nearly half of those who receive outpatient treatment have no mental disorder that can be identified through surveys, although some of these might have disorders that could be identified by clinicians (R. Kessler et al., 2005b). What explains this discrepancy between experiencing symptoms and receiving treatment?

According to Allan Horwitz, "Symptoms of mental disorder are usually vague, ambiguous, and open to a number of varying interpretations. . . . Labels of 'mental illness,' 'madness,' or 'psychological disturbance' are applied only after alternative interpretations have failed to make sense of the behavior" (1982: 31). The key question, then, is how does this happen?

Self-Labeling

Regardless of how others define their situation, at least initially individuals usually define themselves as mentally healthy, using a process Whitt and Meile (1985) refer to as **aligning actions,** or actions taken to align one's behavior with social expectations. If individuals' problems increase, however, these aligning actions become less convincing. In a process Whitt and Meile refer to as **snowballing,** each additional problem becomes more difficult to deal with than the previous one, so a person with four problems experiences more than twice the difficulty of a person with two problems. As this snowballing occurs, individuals become more likely to define themselves as mentally ill and to seek care.

Peggy Thoits (1985) has provided a more detailed model of how self-labeling works among those—the majority—who experience only acute or mild problems. Her model, like that of Erving Goffman, draws on the theory of symbolic interactionism. Thoits applies this to mental illness by hypothesizing that well-socialized individuals sometimes label themselves as mentally ill when their behavior departs from social expectations, even if others do not consider their behavior disturbed or disturbing.

Because individuals recognize the stigma attached to mental illness, however, they work to avoid this label. According to Thoits, and as described earlier, most of the behavior that can lead to the label of mental illness involves inappropriate feelings or expressions of feelings. To avoid the label of mental illness, therefore, individuals can attempt to make their emotions match social expectations, through what Arlie Hochschild (1983) refers to as **feeling work.**

Feeling work can take four forms. First, individuals can change or reinterpret the situation that is causing them to have feelings others consider inappropriate. For example, a working woman distracted from her work by worries about how to care for an ill parent—and distracted while with her parent by worries about her work—can quit her job. Second, individuals can change their emotions physiologically, through drugs, meditation, biofeedback, or other methods. The woman with the ill parent, for example, could drink alcohol or take Prozac to control anxiety. Third, individuals can change their behavior, acting as if they feel more appropriate emotions than they really do. Fourth, individuals can reinterpret their feelings, telling themselves, for example, that they only feel tired rather than anxious.

When feeling work succeeds, individuals can avoid labeling themselves mentally ill. This is most likely to happen when the situations causing the emotions are temporary and brief and when supportive others legitimize their emotions. If, for example, the woman with the ill parent has similarly situated friends who describe similar emotions, she might conclude that her emotions are understandable and acceptable. If, on the other hand, her colleagues do not sympathize with her concerns and continually tell her to put her work first, her attempts at feeling work could fail, and she might conclude that she has a mental problem.

Ironically, some individuals label themselves mentally ill or are labeled by others because they succeed too well at feeling work. For example, those who rely too heavily on drugs to manage their feelings can lose control of their lives, and those who consistently reinterpret their emotions—telling themselves that they are not angry, for example, even while punching a wall or a spouse—can find that others label them crazy when their emotions and behavior don't match. In addition, those who consistently engage in feeling work can lose the ability to interpret their feelings accurately and experience them fully. The resulting sense of numbness and alienation eventually can lead individuals to define themselves as mentally ill.

Labeling by Family, Friends, and the Public

Like individuals, families only reluctantly label their members mentally ill (Horwitz, 1982). Instead, families can deny that a problem exists by convincing themselves that their relative's behavior does not depart greatly from the norm. If they do recognize that a problem exists, they can convince themselves that their relative is lazy, a drunkard, "nervous," responding normally to stress, or experiencing physical problems rather than mental illness. Finally, families might recognize that their relative is experiencing mental problems but define those problems as temporary or unimportant.

Two factors explain how and why families can ignore for so long behavior that others would label mental illness. First, those who share cultural values, close personal relationships, and similar behavior patterns have a context for interpreting unusual behavior and therefore can interpret behavior as meaningful more easily than outsiders could. Second, families often hesitate to label one of their own for fear others can reject or devalue both the individual and the family. As a result, families have a strong motive to develop alternative and less stigmatizing explanations for problematic behavior.

Surprisingly, strangers as well as intimates tend to avoid interpreting behavior as mental illness. In one study, for example, researchers had subjects read vignettes describing individuals who met the criteria for various psychiatric diagnoses (D'Arcy and Brockman, 1976). The researchers found that the proportion of subjects who defined the described individuals as mentally ill declined from 70 percent for the vignettes of paranoid schizophrenics to 34 percent for the vignettes of simple schizophrenics, 25 percent for the vignettes of alcoholics, and less than 10 percent for the vignettes of neurotics (that is, persons who experience psychological distress but are in touch with reality and able to function). This evidence suggests that the public applies the label of mental illness only when disordered behavior is public, violent, dramatic, or otherwise unignorable.

Moreover, even when relatives and other intimates define an individual as mentally ill, they do not necessarily bring the individual to treatment. Instead, they can continue to protect the individual against social sanctions through a process Lynch (1983) refers to as **accommodation.** Accommodation refers to "interactional techniques that people use to manage persons they view as persistent sources of trouble" and to avoid conflict (Lynch, 1983: 152).

Based on analyzing essays in which college students described how they handled family members, workmates, fraternity brothers, and others whom they regarded as disturbed, Lynch identified three forms of accommodation. First, students could minimize contact with problematic individuals—avoiding them, ignoring them when they could not be avoided, or restricting interactions to a minimal and superficial level when they could not be ignored. Second, students could limit the trouble individuals could cause through such actions as taking over the individuals' responsibilities or humoring their wishes and beliefs. Third, they could manage the *reactions* to the problematic individual through such actions as providing excuses when the individual did not meet social expectations or hiding the individual from others' view—for example, keeping a "crazy" fraternity brother out of sight when outsiders were present during parties.

Nevertheless, despite these attempts to normalize and accommodate mental illness, families and friends may eventually conclude that an individual needs treatment. At that point, they must either get the individual to agree or coerce the individual into getting treatment despite his or her active resistance. One study of all persons seeking care for a serious mental illness for the first time found that 42 percent had actively sought care and 23 percent had been coerced (Pescosolido, Gardner, and Lubell, 1998). Coercion was most common among those with bipolar disorder, who often enjoyed the "highs" of mania even though others regarded them as seriously disturbed, and among those with large, tight social networks. In another 31 percent of cases, families "muddled through"; either the individuals went along with treatment decisions made by others without accepting or rejecting those decisions, or no one in the family seemed to have been in charge of the decision-making process.

Labeling by the Psychiatric Establishment

Once individuals enter treatment, a different set of rules applies, for whereas the public tends to normalize behavior, mental health professionals tend to assume illness. First, because the medical model of mental illness stresses that treatment usually helps and rarely harms, it encourages mental health workers to define mental illness broadly. Second, because mental health workers see prospective patients outside of any social context, behavior that might seem reasonable in context often seems incomprehensible. This is especially likely when mental health workers and prospective patients come from different social worlds, whether because they differ in gender, ethnicity, social class, or some other factor. Third, mental health workers assume that individuals would not have been brought to their attention if they did not need care. Finally, because normalization and accommodation are so common, mental health workers often do not see individuals until the situation has reached a crisis, making it relatively easy to conclude that the individuals are mentally ill.

The Post-Patient Experience

Research on the post-patient experience has focused on the sources, conse-quences, and extent of stigma experienced by former patients. This is a crit-ical issue, for it challenges the medical model's assumption that psychiatric treatment is benign.

Those who support a medical model of illness point to several studies suggesting that the public stigmatizes only those former patients who continue to engage in problematic behavior (Link et al., 1987). Yet nation-ally representative surveys continue to find that persons with mental ill-nesses evoke substantial fear and social rejection from others (Link et al., 1999).

To explain why some studies find high rates of stigma toward former mental patients and others do not, Bruce Link and his colleagues (1987) asked a random sample of survey respondents to fill out questionnaires regarding their attitudes toward persons with mental illness and to respond to a descrip-tion of a person whose behavior met the definition of mental illness. None of the respondents was told that the person was mentally ill, but half were told that he was a former mental patient. Respondents who believed mentally ill persons are dangerous proved *more* likely to reject a person who was described as a former mental patient, whereas those who believe persons with mental illness are generally harmless proved *less* likely to reject the former patient. The authors conclude that previous studies found no evidence of stigma because they unintentionally had combined these two groups.

In two further studies, Link and his colleagues argued that labeling an individual mentally ill has negative effects not only because of how the general public responds but also because of how the labeled individual responds (Link, 1987; Link et al., 1989). According to these studies, former patients believe that most people devalue and reject former mental patients. As a result, former patients devalue themselves, which damages their self-esteem and their work performance. In addition, because former patients expect rejection, they often engage in defensive behaviors such as secrecy and emotional withdrawal, which further harms their social relationships.

These findings, of course, do not necessarily mean that the hazards of stigma outweigh the benefits of treatment. Substantial evidence suggests that both psychotherapy and drug treatment can reduce symptoms and prevent relapse, at least in the short term (Link et al., 1997). Other research, however, suggests that the negative effects of stigma coexist with the benefits of treat-ment, partially canceling each other out (Link et al., 1997; Rosenfield, 1997). These results led Bruce Link and his colleagues to conclude that

> stigma has important effects, effects that remain even when people improve while participating in treatment programs. Health care providers are therefore faced with the challenge of how to address stigma in its own right if they want to maximize the quality of life for those they treat and maintain the benefits of treatment beyond the short term. (1997: 187)

Box 7.3　　　*Ethical Debate: Confidentiality and the Duty to Warn*

In the fall of 1969, Prosenjit Poddar entered outpatient psychotherapy at the University of California–Berkeley Student Health Center. During the course of therapy, he told his therapist, Dr. Lawrence Moore, that he planned to kill his girlfriend, fellow student Tatiana Tarasoff.

Therapists, like medical doctors and clergy, always have regarded their discussions with patients as privileged communication in which, both legally and morally, confidentiality must be safeguarded. In a situation such as this one, however, therapists must weigh the danger to their patients if they breach confidentiality against the danger to others if they do not.

Dr. Moore's first response was to consult with his two supervisors. All three concurred that Poddar needed to be hospitalized for observation. Moore's supervisor then notified the campus police and asked them to bring in Poddar. When the police detained and interviewed him, however, they concluded that he was rational and not dangerous. As a result, Moore's supervisor rescinded the original commitment order.

Not surprisingly, Poddar felt betrayed by his therapist's breach of confidence and broke off therapy. Two months later, when Tarasoff returned from a long trip, Poddar killed her.

After Tarasoff's death, her parents learned that Poddar had told his therapist of his intentions. In Tarasoff v. Regents of the University of California (131 California Reporter 14, July 1, 1976), the parents successfully sued Dr. Moore and the university on the grounds that therapists must abandon confidentiality when another life is endangered and that, specifically, they must inform intended victims as well as legal authorities.

At first reading, the message of the *Tarasoff* case seems obvious: If a therapist reasonably suspects a client is dangerous, the therapist must warn both the legal authorities and the intended victims. This same reasoning has been applied to clients who tell their therapists of suicidal thoughts. More recently and in a somewhat different vein, some have argued that health care workers must breach confidentiality when they learn of **HIV**-infected clients having unprotected sex without informing their sexual partners of their infection. The codes of ethics of both the American Medical Association and the American Psychiatric Association, as well as various legal

---

The potential for stigmatizing mental patients and the problems that arise when the interests of mental patients conflict with the interests of others are discussed in this chapter's ethical debate (Box 7.3).

Conclusion

In this chapter we have compared the sociological and medical models of mental illness. As with the medical models of physical illness and disability discussed in Chapters 5 and 6, the medical model of mental illness asserts that mental illness is a scientifically measurable, objective reality, requiring

decisions, declare that doctors must breach confidentiality when the health or welfare of either a client or others in the community is endangered.

A closer look at the *Tarasoff* case, however, reveals some of the difficulties of reaching any simple conclusion. On the one hand, it could be argued that if Tatiana had been informed, she could have protected herself. Yet women are killed daily who know full well that their husbands or lovers want to kill them. Police often offer little protection to these women, and the women often can do little to protect themselves.

In addition, in the *Tarasoff* case, the one documented result of informing the police was that Poddar ended therapy. It could be argued, therefore, that far from protecting the intended victim, breaching confidentiality placed her in greater danger by convincing Poddar to end therapy, thus reducing the chances that he would find a nonviolent way of managing his anger.

Finally, the argument that therapists must breach confidentiality regarding dangerous clients assumes that therapists know which clients are dangerous. Yet, as various studies have shown and as the American Psychiatric Association and several other professional organizations argued in briefs filed on behalf of the therapists in the *Tarasoff* case, this assumption is far from true. Moreover, if psychiatrists wrongly conclude that clients are dangerous and therefore breach confidentiality, they can subject the clients to substantial stigma, sometimes with permanent consequences. Indeed, with the growth of large, all-too-accessible, computerized data banks of medical records and the growth in access to those records by insurers, peer review organizations, and the like, the more serious issue facing therapists in the future may be how to *protect* confidentiality, not when to breach it.

Sociological Questions

1. What social views and values about medicine, society, and the body are reflected in this policy? Whose views are these?

2. Which social groups are in conflict over this issue? Whose interests are served by the different sides of this issue?

3. Which of these groups has more power to enforce its view? What kinds of power do they have?

4. What are the intended consequences of this policy? What are the unintended social, economic, political, and health consequences of this policy?

prompt treatment by scientifically trained personnel. As such, this model downplays the role of social and moral values in the definition and treatment of mental illness and the effect of mortification and stigma on those who receive treatment.

Entering the twenty-first century, we find ourselves facing a situation uncomfortably similar to that of past centuries. As in the years before the Great Confinement, thousands of persons who have mental illnesses now live on the streets and support themselves at least partly by begging. Many more—along with others who experience social rejection—are confined in nursing homes, board and care homes, or prisons, in the same way that earlier societies confined persons with mental illness in almshouses along with the poor, the disabled, and those without families. Although drugs largely have replaced shackles, society still allocates far too few resources to provide humanely for those who suffer mental illnesses. We can only hope that, in the future, with a greater understanding of the nature of mental illness and of the social response to it, we can develop more compassionate and effective means of coping with mental illness.

Suggested Readings

Goffman, Erving. 1961. *Asylums.* Garden City, NY: Doubleday. The classic text on the nature of mental hospitals and other total institutions. Still fascinating reading.

Kaysen, Susan. 1993. *Girl, Interrupted.* New York: Random House. A memoir of mental illness and its treatment.

Wagner, Pamela Spiro, and Carolyn S. Spiro. 2005. *Divided Minds: Twin Sisters and Their Journey Through Schizophrenia.* New York: St. Martin's Press. Two sisters, one who has fought a lifelong battle with schizophrenia and the other a psychiatrist, offer their joint memoir of how schizophrenia has affected their lives and relationship.

Getting Involved

American Civil Liberties Union. 132 W. 43rd Street, New York, NY 10004. (212) 944-9800. www.aclu.org. Among other things, works for the civil rights of mental patients.

Bazelon Center for Mental Health Law. 1101 15th Street NW, Suite 1212, Washington, DC 20005. (202) 467-5730. www.bazelon.org. Works to advance and preserve the rights of people with mental illnesses and developmental disabilities. Provides extensive information about current issues and late-breaking news in this area.

MindFreedom Support Coalition International. 454 Willamette, Suite 216, PO Box 11284, Eugene, OR 97440. (877) MAD-PRIDE. http://mindfreedom.org. A grassroots organization of self-described survivors of psychiatric treatment, which, among other things, has led campaigns to end involuntary electroshock and psychiatric drugging.

National Alliance for the Mentally Ill. 2107 Wilson Blvd., Suite 300, Arlington, VA 22201. (703) 524-7600. www.nami.org. The nation's leading grassroots, self-help, and family advocacy organization devoted to improving the lives and treatment of persons with severe mental illnesses. Supports the medicalization of mental illness.

Review Questions

How and why do ethnicity, gender, and social class affect rates of mental illness?

What is the relationship between life events and mental illness?

What are the differences between the medical and sociological models of mental illness?

What are the problems embedded in psychiatric diagnoses?

What was moral treatment, and why did it fail?

What was the antipsychiatry critique?

What were the sources and consequences of deinstitutionalization?

What is the remedicalization of mental illness?

How is managed care affecting the treatment and experience of mental illness?

How do individuals become mental patients?

What are the consequences of labeling an individual mentally ill?

Internet Exercises

1. Browse the website for the National Alliance for the Mentally Ill (NAMI) (www.nami.org), the major organization promoting the interests of persons with mental illness and their families. What is NAMI's approach to mental illness? How is it similar to or different from the perspective presented in this chapter?

2. To ascertain the extent to which Freudian ideas now permeate American culture, obtain access through your library or the Internet to *Periodical Abstracts,* the *Readers Guide to Periodical Literature,* or another index of popular magazine articles. Then search for all English language articles from the last two years that use the word *Freudian.* In what ways is the term now used, by what sorts of persons and organizations, and for what purposes?

3

Health Care Systems, Settings, and Technologies

In Part Two, we looked at illness primarily from the perspective of the ill individual. In this part, we move to a macrosociological level, looking at health care systems and settings. In Chapter 8, we consider the history and current nature of the U.S. health care system, examining why and how millions of Americans have found themselves uninsured, underinsured, or precariously insured—threatened with the loss of health insurance at any moment. Chapter 9 begins by presenting a series of measures useful for evaluating any health care system, and then uses these measures to explore four alternative health care systems—those of Canada, Great Britain, the People's Republic of China, and Mexico. With this as a basis, the chapter concludes with a look at the prospects for reforming the U.S. health care system. Finally, in Chapter 10, we investigate the major settings in which health care is offered in the United States (other than individual doctors' offices), and the increasingly important role technology plays in those settings, as it helps solve old problems and creates new problems.

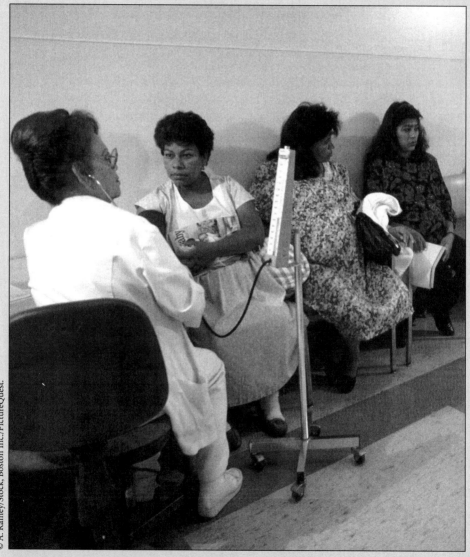

The U.S. Health Care System and the Need for Reform

Health care in the United States is a system in crisis. Consider, for example, Kim's story:

> Born in Chicago and raised in the city's housing projects, Kim had few advantages in life other than having a father in the U.S. military. Though he did not live with the family, he did list her on his health insurance policy until she turned eighteen. At that point, his plan would no longer cover her.
>
> After high school, Kim went to community college to study early childhood education. Like many students, she assumed that her degree would lead to a permanent job and benefits. Since graduating from (community) college, however, she has been working part-time at a day care center. She would like to work full-time, but the center isn't hiring full-time employees. She also works part-time at a Walgreen's drugstore. Though she isn't thrilled with the work (which doesn't utilize her college training), she would agree to work full-time, except that Walgreen's isn't hiring full-time employees either. Kim explained that she tried working more hours there after her boss told her that she would need to work full-time for twelve weeks in order to be eligible for insurance. But when she approached the twelve-week mark, her hours were cut, making her ineligible for insurance. . . .
>
> Kim knows that she has serious health problems and that it is dangerous for her to go without medical care and medication. Since late childhood, she has had diabetes. She needs to take insulin and Glucophage, and she must test her blood sugar several times each day. The medicine and testing equipment cost far more than she can afford on her minimum-wage salary (she earns about $1,000 a month), and she has not been to the doctor for longer than she can remember. [As a result, she says,] "I haven't been taking my medicine like I was supposed to, because I couldn't afford it." . . .

Untreated diabetes not only makes her feel worse day by day but also hastens the onset of the serious complications the disease can cause. Because she is unable to monitor and manage her blood sugars and get recommended preventive care, she is at high risk for premature blindness, heart disease, limb amputations, and kidney failure. Standard medical treatment aims to prevent or at least significantly forestall such outcomes, but Kim does not see a way to access standard treatment (Sered and Fernandopulle, 2005: 137–138).

In desperation, Kim went to a diabetes clinic she had used while still insured and asked if she could arrange for a reduced fee. The answer was no. She then applied for Medicaid, the federal program for health care for the poor, but earned too much to get on the program unless she was pregnant.

The most basic element in any nation's health care system is how it provides and pays for health care. As Kim's story illustrates, the United States has no mechanism for guaranteeing health care to its citizens. Nor, despite this chapter's title, does it really have a health care system. Instead, an agglomeration of public and private health care insurers (such as Medicaid and Aetna), health care providers (such as doctors and nurses), and health care settings (such as hospitals and nursing homes) function autonomously in myriad and often competing ways. In this chapter we look at how health insurance is structured in the United States, how pharmaceutical companies increasingly affect the costs and nature of U.S. health care, and the growing crisis in U.S. health care.

Health Insurance in the United States

Until at least the 1930s, most Americans paid for their health care out of pocket. The wealthy could buy whatever health care they desired, the middle class could afford most needed health care, and the poor mostly went without.

There still are some Americans who can afford to purchase whatever care they want, as well as many, like Kim, who cannot afford needed care, for the United States is the only **industrialized nation** that does not guarantee health care to its citizens. (The problems faced by the uninsured are discussed later in this chapter.) Most Americans, however, rely on health insurance to make health care affordable. In this section, we first look at the two main health insurance *models* that historically existed in the United States, fee-for-service insurance and health maintenance organizations (which are a form of managed care). Key Concepts 8.1 compares these models. Although both models have changed considerably over the years, understanding them makes it easier to understand the newer models that have emerged more recently. After looking at these two models, we look at how U.S. health insurance overall has moved toward managed care. Finally, we look briefly at the two main government-provided health insurance *programs* in the United States, Medicare and Medicaid, each of which offers insurance based on both the fee-for-service and health maintenance models.

Key Concepts 8.1	*Comparing Insurance Models*	
MODEL	FEE-FOR-SERVICE INSURANCE	HEALTH MAINTENANCE ORGANIZATIONS
Examples	Blue Cross/Blue Shield	Kaiser Permanente HMO
Underlying purpose	Protect doctors and hospitals.	Provide health care to all.
Historically restrained costs through . . .	Community ratings: insure entire, largely healthy, communities to reduce risk and spread costs.	Community ratings, plus emphasis on maintaining health and preventing costly illness.
Doctors paid	Fee-for-service	Salary
Typical coverage	Open choice of and access to doctors.	Limited choice of doctors and limited access to specialists.
	Many bills not covered (deductibles, preexisting conditions, prescription drugs, limits on yearly and lifetime coverage).	Almost all bills covered.
	Preventive care not covered.	Preventive care emphasized.
Changes in model over time	Commercial fee-for-service insurers emphasize generating profits for stockholders.	Commercial HMOs emphasize generating profits for stockholders.
	Move to actuarial risk rating.	Increased use of copayments and restriction of HMO membership to healthier populations.
	Doctors pressed to accept negotiated fee schedules (in PPOs).	Doctors paid on capitation or fee-for-service.
	Choice of doctors limited in preferred provider options.	Choice of doctors expanded by preferred provider options.
		Access to specialists expanded with elimination of "gatekeepers."
	Managed care strategies become common (utilization review, etc.).	Managed care strategies become more common.

Health Insurance Models

Both fee-for-service insurance and health maintenance organizations first appeared during the Great Depression of the 1930s, when millions of Americans were out of work and few could afford to pay for medical care. But the two forms differed dramatically in their origins and goals.

Fee-for-Service Insurance

The first major fee-for-service insurance program, **Blue Cross,** was founded by the American Hospital Association. Through selling insurance to cover individuals' hospital bills, the association hoped to preserve hospitals' income and protect them from bankruptcy. The success of Blue Cross led the American Medical Association (AMA) to found **Blue Shield** shortly thereafter. Whereas the purpose of Blue Cross was to protect hospitals' incomes, the purpose of Blue Shield, which provides coverage for medical bills, was to protect doctors' incomes, by ensuring that middle-class Americans would be able to afford medical care. These two nonprofit plans (collectively known as "the Blues") continue to play an important role in the U.S. health care system; during 2004, 92.3 million Americans belonged to these plans (Blue Cross and Blue Shield Association, 2005).

Historically, individuals who had Blue Cross/Blue Shield insurance could seek care from whatever hospitals and doctors they chose. In turn, hospitals and doctors charged Blue Cross/Blue Shield patients on a **fee-for-service** basis; that is, patients were billed a fee for each office visit, test, or other service they received. For this reason, Blue Cross/Blue Shield is known as **fee-for-service insurance.** Under such insurance, individuals must first pay their medical bills and then request reimbursement from their insurance providers. However, individuals typically must pay on their own the first $100 to $500 in bills they receive each year (known as the **deductible**), 20 percent or more of their hospital bills, and all costs for preventive medical care. To keep Blue Cross/Blue Shield premiums low, many plans now offer **preferred provider organizations (PPOs),** in which doctors agree to charge lower, preset fees in exchange for the additional business, and consumers agree to obtain care from these doctors in exchange for lower premiums and deductibles.

Both Blue Cross and Blue Shield usually establish lifetime and sometimes annual maximums. Individuals who exceed their maximums must pay their remaining bills themselves, a serious problem for those with **chronic illnesses** or serious injuries.

Until the 1980s, both Blue Cross and Blue Shield established their fees based on **community rating.** Under community rating, each individual pays a "group rate" premium (or yearly fee) based on the average risk level of his or her community as a whole. Even if a particular individual is a bad insurance risk because of a preexisting illness, a dangerous job, or a family history of illness, the insurer need not charge that individual a high premium because most members of the community will have much lower risks, keeping the average costs to the insurer low. This explains why those who purchase insurance as part of a large group, such as all employees of IBM, pay far lower premiums than do those who purchase insurance individually.

In contrast, fee-for-service insurance offered by **commercial insurance companies** (i.e., insurers that function on a for-profit basis) is based on **actuarial risk rating** rather than community rating. Under actuarial risk rating, insurers maximize their profits by insuring only individuals whose health risks are low or by charging very high premiums to those whose health risks are high. For example, commercial insurers typically charge higher premiums to those who have allergies, back strain, kidney stones, or ulcers; typically deny coverage to those who have ulcerative colitis, diabetes, or severe obesity; and often deny coverage to individuals who work in high-risk fields or in fields that attract risk takers, such as aviation, auto sales, construction, and law.

Conversely, to attract a low-risk clientele, commercial insurers charge lower rates to such individuals. As a result, they have successfully lured many low-risk individuals away from the Blues, leaving the Blues with a sicker clientele overall. To avoid having to raise their rates for all members to cover the bills of their sicker members, many Blue Cross/Blue Shield companies now use actuarial risk rating.

Health Maintenance Organizations

The 1930s and 1940s also saw the rise of a very different type of health insurance program: **health maintenance organizations (HMOs).** Unlike the Blues and the commercial fee-for-service insurers, the first HMOs to attract national attention—Kaiser Permanente and the Group Health Cooperative of Puget Sound—were organized by individuals whose primary aim was providing affordable, high-quality health care to their communities. Like the Blues, these HMOs based their fees on community rating. But whereas the Blues and the commercial insurers used **retrospective reimbursement,** reimbursing individuals for health care costs after they fell ill, the HMOs used **prospective reimbursement** in an attempt to keep people from falling ill in the first place.

Under prospective reimbursement, HMOs paid doctors a salary, rather than paying them on a fee-for-service basis. Because doctors received the same salary regardless of how many times they saw their patients or how many procedures they performed, they could not increase their income by providing unnecessary medical care. Instead, doctors would earn the highest net income by keeping patients healthy so the patients would require less of their time and resources in the long run.

In line with their emphasis on restraining costs by keeping members healthy, HMOs, unlike the Blues and commercial insurers, paid the full cost of preventive care. Patients, meanwhile, paid nothing beyond the cost of their insurance premiums as long as they used only doctors affiliated with their HMO and saw specialists only if referred by their **primary care doctor** (known as a **gatekeeper** in systems of this sort).

As research increasingly suggested that HMOs could provide care at least as good as that offered by fee-for-service insurance but at lower cost (e.g., Leape, 1992), interest in developing HMOs to generate corporate profits began to grow. As a result, by 2002, 31 percent of privately insured Americans belonged to HMOs (National Center for Health Statistics, 2004: 355), with most of these belonging to for-profit HMOs.

The interest in HMOs as cost-saving and profit-generating mechanisms has altered the structure of HMOs substantially. Like commercial fee-for-service insurers, commercial HMOs work to enroll as healthy a population as they can. To discourage unnecessary medical visits by members, most HMOs now charge **copayments**—small fees consumers must pay each time they see a care provider. To discourage primary care doctors from unnecessarily referring patients to specialists, HMOs began setting aside annually a pool of money to pay for referrals to specialists and allowing primary care doctors to divide among themselves any money left over at the end of the year. A California survey of primary care HMO doctors found that 57 percent felt pressured to limit referrals (Bodenheimer, 1999); those who do not limit referrals are less likely than others to have their contracts renewed. To further increase doctors' incentives to control the costs of health care, most HMOs no longer pay doctors on salary. Instead, HMOs typically pay primary care doctors by **capitation,** paying them a set annual fee to cover all care (both primary and specialty) per patient in their practice, and pay specialists (and occasionally primary care doctors) on a fee-for-service basis. Like doctors in PPOs, however, HMO doctors paid fee-for-service must abide by a schedule of fees negotiated in advance with the HMO.

The Managed Care Revolution

The most striking change in the U.S. health care system over the last quarter-century has been the dramatic rise of **managed care.** Managed care refers to any system that controls costs through closely monitoring and controlling the decisions of health care providers. Most commonly, managed care organizations (MCOs) monitor and control costs through utilization review, in which doctors must obtain approval from the insurer before they can hospitalize a patient, perform surgery, order an expensive diagnostic test, or refer to a specialist outside the insurance plan. In addition, MCOs typically organize panels of doctors, pharmacists, and administrators to create lists (known as **formularies**) of the most cost-effective drugs for treating specific conditions. Doctors who work for an MCO must get special permission to prescribe any drugs not on that MCO's formulary.

Although the terms *HMO* and *managed care* increasingly are used interchangeably, HMOs represent only one form of managed care, and most fee-for-service insurers now also use managed care. Most Americans who have private insurance now belong to some form of managed care plan.

The Rise of Managed Care

The use of managed care spread rapidly around the country during the 1980s and 1990s in an effort to restrain spiraling health care costs. This explosive growth led to many questions regarding whether MCOs cut quality of care along with costs. Research suggests that in at least some circumstances, managed care can reduce costs while maintaining or even improving quality of care. For example, one study tracked, for seven years, almost 2,000 patients who had high blood pressure or adult diabetes; no differences were found in outcomes between managed care patients and other patients, even though the managed care patients received fewer tests, had fewer hospitalizations, and thus had lower bills overall (Greenfield et al., 1995). Similarly, another study found that older women with breast cancer who received managed care through nonprofit HMOs were more likely than those who had fee-for-service insurance without managed care to have their cancers diagnosed at earlier stages and to receive all treatments currently recommended by medical experts (Riley et al., 1999).

Overall, however, most studies have found few significant differences between managed care and other plans in access to care, quality of care, or patient satisfaction (Mechanic, 2004; R. Miller and Luft, 1997). At any rate, current research provides a poor basis for predicting the economic or health impact of MCOs in the future. As the use of MCOs has spread, they have attracted a more typical and less-healthy population than in the past. For these less-healthy patients, MCOs' emphasis on preventive, primary care rather than on interventionist and specialty care may not be the best choice, and so the health benefits of MCOs are diminishing (Draper et al., 2002).

Perhaps the more important issue, though, is not the impact of managed care per se but the impact of the for-profit motive. Importantly, although both for-profit and nonprofit HMOs use managed care to control costs, the former due so to generate profits, while the latter do so to free the funds needed to improve services for their members. Data collected in 1997 from most HMOs in the country found that for-profit HMOs scored lower than nonprofit HMOs on all fourteen indicators of quality of care, including rates of childhood immunization, mammograms, prenatal care, and appropriate treatment of persons who had diabetes or heart attacks (Himmelstein et al., 1999). (This chapter's ethical debate, Box 8.1, similarly discusses the impact of profit incentives on pharmacists' services.)

The Backlash Against Managed Care

Despite evidence suggesting that managed care makes little differences in either patient outcomes or patient satisfaction, there has been a substantial backlash against the managed care revolution. A string of legislative and legal moves—often framed as "Patients' Bills of Rights"—have pressed insurers to drop some of the less popular aspects of managed care. For example, legislators have opposed the early release of women from hospitals soon after

---

Box 8.1   *Ethical Debate: Pharmacists and Conflicts of Interest*

In the same way that the interests of doctors and patients clash when doctors have a vested economic interest in referring patients for particular tests at particular laboratories, many pharmacists now have a vested economic interest in selling certain drugs rather than others (Kolata, 1994).

In 1992, Merck Pharmaceuticals bought Medco, a nationwide drug supply company that buys drugs from manufacturers and sells them at discounts to its 38 million U.S. members through pharmacies. Since then, two other major pharmaceutical companies, SmithKline Beecham and Eli Lilly, have bought drug supply companies.

Since Merck bought Medco, it has offered cash commissions to pharmacists who convince customers to buy Merck products rather than competing drugs. For example, if a customer who belongs to Medco brings in a prescription for an ulcer medication not produced by Merck, the pharmacist may tell the customer that, under their Medco coverage, they can purchase a similar and equally effective drug more cheaply. The pharmacist then offers to call the customer's doctor to request that the doctor approve switching drugs. What the pharmacist will not tell either the customer or the doctor is that Merck makes the recommended drug and that the pharmacist will benefit financially from this switch.

Because in the past pharmacists had no financial links to pharmaceutical companies, doctors generally assume that pharmacists' suggestions are both educated and impartial. Doctors therefore agree to switch drugs in about 80 percent of cases (Kolata, 1994). After several such phone calls from pharmacists, doctors may begin routinely prescribing the recommended drug instead of the drug that they used to prescribe.

Is it unethical for pharmaceutical companies to offer financial rewards to pharmacists who sell certain drugs, or for pharmacists to accept those rewards? Those who participate in these arrangements, of course, consider them merely an extension of normal business practices. Because many of the most popular drugs

---

giving birth (labeled "drive-by deliveries" by the media) even though in general, early release is safer because it reduces women's chances of contracting an infection in the hospital. Similarly, legislators have fought to get patients access to experimental treatments, although patients are more likely to be harmed than helped by them. Even in the absence of legislative pressure, the need to keep both consumers and contracted doctors happy has led insurers to scale back the use of formularies and utilization review, and to virtually abandon the use of primary care gatekeepers (Bodenheimer, 1999).

Why has this backlash been so large and effective? The answer lies in American culture, media, and politics (Mechanic, 2004). A central theme in American culture is an emphasis on individual autonomy and independence. (In contrast, the countries of northern and western Europe have a far stronger emphasis on community and social solidarity, leaving them far more willing to support social ventures such as universal health care.) By its very nature, managed care reduces individual choices for both consumers

on the market are virtually identical to competing drugs, supporters argue, customers lose nothing by switching drugs and gain if the new drugs are cheaper. Moreover, they claim, if drugs do differ significantly, it is the doctor's responsibility—not the pharmaceutical company's or pharmacist's—to know that and to protect his or her patients. In essence, proponents argue, drugs are like any other consumer good; no ethical rules apply beyond the normal rules of the marketplace, such as not advertising a product's effects falsely.

Opponents of these arrangements, on the other hand, argue that such practices necessarily produce unethical conflicts of interest. A pharmacist who can earn extra money by recommending certain drugs over others is more likely to recommend that drug, whether or not it really is the best drug for the customer. Moreover, the entire transaction is grounded in dishonesty, for neither customer nor doctor knows that the pharmacist has a vested interest in selling certain products. Rather, both customer and doctor reasonably assume that

pharmacists, as professionals, are bound by a code of ethics that restrains any tendency to place their economic self-interest ahead of customer welfare. These problems led the federal government in 2002 to release new guidelines that identify these practices as illegal frauds and kickbacks. It remains to be seen how much effect the guidelines will have.

Sociological Questions

1. What social views and values about medicine, society, and the body are reflected in this policy? Whose views are these?

2. Which social groups are in conflict over this issue? Whose interests are served by the different sides of this issue?

3. Which of these groups has more power to enforce its view? What kinds of power do they have?

4. What are the intended consequences of this policy? What are the unintended social, economic, political, and health consequences of this policy?

and health care providers. As a result, media and political attacks on managed care resonated well with popular sentiment.

The media and politicians also found managed care an easy target simply because its size made it so visible. As we saw in Chapter 2, medical errors are rife throughout the health care system. Yet when fee-for-service doctors working outside of managed care plans are identified as dangerous, we think of them as individuals, not as representatives of the fee-for-service system. In contrast, because managed care doctors belong to huge, visible, corporations, it is far easier for opponents to generalize concern about problematic doctors or clinics to managed care as a whole.

Similarly, the belief that more health care is better health care is long-standing in American culture. Under the fee-for-service system *without* managed care, doctors have an incentive to provide as much treatment and testing as their patients' insurance or budget will cover, leading at least in some circumstances to dangerous overtreatment (Leape, 1992). For example, mortality rates are *higher* in geographic regions where Americans receive more extensive medical care, apparently because the extra medical treatment is more dangerous than helpful (E. Fisher et al., 2003). Yet because of our belief that more is better in health care, the public rarely questions whether the ease of access to care under the fee-for-service system might be dangerous.

With the rise of managed care, the inherent financial incentives of the health care system have reversed, so that now doctors can increase their incomes by *restricting* the treatments they provide or the drugs they prescribe. Because this system goes against the American cultural belief that more health care is better, it is far easier for patients to see the dangers of undertreatment inherent in managed care than the dangers of overtreatment inherent in fee-for-service medicine prior to managed care. Similarly, although the time doctors spent with each patient actually *increased* slightly between 1989 and 1999, most Americans believe it *decreased* due to managed care, which has further eroded their trust in the health care they receive (Mechanic, 2001a). More broadly, some patients now think of their doctors as "double agents," whose loyalties are split between serving their patients and serving the MCOs that pay their bills (Shortell et al., 1998). Such patients are less likely to trust their doctors and, as a result, more likely to decline treatment, participate in treatment only halfheartedly, or withhold needed information about their health from health care providers (Mechanic, 1999).

These cultural factors made managed care an easy target for the mass media, politicians who wanted to spruce up their image with the public, medical groups that wanted to regain some of their former independence, and pharmaceutical companies that wanted to reduce the power of MCOs over drug prescribing or prices. As a result, the managed care revolution has been substantially curtailed.

Government-Funded Health Insurance Programs

Although the United States does not offer a national health insurance program to cover all citizens, it does offer smaller programs for specified subgroups. For example, the Veterans Administration offers health coverage to veterans, TRICARE (formerly CHAMPUS) offers coverage to active and retired members of the armed forces and their families, and the Federal Employees' Health Benefits Program offers coverage to federal employees and their families. In this section we focus on the two best-known, government-funded health insurance programs, Medicare and Medicaid. Both programs started as traditional fee-for-service without managed care but increasingly are based on managed care principles.

Medicare

Medicare covers more persons than any other single insurance program in the nation. Virtually all Americans over age 65 receive Medicare, as do some permanently disabled persons. All persons eligible for Medicare receive, at no cost, coverage for as many as 150 days of hospital care, although these persons must pay substantial deductibles and copayments. In addition, they receive limited coverage for posthospital nursing services, home health care, and hospice care. Medicare also offers fee-for-service insurance for outpatient medical costs, at a monthly premium of $66 as of 2004. This insurance, too, has substantial deductibles and does not cover many medical costs, such as prescription drugs, long-term nursing home care, and routine eye care. Adding together the costs of copayments, deductibles, premiums, and items not covered by insurance, 60 percent of Medicare recipients over age 65 spend more than 20 percent of their income on health care (Health Care Financing Administration, 2000: 27). To keep their costs to a minimum, almost all Medicare recipients purchase (or receive from their former employers) additional insurance known as **medigap policies.** (The poorest Medicare recipients may receive additional coverage through Medicaid, the government's program for indigent health care.)

Medicare faces increasing economic pressures from all sides. Medicare is primarily funded through federal Social Security taxes. Essentially, working adults pay taxes into a trust fund that pays the health care bills of the elderly. Because of our aging population, this financial structure cannot work in the long run (Health Care Financing Administration, 2000). In 2004, 14 percent of Americans received Medicare (DeNavas-Walt, Proctor, and Mills, 2004). If current trends continue, by 2030, when 22 percent of Americans will be eligible, federal researchers expect the system to go bankrupt, because there will be too few workers paying into the system to support it. Responding to this problem, Congress has instituted a long-term program for increasing Medicare premiums paid by consumers and reducing fees paid to health care providers.

Medicaid

Whereas Medicare provides coverage to individuals based primarily on age, **Medicaid** (and S-CHIP, the associated the State Children's Health Insurance Program) provide coverage based on income and physical vulnerability. To receive Medicaid, adults must be both poor and either aged, blind, disabled, pregnant, or the parent (almost always the mother) of a dependent child. About 13 percent of Americans have Medicaid insurance, most of which comes through some form of managed care organization.

Medicaid is funded through a combination of federal and state taxes. States have considerable leeway to determine eligibility and benefits, however. In the last few years, as the current economic recession combined with political pressure to reduce taxes have reduced states' income, states have found it increasingly difficult to pay the expenses of running Medicaid programs (Pear and Toner, 2002). As a result, about one-quarter of poor children and half of poor adults are not covered (Kaiser Commission on Medicaid and the Uninsured, 2004). States also have reduced the amounts they pay to health care providers (physicians, hospitals, and nursing homes) who work with Medicaid patients, leading many doctors to refuse to treat Medicaid patients except in life-threatening emergencies.

"Big Pharma": Pharmaceutical Companies and U.S. Health Care

In addition to the health insurance system, the other "big player" in the U.S. health care world is the pharmaceutical industry, or "Big Pharma," as it is often known. Because it is a for-profit enterprise, Big Pharma's goal is not only to develop drugs but to sell those drugs. As a result, the pharmaceutical industry plays a major role in determining how doctors and the public think about illnesses and treatments and in the rising costs of health care.

Big Pharma Comes of Age

The pharmaceutical industry is an enormous—and enormously profitable—enterprise. Indeed, it has been the most profitable industry in the United States since the early 1980s. In 2001, for example, the combined profits of the ten pharmaceutical companies in the Fortune 500 surpassed the profits of the other 490 companies on the list combined (Angell, 2004). Although the pharmaceutical industry routinely argues that their high profits merely reflect the high cost of researching and developing new drugs, such work accounts for only 14 percent of their budgets. In contrast, marketing accounts for about 50 percent (Angell, 2004). Due largely to this marketing, American citizens now spend a total of about $200 billion per year on prescription drugs, not including drugs purchased by doctors, nursing homes, hospitals, and other institutions (Angell, 2004: 3). Americans are buying *more* drugs, buying more *expensive* drugs, and seeing the *prices* of the most popular drugs rise more often than ever before. (The price of the popular antihistamine Claritin, for example, rose 13 times in 5 years.) Prescription drugs now account for more than one-quarter of all U.S. health care expenses (National Institute for Health Care Management Foundation, 2002).

The pharmaceutical industry has not always been this profitable. Profits only began soaring in the early 1980s, following a series of legal changes reflecting both the increasingly "business-friendly" atmosphere in the federal government and the increased influence of the pharmaceutical industry lobby—now the largest lobby in Washington (Angell, 2004). First, new laws allowed researchers whose work was funded by federal agencies (including medical school faculty, university professors, researchers working for small biotech companies, and some federal employees) to patent their discoveries and license those patents to pharmaceutical companies. This change gave these researchers a vested interest in supporting the pharmaceutical industry, and made it possible for the industry to dramatically decrease its own costs for research. Second, new laws almost doubled the life of drug patents. As long as a drug is under patent, the company owning that patent has the sole right to sell that drug. As a result, the company can set the price for that drug

---

Box 8.2   ***Making a Difference: No Free Lunch***

No Free Lunch was founded by Dr. Bob Goodman (Koerner, 2003). From the start of his medical training, Dr. Goodman had questioned the influence of pharmaceutical companies on doctors' prescribing practices. When in 1993 he opened a clinic for low-income patients in a poor New York City neighborhood, Dr. Goodman decided he would no longer accept samples or other "goodies" from pharmaceutical salespeople. But like most doctors, he had come to depend on samples for treating patients who could not afford to buy drugs. To help pay for the drugs his patients needed, Dr. Goodman started a website, www.nofreelunch.org, to provide up-to-date information on the nature, extent, and consequences of pharmaceutical advertising to doctors while selling mugs and pens with the "No Free Lunch" logo he had devised. The website also features a list of doctors who have signed a pledge "to accept no money, gifts, or hospitality

from the pharmaceutical industry; to seek unbiased sources of information and not rely on information disseminated by drug companies; and to avoid conflicts of interest in my practice, teaching, and/or research."

No Free Lunch remains mostly a one-man (money-losing) operation, although it now has many members and other supporters around the country. Physician members have organized talks at their hospitals and medical schools on the impact of pharmaceutical advertising on medical behavior. Medical student members have held "pen amnesty days," in which students and doctors are encouraged to turn in their drug company pens and other paraphernalia for No Free Lunch pens. Pen Amnesty Days often are accompanied by lectures, other events, and media coverage to help spread the word about the dangers of relying on pharmaceutical companies for medical information.

---

as high as the market will bear, with no concern about competition. In addition, current regulations make it easy for companies to extend their patents by developing "me-too" drugs, which differ only slightly from existing drugs in their dosage, formula, or advertised target market. Me-too drugs now account for about 75 percent of all new drugs on the market (Angell, 2004). Third, the pharmaceutical industry won the right to market drugs direct to consumers, on television as well as in print media. Direct-to-consumer advertising—a $3.8 billion business in 2005—has proven highly effective. According to a nationally representative survey conducted in 2001 for the nonprofit Kaiser Family Foundation, 30 percent of American adults have asked their doctors about drugs they've seen advertised, and 40 percent of these received prescriptions for the drugs as a result (Brodie, 2001). Similarly, in one experimental study, pseudo-patients were sent to doctors' offices to request specific prescriptions, and more than half received them (Kravitz et al., 2005). Box 8.2 describes the work of No Free Lunch, a group dedicated to weaning doctors from their dependence on the pharmaceutical industry.

Passage of the Medicare drug benefit program, which goes into effect in 2006, is expected to raise pharmaceutical profits even higher. The pharmaceutical industry was heavily involved in the drafting and passage of this program, under which Medicare recipients can choose to buy supplemental insurance to cover some of their prescription drug costs (Abramson, 2004; Angell, 2004). However, most Medicare recipients will pay more in premiums and deductibles for the drug program than they will save by enrolling in it. In addition, the program is so complex that few consumers will be able to make informed decisions about whether to purchase the insurance. The pharmaceutical industry, meanwhile, is guaranteed to earn high profits from the program, for under the new law, Medicare (unlike private insurance programs) cannot restrict which drugs will be purchased and cannot negotiate with pharmaceutical companies to purchase drugs at bulk rates.

Developing New Drugs

Whenever a new drug is developed, the crucial question for health care providers and patients is whether its benefits outweigh its dangers. For this reason, it is crucial that any new drug be extensively tested to determine whether it works better than already available drugs (which almost certainly are cheaper), whether it works differently in different populations (does it help men as well as women? adults as well as children? persons with early as well as late-stage disease?), what are appropriate dosages, and what are the potential side effects? But because pharmaceutical companies earn their profits by selling drugs, they have a vested interest in overstating benefits and understating dangers. And increasingly, these companies are both willing and able to manipulate the data available to outside researchers, doctors, federal regulators, and consumers (Abramson, 2004; Angell, 2004).

In the past, university-based drug researchers provided at least a partial check on the drug research process, because these researchers could bring a more objective eye to their research. Between 1980 and 2000, however, pharmaceutical industry funding for research by university-based scientists increased almost nine times (Lemmens, 2004). That funding comes in many forms, from research grants, to stock options, to all-expenses-paid conferences in Hawaii. Moreover, as other federal funding for universities declined over the past quarter-century, university administrators came to expect their faculty to seek pharmaceutical funding. Importantly, when the pharmaceutical industry funds university-based research, it often retains the rights to the findings of that research, and can keep university researchers from publishing any studies suggesting that a particular drug is ineffective or dangerous (Angell, 2004; Lemmens, 2004).

At the same time that the pharmaceutical industry has increased its funding to university-based researchers, it has even more dramatically increased funding to *commercial* research organizations (Lemmens, 2004). These organizations are paid not only to conduct research but also to promote it. To keep on the good side of the companies that fund them, these research organizations must make drugs look as effective and safe as possible by, for example, selecting research subjects who are least likely to suffer side effects, studying drugs' effects only briefly before side effects can appear, underestimating the severity of side effects that do appear, and choosing not to publish any studies suggesting that a drug is ineffective or dangerous.

Doctors, medical researchers, sociologists, and others have raised concerns about the impact of bias on research publications (Bodenheimer, 2000). Researchers have found that articles published in medical journals and written by individuals who received pharmaceutical industry funding are four to five times more likely to recommend the tested drug than are articles written by those without such funding (Abramson, 2004: 97). Concern about such biases led the *New England Journal of Medicine* (one of the top two medical journals in the United States) to briefly adopt a policy forbidding authors who have financial interest in a drug from writing editorials or review articles on that drug. This policy was dropped quickly because it was virtually impossible to find authors who did not have such financial interests (Lemmens, 2004).

Even more astonishing than pharmaceutical industry funding of university-based researchers is the growing practice of paying such researchers to sign their names to articles actually written by industry employees (Elliott, 2004). For example, between 1988 and 2000, ninety-six articles were published in medical journals on the popular antidepressant Zoloft. Just over half of these were written by pharmaceutical industry employees but published under the names of university-based researchers. Moreover, these ghost-written articles were *more* likely than other articles to be published in prestigious medical journals.

Regulating Drugs

In the United States, ensuring the safety of pharmaceutical drugs falls to the Food and Drug Administration (FDA). But during the same time period that the profits and power of the pharmaceutical industry grew, the FDA's power and funding declined, as part of a broader public and political movement away from "big government." These two changes are not unrelated: The pharmaceutical industry now routinely provides funding of various sorts to staff members at government advisory agencies, doctors who serve on FDA advisory panels, and legislators who support reducing the FDA's powers (Lemmens, 2004).

Under current regulations, the FDA must make its decisions based primarily on data reported to it by the pharmaceutical industry. Yet the industry is required to report only a small fraction of the research it conducts. For example, the company that produced the antidepressant Paxil had considerable data indicating that, among teenagers, Paxil did *not* reduce depression but *could* lead to suicide. To avoid making this information public, the company submitted to the FDA only its data from studies on adults (Lemmens, 2004). Similarly, drug companies must demonstrate only that new drugs work better than **placebos.** In contrast, in Europe drug companies must demonstrate that new drugs work better than older, less expensive drugs—a standard few new drugs attain. For example, the painkiller Vioxx, at $4 per pill, was found to be no more effective than ibuprofen, at 50 cents per pill. Yet Vioxx quickly became one of the most popular drugs worldwide before it was with-drawn from the market because of its sometimes-fatal side effects. Much of the recent rise in health care costs in the United States comes from the shift to new drugs; as of 2004, spending on drugs stands at $162.4 billion—more than double the amount spent in 1997 (Harris, 2004).

Marketing Drugs

Once the pharmaceutical industry develops a drug and gets FDA approval, the next step is to market the drug. One of the most important limitations to the FDA's power is that, once it approves a drug for a single use in a single population, doctors legally can prescribe it for *any* purpose to *any* population. For example, some doctors prescribe human growth hormone to middle-aged men to stimulate muscle growth, even though the FDA has approved its use only for children with genetic pituitary defects that produce short stature.

Drug marketing has two major audiences, doctors and the public. Marketing to doctors begins during medical school, as students quickly learn that pharmaceutical companies provide a ready source not only of drug samples and information but also of pens, notepads, lunches, and all-expense-paid "educational" conferences at major resorts. Once doctors graduate, the pharmaceutical industry continues to serve as their main source of information about drugs. The *Physicians' Desk Reference* (or *PDR*), the main reference doctors turn to for drug information, is solely comprised of drug descriptions written by drug manufacturers. In addition, the pharmaceutical industry spends $6,000 to $11,000 (depending on medical specialty) per doctor per year to send salespeople to doctors' offices, this on top of the money it spends advertising drugs to doctors in other ways. Most doctors meet with pharmaceutical salespeople at least four times per month and believe their behavior is unaffected by these salespeople. Yet doctors who meet with drug salespeople prescribe promoted drugs more often than other doctors do, even when the promoted drugs are more costly and less effective than the alternatives (Angell, 2004; D. Shapiro, 2004). In addition to these personal meetings with doctors, the pharmaceutical companies now pay for much of the "continuing education courses" doctors must take each year. To hide their role, however, pharmaceutical companies now typically pay for-profit firms to organize these courses, and these firms in turn pay universities to accredit their courses (Angell, 2004).

In recent years, and as noted earlier, marketing directly to consumers has become as important as marketing to doctors. Since 1997, when pharmaceutical companies won the right to advertise brand-name prescription drugs on television, such advertising has skyrocketed. To the companies, such advertising is simply an extension of normal business practices, no different from any other form of advertising. Moreover, they argue, advertising to consumers is a public service, because it can encourage consumers to seek medical care for problems they otherwise might have ignored. Finally, companies have argued that these advertisements pose no health risks because consumers still must get prescriptions before they can purchase drugs, thus leaving the final decisions in doctors' hands. Those who oppose such advertisements, on the other hand, argue that consumers lack the expertise to evaluate the (frequently misleading) advertisements (*Consumer Reports,* 1996). And because the purpose of these advertisements is to encourage consumers to press their doctors for prescriptions, it is disingenuous of advertisers to argue that doctors will protect consumers from making poor drug choices. At any rate, companies increasingly encourage consumers to obtain prescriptions and drugs on the Internet, guaranteeing that they will do so without a doctor's advice.

Marketing Diseases

As part of its marketing, the pharmaceutical industry "sells" not only treatments for diseases, but the diseases themselves. In some cases, drug companies have encouraged doctors and the public to define disease *risks* (such as high blood pressure) as *diseases* (such as hypertensive disease). In other cases, drug companies have defined symptoms as diseases. For example, a variety of neurological conditions (such as head trauma, stroke, Lou Gehrig's disease) can cause uncontrollable laughing or crying unrelated to individuals' emotional state. Avanir Pharmaceuticals markets the drug Neurodex to reduce these symptoms (Pollack, 2005). Although Neurodex seems to help some patients, its side effects are serious enough to cause at least one-quarter of users to stop taking the drug. Critics have questioned whether it is worth promoting a new, under-studied drug to individuals who have far more serious problems and must take numerous other medications.

In addition to marketing Neurodex, Avanir is marketing the symptom of uncontrollable laughing or crying as a condition it has named pseudobulbar affect, or PBA. To convince doctors that uncontrollable laughing and crying is a disease, Avanir has advertised in medical journals and sponsored continuing education courses, conferences, and a PBA newsletter. Because the drug does not yet have FDA approval, none of this marketing can mention Neurodex by name, but it can talk about the need to treat PBA and mention that Avanir has a new treatment for this new "disease."

Avanir is also marketing the concept of PBA directly to consumers. It has targeted consumers through its PBA website and by giving educational grants to stroke and multiple sclerosis patient advocacy groups. For drugs that have FDA approval, direct-to-consumer advertising can go much farther, describing drugs by name and suggesting that consumers mention the drugs to their doctors.

The Crisis in Health Care

Whereas the rise of managed care and the increasing power of the pharmaceutical industry have raised concern about the quality of care available in the United States, the increased costs of health care and the resulting decrease in access to it have challenged the very basis of the U.S. health care system.

Figure 8.1   **Expenditures on Health Care as Percentage of Gross Domestic Product, 1960–2002**

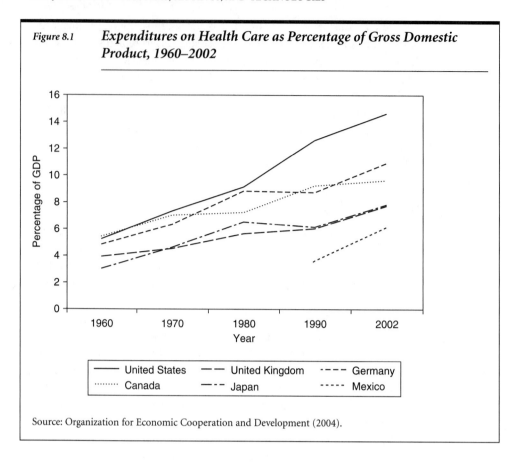

Source: Organization for Economic Cooperation and Development (2004).

Rising Health Care Costs

According to federal researchers, in the United States average costs per capita in 2003 for medical care, drugs, supplies, and insurance was $5,241, with expenditures expected to double by 2013 (U.S. Bureau of the Census, 2004: Tables 114 and 117). These costs continue to be higher and to rise more quickly than in other industrialized nations—far outpacing inflation, as Figures 8.1 and 8.2 show (Organization for Economic Cooperation and Development, 2004).

What accounts for the rising costs of health care? If you ask the typical American—or member of Congress—he or she is likely to respond with one of four popular "myths" about U.S. health care (Starr, 1994).

The first myth is that Americans receive more care than do citizens of other nations. Yet on average, the reverse is true.

The second myth attributes our high health care costs to our unique propensity for filing malpractice suits. Yet malpractice insurance accounts for less than 1 percent of total U.S. health care costs (De Lew, Greenberg, and Kinchen, 1992). Even if we add the estimated costs of **defensive medicine**— tests and procedures doctors perform primarily to protect themselves against

Figure 8.2   *Per Capita Expenditures on Health, 1960–2002*

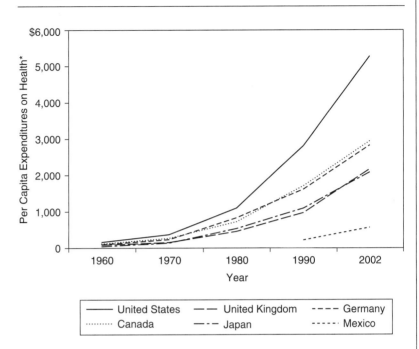

*Dollar amounts adjusted for purchasing power parity. This strategy controls for differences over time and across countries in the worth of a nation's currency by factoring in the number of units of a nation's currency required to buy the same amount of goods and services that $1 would buy in the United States.

Source: Organization for Economic Cooperation and Development (2004).

lawsuits—these expenses increase to only 4 percent of total health care costs. Moreover, those tests and procedures would offer doctors no legal protection if they were obviously unnecessary (Starr, 1994). Consequently, doctors might do the tests and procedures even if not pressured by fear of lawsuits. Nor would health care costs necessarily decline if doctors stopped doing defensive medicine, because they could still maintain their incomes by increasing the number of other services they provided.

The third myth attributes our rising health care costs to our aging population. Yet the population of the United States is no older than that of any of the other top industrialized nations (Population Reference Bureau, 2004).

The fourth myth is that health care costs are so high in the United States because of our advanced technologies. Although these technologies certainly play a role in health care costs, they account for only a small fraction of all health care costs. Moreover, the same technologies exist in the other industrialized nations without producing equally high health care costs. Thus the mere existence of technology cannot explain these costs.

If patient demand, malpractice costs, the aging population, and advanced technology do not explain the rising costs of health care, what does? Research points to two underlying factors: a fragmented system that multiplies administrative costs, and the fact that health care providers (doctors, hospitals, pharmaceutical companies, and so on) have greater power to set prices than do health care consumers, whether individuals, the government, or insurers (Reinhardt, Hussey, and Anderson, 2004).

Because Canadian society is probably the most similar to U.S. society, comparing these two countries helps to illustrate why costs are so high in this country. In the next chapter we examine the Canadian health care system in detail. At this point, we need only note a few major points. Most important, Canadians receive their health insurance directly from the government. Similarly, hospitals receive an annual sum each year from the government to cover all costs. Those costs are restrained because, unlike in the United States, Canadian hospitals do not need an expensive administrative system to track patient expenses and submit bills to multiple insurers. Costs are also restrained by government oversight on major capital development: If a Canadian hospital wants to add new beds or purchase new advanced technologies, it must first convince the government that such services are needed. As a result, hospital costs are considerably lower in Canada than in the United States, even though admission rates are about equal and average stays are longer.

Similar forces keep medical and drug costs down. Like hospitals, doctors need submit their bills only to the national insurance system, rather than filing myriad different forms with different insurers. Meanwhile, no one need spend money on advertising or selling insurance, trying to collect unpaid bills, or covering the costs of unpaid bills. Drug costs are limited because provincial health administrators can develop formularies of cost-effective drugs and negotiate with pharmaceutical companies to buy those drugs at discount. Similarly, the national health care system has the economic "muscle" to control the prices it pays doctors, technology companies, and other health care providers.

The second major reason health care costs are higher in the United States than in Canada is that U.S. health care providers have proportionately more market power than do U.S. health care consumers. This results from the fact that profit-making—by doctors, hospitals, insurers, pharmaceutical companies, and others—lies at the heart of the U.S. health care system.

As we have seen, in the United States, pharmaceutical companies are largely able to control which drugs come to market, how they are advertised, and at what prices, with few constraints imposed by any national consumer or government forces. Similarly, because no national health care system effectively controls the number or distribution of doctors in the United States, there are far too many specialists here. Because health care consumers typically purchase whatever medical services their doctors recommend, when an oversupply of doctors increases competition for patients, doctors can protect their incomes by increasing their charges for services or the number of services they perform. As a result, persons living in areas with the greatest numbers of doctors per capita receive more medical tests, surgeries, and other procedures and pay more for those services, with *worse* health outcomes as a result (Center for the Evaluative Clinical Sciences, 1996). The rise of managed care has constrained doctors' incomes only slightly, because the primary goal of most MCOs is to increase their profits, not to restrict costs to consumers.

Like U.S. doctors, U.S. hospitals are both free of the sort of national oversight that lies at the heart of the Canadian system and forced to compete for patients to pay their bills, let alone earn a profit. As a result, hospitals can and must create demand for their services. To do so, hospitals have added more beds, units (such as heart transplant units), and expensive technologies (such as CT scan machines), whether or not they are needed in their communities.

Unfortunately, whereas in any other field low demand leads to lowered prices, the reverse is true in medical technology. For example, as sociologist Paul Starr explains (1994: 25):

> With fully utilized mammography machines, a screening mammography examination [for early breast cancer detection] should cost no more than $55, according to studies by the GAO [U.S. General Accounting Office] and Physician Payment Review Commission. But because machines are typically used far beneath capacity, prices run double that amount [so that hospitals can recoup their investment]. With prices so high, many women cannot afford a mammogram. . . . In other words, *because* we have too many mammography machines, we have too little breast cancer screening. Only in America are poor women denied a mammogram because there is too much equipment. [Emphasis in original.]

Moreover, when equipment is underutilized, health care providers cannot maintain their skills, so rates of complications and death rise significantly. To maintain skills—and profits—hospitals and doctors tend to overuse any technologies they have at their disposal, leading to wide regional variations in usage (Leape, 1992). In sum, whereas under the normal laws of the marketplace, greater supply leads to lower prices, in health care, greater supply leads to *higher* prices.

In addition, Canada has succeeded at cost control better than the United States because attempts at cost control occur in a unified system where everyone shares the same goal. In contrast, those who have attempted in the past to control the costs of medical and hospital care in the United States have failed because they did not take into account the broader, hostile, profit-driven system in which those costs were generated. For example, faced with rising costs under the Medicaid and Medicare programs, the government since 1983 has used a system of **diagnosis-related groups (DRGs)** that sets an average length of hospital stay and cost of inpatient treatment for each possible diagnosis. Under this prospective reimbursement system, the government determines in advance each year the amount it will pay hospitals per patient based on the average cost of treating someone with a given DRG. If the hospital spends less than this amount, it earns money; if it spends more, it loses money. Theoretically, then, the DRG system should have limited the costs of providing care under Medicaid and Medicare. Instead, and taking advantage of the fact that patients often have multiple illnesses and that the same symptoms often suggest more than one diagnosis, doctors and hospitals now sometimes use sophisticated computer software to identify the most remunerative, but still plausible, diagnosis for a given patient—a process known as DRG creep. In addition, hospitals responded to the adoption of the DRG system by shifting services to outpatient units (where the DRG system does not apply) and by increasing the number of patients they admitted. As a result, the DRG system only marginally reduced government costs for hospital care. Similarly, when the government restricted the fees it would pay health care providers for treating Medicare and Medicaid patients, providers increased the fees they charged other patients.

Declining Coverage

Uninsured Americans

The rising costs of care have led directly to declining coverage. Whereas in 2000 about 40 million Americans were uninsured, by 2004, about 45 million Americans—18 percent of the population under age 65—were uninsured (Kaiser Commission on Medicaid and the Uninsured, 2004.) Moreover, almost two-thirds of these individuals have lacked insurance for two years or more.

Because Medicare covers almost all Americans over age 65, health care coverage is essentially a problem of the young and middle aged. As Figure 8.3 shows, lack of health insurance affects substantial portions of all age groups below age 65, but is especially acute among working-age adults (the population least likely to be covered by government health care programs). Many of these individuals simply cannot afford to purchase health insurance; others are in good health and so do not feel it is worth purchasing insurance if it is expensive.

As described earlier, insurance in the United States is typically linked to employment, with about two-thirds of Americans receiving insurance through their employer or a family member's employer. This system is far from perfect, however. Over the last decade or so, employers have kept profits high by reducing the benefits they offer to full-time employees and hiring more part-time and temporary workers without benefits. Consequently, in 2004, 69 percent of the uninsured live in families with one or more full-time workers (Kaiser Commission on Medicaid and the Uninsured, 2004). Not surprisingly, insurance coverage also varies by income level, with poor and near-poor individuals making up two-thirds of the uninsured (Figure 8.3).

Size of employer also affects the likelihood of insurance coverage. Almost all firms with 200 or more employees, compared to only 37 percent of smaller

Figure 8.3   *Uninsured Americans, 2004*

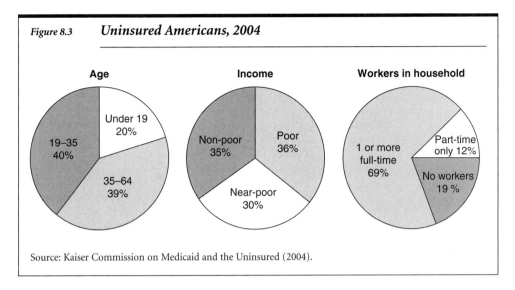

Source: Kaiser Commission on Medicaid and the Uninsured (2004).

firms, offer health insurance (Kaiser Commission on Medicaid and the Uninsured, 2004). Three factors explain why persons who work in small firms or are self-employed are most likely to be uninsured. First, whereas large firms can spread the administrative costs of insurance over many employees, small firms and self-employed persons cannot. Consequently, although those costs pose a minor nuisance for large firms, they can make insurance prohibitive for small firms and self-employed persons. Second, large firms, unlike small firms, have enough ready capital to **self-insure**— putting aside a pool of money from which to pay all health care expenses for their workers rather than purchasing insurance from a commercial provider. Because self-insuring costs less than buying insurance, large firms that self insure can better afford to insure their workers. Third, insurers are more willing to offer lower rates to large firms because they assume that any money they might lose paying for the health care of ill employees will be more than counterbalanced by the money they earn on the many healthy employees in the same firm.

Women and men are equally likely to be uninsured, but ethnicity plays an important role: About 33 percent of Hispanics, 25 percent of Native Americans, 21 percent of African Americans, and 20 percent of Asian Americans are uninsured, compared to 13 percent of white Americans (Kaiser Commission on Medicaid and the Uninsured, 2004).

Insurance coverage also varies by state, with insurance less common in those states that provide less-generous Medicaid coverage, have higher proportions of residents who work for small firms, or have higher proportions of poor residents. The chances of a person being uninsured are about twice as high in parts of the South and Southwest when compared to the Upper Midwest, for example.

Finally, insurance coverage varies by health status. Ironically, health insurance is hardest to get when a person actually needs it. In most jurisdictions, insurers legally may reject any applicants for individual health insurance who do not pass a series of medical tests and have clean health records. Consequently, although most uninsured adults are healthy, a minority is much sicker than the rest of the population.

Paradoxically, not only have rising costs led to declining coverage, but declining coverage has also led to rising costs. As the costs of coverage have increased, many healthy people have concluded that they cannot afford insurance. Those who know they have health problems, however, more often decide that they must purchase insurance regardless of its costs. Consequently, compared with the past, a higher proportion of insured Americans are ill. To maintain their financial stability, therefore, insurance companies must increase prices, driving away still more healthy persons. This process creates a **rate spiral** in which increasing costs and declining coverage each foster the other.

Underinsured Americans

In addition to those who have no coverage, many more Americans have insurance that leaves them with more medical bills than they can afford to pay. These problems stem from required premiums, deductibles, and copayments; long waiting periods before insurance covers preexisting conditions; caps on insurance reimbursement per treatment, per year, or per lifetime; and lack of insurance for certain costs, such as nursing-home care and prescriptions. Data collected during 2003 indicate that 16 million Americans—most either chronically ill or with low to moderate income—are underinsured (Schoen et al., 2005). Just over half of underinsured Americans went without needed medical care during 2003 because they could not afford it, and just under half already have medical bills they cannot pay. Medical bills are responsible for between one-third and one-half of all personal bankruptcies in the United States, even though most people who file for bankruptcy have health insurance (Sered and Fernandopulle, 2005). Another large national survey conducted in 2001 found that 8 percent of Medicare recipients, 8 percent of adults with insurance through their employers, and 26 percent of Medicaid recipients could not afford to purchase a drug their doctors had prescribed (Pear, 2002b). Because of these problems, many who live near Mexico purchase health care or prescription drugs there, and many who live near Canada fraudulently use the Canadian health care system (Rosenau, 1997; Vuckovic and Nichter, 1997).

Other Americans face financial difficulties not because they lack sufficient insurance but because they cannot get their insurers to pay for their care (Light, 1992). For example, in the past, once an individual had belonged to a plan for about six months, his or her insurance generally would cover any medical bills for preexisting conditions. Now, however, insurers sometimes demand new contracts each year, with new lists of preexisting and excluded conditions. In addition, insurers can adopt near-impossible rules and proce-dures to avoid paying individuals' bills, such as requiring individuals to obtain insurer approval within twenty-four hours after receiving emergency care or assigning insufficient personnel to staff claims department telephones.

Precariously Insured Americans

Finally, in addition to the millions of Americans who are uninsured or under-insured, many more are precariously insured—liable to lose their insurance coverage at any time. Those who receive Medicaid lose their coverage once their income rises above a specified ceiling. Those who receive their insur-ance as part of a family plan can lose their insurance following divorce. Those who are covered through their own employment can lose coverage if they change to a job that does not offer insurance or where the insurance does not cover health problems they developed earlier. Finally, those whose employers self-insure (thus avoiding state insurance regulations) or negotiate a new yearly contract with an insurance company may have their insurance dropped if they or a family member becomes ill.

The Consequences of Declining Coverage

The decline in health care coverage in the United States has directly affected the use of health care services and indirectly affected health outcomes among the uninsured and underinsured.

Individuals who do not have health insurance still sometimes can obtain health care. Federal, state, and some local governments provide clinics and public hospitals that offer low-cost or free care. In addition, governments sometimes provide low-cost or free vaccination, cancer screening, and "well child" programs. These facilities and programs, however, are not always geo-graphically accessible to those who need them. In addition, these facilities are continually underfunded, so individuals may have to wait hours for emergency care and weeks or months for nonemergency care.

Uninsured persons also sometimes can obtain health care through the private sector. First, some individuals can find private doctors who will reduce or waive their fees, and some live in communities where nonprofit hospitals offer inexpensive outpatient clinics. Second, uninsured persons can obtain care for both acute and chronic, emergency and nonemergency health problems from hospital emergency rooms; although emergency rooms legally can refuse care to anyone who is medically stable, many provide at least basic treatment to all who present themselves. As a result, emergency rooms around the country have become primary care providers for those who cannot afford care, even though the services they offer only poorly match the needs of these individuals and could be provided at far lower costs elsewhere. Finally, uninsured persons increasingly have volunteered for experimental trials of new drugs as a way of receiving sporadic treatment (Kolata and Eichenwald, 1999). Yet in such experiments some patients will receive **placebos,** some will receive drugs that prove ineffective, and some will receive drugs that prove harmful. Moreover, even if the drugs work well, patients receive only temporary benefit, because the drugs become unavailable once the experiments end.

Depending on where they live, therefore, uninsured persons may have some access to health care. However, this access is substantially less than that available to other Americans. According to a large national random survey by the nonprofit, nonpartisan Kaiser Commission on Medicaid and the Uninsured (2004), 47 percent of the uninsured (compared to 15 percent of the insured) had delayed seeking needed care due to costs. Similarly, 37 of the uninsured (compared to 13 percent of the insured) had not filled needed prescriptions. Uninsured persons are also significantly less likely than others to receive basic preventive health care, such as physical examinations, blood pressure checks, pap smears, and mammograms. Because of these differences in access to care, the health problems of uninsured persons are usually worse and more difficult to treat than those of insured persons.

When uninsured persons do seek health care, they typically receive less care, of lower quality, than do insured persons, even in life-threatening emergencies. For example, a thorough review of published research conducted by the prestigious federal Institute of Medicine (2002) found that compared with other Americans, uninsured Americans injured in car accidents were less often admitted to hospitals, received fewer services when admitted, and were substantially more likely to die from their injuries. Because of both undertreatment and lower quality of care, uninsured Americans are 25 percent more likely than other Americans are to die in any given year (Institute of Medicine, 2002).

Why the United States Lacks National Health Care

Why is the United States the only industrialized nation that does not guarantee access to health care for its citizens? The answer to this question reflects the particular history, politics, and culture of this country.

As Chapter 10 will describe, since the nineteenth century the government has provided free care to indigent persons at hospitals scattered around the country. Many Americans, however, live in areas not served by such hospitals. Moreover, hospitals focus on providing intensive high-technology care, not the primary care individuals more often need.

Concern about the lack of basic health care coverage for the poor (as well as the middle class) first surfaced during the first half of the twentieth century. In 1912, Theodore Roosevelt proposed a national health insurance system as part of a broader package of "Progressive Era" programs during his unsuccessful presidential campaign. Twenty years later, when poverty rates soared during the Great Depression and fears of a socialist uprising were rampant, President Franklin D. Roosevelt supported including national health insurance in the new Social Security program. His successor, Harry Truman, supported a similar plan. In each case, however, **stakeholder mobilization**—organized political opposition by groups with vested interest in the outcome—stymied the proposals (Quadagno, 2005).

Opposition to national health care came from numerous sources, each of which benefited from having organizational strength at the local, state, and national levels (Quadagno, 2005). During the first half of the twentieth century, probably the most important opponent of national health care proposals was the AMA, which feared that such proposals might reduce doctors' incomes or autonomy. More surprisingly, labor unions opposed national health insurance because it would eliminate one of the major benefits they could offer members: the ability to press employers to offer health insurance. In addition, national health care was opposed by conservative politicians who considered it socialistic and by Southern politicians who feared it would force racial integration of health care facilities. Meanwhile, the development of Blue Cross and Blue Shield in the mid-1930s freed most middle-class Americans from worrying about paying their health care bills. As a result, popular support for national health care among this important segment of the voting public declined, leaving insufficient stakeholder mobilization in favor of national health care to defeat its opponents (Quadagno, 2005; D. Rothman, 1997).

By the 1960s, however, it had become apparent that access to health care was a major problem among the poor and the elderly, including those who had enjoyed middle-class status earlier in life. Reflecting the rise of the civil rights movement, the growing belief in the power and obligation of government to improve Americans' lives, and the shift of labor unions toward supporting national health care, Congress in 1965 authorized the Medicare and Medicaid programs. These programs, however, only partially and temporarily solved the problem. But by alleviating middle-class Americans' guilt over the suffering of the poor and fears of being impoverished by medical bills in old age, passage of these programs reduced public pressure for national health care.

Such pressures began simmering again during the late 1980s and early 1990s, as more and more Americans found themselves uninsured or otherwise unable to pay their health care bills. These problems led President William J. Clinton to propose his Health Care Security Act (HCSA) in 1993.

The HCSA represented a liberal approach to health care reform. If adopted, the act would have broadened access to care without seriously threatening the basically entrepreneurial nature of the U.S. health care system or the power of the "big players" in health care. Under the HCSA, Americans still would have received health insurance from many different insurers, retaining the complexity and costs of the current system. Wealthier Americans would have retained the right to purchase health care options unavailable to others, and so health care would have remained a two-class system. And the proposal included no oversight mechanisms to restrain the costs (and profits) of hospital, drug, or medical care.

Nevertheless, opposition to the plan was fierce, especially from the insurance and pharmaceutical industries. Even though the HCSA was designed to limit the threat to these industries, they still feared government oversight and price controls. In addition, small businesses feared that the plan would shift too many costs to their shoulders. These groups poured millions into fighting the bill, outspending those who favored it by a ratio of 4 to 1 (Quadagno, 2005: 189). In the end, Congress rejected it without even a floor vote.

The defeat of the HCSA showed once again the importance of stakeholder mobilization, even though the stakeholders were different from those in previous battles. In addition, this defeat illustrated the difficulties of developing a coherent and acceptable plan for completely overhauling a complex health care system. It also illustrated how antitax sentiment and distrust of "big government" has become a powerful force in U.S. politics, making it difficult to generate support for governmental programs (D. Rothman, 1997; Skocpol, 1996). Nevertheless, surveys consistently find that most Americans support health care reform, are willing to pay more taxes to fund health care, and believe that the government should play an important role in providing care to citizens.

Conclusion

As we have seen, Americans obtain their health care through a wide range of funding mechanisms, from publicly subsidized health care programs to private fee-for-service insurance to nonprofit health maintenance organizations (HMOs). Although some Americans have nearly unlimited access to health care—including unneeded and potentially dangerous care—others lack access to even the most basic health care. As a result, the United States must cope simultaneously with economic and health problems caused by both overuse and underuse of health care services.

Whether we choose to tackle these dilemmas depends on how we—both individually and as a nation—define the situation. If we view obtaining health care as an individual responsibility, we are likely to oppose any attempts to extend government sponsorship of health care. However, if we view health care as a basic human right, we are likely to support extending health care to all. At the same time, regardless of whether we view health care as a right, we may support health care reform as a means of protecting the nation's economy; many corporations, for example, have begun lobbying for health care reform because they believe the money they spend on insuring their employees places them at a disadvantage compared with manufacturers in other nations that have national health care systems.

For those who believe reform is necessary, the question of *how* to reform the system becomes paramount. In the next chapter we grapple with this question.

Suggested Readings

Angell, Marcia. 2004. *The Truth About the Drug Companies: How They Deceive Us and What to Do About It*. New York: Random House. Former *New England Journal of Medicine* editor Angell explains how the pharmaceutical industry has grown so powerful and wealthy, and what consumers can do to protect their health and the health care system.

Himmelstein, David U., and Steffie Woolhandler. 2001. *Bleeding the Patient: The Consequences of Corporate Health Care*. Monroe, ME: Common Courage. A series of charts and tables that succinctly explains and describes many of the problems with health care in the United States.

Sered, Susan Starr, and Rushika Fernandopulle. 2005. *Uninsured in America: Life and Death in the Land of Opportunity*. Berkeley: University of California Press. Explains in gripping detail who the uninsured are and what happens to individuals' health, income, and lives once they lose their health insurance.

Getting Involved

People's Medical Society. 462 Walnut St., Allentown, PA 18102. (610) 770-1670. www.peoplesmed.org. A consumer organization that investigates the cost, quality, and management of health care; promotes self-care and alternative health care procedures; and represents consumer interests in health care.

Review Questions

What is the nature of Blue Cross/Blue Shield insurance, and how does it differ from commercial health insurance?

Why did the originators of health maintenance organizations believe HMOs would provide better health care at lower cost than would traditional insurers?

What is managed care? How can it restrain health care costs, and how can it harm individuals' health?

What are Medicaid and Medicare?

Why have health care costs in the United States risen?

Who are the uninsured?

Why do individuals who have health insurance still sometimes face financial difficulties in paying their health care bills?

How can individuals lose their health insurance?

How does lack of insurance affect health care and health status?

Internet Exercises

1. Find the website for the nonprofit Kaiser Family Foundation. Search the website for information on the characteristics of uninsured children.

2. Find the website for the nonprofit Consumers Union, and then find its section on health care. What does Consumers Union believe are the most serious problems in the U.S. health care system? What sorts of strategies does Consumers Union propose for relieving those problems?

3. Go to the website for the University of California's Survey Documentation and Analysis (SDA) Archive. This archive contains data from several national random surveys. Enter the archive, and then click on the GSS Cumulative Datafile, 1972–2002, full analysis. *Bookmark this page.* Select "browse codebook," and then click on "start." Next, click on "Standard codebook." On the left side of your screen, under Indexes, click on "Alphabetical Index." Once you get to the alphabetical index screen, you will see that the left-hand side of that window shows the mnemonic names for all the variables in the General Social Survey, with a brief description of the variable to its right. Click on each of the questions on opinions of HMOs (HMO1, HMO2, and so on). At this point, you'll be able to see what percentage of respondents answered each of the different questions.

   To find out how different groups felt about these questions, go back to the page you bookmarked. This time, select "Frequencies or Crosstabulations." Then click on "Start." A form with several blank spaces will appear on your screen. For row variable, type *HMO1*. For column variable, type *class*. Click on the boxes to the left of "Column Percentaging," "Statistics," and "Question Text." Then click the button to "Run the Table." Repeat, using first *sex* and then *health* as the column variables. Do the same thing, using as your row variable *HMO2* and then *HMO3*. Which groups have the most positive opinions of HMOs? Which groups have the least positive opinions?

CHAPTER **9**

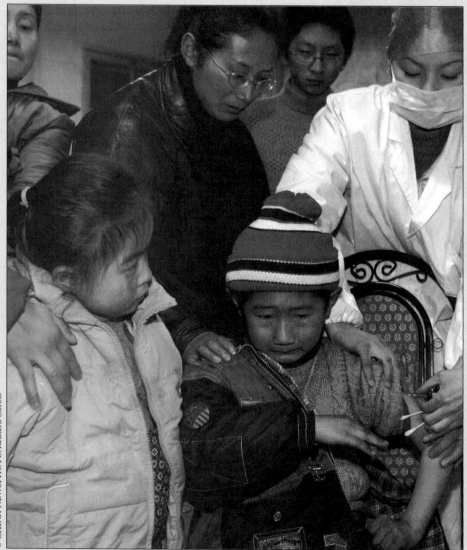

Alternative Health Care Systems

Joan Brooks, a 58-year-old grandmother, lives in Toronto. Her husband died a year and a half ago after suffering from cancer and kidney failure. He spent his last nine months in the hospital. The Ontario Health Insurance Plan covered all his medical expenses, leaving her no unpaid bills.

Brooks' only income is her husband's veterans pension—about $15,000 in U.S. dollars. But paying for medical care is not one of her worries. The Ontario health plan, to which every Ontarian belongs, covers those expenses.

She has severe arthritis and gout in both ankles and is unable to walk unless she takes prescription medicine. Not long ago, she was experiencing dizziness; her doctor suspected a drug toxicity affecting her liver and ordered a diagnostic ultrasound procedure. Brooks had the procedure within one week. She says it could have been done sooner but her schedule didn't permit an earlier appointment.

When the ultrasound revealed an enlarged liver, her doctor referred her to a specialist. Within days, the specialist admitted her to the hospital's outpatient unit and performed a needle biopsy. The Ontario plan paid the doctor about $54 for his work. Under the rules of the Canada Health Act, the doctor must accept the plan's payment, which is negotiated by the province and the provincial medical association. He cannot bill Brooks any additional amount.

Across Lake Ontario, in Buffalo, New York, insurance carriers pay doctors about $139 to perform the same procedure, and doctors can "balance bill"—that is, charge their patients more than they receive from insurers.

The biopsy shows that Brooks was suffering from excessive fat in her liver. . . . Her doctor has referred her to a nutritionist at the hospital who is helping her plan low-fat menus.

Right now she takes a drug for arthritis that costs $18 a month. The Ontario plan doesn't cover prescription drugs for Ontario residents unless they are over 65 or on welfare. But many people who are employed have drug coverage through private insurance provided by their employers. . . .

Because of the medication, Brooks' doctor checks her blood every three months to see if the dosage needs fine tuning. Ontario's health plan pays for the lab tests. When she needs to have her eyes examined, the plan pays for that checkup as well. The insurance plan covers the cost of one complete eye exam each year. Recently she had to return to the optometrist because the glasses he prescribed weren't adequate. The plan also covered the second visit, since it pays for subsequent visits if they are necessary.

Brooks . . . isn't interested in trading in Canadian health care for treatment in the U.S. "You can't buy the kindness and caring of this system," she says. *"I have no dark tales to tell."* (Consumer Reports, 1992: 585)*

On television, in newspapers, and in public discussions, we often hear that the United States offers the best health care in the world. Yet other countries—both Western and non-Western, rich and not so rich—provide far better access to care for their citizenry, at lower costs, and with better health outcomes. In this chapter, we look at alternatives to the U.S. health care system, beginning with some basic measures for evaluating any health care system and then exploring the systems in four other countries—Canada, Great Britain, China, and Mexico. The health care systems in Canada and Great Britain are ranked higher than the U.S. health care system by the World Health Organization (2000b) and are often cited as possible alternative models for the United States (Table 9.1).

Table 9.1	**Health Care Systems as Ranked by the World Health Organization**
COUNTRY	RANK
Great Britain	18
Canada	30
United States	37
Mexico	61
China	144

Source: World Health Organization (2000b).

The health care systems in China and Mexico demonstrate how poorer countries have struggled to improve their nation's health despite limited resources. Finally, we look at the prospects for reforming the U.S. health care system.

Evaluating Health Care Systems

Universal Coverage

The most basic measure of any nation's health care system is whether it provides **universal coverage,** guaranteeing health care to all citizens and legal residents of a country. The United States is the only **industrialized nation** that neither does so nor recognizes a right to health care (the topic of this chapter's ethical debate, Box 9.1). Instead, the U.S. government provides insurance to a small percentage of the population, and private insurers have nearly free rein to choose whom they will insure and at what prices. In contrast, any legal resident of Great Britain or Canada, regardless of income, place of residence, employment status, age, or any other demographic characteristic, can obtain state-supported health care—although not everything they want when they want it.

In the absence of universal coverage, uninsured U.S. citizens must do without needed care, rely on charity, or try to obtain government-funded health care. When individuals are not eligible for government-funded care, hospitals and doctors may provide care, but must make up the financial losses they incur by raising the prices they charge others through a process known as **cost shifting.** Consequently, from the perspective of the system as a whole, it is more cost-effective to plan to provide care to everyone who needs it and budget accordingly than to have to find ways to pay for that care after the fact.

Portability

A second important measure of health care systems is whether they offer portable benefits. As described in Chapter 8, most U.S. citizens receive their health insurance through their jobs, their spouses' jobs, or their parents' jobs, leaving them vulnerable to losing their insurance if their family or work situation changes. Similarly, individuals who receive **Medicaid** can lose this coverage if they move to another state or if their income rises above the legal maximum; and those who retire or go on disability often find that they cannot move to another area, because the health insurance they receive from their former employer will not cover them elsewhere. In contrast, in other developed nations individuals need not worry about losing their insurance no matter what changes occur in their personal lives.

Box 9.1   ***Ethical Debate: Is There a Right to Health Care?***

Every industrialized nation in the world other than the United States considers health care a basic right and provides all its citizens with access to health care. In the United States, on the other hand, many question whether individuals have a right to health care, and no court has ever recognized such a right.

Those who argue against a right to health care draw on the language of autonomy and individualism, stressing the rights of individuals over any socially imposed rights accruing to all members of a society (Sade, 1971; Engelhardt, 1986). Those who take this position note that in asserting individual's rights to health care, we implicitly assert that health care workers have a duty to provide that care. In so doing, therefore, we restrict the rights of health care workers to control their time and resources. If we would not force a baker to give bread to the hungry, how can we force doctors to give their services away, or restrict what patients doctors see, what services they provide, and what charges they assess?

Similarly, in asserting a right to health care, we implicitly assert that all members of a society have a duty to pay the costs of that care. When we subsequently use tax dollars to pay for health care, we restrict the rights of individuals to spend their money as they please. Some individuals, both rich and poor, might consider this a good investment, but many others would prefer to choose for themselves how to spend their money.

Moreover, according to those who take this position, asserting a right to health care fails to differentiate between unfortunate circumstances and unfair ones (Engelhardt, 1986). Although it is certainly unfortunate that some individuals suffer pain, illness, and disability, it is not necessarily unfair. Society may have an obligation to intervene when an individual unfairly suffers disability because another acted negligently, but society cannot be expected to take responsibility for correcting all inequities caused by biological or social differences in fortune.

Finally, if we assert that individuals have a right to demand certain social goods from a society, where do we draw the line? Do individuals have a right only to a minimum level of health care, or do they have a right to all forms of health care available in a given society? And, if we grant individuals a right to health care, how can we deny them a right to decent housing, education, transportation, and so on?

Those who argue in favor of a right to health care, on the other hand, draw on the

---

Geographic Accessibility

Even those who have health insurance can face obstacles to receiving care depending on where they live. Both rural areas and poor inner-city neighborhoods in the United States typically have relatively few health care providers per capita. Meanwhile, other areas have an excess of doctors—a situation that can pressure doctors to increase their prices or perform perhaps unnecessary procedures to maintain their incomes despite the competition for patients. These problems suggest that, for both economic and medical reasons, we should also evaluate health care systems according to whether they include mechanisms for encouraging an equitable distribution of doctors, such as providing low-cost loans to doctors who work in

language of social justice (Rawls, 1971). Believing each individual has inherent worth, they reject the distinction between unfortunate and unfair circumstances and the idea that health care is a privilege, dependent on charity or benevolence. Instead, they argue that each individual has a right to at least a minimum level of health care. Moreover, they argue, all members of a society are interdependent in ways that a rhetoric of individualism fails to recognize. For example, doctors who believe they should have full control over how and to whom they provide services fail to recognize the many ways they have benefited from social generosity. Medical training relies heavily on tax dollars, as do medical research projects, technological developments, hospitals, and other health care facilities. In accepting these benefits of tax support, therefore, doctors implicitly accept an obligation to repay society through the health care they provide.

Similarly, those who support a right to health care argue that to consider the decision to purchase health care as simply an individual choice misrepresents the nature of this decision, for it hardly makes sense to define something as a choice when the alternative is death or disability. Nor does it make sense to

talk about the purchase of health care as a choice when individuals can do so only by giving up other essentials such as housing or food.

Finally, those who support a right to health care recognize that society could never afford to provide all available health services to everyone, but argue that this should not limit society's obligation to provide a decent minimum of care to all. Doing any less denies the basic worth of all humans.

Sociological Questions

1. What social views and values about medicine, society, and the body are reflected in this debate? Whose views are these?

2. Which social groups are in conflict over this issue? Whose interests are served by the different sides of this issue?

3. Which of these groups has more power to enforce its view? What kinds of power do they have?

4. What are the intended consequences of the various policies under consideration? What are the unintended social, economic, political, and health consequences of these policies?

underserved areas or refusing permission for doctors to open practices in over-served areas.

Comprehensive Benefits

Another important measure of health care systems is whether they offer all the essential services individuals need. The difficulty lies in defining what is essential. Although all observers would agree that comprehensive health care must include coverage for **primary care,** agreement breaks down quickly once we begin discussing specialty care. Some individuals, for example, consider coronary bypass surgery an essential service, whereas others consider it an overpriced and overhyped luxury. Similarly, some favor offering only proce-dures necessary to keep patients alive, whereas others support offering proce-dures or technologies like hip replacement surgery, home health care, hearing aids, or dental care, which all improve quality of life but do not extend life.

Any system that does not provide comprehensive benefits runs the risk of devolving into a two-class system, in which some individuals can buy more care than others can. To those who believe health care is a human right, such a system seems ethically unjustifiable. Others object to such systems on practi-cal economic grounds, arguing that it costs less in the long run to plan on pro-viding care for everyone rather than to haphazardly shift costs to the general public when individuals who cannot afford care eventually seek care anyway.

Affordability

Guaranteeing *access* to health care does not help those who cannot afford to *purchase* it. Consequently, we also must evaluate health care systems accord-ing to whether they make health care coverage affordable, restraining the costs not only of insurance premiums but also of **co-payments, deductibles,** and other health care services such as prescription drugs.

For health care to be affordable, individual costs must parallel individ-ual incomes. As noted earlier, most insured Americans receive their insur-ance through employers. Employers typically pay a proportion of the costs for premiums and deduct the remainder from individuals' wages. To pay their share of the premiums, employers typically pass their costs on to their employees, dividing the costs equally among all employees and reducing salaries accordingly (Iglehart, 1999). As a result, low-wage and high-wage workers in essence pay (through reductions in salary) the same dollar amount for their health insurance, even though that dollar amount repre-sents a much higher percentage of income for the low-wage worker. Consequently, paying for health insurance imposes a far heavier burden on poorer persons than on wealthier persons; having to pay $3,000 per year for health insurance, for example, might force wealthier persons to scale back their vacation plans but might force poorer persons to put off reroof-ing their houses. In contrast, when, as in Great Britain and Canada, health coverage is paid for through graduated income taxes, poorer persons pay a lower percentage of their income for taxes and therefore for health care than wealthier persons. Either way—whether through taxes or lowered wages—citizens pay all costs of health care. The only difference is who pays how much.

Financial Efficiency

Another critical measure of a health care system is whether it operates in a financially efficient manner. Currently, the multitude of private and public insurers in the United States substantially drives up the administrative costs of the health care system. At the same time, the atomized and essentially entrepreneurial nature of our health care system makes it virtually impossible to impose effective cost controls. For example, in Chapter 8 we saw how the federal government now tries to restrict costs by paying hospitals prospectively for patient care based on **diagnostic-related groups (DRGs)**—paying the same fee for all patients with the same diagnosis. To maintain profits despite the DRG system, hospitals have shifted patient care from inpatient to outpatient settings where DRGs do not apply. Similarly, doctors have responded to financial limits on Medicare payments by raising the fees they charge to non-Medicare patients. For these reasons, true reform probably must include some mechanism for simplifying and centralizing control over the health care system and for restraining entrepreneurial elements.

Consumer Choice

We also need to evaluate health care systems according to whether they offer consumers a reasonable level of choice. Currently, wealthy Americans can purchase any care they want from any willing provider. In addition, Americans who have **fee-for-service insurance** can seek care from any provider as long as they can afford the copayments and deductibles and, if their plan uses **managed care,** as long as their insurer approves the care. Those who belong to **health maintenance organizations (HMOs),** meanwhile, can seek care only from providers affiliated with their plans, unless they have purchased additional coverage and can afford the extra charges. Finally, those who have Medicaid or Medicare coverage can obtain care only from providers willing to accept the relatively low rates of reimbursement offered by these programs, and those who have no health insurance can obtain care only from the few places willing to provide care on a charity basis.

No health care system can afford to grant all individuals full access to all providers. To be acceptable to Americans, however, an alternative health care system probably would need to provide at least the level of consumer choice that **managed care organizations (MCOs)** now offer and that many Americans have come to expect.

Provider Satisfaction

Finally, for a health care system to function smoothly, providers as well as consumers must feel satisfied with the system. Consequently, we must evaluate health care systems according to whether they offer health care providers an acceptable level of clinical autonomy, an income commensurate with providers' education and experience, and some control over the nature of their practices.

Health Care in Other Countries

With these measures in mind, we can now look at the health care systems in Canada, Great Britain, China, and Mexico. Canada and Great Britain guarantee portable, affordable, and universal health care coverage to their citizens. China has a stunning but now fraying record of increasing coverage, while Mexico continues its struggle to improve access to care.

Health care in the United States is primarily organized through an **entrepreneurial system,** that is, a system based on private enterprise and the search for profit. In contrast, Canadian primary care doctors, although also functioning as private practitioners paid on a fee-for-service basis, receive their payments through provincial government insurance programs. In Great Britain, meanwhile, most primary care doctors are paid through a mix of **capitation** and fee-for-service payments directly from the government, with insurance companies playing little role in health care. Finally, the Mexican and Chinese systems combine socialistic and entrepreneurial elements. (Table 9.2 summarizes the characteristics of these systems.)

Not surprisingly, each of these systems has changed over time. More interestingly, the changes seem to have moved these and other health care systems toward increasing convergence. This observation led David Mechanic and David Rochefort to propose a **convergence hypothesis,** which argues that health care systems become increasingly similar over time due to a combination of "scientific, technological, economic, and epidemiological imperatives" (1996: 242).

First, Mechanic and Rochefort argue, doctors always seek the most current medical knowledge and technology, both to improve the services they offer and to increase their incomes and prestige. In recent decades **globalization** has expanded access to such knowledge, as doctors increasingly use medical journals and Internet resources from around the world and as medical and pharmaceutical corporations market new technologies internationally. Thus doctors in many different countries are adopting the same technologies and placing similar economic pressures on their health care systems. In turn, those systems have adopted similar strictures to limit both specialization and the use of technological interventions whose benefits do not justify their costs.

Broader economic shifts also can push health care systems inadvertently toward convergence. In countries with booming economies and largely capitalist health care systems, expenditures on health care typically rise steeply, eventually resulting in efforts to contain costs through *restricting* the role of the market in health care. Conversely, countries with weakening economies and largely socialistic health care systems find it increasingly difficult to support universal health care and typically respond by adopting measures designed to *increase* the role of the market, such as allowing wealthier persons to purchase health care outside of a national health care system. Thus, both sets of countries gradually move toward health care systems in which market forces play a role, but that role is restricted by the state.

Table 9.2 **Characteristics of Health Care Systems in Other Countries**

CHARACTERISTICS	UNITED STATES	CANADA	GREAT BRITAIN	CHINA	MEXICO
Nature of system	Entrepreneurial	National Health Insurance	National Health System	In flux from national health system to entrepreneurial	Multiple options, providing unequal access
Role of private enterprise	Very high	Moderate	Low but rising	Moderate and rising	Moderate
Payment mechanism for primary care	Wide variety of mechanisms	Fee-for-service	Capitation	Primarily fee-for-service	Primarily salaried
Payment source for primary care	Mix of private, nonprofit, and government insurers	Government	Government	Primarily individuals	Primarily government
Universal coverage	No	Yes	Yes	No, but good access to primary care	Yes, but with inequitable services
Payment mechanism for hospital doctors	Salaried and fee-for-service	Salaried	Salaried	Salaried	Salaried
Payment mechanism for hospital expenses	Varied	Lump sum from government	Lump sum from government	Lump sum from government	Lump sum from government

Epidemiological changes also promote convergence. As populations have aged around the world, health care systems have had to shift more toward treating chronic degenerative diseases rather than treating acute diseases. At the same time, the globalization of knowledge has increased people's expectations regarding health and health care because they now compare themselves not only to their neighbors but also to those they see in the mass media. This shift has forced health care systems to pay greater attention to patient satisfaction and choice, while providing support for parallel systems that allow the wealthy to buy care unavailable to others.

Canada: National Health Insurance

Like the United States, Canada is an industrialized democracy made up of various provinces and territories more or less equivalent to U.S. states. Although its 2005 gross national income (GNI) per capita of $28,930 is almost 20 percent lower than in the United States, its economy is strong. In addition, because of steady immigration, Canada has a younger population than those in most industrialized nations, which guarantees it a relatively healthy population.

Canada is also, however, the second largest country in the world, with vast social differences reflecting its vast geographic spaces. Its population is highly concentrated along its southern border, as are most health care personnel and facilities. Neither health status nor health access is as good in rural areas or in its remote northern regions, where many of the residents are poor Native Americans (known in Canada as "First Nations").

Structure of the Health Care System

All Canadian health insurance is obtained through one source—the federal government—and is coordinated through the Canada Health Act. For this reason, the Canadian system is referred to as **national health insurance,** or a **single-payer system.** In fact, however, the Canadian system is a decentralized one, with each province retaining some autonomy and offering a somewhat different health care system. This brief discussion of the Canadian health care system necessarily obscures some of these differences.

The national health insurance system has evolved gradually since the late 1940s (P. Armstrong and H. Armstrong, 1998; Woodward and Charles, 2002). Underpinning the system are payments that the federal government gives the provinces yearly to run their health care systems. To receive these payments, provinces must offer comprehensive medical coverage to all residents through a public, nonprofit agency. Provinces must charge residents no more than minimal fees and must allow residents to move from one province to another without losing their coverage.

Purchasing Care

Unlike patients in the United States, most Canadians rarely see a medical bill, an insurance form, or any other paperwork related to their health care. Through a combination of federal and provincial taxes, the health insurance systems cover the costs of hospital care, medical (but not dental) care, and prescription drugs for the elderly (and, in some provinces, for younger persons). It also partially covers the costs of long-term care and mental health services. Because the system is based primarily on graduated income taxes, it is **financially progressive,** placing the heaviest financial burdens on those who can best afford it: Those who earn more money pay a higher proportion of their income in taxes and therefore pay more toward health care than do those who earn less money.

Increasingly, though, costs are being shifted to individuals. Two provinces now charge insurance premiums. (Unlike in the United States, however, those premiums are charged equally to all citizens, rather than charging higher premiums for those who have more health risks.) A growing list of services (such as in vitro fertilization and routine circumcision) are no longer considered medically necessary and so are no longer covered by the insurance system. And, as in the United States, patients are now released quicker from hospitals (where all costs are covered) to their homes (where they must pay some costs on their own).

Paying Doctors

Most Canadian doctors work in private practices and are paid on a fee-for-service basis by the government insurance systems. Doctors submit their bills directly to the health insurance system using fee schedules negotiated annually between the provincial medical associations and provincial governments. Unlike in the United States, in Canada doctors who consider these fees too low cannot **balance bill** (that is, bill their patients directly for the difference between what the patients' insurance will pay and what the doctor wants to charge). In addition, some provinces control costs by setting annual caps on the total amounts they will reimburse either each doctor or the doctors as a group. In practice, this means reimbursing doctors less for each service rendered as the number of services they bill for rises.

Although recent years have seen increasing grumbling among Canadian doctors about their incomes (Woodward and Charles, 2002), overall they express strong support for their country's health care system. Of a random sample of 3,387 Canadian doctors who participated in a nationwide survey in 1992 (the latest data available), about 85 percent preferred the Canadian system to the U.S. system (Himmelstein and Woolhandler, 1994: 265).

Several factors explain Canadian doctors' support for their system. First, Canadian doctors have retained considerably more clinical autonomy than have U.S. doctors. In addition, most Canadian doctors work in solo private practice, free from the constraints of group settings or regulations. At the same time, doctors' workloads have remained essentially unchanged since the start of national health insurance, and their incomes have remained high. Primary care doctors (about 60 percent of all doctors in Canada, compared with 13 percent in the United States) earn approximately the same net incomes in Canada and the United States, although specialists earn considerably more in the United States. Moreover, because medical education is highly subsidized by the Canadian government, Canadian doctors do not enter practice burdened by heavy debts, so their incomes go farther. (However, this could change quickly, because some provinces are about to dramatically increase medical school tuition.)

Paying Hospitals

To cover their operating costs, Canadian hospitals (most of which are privately owned) receive an annual operating budget from their provincial insurance system. Hospitals can spend their budgets as they like, with no controls imposed by the government, as long as they provide care to anyone in their region who needs services. In addition, hospitals annually receive a capital expenditure budget. Because the government controls both operating and capital budgets, it can limit both unneeded hospital growth and the proliferation of high-cost technologies.

Access to Care

Despite having national health insurance, on average Canadians retain more control over their health care than do most U.S. residents (P. Armstrong and H. Armstrong, 1998; Woodward and Charles, 2002). Whereas most Americans can receive care only from the doctors affiliated with one particular health care plan, Canadians can choose any primary care doctor they want and theoretically can switch doctors at will (although often doctors will not accept new patients, particularly if they practice in underserved areas). However, as in U.S. HMOs in the past, individuals typically must get a referral from their primary practitioner before seeing a specialist.

Nevertheless, access to care has decreased since the early 1990s, as budgetary pressure has led to reductions in federal subsidies for health care (P. Armstrong and H. Armstrong, 1996). As a result, some Canadian provinces now purchase certain high-technology services from providers in the United States (for example, sending persons from Toronto to Buffalo for chemotherapy), and some now offer a more limited package of benefits than in the past. In addition, waiting times for some procedures have grown unacceptably longer. Consequently, some Canadians now purchase certain medical or surgical services out of pocket in Canada or in the United States. Meanwhile, provinces have closed some hospitals, sparking interest among U.S. investors in buying these hospitals and turning them into fee-for-service facilities.

These problems, though real, have been blown far out of proportion by the U.S. media, leading many U.S. citizens to conclude that Canadians have far less access to health care than do persons living in the United States (Brundin, 1993). This image is almost totally inaccurate. It is true that Canadians sometimes cross the border to seek health care, but so do many U.S. residents, for both Canadian and U.S. insurers sometimes find it cheaper to pay to send their patients to the other country for services than to provide those services themselves (Lassey et al., 1997). On the other hand, no Canadian is forced to come to the United States because he or she can't afford needed medical care, whereas many U.S. citizens fraudulently claim to be Canadians to receive medical care they otherwise couldn't afford. For the same reason, many charter buses now regularly go from the United States to Canada (and Mexico) solely to allow individuals to purchase prescription drugs more cheaply. Finally, Canada, like the United States, had in past years permitted the building of unneeded hospital beds, driving up the cost of health care. By closing some of these beds and centralizing services to a smaller number of locations where staff constantly practice their skills, Canada has both lowered costs and improved the quality of care. (However, decisions regarding which hospitals to close are partially shaped by political rather than by strictly health concerns and have increased problems with accessibility in rural areas.)

It is true that affluent U.S. citizens can obtain better (or at least more) health care than the average Canadian can. When we look at the two populations overall, however, Canadians have the same, or better, access to care as U.S. citizens have on almost every measure, such as number of doctor visits per capita, number of hospital admissions per capita, and average length of hospital stay (Himmelstein and Woolhandler, 1994). Canadians do wait longer than Americans do for some forms of high-technology care, but rarely do so in life-threatening situations. Canadians also are less likely to receive some (although not all) high-technology procedures, such as coronary artery bypass graft surgery. However, this more likely reflects *overuse* in the United States than *underuse* in Canada (Himmelstein and Woolhandler, 1994). Moreover, a 1994 national random survey (Donelan et al., 1996) found U.S. citizens slightly more likely than Canadians (or Germans) to report that they were unable to get needed medical care, had postponed getting needed medical care, or had serious problems paying their medical bills (although Canadians and Germans more often reported long waits to get appointments with specialists); Table 9.3 provides details. Finally, in both Canada and the United States poorer residents have worse health and so need more surgeries performed than more affluent residents do. It is therefore not surprising that in Canada poorer persons receive more surgical procedures than do affluent persons. In the United States, on the other hand, poorer residents receive *fewer* surgeries than do more affluent persons, even though poorer persons' needs are greater (P. Armstrong and H. Armstrong, 1998: 47).

Table 9.3   ***Consumers' Self-Reported Experiences with Health Care in the United States and Canada***

	UNITED STATES (%)	CANADA (%)
Not able to get needed medical care	12	8*
Postponed needed medical care	30	16*
Had serious problems paying medical bills	20	6*
Long waits to get appointments with specialists	20	34*

*p < .05

Source: Donelan et al. (1996).

Costs of Care

In addition to improving access while maintaining quality of care, the Canadian health care system has at least partially restrained health care costs. As of 2002, the United States spent $5,267 per capita, or 14.6 percent of its gross national product, on health care; Canada spent $2,931 per capita, or 9.6 percent of its national product, on health care (Organization for Economic Cooperation and Development, 2004).

How does the Canadian system restrain health care costs? Most important, a single-payer system dramatically reduces administrative overhead. In a single-payer, nonprofit system, no one need spend money on selling insurance, advertising insurance, or paying profits to stockholders. Nor is money spent on collecting funds to run the system, for those funds are collected from the public through already-existing taxation systems. Doctors, too, have fewer expenses because they need to submit bills to only one insurer using one standard form. Hospitals, meanwhile, need not spend money tracking or collecting bills for each patient, because they receive a lump budget for the year regardless of how many patients they treat or what services those patients receive. As a result, the Canadian insurance system spends only 1 percent of its budget on overhead, compared to the 20 to 23 percent spent by the largest MCOs in the United States (Himmelstein and Woolhandler, 2003).

According to the U.S. General Accounting Office,

> if the universal coverage and single-payer features of the Canadian system were applied in the United States, the savings in administrative costs alone would be more than enough to finance insurance coverage for the millions of Americans

who are currently uninsured. [In addition,] there would be enough left over to permit a reduction, or possibly even the elimination, of co-payments and deductibles. (1991: 3)

The single-payer system also saves money by centralizing purchasing power. As the sole purchaser of drugs in Canada, the Canadian government has substantial leverage to negotiate with pharmaceutical companies regarding drug prices. As a result, Canadians pay an average of 38 percent less than Americans do for identical drugs (Himmelstein and Woolhandler, 1994: 138). Similarly, Canadian doctors, like fee-for-service doctors in the United States, can increase their incomes by increasing the number of services they perform. But as the sole payer, the Canadian government can control how much it will reimburse the doctors per service. Finally, the single-payer system restrains costs by enabling Canada to implement efficient regional planning and to avoid unnecessary duplication of expensive facilities and technologies.

The Canadian system is not, however, free of problems (P. Armstrong and H. Armstrong, 1998: 124–138). Payment of doctors on a fee-for-service basis makes it more difficult for Canada to control medical costs. When, for example, the provinces banned balance billing, doctors responded by increasing the number of services they performed (with the provinces responding by reducing the amount they reimbursed for each service). Similarly, the provinces have instituted **utilization review** boards to identify any doctors who perform medically unjustifiable tests and procedures, but have given these boards little authority to sanction doctors. Finally, as noted earlier, declining budgets have led to declining benefits across the country and have led Canadian hospitals, like U.S. hospitals, toward outpatient services and shorter patient stays, thus moving some costs from the health care system to family caregivers.

Concern about increasing costs, longer waits, and declining benefits, plus political pressure from political conservatives and corporations interested in profiting from health care provision, has resulted in the increasing incorporation of market forces into the Canadian health care system (P. Armstrong and H. Armstrong, 1998: 138–142). A major report published in 2002 by an influential Canadian Senate committee argued that an additional $5 billion is needed to improve Canadians' access to advanced medical technologies. The report argues that Canada must either raise taxes to cover these costs or allow the development of a parallel tier of health services accessible only to those who can pay out of pocket. It also recommended increasing market competition by having private, for-profit clinics compete to provide certain publicly funded services. This recommendation won support from a 2005 court decision in Quebec that upheld the right to privately purchase health insurance. Although support for the national health insurance system remains strong, that system nonetheless seems increasingly fragile.

Health Outcomes

Perhaps the most important question to ask about the Canadian health care system is how health outcomes compare with those found in the United States. The data suggest that outcomes are at least as good if not better in Canada. Canadians have lower infant and maternal mortality rates, live two years longer on average, and enjoy more years free of disability, even when ethnic differences are controlled (P. Armstrong and H. Armstrong, 1998: 79–80). Of course, these health outcomes tell us more about social conditions than about the quality of health care. Nevertheless, studies that have looked more directly at health care have reached similar conclusions. For example, a study that compared matched populations of elderly persons who received surgery in Manitoba and New England found that long-term survival rates were higher in Manitoba for nine of the ten studied surgical procedures (Roos et al., 1992). These data suggest that the Canadian health care system, although certainly not perfect, is superior to the U.S. system.

Great Britain: National Health Service

As the home of the Industrial Revolution, Britain for many decades was a leading industrial power. Along with its industrial strength came a strong labor movement, as workers united to gain political power within Britain's parliamentary government. As a result, a commitment to protecting its citizens, including a commitment to universal health care coverage, has long been a central part of Britain's identity. During the 1980s and into the 1990s, however, the nation's economy declined and conservatives took over the national government. Both these factors put Britain's health care system in jeopardy, although since 1997 a more liberal government has reinstated the nation's social and economic commitment to the national health care system. Currently, GNI per capita is $26,580, significantly lower than in the United States.

Structure of the Health Care System

Since 1911, Great Britain has provided low-income workers with subsidized care from general practitioners. Due to the sacrifices made by the British people during World War II, however, popular sentiment increasingly held that all Britons had earned the right to a decent quality of living, including access to health care. This sentiment, coupled with other social forces, resulted in the creation of the **National Health Service (NHS)** in 1946.

Whereas Canada provides its citizens with national health *insurance,* Great Britain provides a national health *service.* In Canada, the government provides insurance so individuals can purchase health care from private practitioners. In Great Britain, on the other hand, no individual need purchase health care or health insurance because the government directly pays virtually all health care costs. As a result, the two systems look quite similar to health care consumers, but differ substantially from the perspective of hospitals, health care workers, and the government.

Purchasing Care

As in Canada, British citizens can obtain comprehensive health care unburdened by bills or bureaucratic forms. The NHS uses tax revenues to pay virtually all costs for a wide range of health care services, including medical, dental, and optical care. In turn, the NHS receives its funds almost solely through general taxation, with small supplements from employers and employees who make national insurance contributions. As in Canada, because the health care system is paid for through taxes it is financially progressive.

To obtain care through the NHS, individuals first must choose a general practitioner. As in Canada, individuals can choose any general practitioner in their area who is taking new patients. But after registering with a general practitioner, they can see only that doctor (or others in his or her practice). Individuals can, however, change their general practitioner at any time, although few choose to do so. Individuals can see specialists only if referred by their general practitioner. However, individuals are free to go to emergency rooms if they feel it is needed, whether or not they see their general practitioner first. In addition, large primary care group practices as well as local governments offer a wide range of public health services, including visiting nurses for the homebound, homemakers for chronically ill persons, and long-term care.

Paying Doctors

British doctors divide sharply into ambulatory care doctors (almost all of whom work in primary care) and hospital-based doctors (all of whom are specialists). Most British general practitioners work as private contractors, although a growing number are choosing salaried government work. Those who work in private practice are paid by **capitation,** receiving a set annual fee from the government for each patient in their practice regardless of how many times they see the patient or how many procedures they perform. In addition, general practitioners receive additional payments for low-income and elderly patients to compensate for the extra expenses of caring for such patients. Doctors also receive allowances to pay for office expenses and bonus payments if they meet targets set by the government for preventive services, such as immunizing more than a certain percentage of children in their practices. In addition, to encourage access to health care, the NHS offers financial supplements to doctors who practice in medically under-served areas.

When the NHS began, most British general practitioners worked in solo practice. Over time, NHS administrators became convinced that working in group practices improved quality of care while reducing costs by enabling doctors to learn from each other and to share clerical and nursing staff. As a result, the NHS offered financial incentives to those who practiced in groups, and most general practitioners now work in groups of three or more.

To encourage doctors to enter primary care, the NHS increased capitation payments to general practitioners and added supplemental payments for house calls and other services offered by general practitioners. As a result, general practitioners now earn approximately the same incomes as specialists.

Unlike general practitioners, specialists almost always work as salaried employees of the NHS at hospitals or other health care facilities. All specialists, regardless of field, receive the same annually negotiated salary from the NHS. Specialists can, however, earn extra income through merit bonuses (usually given toward the end of a person's career) and through seeing private patients.

Income for all doctors remains far higher than for other occupations in Great Britain. Those incomes go farther than they would in the United States because municipal governments pay most costs of medical training, so British doctors enter practice virtually debt-free.

Paying Hospitals

The vast majority of hospitals in Britain, including mental hospitals, chronic disease facilities, and tuberculosis hospitals, belong to the government (although some of these include beds for private patients). Until recently, hospitals received their funds in an annual budget allocated by their regional health authority. To control costs and increase the role of market forces, the conservative government in the early 1990s encouraged hospitals to compete for patients. Specifically, hospitals were encouraged to bid against each other on contracts to cover hospital care for any patients from a given local Health Authority or general practice. To fund new hospital construction, the conservative government gave hospitals the authority to consolidate, sell land, or relocate to cheaper sites, as well as to solicit and use private funding. These policies continue under the current, more liberal government.

Access to Care

Under the NHS, individual financial difficulties no longer keep Britons from receiving necessary care. In addition, the NHS has reduced substantially the geographic inequities that for generations made medical care inaccessible to many rural dwellers, although serious deficits remain in access to care in inner-city areas.

Although Britons' access to primary care is excellent, their access to high-technology care is somewhat limited. Britain's economic decline during the last few decades has left few funds available for new hospital construction. In addition, during the 1980s and into the 1990s, conservative politicians successfully fought to keep health care funding levels significantly below that in other European countries, a problem that still continues. Consequently, although the quality of health care offered in Britain remains high, both the quality of hospital facilities and the number of hospital beds fell. British citizens thus continue to receive fewer days of hospital care per capita than do citizens of almost any other country in Europe. At the same time, the government has restricted the purchase of advanced technologies. Nevertheless, although individuals sometimes experience long waits before receiving elective surgery, no one must wait for emergency care. And whereas in the United States, those with the best ability to pay receive surgery first, in Britain those who need it the most receive it first. As a result—and as reflected in recent election results—most Britons have little interest in developing a privatized, U.S.-style health care system and instead favor a return to a strong, well-funded NHS, committed to affordable, publicly sponsored health care. Reflecting these sentiments, the British government has committed to increasing funding to the levels of the best-funded European nations by 2007 (S. Stevens, 2004).

Controlling the Costs of Care

Even with these new increases, Great Britain will be spending (as a percentage of gross domestic product) about one-third less than the United States spends on health care (S. Stevens, 2004). Like Canada, Britain has made its health funds go farther than they otherwise would through national and regional planning, and through keeping salaries relatively low. Because the government owns a large proportion of health care facilities and employs a large proportion of health care personnel, it can base decisions about developing, expanding, and locating high-technology facilities on a rational assessment of how best to use available resources and can avoid the unnecessary proliferation of expensive facilities.

Great Britain also has restrained government health care expenditures by increasing the role of market forces and shifting costs and services from the NHS to the private sector (Lassey et al., 1997). During the 1980s, the then-ruling Conservative Party refused to grant salary increases to specialists to encourage them to develop private practices. As specialists increased their private practices, they had less time for NHS patients, who soon complained of having to wait longer for specialized care. (However, it remains unclear whether patients actually had to wait longer or whether general practitioners had begun putting patients on specialists' waiting lists sooner.) Due to these problems (whether perceived or real), a small number of Britons began buying health insurance so they could buy their way more quickly into a specialist's office.

Similarly, during the 1980s the Conservative Party increased the number of beds set aside for private patients in NHS-owned hospitals. During those same years, private corporations began building private hospitals in Britain. Although these hospitals contain only a small fraction of all beds in the country, they threatened to drain personnel from the NHS by offering higher salaries and better working conditions. Meanwhile, underfunding of the NHS increased staffing pressures and waiting times at NHS hospitals, contributing to public dissatisfaction. Recent increases in funding for the NHS are expected to significantly alleviate these problems, even while political support for private hospitals continues. However, both public and private hospitals will now have to control costs under a system similar to the DRG system used in the United States (S. Stevens, 2004).

The major change in the NHS since the Labor Party took over is the change from primary care group practices to "Primary Care Trusts" (U.K. Department of Health, 2005). A Primary Care Trust is an integrated group of doctors, nurses, and other health workers involved in primary care in a given community. In theory, the development of these trusts promised a major shift of responsibility and authority from centralized control to local control. Whereas in 1997 primary care doctors controlled about 15 percent of NHS funding, by 2004 they controlled 75 percent. In exchange, the trusts are now responsible for deciding what services should be offered in their areas and how they should be structured. The goal of these changes is to move decision making closer to patients and communities and to provide better, more accessible, and more integrated patient care.

Health Outcomes

Despite the problems in the NHS, health outcomes have remained good. Infant mortality (5.3 per 1,000 live births) is lower than in the United States, and life expectancy is one year higher.

China: Good Health at Low Cost

Although many observers have proposed using the health care systems of Canada and Great Britain as models for a restructured U.S. health care system, few would seriously propose China as a viable model. China's culture differs greatly from that of the United States, and so its citizenry has very different values regarding what constitutes an acceptable health care system. In addition, China's GNI per capita of only $4,520 (Population Reference Bureau, 2004) severely limits its options, and the remaining communistic underpinnings of its economy make some health care options more feasible and some less feasible than in the United States. Nevertheless, China's story provides useful clues regarding how to provide good health to the citizenry of a poor country.

China's health care system reflects its unique history and situation (Lassey et al., 1997). When, after many years of civil war, the Communist Party in 1949 won control of mainland China, it found itself in charge of a vast, poverty-stricken, largely agricultural, and densely populated nation of about 1 billion persons. Most people lived in abject misery while a small few enjoyed great wealth. Malnutrition and famines occurred periodically, life expectancies for both men and women were low, and infant and maternal mortality were shockingly high. In urban areas, only the elite typically could afford medical care, whereas in rural areas, where most of the population lived, Western medical care barely existed.

Structure of the Health Care System

In 1950, one year after winning control of mainland China, the Communist government announced four basic principles for the new nation's health care system. First, the primary goal of the health care system would be to improve the health of the masses rather than of the elite. Second, the health care system would emphasize prevention rather than cure. Third, to attain health for all, the country would rely heavily on mass campaigns. Fourth, the health care system would integrate Western medicine with traditional Chinese medicine.

These principles reflected both the political climate and the practical realities of the new People's Republic of China. The first goal—improving the health of the masses—stemmed directly from the communist political philosophy underpinning the revolution. The years of bloodshed were to be justified by a new system that would more equitably redistribute the nation's wealth and raise the living standards and health status of China's people. The second and third goals reflected unignorable facts about China's situation. Lacking both a developed technological base and an educated citizenry, China's greatest resource was the sheer labor power of its enormous population, which could be efficiently mobilized because of its now-centralized economy. Focusing on prevention through mass campaigns promised to deliver the quickest improvements in the nation's health. Finally, the decision to encourage both Western and traditional medicine similarly recognized the difficulties China would face in developing a Western health care system, as well as the benefits of including traditional medicine in any new system. By encouraging traditional as well as Western medicine, China could take advantage of its existing health care resources and gain the support of the peasantry, who remained skeptical of Western medicine. At the same time, incorporating traditional medicine into the new, modernized Chinese health care system offered a powerful statement to the world regarding the new nation's pride in its traditional culture. Simultaneously encouraging the growth of Western medicine, meanwhile, would help bring China into the scientific mainstream.

Given its large and poverty-stricken population and its lack of financial resources and medically trained personnel, China needed to adopt innovative strategies if it were to meet its goal of improving the health of the common people. Two of these strategies are the use of mass campaigns and the development of **physician extenders**—individuals (such as nurse practitioners and physician assistants in the United States) who can substitute for doctors in certain circumstances.

One of the more unusual aspects of China's health care policy has been its emphasis, especially in the early years of the People's Republic, on mass campaigns (Horn, 1969). For example, to combat syphilis, which was **endemic** in much of China when the Communists came to power, the government first closed all brothels, outlawed prostitution, and retrained former prostitutes for other work. Second, the government began the process of redistributing income and shifting to a socialist economy so that no young women would need to enter prostitution to survive. During the next decade, the government trained thousands of physician extenders to identify persons likely to have syphilis by asking ten simple questions, such as whether the person had ever had a genital sore. By so doing, the government made manageable the task of finding, in a population of 1 billion, the small percentage that needed to be tested and treated for syphilis.

To convince people to come to health centers for testing, these physician extenders posted notices in villages, performed educational plays in marketplaces, and gave talks around the country, explaining the importance of eradicating syphilis and attempting to reduce the stigma of seeking treatment for syphilis by defining the disease as a product of the corrupt former regime rather than a matter of individual guilt. Those identified as likely to have syphilis were tested and treated if needed. These methods—coupled with testing, among others, persons applying for marriage licenses, newly drafted soldiers, and entire populations in areas where syphilis was especially common—dramatically reduced the **prevalence** of syphilis in China.

Health Care Providers

The second innovative strategy for which China has won acclaim is its use of physician extenders. In urban areas, **street doctors** (sometimes known as **Red Cross health workers**) offer both primary care and basic emergency care, as well as health education, immunization, and assistance with birth control. Street doctors have little formal training and work in outpatient clinics under doctors' supervision.

In rural areas, **village doctors** (formerly known as **barefoot doctors**) play a similar role. Village doctors were first used in 1965 during China's Cultural Revolution, a political movement started by students and fostered by some members of the national government to uproot the last vestiges of the old class structure (as well as to eliminate political dissidents). Village doctors, it was hoped, would alleviate the continued lack of health care providers in rural areas as well as reduce the political power of urban medical doctors, who remained a reminder of the precommunist elites. Novice village doctors were selected for health care training by their fellow workers based on their aptitude for health work, personal qualities, and political "purity." Following about three months of training (supplemented yearly by continuing education), village doctors returned to their rural communes, where they divided their time between agricultural labor and health care. Since the end of the Cultural Revolution in 1976, the number of village doctors has declined substantially. Training is now more rigorous, and individuals must pass an exam before entering practice, but they still receive relatively little supervision from better-trained health care workers.

Above village doctors in the Chinese health care hierarchy are **assistant doctors.** These individuals receive three years of postsecondary training similar to that received by medical doctors, during which they learn both Western and traditional Chinese medicine. Finally, at the top of the hierarchy are medical doctors. Individuals must complete a minimum of five to eight years of postsecondary training to become doctors, plus a supervised residency program to become specialists. All doctors receive training in both Western and traditional Chinese medicine and may focus on either field, although relatively few choose traditional medicine.

Purchasing Care

As China's economy has changed from a largely socialized and centrally controlled system toward a more decentralized, economically heterogeneous model, so has its health care system (Chen, 2001; Lassey et al., 1997). For the majority of urban residents, these shifts have brought few changes. As in the past, the government pays most costs of health insurance and health care for government employees, military personnel, and students. Public industries and urban industrial collectives also pay for care for their workers. The growing and now significant numbers of urban residents who work in private enterprises, however, often lack any health insurance.

For rural Chinese—about 78 percent of China's population—recent years have dramatically changed the nature of health care. Before the 1980s, rural residents received their care at little or no cost through the agricultural communes where they lived and worked. Within these communes, members shared all profits and costs, including those for health care. Each commune had between 15,000 and 50,000 members and offered its own clinic staffed by assistant doctors (also commune members) who provided both primary care and minor surgery. In addition, communes were divided into production teams of 250 to 800 people, each including a village doctor.

Beginning in the early 1980s, most agricultural communes reverted to their original non-communal village structures, with each family given land to farm by the village. Families now keep their profits, but are responsible for their own welfare should costs exceed profits. Due to this shift in financing, the former communes no longer earn sufficient revenues to continue providing health care. Many village doctors returned to full-time agricultural work, and most rural assistant doctors moved to township or city clinics. Almost all rural residents now receive their primary health care on a fee-for-service basis, and financial difficulties have forced some to cut back on needed care. In addition, waning government support for large-scale public health activities has allowed previously conquered diseases to reemerge. For example, schistosomiasis, a debilitating and sometimes deadly disease once eradicated by mass campaigns to kill the snails that carry it, is again **endemic** in some rural areas (Yardley, 2005).

Paying Doctors

Currently, ambulatory care doctors in China work primarily on a fee-for-service basis and hospital doctors work on salary. In addition, many townships (made up of six or more rural villages) have a clinic where doctors work on salary. As in many HMOs in the United States, however, these doctors can divide among themselves any profits generated by the clinic and not needed for new equipment or facilities, thus encouraging market forces to play a role in controlling costs.

Paying Hospitals

Unlike primary care, hospital care has remained largely a public enterprise. Almost all hospitals receive their operating and capital budgets from federal or local governments. In recent years, however, budgets have been cut and great pressure has been placed on hospitals to generate income through selling services and starting other enterprises.

Access to Care

Because of the changes in China's health care system, prices for health care have risen and access has diminished, especially in rural areas, where fewer hospital beds and doctors are available per capita. Although primary care remains affordable, even for those who lack health insurance, hospital care is not. To equalize access to care, the government has established a national fund to supplement the health care budgets of poorer regions and an insurance program for childhood immunizations. Those who, for a small premium, purchase this insurance receive free immunization for children to age 7 and free treatment if a child develops one of the infectious diseases the immunization program is supposed to prevent. More than half of all children in the country belong to this program. Finally, a similar insurance program offers prenatal and postnatal care to women and infants; it is not known how many are covered by this program.

Health Outcomes

As a poor country, China spends only 5.8 percent of its GNP on health care, compared with the 14.6 percent spent by the United States (World Health Organization, 2005b). Nevertheless, China's commitment to equalizing both income and health care has allowed it to attain health outcomes far greater than its economic status or investment in health care might predict. Although median income in China remains similar to that in many other developing nations, China boasts health outcomes only slightly below those of the industrialized nations. Whereas in 1960 infant mortality was 150 deaths per 1,000 and life expectancy was 47 years, as of 2004 infant mortality is

32 per 1,000 and life expectancy is 71, only 6 years lower than in the United States (Population Reference Bureau, 2004). Although large and increasing differences in health status remain between rural and urban dwellers, China now stands on the cusp of the epidemiological transition, with chronic and degenerative diseases increasingly outpacing infectious diseases as the leading causes of death. (Lung cancer, especially, is a growing problem because the government relies on tobacco products as major sources of tax revenue and export dollars and so has invested almost no funds in smoking prevention efforts.)

Nevertheless, some regions of China continue to face health problems common in developing nations, such as insufficient access to clean drinking water. The rise of a market economy has contributed to these problems (Chen, 2001). The pressure to develop profitable industries has increased water and air pollution and decreased occupational safety, especially in rural areas. Similarly, pressures on the health care system to control costs and generate profits has led to a decreased emphasis on preventive care and increased emphasis on profit-generating treatments.

Despite these problems, China's great accomplishments in improving the health of its people deserve recognition. To find the key to China's successes, we need to look beyond the nature of its health care system. This topic has been investigated through a series of studies begun by the Rockefeller Foundation (Caldwell, 1993). These studies explored how China, along with Sri Lanka, Costa Rica, Vietnam, Cuba, and several other countries, has achieved substantially better health outcomes at lower cost than have countries that spend more and have higher per capita incomes. Three factors seemed to account for these outcomes. First, health outcomes in these countries improved somewhat when access to medical care improved. Second, and more important, health outcomes improved when nations encouraged education for men and emphasized family planning for both men and women. Finally, and as explained in Chapter 4 health outcomes improved most dramatically when nations made a commitment to educating women. Once women's educational levels increased, their status increased as well, and they gained greater power to control or delay reproduction. Women's lives thus were less often cut short by childbirth, and their babies were born healthier. A rise in women's status also brought a more equitable distribution of food between women and men, so that both women and the children who relied on them for food were less likely to suffer malnourishment and more likely to survive.

Mexico: Struggling to Provide Health Care Equitably

Understanding Mexico's health care system is particularly important for U.S. citizens because Mexico shares a long and permeable border with the United States. As a result, health issues in Mexico directly affect the United States, as people (and often diseases) travel across the border in both directions to seek work or pleasure (Skolnick, 1995). In addition, both Mexicans *and* U.S. citizens sometimes cross the border to the other country to seek health care, although Mexicans more often seek basic medical care for life-threatening health conditions, whereas U.S. citizens more often seek inexpensive cosmetic surgery, dental work, or pharmaceutical products.

Mexico stands on the cusp between being an industrialized and a developing nation. As Mexican industry has developed, many have moved off the land, and now more than three-quarters of Mexico's population live in cities. Those cities contain both middle-class neighborhoods, which enjoy health and living conditions similar to those found in the industrialized nations, and impoverished slums that lack such basic facilities as running water and sewer systems. These slums are inhabited primarily by migrants from rural areas. Rural areas, especially those inhabited primarily by Indians, generally are poor, with only 37 percent having sewer systems (Pan American Health Organization, 2005). Mean GNI per capita remains only $8,800—far higher than in China, but far lower than in any of the other nations discussed in this chapter (Population Reference Bureau, 2004).

Structure of the Health Care System

Unlike any of the other countries described in this chapter, Mexico has a three-tiered system for health care: private health care for the wealthy, high quality government-provided insurance for the middle third of the population, and lower-quality government-provided services for the poor (Durán-Arenas et al., 2002; Lassey et al., 1997). This three-tiered system is a product of Mexico's unique history, in which revolutionary fervor and conservative sentiments have always counterbalanced each other and in which the social and economic division between Indians (who now make up less than 10 percent of the population) and others (who are primarily a mix of Spanish and Indian) has remained important.

Over the centuries, Mexico has experienced several revolutions—some violent and some at the ballot box. Throughout the twentieth century, these revolutions resulted in gradual improvements in the health care available to Mexico's citizens. In 1917, Mexico's new constitution first gave the federal government responsibility for health care. Simultaneously, many large estates were taken out of private control and divided into small cooperatives owned by the local peasantry. These rural cooperatives subsequently received funding from the federal government to establish local clinics, typically run by minimally trained health aides. Staffing improved during the 1930s when, responding to the revolutionary spirit of the times, the federal government established a continuing program under which all new physicians must work for one year in a rural community.

The next major change in the health care system occurred in 1942, when the government established the Social Security program and opened a network of modern health clinics and hospitals around the country. However, only salaried workers employed by private industries in Mexico's cities were eligible for Social Security and allowed to use these facilities. Since then, the system has expanded to include government employees and salaried agricultural workers, covering about half of the population by 2001 (Durán, 2002). In addition, other individuals now purchase Social Security insurance—some using their own funds and others under an experimental governmental program that subsidizes these costs for the poor.

Social Security provides a comprehensive package of ambulatory and inpatient benefits. However, some Mexicans receive considerably more and better quality benefits than others do because benefits are allocated through several separate Social Security organizations with separate clientele and budgets. For example, the Social Security organization responsible for the health care of workers in the oil industry spends twice as much per capita as does the organization responsible for the health of workers in the private sector.

Mexicans who are not eligible for health care under Social Security receive a less comprehensive package of coverage through the Ministry of Health. The ministry has expanded access to health care steadily, building clinics and hospitals in both rural and urban areas. In general, however, these facilities are inferior to facilities run by Social Security. On the other hand, the ministry also runs some of the country's best specialized hospitals, used by private patients as well as by ministry patients. In addition, the ministry and other governmental agencies have funded widespread improvements in living conditions—food subsidies, new school construction, fluoridation of water, home improvements, and sanitary water systems—which have improved the health of the population. Between Social Security and the Ministry of Health, 99.5 percent of Mexicans now have regular access to modern health care.

Despite this coverage, affluent Mexicans sometimes choose to purchase private insurance or care from private doctors on a fee-for-service basis. Although most Mexican doctors work as salaried government employees, most also take private, fee-for-service patients on a part-time basis and some work solely for private patients. Because the government does not regulate the private purchase of medicine, little is known about this sector of the health care system.

Purchasing Care

Individuals who purchase health care in the private sector have, of course, a wide choice of doctors and hospitals. Other Mexicans, however, must use the doctor or the clinic to which they are assigned for primary care (although in theory they have some choice). Copayments vary by source and type of service, but range from nominal to nonexistent.

To obtain specialty care, patients must first get referrals from their primary care doctors. Such referrals can be difficult to get, however, because of government cost controls that restrict the number of practicing specialists. For the same reason, patients who do get referrals typically have long waits before they can get appointments with specialists. As a result, many patients subvert the system by instead seeking specialty care at emergency clinics or from private doctors, if they can afford to do so.

Mexicans' access to technologically intensive care remains limited. In addition, these services are haphazardly distributed, with more services available in cities compared with rural areas and in northern regions of the country compared with the south. Consequently, some hospitals and clinics are underutilized whereas others are overburdened.

Health Outcomes

Although Mexico remains rife with social and economic inequities and resulting inequities in health, it has nevertheless achieved notable improvements in health outcomes for much of its population. Consequently, by some measures Mexico appears to have completed the epidemiological transition—cancer and heart disease now kill more Mexicans than do infectious diseases, and life expectancy is 75 (Population Reference Bureau, 2004). On the other hand, the infant mortality rate remains high (25 per 1,000 live births). In addition, poor rural Mexicans still experience health conditions characteristic of developing nations, and rates of some infectious diseases, such as malaria and tuberculosis, are rising. Nevertheless, preventive health campaigns have improved health throughout the nation: A massive vaccination program eradicated polio in 1991, and, as of 2004, 95 percent of children receive all recommended vaccinations by age 1 (Pan American Health Organization, 2005).

These health outcomes have been achieved at relatively little cost. As of 2000, Mexico spent 6.1 percent of its gross domestic product (GDP)—compared with the 14.6 percent spent by the United States—on health care (World Health Organization, 2005b).

Reforming Health Care in the United States

According to the World Health Organization (2000b), the United States spends a higher percentage of its gross domestic product on health care than do any of the other 191 member countries, but it ranks only thirty-seventh in performance in 2000 (the latest data available). Clearly, this system needs reform.

As Box 9.2 describes, Physicians for a National Health Program (along with numerous other organizations and individuals) continue to fight for a single-payer system. Even if they don't succeed, their efforts add to the political pressures that may eventually result in the incremental reform that most observers believe is more likely.

Since the defeat of President Clinton's 1993 attempt to overhaul the health care system, numerous proposals have been presented at the state and federal level to incrementally expand health insurance coverage. These proposals have

---

Box 9.2   *Making a Difference: Physicians for a National Health Program*

Physicians for a National Health Program (PNHP) has been at the forefront of the U.S. movement for universal health care coverage under a single-payer plan. More than 10,000 doctors, medical students, and other health care providers have joined the nonpartisan, nonprofit organization since it started in 1987.

To the members of PNHP, the corporate control of our current health care system makes it impossible to carry out what they consider the primary mission of physicians: to act as advocates for their patients, providing the best care they can. A national health care system, they argue, run on a nonprofit basis and funded by tax dollars, would allow physicians to provide high-quality care to all patients, rather than forcing physicians to make decisions about who they treat and how based on what will best protect their profits or their corporate employers' profits.

The core mission of PNHP is to educate health care workers and the general public about the need for universal health care as well as the need for a single-payer system to make

such care economically feasible. Because the core of its leadership is comprised of respected, nationally prominent physicians, PNHP brings considerable credibility to its arguments, which it presents often in town hall meetings, debates, conferences, medical journal articles, popular books, newspapers articles and editorials, and television and radio presentations across the nation. The PNHP website (www.pnhc.org) provides access to a speakers bureau as well as to a wealth of material on the need for a single-payer system, including press releases, articles, and PowerPoint presentations. Members engage in such activities as writing letters to newspapers and medical specialty journals, giving or arranging for lectures on health care reform at pizza parties for medical students or local medical society meetings, and lobbying legislators regarding proposed health care legislation. Through all these activities, PNHP members encourage both their fellow health care workers and other Americans to think deeply about the underlying ethical, medical, and economic issues involved in health care.

---

generally taken two forms: expanding eligibility for already-existing government-run health insurance programs or combining tax incentives with other options and regulations to make commercial insurance more affordable.

Those who favor expanding government programs have proposed, for example, extending Medicaid to children who are near-poor or to disabled persons with middle-class incomes. Such proposals have the benefit of taking advantage of existing structures rather than requiring new bureaucracies, but run the risk of straining already overburdened programs.

Those who favor making commercial insurance more affordable have proposed such tactics as providing tax credits or tax deductions to individuals to subsidize the cost of insurance or requiring all employers to provide health insurance, coupled with developing statewide insurance purchasing pools that would provide affordable insurance for small firms. These proposals present a different set of problems. With tax credits, individuals can reduce their federal taxes by the amount they have spent on health insurance, up to a set limit. All the proposals so far, however, have set limits so low that they would cover only a small portion of the cost of health insurance. As a result, these proposals seem more likely to benefit those who already have health insurance than those who currently find insurance unaffordable. Proposals offering tax deductions, which allow individuals to deduct part of the cost of health insurance from their income before calculating their federal taxes, offer even less benefit, especially to poorer persons who are in low tax brackets anyway. Moreover, the existence of tax credits or tax deductions might make it easier for employers to justify not offering health insurance to workers, thus increasing the number of uninsured Americans. Proposals to require employers to provide insurance, on the other hand, will do nothing to reduce the administrative inefficiencies built into our current system with its hundreds of insurance providers. And in either event, if more people do start purchasing private health insurance, insurance companies would likely respond to this increased demand by raising prices.

Incremental change in the health care system could also come about through state-level reforms. During the last decade, the federal government has supported innovation at the state level, allowing states to develop their own programs to serve persons eligible for Medicaid and Medicare. Some of these programs could eventually serve as models for the nation as a whole.

Hawaii's program has generated especially great interest. In 1974, Hawaii's legislators passed the Prepaid Health Care Act, which required employers to pay at least 50 percent of the cost of health insurance for all full-time employees (Neubauer, 1997). Small businesses that cannot afford to pay their share of premiums can draw subsidies from a special fund established under the act, although very few have done so. Because of Hawaii's booming economy and the resulting competition for workers, most employers voluntarily insure employees' families as well as their employees and pay more than their required 50 percent of the costs.

As in other states, elderly persons and very poor persons receive their health insurance from Medicaid or Medicare. To provide insurance coverage for the "gap group" of unemployed persons and part-time workers who earn too much to receive Medicaid but too little to purchase insurance on their own, Hawaii in 1989 established a state health insurance program (SHIP), which purchases insurance from HMOs for these individuals. By closing the insurance gap, Hawaii secured health insurance for 90.5 percent of its residents (R. Mills, 2002). Because such a high proportion of the state's population is insured, insurers can use community ratings rather than risk ratings—keeping rates affordable for all purchasers—and still remain financially viable.

In addition to ensuring a high level of coverage, the new system enabled Hawaii to achieve unusual success in restraining health care costs. In part, this success resulted from the unintended development of monopolistic, nonprofit insurance plans. About 70 percent of Hawaiians receive their insurance from one of two nonprofit insurers: the Hawaii Medical Service

Association (a **Blue Cross/Blue Shield** plan) and Kaiser Permanente (an HMO that still uses a salaried staff). Because they control such a large share of the market, these two insurers exert considerable control over medical costs. Doctors who refuse to accept their reimbursement schedules or salaries can attempt to seek patients elsewhere, but will find few patients who do not belong to these plans.

More important, Hawaii restrained costs through reducing hospital use and costs. Neither of the major insurers charges deductibles, so individuals have less incentive to put off needed care. As a result, health problems more often are caught at early stages, when treatment is relatively inexpensive. In addition, and unlike most U.S. insurers, both of these insurers pay only for stays in hospital wards, not in semiprivate rooms. Finally, Hawaii has implemented a strict system for prospectively reviewing any hospital capital expenses. Hospitals cannot purchase major equipment or construct new facilities unless they can demonstrate need for those services. Therefore, consumers need not pay the costs of maintaining unused hospital beds or duplicative technologies.

Conversely, the continued existence of Medicare and Medicaid has hampered Hawaii's ability to restrain health care costs. Because these plans do not reimburse hospitals at rates high enough to cover the actual costs of providing care, hospitals have shifted costs to patients with private health insurance. At the same time, Medicaid's and Medicare's low reimbursement schedules have produced problems in access to health care because many doctors will not accept patients who belong to these plans. To control costs, and to equalize the benefits available under SHIP and Medicaid, Hawaii in 1994 merged Medicaid into SHIP (now renamed "QUEST"). Nevertheless, costs have continued to climb (although not as steeply as in other states), largely because of nationwide economic shifts resulting in a larger pool of part-time workers who fall into the gap group. These cost increases have forced Hawaii to reduce the benefits available through its insurance program.

In sum, the Hawaii experiment demonstrates both the advantages of moving toward a single-payer, nonprofit system with strong centralized control and the problems when multiple payers—in this case, public and private insurers—continue to function in the same economic sphere. It also demonstrates the benefits available from a reasonably unified managed care system, and the difficulties of sustaining a strong system in the face of external economic pressures.

Whether a Hawaii-type program or any other program for reforming health care is adopted will likely depend on **stakeholder mobilization,** and especially on whether powerful stakeholders line up in favor of change. At this point, the most important indicator that change might come is the growing support for health care reform among major corporations, which have come to view reform as essential to controlling their costs; in a recent survey, 96 percent of corporate executives identified health care costs as a significant or critical concern (National Coalition on Health Care, 2005: 6).

As the National Coalition on Health Care (2005: 6), a nonprofit alliance that includes corporations as well as labor, consumer, and medical groups, explains:

> The escalation of health care costs is not only a health care issue; it is also a major national economic problem. As these costs rise, they eat into corporate margins, reducing the capacity of firms across the economy to grow their businesses by investing in research, new plants and equipment, and product development. Health care cost increases slow the rate of job growth by making it more expensive for firms to add new workers. . . . And double-digit premium increases—on top of what are already the highest per-worker health care costs in the world—put American firms at a steep and growing disadvantage in global markets, where they must compete against companies with much lower health care costs.

Conclusion

A critical approach to health care reform suggests that for meaningful reform to occur in the U.S. health care system, we must be willing to challenge the power dynamics underlying the current system. Once we do so, the way becomes clearer for us to learn from the experiences of countries that have reformed their health care systems. Canada's history, for example, suggests that eliminating private insurers—major power holders in the current system—can reduce costs substantially by eliminating the costs of selling, advertising, and administering the various insurance plans. Eliminating private insurers also eliminates the costs that accrue when doctors, hospitals, and other health care providers must track and submit bills for each client to each insurance company. Similarly, moving hospitals and other health care centers from private to public control, as Britain has done, and placing them under a single national authority (probably with some decision making reserved for local authorities) would give the government control over both operating and capital budgets for these facilities. As a result, centralizing control of health facilities would allow the government to restrict the duplication of services and proliferation of technologies that have driven up the costs of the existing system. By the same token, establishing a national fee schedule for service providers, such as Canada uses, would enable the government to restrict the rise of those fees. Even more control is possible if the government, like Britain's, restricts doctors to salaried practices so that doctors cannot increase their incomes by increasing the number of procedures they perform. At the same time, mandating national health coverage would guarantee a large enough risk pool to make community rates feasible and affordable while eliminating the possibility of a **rate spiral.** Finally, using income taxes to pay for health care would more equitably distribute the costs of financing the system.

Of course, any proposals incorporating a critical approach would meet major opposition from those who benefit from the current system. Such a proposal, however, would be worth fighting for.

Suggested Readings

Twaddle, Andrew C. 2002. *Health Care Reform Around the World*. Westport, CT: Auburn House. An excellent overview of fifteen health care systems, covering industrialized nations, developing nations, and formerly Communist nations.

Getting Involved

Physicians for a National Health Program. 332 South Michigan Avenue, Suite 500, Chicago, IL 60604. (312) 782–6006. www.pnhp.org. Organization of U.S. physicians for a Canadian-style health care system.

Review Questions

Define the eight measures of health care systems and explain why each is important.

What is the convergence hypothesis? What evidence of convergence can be found in the histories of health care in Great Britain and China?

How are doctors and hospitals paid in Canada? in Great Britain?

What is the difference between national health insurance and a national health service?

How does access to primary and hospital care in Canada compare with access to care in the United States?

What aspects of the health care systems in Canada and Great Britain have helped them to restrain costs? What aspects have kept costs high?

How has the rise of market forces affected health care in Great Britain?

What aspects of its health care system have enabled China to provide good health at low cost to its people?

In what ways is health care in Mexico a two-class system?

Internet Exercises

1. Choose a country you are interested in. Then use the Internet to see what you can find out about its health care system, looking for information comparable with that presented for other countries in this chapter.

2. Using the website for Health Hippo, an online archive of health law materials, find information on a current health policy issue of interest to you. What are some of the current proposals on this issue, and what are some of the arguments that have been offered for or against those proposals?

Health Care Settings and Technologies

Timothy Diamond, a sociologist who spent several years working as a nursing aide in a variety of nursing homes, recounts the following experience:

Mary Ryan, like many others, spent all day in the day room, secured to her chair with a restraint vest. "How y' doin' today, Mary?" I once asked in passing.

She answered the question with a question. "Why do I have to sit here with this thing on?"

I responded automatically with a trained answer, "That's so you won't fall. You know that."

"Oh, get away from me," she reacted with disgust. "I don't trust anyone in white anymore."

Stunned by her rejection, and not completely confident of my own answer, I passed the question on to Beulah Fedders, the LPN [licensed practical nurse] in charge.

"Beulah, why does she have to wear that thing all the time?" Beulah accompanied her quick comeback with a chuckle. "That's so they don't have to hire any more of you."

We snickered together at the humor of her explanation, but an explanation it was, and more penetrating than mine to Mary. It posed a relationship between technology and labor, and in that connection Beulah explained that the use of one could mitigate the need for the other. A different kind of answer to the same question was given during our orientation [by the home's administrator]. "The restraint vests save on incidents. . . ."

Beulah's answer was more accurate than "so you won't fall" and "vests save on incidents," because she connected them both to a common denominator—available labor. If no nursing assistant was there to be with Mary, to walk with her or anticipate her dizziness, and if she sat in the chair without a restraint and without anyone to keep an eye on her, she might have fallen, thus generating an incident. Her restraint vest saved on incidents while it saved on labor costs. (Diamond, 1992: 182)

As this story suggests, a central dilemma of the American health care system is how to provide care in profit-driven institutions, as well as in nonprofit institutions that function within a broader, entrepreneurial system. In this chapter, we look at several settings where Americans obtain health care: hospitals, nursing homes, board and care homes, assisted living facilities, hospices, and family homes. We also consider a sociological analysis of the technologies that have become such a central part of care in these different settings.

The Hospital

The Premodern Hospital

The hospital as we know it is a modern invention. Before the twentieth century, almost all Americans, whether rich or poor, received their health care at home, from friends, relatives, and assorted health care providers. Because these providers used only a few small and portable tools, hospitals were unnecessary.

Some form of institution, however, was needed for those Americans too destitute to pay for care at home and for those who had no friends or relatives who could provide care. For these individuals, the only potential source of care was the **almshouse.** Here they—along with orphans, criminals, the disabled, the insane, and other public wards—would receive essentially custodial care. Conditions in almshouses generally were appalling. Inmates often had to share beds or sleep on the floor, and rats often outnumbered humans. Hunger was common and blankets and clothing scarce. These conditions, coupled with the lack of basic sanitation, made almshouses ideal breeding grounds for disease (Rosenberg, 1987: 31–32).

Wealthy Americans considered almshouse conditions quite acceptable for those they regarded as lazy, insolent, alcoholic, promiscuous, or incurable (categories they believed included all nonwhites). By the end of the eighteenth century, however, wealthy Americans began to view these conditions as unacceptable for those they considered the "deserving" poor—the respectable widow, the worker crippled by accident, the sailor struck by illness far from home. With such individuals in mind, philanthropists decided to develop a new form of institution, the hospital, devoted solely or primarily to **inpatient** care of the "deserving" sick. These hospitals would function as nonprofit, or **voluntary, hospitals,** so named because they reflected a spirit of voluntarism, or charity, rather than a profit motive. Such institutions would protect the morally worthy poor from the degradations of living in an almshouse and associating with the morally unworthy poor.

The first two American hospitals were founded in the late eighteenth century, and a trickle of others appeared during the first half of the nineteenth century. Reflecting their origins in social rather than medical concerns, these early hospitals accepted only patients certified as deserving. Hospitals often required those seeking care to provide letters of reference from their employers or ministers (Rosenberg, 1987: 19–20). In addition, hospitals generally refused patients with chronic, contagious, or mental illnesses, making exceptions only rarely for the few who could pay for care.

Not surprisingly, given the essentially moral concerns of hospital founders, doctors played only a small role in hospital care and an even smaller role in hospital administration. Instead, hospitals relied on lay administrators or trustees, appointed more for their social status and charitable donations than for their medical knowledge (Rosenberg, 1987: 47–68). From the beginning, though, hospitals partially justified their existence by pointing to their role in medical education, and the few elite doctors who worked in hospitals derived both status and financial profit from that association.

Early nineteenth-century hospitals differed dramatically from modern hospitals. Until after the Civil War, the large ward remained the center of all hospital activity. Admissions, diagnostic examinations, surgical operations, the last moans of the dying, and ministrations for the dead all occurred on the ward in full view of other patients and staff.

Although conditions in hospitals were better than in almshouses, they remained unpleasant. Throughout most of the nineteenth century, hospitals were chaotic and dirty places. According to historian Charles Rosenberg:

> Nurses were often absent from assigned wards and servants insolent or evasive. Chamber pots [used for urinating and defecating] remained unemptied for hours under wooden bedsteads, and mattresses were still made of coarse straw packed tightly inside rough ticking. Vermin continued to be almost a condition of life among the poor and working people who populated the hospital's beds, and lice, bedbugs, flies, and even rats were tenacious realities of hospital life. (1987: 287)

These conditions, plus the severe limitations of contemporary medicine, kept **mortality** rates high and taught the public to associate hospitals with death rather than treatment.

Hospitals functioned as **total institutions** (described in Chapter 7), in which patients traded individual rights for health care (Rosenberg, 1987: 34–46). Hospital rules regulated patients' every hour, including mandating work schedules for all who were physically capable. Patients who did not follow the rules could find themselves thrown into punishment cells or frigid showers.

Engraving from *Harper's Weekly*, 1860. Museum of the City of New York. Reprinted by permission.

A ward overrun by rats in New York's Bellevue Hospital. This woman's baby was eaten by rats.

Beginnings of the Modern Hospital

Given the rigors of hospital life, the **stigma** of charity that accompanied hospital care, and the association of hospitals with death, early nineteenth-century Americans entered hospitals only as a last resort. The Civil War, however, began to change this (Rosenberg, 1987: 98–99). During the war, the need to care for sick and wounded soldiers exposed middle- and upper-class Americans to hospital care for the first time, as both patients and health care workers. Of necessity, during the course of the war, hospital organization and care improved, at least for the better-financed Union Army. These changes demonstrated that hospitals need not be either deadly or dehumanizing.

Following the war, widespread adoption of new ideas about the dangers of germs and the importance of cleanliness helped to make hospitals safer and more pleasant, as did technological changes including the development of disposable gauze and cheaper linens, which made cleanliness feasible (Rosenberg, 1987: 122–141). Concurrently, demographic changes made hospitals more necessary. The tremendous spurt in immigration, the growth of cities, and the resulting overcrowding and dire poverty made it impossible for many Americans to recuperate from serious illnesses or injuries at home. Meanwhile, the growth of industry and technology fostered accidental injuries, and poor and crowded living conditions bred contagious diseases that required hospital treatment. Medical changes, too, made hospital care more necessary, as doctors came to value the technologies and germ-free surgical conditions available only in hospitals (Rosenberg, 1987: 149).

Yet affluent Americans remained generally unwilling to tolerate the conditions on even the cleanest hospital wards. As a result, and to compete with the **for-profit, private hospitals** that began appearing during the second half of the nineteenth century, voluntary hospitals developed a class-based system of services (Rosenberg, 1987: 293–294). Those who could pay for private accommodations received better heating and furnishings, exemption from many hospital rules, and privileges such as more anesthesia during operations. In addition, as hospitals increasingly became involved in medical education, private patients retained the right to treatment by their private doctors, while charity patients endured treatment by inexperienced medical students or **residents.** Through these changes, voluntary hospitals began to lose their ethos of service and became increasingly like their for-profit competitors.

The Rise of the Modern Hospital

By the early twentieth century, the hospital as we now know it had become an important American institution and a major site for medical education and research. In the 50 years between 1873 and 1923, the number of hospitals increased from 178 to almost 5,000 (Rosenberg, 1987: 341). These new hospitals also included **government hospitals,** established to provide services to those groups—the insane, the chronically ill, and the "undeserving poor"—that voluntary hospitals considered unworthy and for-profit hospitals considered money losers. However, African Americans still could obtain care only in a few segregated, poorly staffed, and poorly funded wards and hospitals; in municipal hospitals where medical students and residents could learn skills by practicing on African American patients; and sometimes in other hospitals for emergency care (R. Stevens, 1989: 137).

This hospital building boom reinforced the class division within voluntary hospitals. According to Rosemary Stevens (1989: 112), the voluntary hospital of the early twentieth century "was like a multiclass hotel or ship, offering different facilities for different prices. The grade of semiprivate patients, tucked in between private patients and the wards, seemed the logical development of a new 'cabin class' between 'first class' and 'steerage.'" Thus by the 1920s, voluntary hospitals had abandoned much of their original charitable mission and become big businesses. As such, they had come to reflect the American ideology that individuals should get only what they pay for, in health care as in other areas (R. Stevens, 1989: 112).

By this time, surgical admissions to hospitals far surpassed medical admissions (Rosenberg, 1987: 150). Most patients went to a hospital to have their tonsils, adenoids, or appendixes removed; their babies delivered; or their injuries treated (R. Stevens, 1989: 106). The emphasis on technology as a defining aspect of modern hospitals further reinforced hospitals' tendency to focus on the care of **acute** rather than **chronic illness.**

This emphasis, coupled with hospitals' desire to maintain their image as proper middle-class institutions, created problems in the years following World War I, when hospitals proved extremely loath to deal with the chronic health problems of veterans (R. Stevens, 1989: 126–128). Many veterans were poor and suffered from crippling or disfiguring problems not amenable to the acute or surgical care that hospitals emphasized. Yet Americans generally believed that veterans had earned the right to health care. As a result, in 1921, Congress voted to establish a national system of **veterans hospitals.**

By initiating a federal system of veterans hospitals, the government gained a chance to set national norms for health care, overriding local norms of racial segregation (R. Stevens, 1989). Instead, however, the federal government bowed to local political pressure and decided to allow African American veterans to use veterans hospitals only in emergencies or in segregated wards. These policies did not change until after the civil rights struggles of the 1960s (R. Stevens, 1989: 222).

The number of hospitals increased dramatically following passage of the 1946 Hill-Burton Act, which provided funding for hospital construction. During the next 14 years, 707 voluntary hospitals and 475 state and local hospitals were built, and the rate of hospital admissions increased substantially. As with the development of the veterans hospital system, however, the federal government did not use this opportunity to develop a rational and national health care system. Instead of tying funding to regional health needs, the government allowed hospitals to pursue their private financial interests: focusing on acute rather than chronic illness; discouraging non-paying patients; reinforcing local norms of racial segregation; and buying expensive, esoteric technology even if it duplicated that owned by nearby hospitals (R. Stevens, 1989: 200–232).

Hospitals Today

Federal subsidies for hospitals expanded substantially following the implementation in 1965 of **Medicaid** and **Medicare.** These plans dramatically increased the profits available to hospitals and spurred the merger of hospitals into for-profit and voluntary hospital chains (such as Humana and Sisters of Charity, respectively). Chains controlled 45 percent of U.S. hospitals in 2000 (American Hospital Association, 2002).

As hospital profits grew, so did costs to the federal government via Medicaid and Medicare. As a result, the government for the first time developed a vested interest in controlling hospital costs. Ironically, the resulting price-control programs (described in Chapter 8) such as **diagnosis-related groups (DRGs)** have pressured hospitals to pay more attention to the bottom line and therefore encouraged voluntary hospitals, which remain the center of the hospital system, to act more like for-profit hospitals (R. Stevens, 1989: 305).

More recent cost-containment programs have especially squeezed funding for public hospitals. Under any circumstances, it is difficult for public hospitals to make ends meet, because about one-third of their patients cannot pay their hospital bills (Andrulis et al., 1996). Until recently, however, public hospitals could subsidize these patients through "disproportionate share funds" given by each state to hospitals that serve a disproportionate share of poor persons. In addition, public hospitals could subsidize nonpaying patients using grants received from the federal government for training medical residents. In the last few years, however, states instead have given some of their disproportionate share funds to **managed care organizations (MCOs)** in exchange for providing insurance coverage to Medicaid recipients. Meanwhile, the federal government has cut funding for medical residencies as a means of decreasing the oversupply of physicians. Taken in combination, these two changes have reduced budgets substantially at public hospitals, resulting in cutbacks, hospital closings, and, particularly, the closing of emergency rooms (which typically lose money for hospitals).

Concern about costs and profits also has affected the mix of services offered by hospitals (R. Stevens, 1989: 334). Hoping to increase profits by offering services that patients would pay for out of pocket (avoiding managed care restrictions altogether), a growing proportion of hospitals now offer alternative therapies such as yoga, meditation, and massage (Abelson and Brown, 2002). Similarly, because insurers (including Medicare under the DRG system) typically pay only preset amounts for inpatient surgery but give hospitals more leeway in setting prices for outpatient surgery (that is, surgery given without formally admitting the patient to the hospital or requiring an overnight stay), hospitals now offer **outpatient** surgery whenever technically feasible. As a result, outpatient surgery increased from 20 percent of all hospital surgeries in 1981 to 60 percent in 1996 (American Hospital Association, 1998). At the same time, the competitive market environment has encouraged hospitals to offer new, technologically intensive treatments even if other nearby hospitals already do so. The result has been a proliferation of technology, as Table 10.1 demonstrates. Similarly, intensive care units, almost unknown in the 1960s, were found in 66 percent of hospitals by 1998 (American Hospital Association, 1998: 151). Because of these changes, hospitals now treat an older and sicker mix of patients, most of whom suffer from the acute complications of chronic illnesses.

Conversely, as hospitals have shifted toward providing more intensive care for middle-class Americans, some (especially government hospitals) have moved, if unwillingly, toward becoming **primary care** providers for the poor. Patients who have neither health insurance nor money to pay for care will sometimes turn to hospital outpatient clinics and emergency rooms not only for treatment of acute problems, such as gunshot wounds, but also for chronic problems, such as backaches. This **emergency room abuse,** as it is defined by hospitals, aggravates exhausted medical staff and worries hospital administrators concerned about budgets. In turn, it has fostered **patient dumping,** in which voluntary and for-profit hospitals place patients, sometimes in serious medical distress, in ambulances and deliver

Table 10.1	*Proportion of U.S. Nonfederal Hospitals Owning Various Technologies, 1984 and 1998*		
TECHNOLOGY		1984 (%)	1998 (%)
Angioplasty		0	21
CT scanner		48	76
Magnetic resonance imaging (MRI)		3	44
Open-heart surgery facilities		12	18

Source: American Hospital Association (1998: 151–160).

them to the emergency rooms of government hospitals—often without informing either the patient or the receiving hospital beforehand.

In response to this problem, Congress in 1985 passed the Combined Omnibus Budget Reconciliation Act (COBRA), which made it illegal for hospitals to transfer physically unstable patients. This law, however, has not ended the problem. Between 1997 and 1999, federal investigators confirmed reports implicating 500 U.S. hospitals in patient dumping (Blalock and Wolfe, 2001). For-profit hospitals were 1.7 times more likely to dump patients than were nonprofit hospitals. These numbers undoubtedly underestimate such incidents because the groups most likely to be dumped—the poor and the powerless—are the groups least likely to file complaints.

The Hospital-Patient Experience

For many patients, a hospital stay is now a matter of only a few hours or days. For example, before World War II women typically stayed in the hospital for two to three weeks following childbirth; they now stay an average of 2.5 days. Similarly, the average stay for hospital patients overall was 4.9 days in 2002, compared with 12.5 days in 1923 (DeFrances and Hall, 2004; Starr, 1982: 158).

Certainly hospitals no longer terrify and endanger patients as they did in the nineteenth century. Yet, a hospital stay often remains alienating and frightening. The bureaucratic nature and large size of modern hospitals, coupled with the highly technological nature of hospital care, often means that the patient as individual person, rather than just a diseased body, gets lost.

The reasons behind this are obvious and, to some extent, unavoidable. First, increasingly patients enter hospitals needing emergency care. Often, health care workers must respond immediately to their needs and have no time to talk with them to ascertain their preferences—which many are phys-ically incapable of expressing in any case. Second, the highly technical nature of hospital care encourages staff to focus on the machines and the data these machines produce rather than on the patient as a whole person. In the modern obstetric ward, for example, workers often focus much of their attention on the electronic fetal monitor rather than on the laboring woman (E. Martin, 1987: 142–146). Third, as we will see in Chapter 11, medical training encourages doctors to focus on biological issues much more than on patients' psychological or social needs. At the same time, short stays make it less likely that patients will develop a personal relation-ship with either hospital staff or other patients. Fourth, as large institutions necessarily concerned with economic profitability or at least stability, hos-pitals cannot afford to provide individualized care. Instead, hospitals rely on routines and schedules for efficiency. These routines and schedules leave little leeway for individual needs or desires, resulting in such ironies as nurses awakening patients from needed sleep to take their temperature or blood pressure.

Public dissatisfaction with the often dehumanizing nature of hospital care, combined with market pressures, has led hospitals to make at least superficial changes in care. For example, since the early 1990s, most U.S. hospitals have offered people who consider the standard hospital labor and delivery rooms emotionally and physically uncomfortable the option of using a "birthing room," which offers a more home-like environment. Critics, however, note that these rooms are still filled with medical tech-nologies—such as intravenous pumps, fetal monitors, and so on—whose very presence makes their use more likely.

Nursing Homes

From the start, American hospitals focused on caring for acutely ill persons and assumed that families would care for chronically ill persons. During the course of the twentieth century, however, average life expectancy increased; families grew smaller, more geographically dispersed, and less stable; and women less often worked at home. As a result, more and more Americans needed to seek long-term care from strangers, and **nursing homes**—facili-ties that primarily provide nursing and custodial care to groups of individ-uals over a long period of time—became part of the American landscape.

The number of nursing homes has tripled since 1980. Currently there are about 15,000 skilled nursing homes in the United States, with about two-thirds run for profit (U.S. Bureau of the Census, 2004). Skilled nurs-ing homes accept only patients under a doctor's care and provide both medical and trained nursing care. In addition, the many intermediate care nursing homes in the country provide bed and board, but only less-intensive health care.

Who Uses Nursing Homes?

Researchers project that 39 percent of Americans who are now 80 will have to enter a nursing home before they die (Murtaugh et al., 1997: 213). As of 2005, about 1.6 million Americans live in nursing homes. Women comprise 72 percent of these nursing-home residents—not just because women live longer, and thus more often eventually need assistance, but because women less often have a surviving spouse who can and will care for them. Although illness and disability can force individuals into nursing homes at any age, nursing-home residents overwhelmingly are elderly: 78 percent are age 75 or older, and only 10 percent are under age 65.

On average, current nursing-home residents are sicker than were residents a decade ago. This change stems from the economic incentives built into DRGs, which have encouraged hospitals to discharge patients "sicker and quicker"— physically stable but still ill—once their bills and lengths of stay exceed the limits set by Medicare for hospital coverage. Those patients who cannot care for themselves at home often are discharged directly to nursing homes.

Although some people stay in nursing homes for only a few weeks, others stay for several years. A survey of nursing-home residents conducted by federal researchers in 1995 found that the average length of stay for all persons over age 65 was 2.3 years (Dey, 1997).

Financing Nursing-Home Care

As of 2005, nursing-home care costs at least $40,000 per person per year, and more than $100,000 in expensive parts of the country. Few Americans have private insurance that will pay these costs. Although individuals can buy **long-term care insurance** to cover the costs of nursing or custodial care, its steep price and limited benefits make it unaffordable for most. Nor can most Americans rely on Medicare to finance nursing-home care, because Medicare pays only for skilled (rather than custodial) nursing care and only for the first 150 days.

In the absence of comprehensive coverage for long-term care, nursing-home residents rapidly slide toward poverty. Those who survive long enough eventually reach the limits of any private or Medicare coverage. They may then obtain Medicaid or other public aid, but only after selling all their assets (minus their houses if they are married) and spending all their savings (minus the cost of burial expenses and minimum living expenses for their spouses).

As of 2003, Medicare covers 12 percent of all U.S. nursing-home bills, and Medicaid covers 46 percent. These programs pay the homes directly, giving residents only a small monthly stipend from which to purchase all personal items, such as cigarettes, gifts, greeting cards, phone calls, or clothes. Moreover, because Medicaid will pay only a certain amount per month for care, as residents progress from Medicare to Medicaid, nursing homes often move residents to cheaper and lower-quality facilities either within a given home or in another home.

Working in Nursing Homes

Nursing-home care is extremely labor intensive. To provide this care, nursing homes rely almost solely on **nursing assistants** (who often have no training) augmented by **licensed practical nurses** (who have completed approximately one year of classroom and clinical training).

Nationally, nursing assistants (half of whom work in nursing homes and one-quarter in hospitals) form one of the largest and fastest-growing health care occupations (Bureau of Labor Statistics, 2004). Almost all are women, and most are nonwhite. Many come from Africa, Asia, or Latin America and are not native English speakers. Often they obtain their airfare to the United States as loans from nursing agencies in exchange for signing contracts obliging them to work for those agencies until they have repaid their debt (Diamond, 1992). These contracts leave them vulnerable to unscrupulous employers because, as essentially bonded laborers, these women have no legal grounds for requesting better wages or working conditions.

In some states, nursing assistants must complete a seventy-five-hour course and pass a state examination before seeking employment, but in others nursing assistants need neither training nor experience. In 2002 those who worked in nursing homes earned an average of $32,220, in many cases by working two jobs or double shifts (Bureau of Labor Statistics, 2004; Diamond, 1992).

To understand the life of nursing-home residents and the nursing assistants who care for them, sociologist Timothy Diamond (1992) became certified as a nursing assistant and worked for several years in a variety of nursing homes. He soon concluded that the core of working as a nursing assistant is caregiving, but that those who train nursing assistants do not recognize this basic fact. Instead, his instructors taught him to recite biological and anatomical terms, measure vital signs, and perform simple medical procedures. Instructors divorced these skills from any social context or any sense that their patients were people rather than inanimate objects. Moreover, the skills Diamond most needed he was never taught, such as exactly how do you clean an adult who has soiled a diaper in a manner that preserves the individual's sense of dignity? Only by labeling this caregiving as mere physical labor could those who hire nursing assistants label them "unskilled" and treat them so poorly.

Life in Nursing Homes

Diamond's research underlines how the fates of nursing assistants and nursing-home residents intertwine and how even in the best nursing homes, the economics of a profit-driven system produce often intolerable conditions for both. According to Diamond, within nursing homes

> caregiving becomes something that is bought and sold. This process involves both ownership and the construction of goods and services that can be measured and priced so that a bottom line can be brought into being. It entails the enforcement of certain power relations and means of production so that those who live

in nursing homes and those who tend to them can be made into commodities and cost-accountable units. (1992: 172)

In this process of **commodification,** or turning people into commodities, "Mrs. Walsh in Bed 3" becomes simply "Bed 3." To keep down the price of this "commodity," only the most expensive homes provide private rooms or separate areas for residents who are dying, incontinent, smelly, or insane. Privacy, then, also becomes a commodity, which few residents can afford.

Nursing assistants, meanwhile, become budgeted expenses, which homes try to keep to an absolute minimum. According to federal researchers, 91 percent of nursing homes have insufficient staff to provide even the minimum standard of care needed (Pear, 2002c). As a result, patients across the country experience bedsores, malnutrition, pneumonia, and other avoidable health problems. To justify these low staffing levels, nursing-home administrators and owners narrowly define the caregiving that assistants provide and residents need. For example, managers may hire only enough assistants to hurriedly spoon-feed residents rather than enough to allow assistants to chat with residents while feeding them or to help residents retain their dignity by feeding themselves. Similarly, managers can keep residents drugged, strapped to chairs, on a strictly regimented schedule, and in a single central room during the day so that a few assistants can supervise many residents; nationally representative studies have estimated that on any given day, nursing homes physically restrain between 20 and 38 percent of residents (Castle and Mor, 1998). The same logic frequently leads nursing homes to reward aides who work quickly and efficiently (even if the aides must bully or coerce patients to do so) and to penalize aides who spend the time needed to offer true caring (Foner, 1994).

Although all these problems also can occur in nonprofit nursing homes, a review of data collected by federal regulators on all U.S. nursing homes found that both quality of life and quality of nursing and medical care were significantly worse in for-profit homes (Harrington et al., 2001). One reason for this is that within the profit-driven system, managers constantly stress to staff that *providing* care is less important than *documenting* care. As a sign proclaimed in one nursing home where Diamond worked, "If it's not charted, it didn't happen." For example, state regulations where Diamond worked required homes to serve residents certain "units of nutrition" each day. Consequently, each day Diamond collected the cards placed on residents' food trays that named the foods and their nutritional content. Every few months, state regulators would inspect the cards and certify that the homes met state nutritional requirements. Yet these cards bore little relationship to reality, for the appetizing-sounding names given to the foods rarely matched the actual appearance or taste of the food. Nor did the cards note if a resident refused to eat a food because it was cold, tasteless, or too hastily served. Similarly, sanitation regulations required homes to shower residents regularly but did not require that the showers be warm. Nor did they require the homes to hire enough nursing assistants so that residents who used diapers could be cleaned as soon as needed, or so that residents could get the help they needed in using the toilet and avoid the indignity and discomfort of diapers.

Problems such as these led Diamond to conclude:

> It made a certain kind of sense . . . that in the schooling and textbooks there had been no vocabulary of caring. There was no place for it in the records. Words that concerned how to be gentle with Arthur, firm with Anna, delicate with Grace; how to mourn with Elizabeth and mourn for Frances; how to deal with death and dying, loneliness and screaming; how to wait in responding to someone else's slow pace—these constituted much of the work as it went along, but nothing of the job. In the documentation there was nothing relational, no shadow of the passion, only a prescribed set of tasks a doer gave to a receiver. (1992: 163)

Board and Care Homes

Nursing homes were developed to provide long-term care to individuals who did not need hospital care but who required too much medical or nursing care to live on their own. Other individuals, however, require neither medical nor nursing care but do need assistance in routine daily tasks such as bathing, dressing, and meal preparation. This group has grown substantially in recent years, due partly to the aging of the American population, the increasing survival rates of severely disabled infants, and **deinstitutionalization** (described in Chapter 7). Recognition of this market has stimulated the growth since the mid-1980s of **board and care homes**—residential facilities, typically based in private homes with shared baths, that provide assistance in daily living but neither nursing nor medical care. Although some homes serve as many as twenty-five clients, many more are family homes with as few as one client.

Board and care homes remain largely unregulated, and licensure is not required in all states. As a result, only minimal data on these homes are available. The absence of regulation, coupled with the dependence of residents and the emphasis on profits, increases the potential for physical as well as emotional abuse in board and care homes.

Assisted Living Facilities

Like nursing homes and board and care homes, **assisted living facilities** have experienced explosive growth in recent years. Assisted living facilities provide fewer medical and nursing services than do nursing homes but more than do board and care homes, and they offer greater independence and privacy than either of these. More than 500,000 individuals, with an average age of 84, now live in these facilities, the number of which increased 30 percent between 1998 and 2000 alone (*Consumer Reports,* 2001).

Unlike nursing homes, which typically consist of wards, assisted living facilities typically consist of small private or semiprivate apartments. Like nursing homes, they provide help with basic tasks of daily living (such as meal preparation and housecleaning) and with routine nursing tasks (such as administering medications). These facilities also typically offer some medical care, although most states forbid them from caring for persons who have unstable medical conditions or require around-the-clock nursing. In addition, assisted living facilities offer local transportation and social activities for those who are reasonably healthy as well as the opportunity to transfer to nearby units with higher levels of care for those whose health deteriorates.

The promise of assisted living facilities is that they will allow residents to "age in place." In fact, however, residents stay an average of less than three years, with most who leave moving to nursing homes (Chapin and Dobbs-Kepper, 2001).

Assisted living facilities were first developed in response to market demand from upper-income persons, who remain their main clientele. They have grown in number as states increasingly have looked to such facilities as a means of reducing the costs they pay for nursing-home care, which account for about 35 percent of all state Medicaid expenditures. As of 2001, thirty-seven states theoretically cover the costs of assisted living facilities—sometimes more than $4,000 per month—for those who otherwise would be placed in nursing homes at state expense (*Consumer Reports,* 2001). However, obtaining such funding is difficult, and most assisted living residents pay out of pocket.

Hospices

Origins of Hospice

Whereas nursing homes emerged to serve the needs for long-term care not met by hospitals, and board and care homes arose to serve the needs not met by nursing homes, **hospices** emerged out of growing public recognition that none of these options provided appropriate care for the dying.

Only in the last few decades has institutional care for the dying become a public issue. At the beginning of the twentieth century, few individuals experienced a long period during which they were known to be dying. Instead, most succumbed quickly to illnesses such as pneumonia, influenza, tuberculosis, or acute intestinal infections, dying at home and at relatively young ages. Now, however, most Americans live long enough to die from chronic rather than acute illnesses. In addition, as doctors and scientists have developed techniques for detecting illnesses in their earliest stages, they now more often identify individuals as having a fatal illness long before those individuals actually die. Thus, dealing with the dying is to some extent a uniquely modern problem and certainly has taken on a uniquely modern aspect.

Although modern medical care has proved lifesaving for many, its ability to extend life can turn from a blessing to a curse for those who are dying (as this chapter's ethical debate on the right to die, Box 10.1, discusses in more detail). For various reasons, including the **technological imperative** underlying medical care, legal concerns about restricting care, and financial incentives that encourage the use of highly invasive treatments, thousands of Americans each year receive intensive, painful, and tremendously expensive medical care that offers only a small hope of either restoring their quality of life or extending their lives. In nursing homes, on the other hand, the emphasis on profit making and cost cutting often results in dying persons receiving only minimal and depersonalized custodial care.

This lack of appropriate care for the dying led to the development of the hospice movement. The first modern hospice, St. Christopher's, was founded in England in 1968 by Dr. Cicely Saunders, specifically to address the needs of the dying and to provide an alternative to the often alienating and dehumanizing experience of hospital death (Mor, 1987). The hospice admitted only patients expected to die within six months and offered only palliative care (designed to reduce pain and discomfort) rather than treatment or mechanical life supports. The hospice provided care both in St. Christopher's and in patients' homes.

The hospice movement received a substantial boost with the publication of Elizabeth Kübler-Ross's book *On Death and Dying* (1969), which helped to make dying an acceptable topic for public discussion. The first American hospice, which closely resembled St. Christopher's, opened five years later in New Haven, Connecticut. Other hospices soon followed, emerging from grassroots organizations of religious workers, health care workers, and community activists seeking alternatives to hospitals and nursing homes. Public support for hospices was so immediate and so great that in 1982, only eight years after the first American hospice opened, Congress (hoping that supporting hospices would both reduce health care costs and garner votes) approved covering hospice care under Medicare (Mor, 1987: 12–14).

The Hospice Philosophy

The early hospice philosophy differed markedly from mainstream medical philosophy (Abel, 1986; Finn Paradis and Cummings, 1986; Mor, 1987). First, the hospice philosophy asserted that patients should participate in their own care and control as much as possible the process and nature of their dying. Hospices strove to give clients choices over everything from what they ate to where they would die. Most significantly, hospices allowed residents to decide when to receive pain medications, how much, and what kinds. To eliminate pain from the experience of dying, hospices used whatever drugs would work, including opiates such as heroin. In contrast, nursing-home staff do not have the expertise to prescribe or supervise the drugs that dying patients need, and hospital staff often oppose using addictive drugs because their commitment to healing makes it difficult for them to acknowledge that certain patients are dying and therefore cannot be harmed by addictive drugs.

Box 10.1   ***Ethical Debate: A Right to Die?***

In 1983, 26-year-old Elizabeth Bouvia, suffering near-total paralysis from cerebral palsy and near-constant pain from arthritis, presented herself for admission to Riverside General Hospital. In years past, and despite her physical problems, Bouvia had earned a degree in social work, married, and lived independently. However, after her efforts to have children failed, her husband left her, and the state stopped paying for her special transportation needs, she lost interest in living. Her purpose in coming to the hospital, she told the hospital staff soon after her admission, was to obtain basic nursing care and painkilling medication while starving herself to death, cutting short what might otherwise have been a normal life span. The hospital's doctors took her case to court and won the right to force feed her, on the grounds that although individuals have the right to commit suicide they cannot force health care workers to commit **passive euthanasia** (i.e., to allow patients to die through inaction).

In 1990, Janet Adkins, 54 years old and suffering from Alzheimer's disease, killed herself with the assistance of Dr. Jack Kevorkian. A pathologist, Kevorkian had designed a machine that allowed people with severe disabilities to give themselves a fatal dose of sodium pentothal and potassium in the privacy and freedom of their homes. Over the next decade, Kevorkian provided doctor-assisted euthanasia to more than 100 people. He has been charged with murder multiple times, but was first convicted in 1999, after administering a lethal injection himself, rather than having his client do so, and sending a videotape of the death to CBS-TV.

In the Netherlands, meanwhile, doctors legally can practice active voluntary euthanasia so long as they follow established guidelines. Those guidelines restrict active euthanasia, in which a doctor ends a patient's life through action rather than inaction, to cases in which mentally competent but incurably ill individuals suffering intolerable and unrelievable pain authorize their doctors in writing to give them a lethal injection. According to several national surveys conducted over the past 15 years, at least two-thirds of Americans believe that terminally ill people have a right to die, and a right to their doctors' assistance (*Contexts*, 2004). As of 2005, only one state, Oregon, has adopted a legal statute permitting doctor-assisted suicide. (The Bush administration contested the legality of that statute, but the Supreme Court issued a decision upholding it in 2006.) Even in states that lack such laws, however, some U.S. doctors engage in euthanasia or physician-assisted suicide; in a nationwide random survey of oncologists (physicians who treat cancer), 10.7 percent reported having done so at some point in their careers (Emanuel et al., 1998).

Those who support a "right to die" argue that competent adults have the right to make decisions for themselves, including the ultimate decision of dying. They argue that death sometimes can be a rational choice and that forcing individuals to suffer extreme physical or mental anguish is unwarranted cruelty.

If we accept that death can be a rational choice, then harder questions follow. Why is it rational only if one's condition is terminal? Doesn't it make even more sense to end the life of someone like Elizabeth Bouvia, whose agonies may continue for another 50 years, than to end the life of someone who will die soon regardless? Why should this choice be forbidden to individuals simply because they cannot, either physically or emotionally, carry it out themselves? And why should we allow individuals to choose death only through passive euthanasia, leaving them to languish in pain while awaiting death, if instead they could be killed quickly and painlessly?

Opponents of this view argue that the duty to preserve life overrides any other values and that euthanasia is merely a nice word for suicide or murder. They question whether Elizabeth Bouvia would still want to kill herself if she once more had the resources she needs to live independently, and they wonder whether euthanasia is merely an easy way out for a society that wants to avoid responsibility for relieving the burdens imposed by illness and disability. Opponents who have studied the Netherlands suggest that doctors there in fact do not always follow the legal guidelines, but instead sometimes end patients' lives without their consent and without first attempting to make the patients' lives worth living (Hendlin, Rutenfrans, and Zylicz, 1997). In addition, opponents question whether acceptance of euthanasia in the Netherlands explains why there are fewer hospices in the Netherlands than elsewhere in Europe and why Dutch doctors receive relatively little training in pain relief.

In sum, the use of euthanasia, whether active or passive, raises numerous difficult questions: What are the consequences of, in effect, declaring it reasonable for disabled people to choose death? What pressures does this place on individuals to end their own lives rather than burdening others? What responsibilities does this remove from society to make these individuals' lives less burdensome? Finally, given that social factors, such as age, gender, and social class, affect our perceptions of individuals' worth, how do we ensure that health care workers and courts will not be more willing to grant a right to die to those who belong to socially disvalued groups?

Sociological Questions

1. What social views and values about medicine, society, and the body are reflected in this debate? Whose views are these?

2. Which social groups are in conflict over this issue? Whose interests are served by the different sides of this issue?

3. Which of these groups has more power to enforce its view? What kinds of power do they have?

4. What are the intended consequences of the various policies under consideration? What are the unintended social, economic, political, and health consequences of these policies?

Second, the hospice philosophy foreswore regimentation and stressed the importance of integrating hospice care into clients' everyday lives rather than integrating clients into hospice routines. Where possible, hospices would offer services in clients' homes. For those who needed care in the hospice, the hospice would offer a home-like environment, without the regulations regarding schedules, visitors, food, clothing, and so on that rule life in hospitals and nursing homes.

Third, the hospice philosophy emphasized a true team approach. Because hospices provided neither diagnosis nor treatment, doctors could claim little special expertise (Abel, 1986). As a result, within hospices, doctors had little more importance or influence than did social workers, nurses, ministers, psychotherapists, nutritionists, and others. Hospices explicitly worked to minimize the authority of doctors and to increase the role and status of nonprofessional volunteers.

Fourth, hospices focused not only on the dying person but also on his or her friends and relatives. Hospices attempted to involve these others in the process of dying and to meet their social and psychological needs. As a result, hospice care did not end with the client's death but extended to bereavement counseling for survivors.

Finally, hospices viewed dying "as a natural event rather than as technological failure" (Abel, 1986: 71). Workers viewed dying as an important phase of life, suitable for and worthy of open discussion. Neither the dying process nor the disease was to be hidden.

The Cooptation of Hospice

The U.S. hospice movement has proved enormously successful, growing from one hospice in 1974 to 3,300 in 2003 and serving almost 1 million clients annually (National Hospice and Palliative Care Organization, 2001). As the movement has spread, however, it has undergone substantial **cooptation,** exchanging much of its initial philosophy and goals for social acceptance and financial support (Finn Paradis and Cummings, 1986; Mor, 1987: 17).

The history of hospice resembles the history of many other reform movements and organizations. As various sociologists have observed, successful social movements over time often come to resemble the very institutions they sought to reform (DiMaggio and Powell, 1983; McCarthy and Zald, 1973). These changes evolve gradually and naturally. For a movement to survive, it must mobilize people and develop sources of funding. To do so, reformers typically must develop hierarchies and rules, abandon their grassroots and voluntaristic approach, and hire professional staff. Battered women's shelters, for example, initially established by feminists as a radical means of protecting women from violent men, soon came to rely primarily on social workers whose goal is restoring the family unit (Schechter, 1982).

The cooptation of hospice similarly derives from natural developments in that field, especially the need to develop a stable economic base. Initially, many hospice organizers, reflecting the countercultural values of the late 1960s and early 1970s, expressed little concern for financial stability (Abel, 1986: 75). Very quickly, though, and despite qualms among some hospice organizers, hospices began to seek federal funds to support hospice development, as well as **third-party reimbursement** (that is, the ability to bill insurers for services rendered).

To gain support, organizers worked with the federal government and with the American Hospital Association to develop standards for hospice care and accreditation. The resulting standards legitimated hospice care and paved the way for Medicare and, later, Medicaid and private insurance reimbursement. Not surprisingly, they also made hospices more like hospitals.

Medicare funding and the associated federal regulations also have changed hospices and threatened the original hospice philosophy (Finn Paradis and Cummings, 1986). For example, Medicare will not reimburse hospices for costs above its set maximum number of dollars and number of days of care per patient. In addition, it will reimburse hospices for the cost of inpatient care only to the extent that inpatient care comprises no more than 20 percent of all care given. These regulations encourage hospices to accept patients who have sufficient family support to stay at home rather than in the hospice, who are near death, and whose time of death can be predicted with reasonable accuracy. In addition, to obtain reimbursement, hospices must provide services that meet specified standards and must document these services. These requirements have made it difficult for hospices to maintain their commitment to individualized care and to patient control and participation.

Medicare and private insurers also have placed limitations on who can provide care, requiring hospices to reduce their reliance on volunteers, social workers, ministers, and the like, and instead to hire professionally trained health care workers and administrators. These latter individuals often bring with them traditional ideas about health care, about the health care team, and about dying itself. Former hospital nurses, for example, might resist allowing patients to refuse intravenous feeding because that seems an unacceptable admission that health care has failed and might resist allowing patients to choose when to receive medications because the nurses prefer the ease of a hospital-like schedule (Abel, 1986: 77).

Internal pressures have forced other changes in hospice care. The original hospices were freestanding units, unaffiliated with other health care institutions. This model has proved both financially and administratively unfeasible. Freestanding hospices lacked ready access to the support services available at hospitals and other health care institutions. In addition, their independent status hampered efforts to get funding and to get referrals of patients from hospitals. As a result, although two-thirds of hospices remain nonprofit, most no longer are independent, community based, or largely volunteer run. (Box 10.2, however, describes one inspiring exception.) Hospitals or home health care agencies own most of the rest. Yet despite these changes, studies find that hospice clients and their families feel more

Box 10.2   ***Making a Difference: The Human Service Alliance***
by W. Bradford Swift with Kimberly Ridley

The Human Service Alliance (HSA) is an experiment both in delivering free health and social services and in voluntarism as a way of life. In addition to its Care for the Terminally Ill (CTI) facility, the organization runs ... a weekend respite program for families with disabled children ... and a health and wellness program for people with chronic illnesses. Even the administrative jobs here, from accounting to filing, are performed by HSA's twenty-four volunteer board members.

In 1996, HSA volunteers provided 70,000 hours of service work, the equivalent of thirty-five people working full time. They delivered an estimated $926,800 worth of services ... on a total operating budget of just $80,000, which comes from individual contributions and a few grants from area corporations. In the 11 years since the organization's inception, its methodologies have drawn the attention of administrators from nursing homes and schools of medicine and public health. One visiting physician, a cancer specialist, remarked after perusing the caregiving charts and detailed notes on each patient, "In the hospital, we cannot come close to offering this kind of attention, and having the rapport that HSA's caregivers do."

Forty-seven guests spent their final days in the CTI wing, twenty-four families utilized the services of the Respite Care Program, and twenty individuals with chronic health problems were served by the Health and Wellness Project in 1996. ...

Human Service Alliance began in 1986 when a handful of people in remote Boomer, North Carolina, started taking care of one terminally ill neighbor at a time in makeshift quarters in a refurbished trailer. ... By 1988, a core group had evolved, incorporated HSA, and moved the organization to the outskirts of Winston-Salem. They committed to operating debt-free by recruiting volunteers and raising donations before spending money. Within a few years, they had raised $400,000 to build HSA's facility for the terminally ill, which opened in 1991. ...

Perhaps among all of HSA's programs, the Care for the Terminally Ill unit is where some of the most intensive services are provided. The unit, which accommodates up to six terminally ill "guests" in private and comfortable rooms, helps fill an important gap by caring for dying individuals who don't require the medical services of a traditional hospice, but whose families are unable to care for them at home. ...

All guests accepted onto the CTI unit are selected by a committee of board members based upon the organization's ability to care for their specific needs, the guest's willingness to live out his or her final days at HSA, and the

satisfied with their care than do those who receive care from conventional sources (Mor, 1987: 150–156).

Use of Hospice

About one of every four persons who die in the United States uses hospice services, with most of these over age 65 (National Hospice and Palliative Care Organization, 2005). Whites, who make up about 75 percent of the general

family members' willingness to be a part of the process of their loved one's death. . . .

Family members are expected to visit regularly and are encouraged to volunteer some of their time serving at HSA, not because more volunteers are needed, but because it's been found that volunteering is often therapeutic for the family. Using volunteer activity in a therapeutic manner has also worked well in HSA's Health and Wellness program for clients with chronic illness. Todd Thornburg, a board member who started the program in 1988, says volunteering seems to be some of the best medicine the organization has to offer. He describes one young woman who entered the Health and Wellness program a few years ago with the complaint that her physician had ruined her knee and her life [through botched surgery]. Volunteering allowed her to redirect her focus, Thornburg says, adding that when she completed the program approximately a year later, she had a new life before her, even though she still had a knee that didn't work properly. . . .

Inspired by their experiences at HSA, a few volunteers have begun developing their own projects back home. Two free, volunteer-run hospices have opened in Jamesville and Fredericksburg, Virginia, [while] in Blue Hill, Maine, writer and former HSA volunteer

Maggie Davis launched Neighborcare, a program in which local volunteers clean, cook, run errands, and provide other help for the sick, elderly, injured, or overwhelmed in their community. "We see ourselves as filling in the gaps where people don't have what they need," Davis says. At first, Davis had in mind a center for the terminally ill, but when she met with representatives from area hospitals and social service organizations, they described more basic needs like rides to and from appointments, companionship, and simple caring. Davis put out the word, and a year and a half later, approximately seventy volunteers are ready to assist their neighbors in a handful of surrounding towns. . . .

The board members and founders of HSA hope to inspire other efforts around the nation and in other countries. But how does the average person find time in a busy life for this kind of work? "Serve in a group," suggests board member Danziger. "If eight people get together and want to serve a respite child, each could do two hours of work a week to give their parents a substantial break. The important thing is to start small and start now." . . .

Source: "Where Care Is Free." *Hope Magazine,* November– December 1997.

population, make up 81 percent of hospice users. Median length of services for hospice clients is only 22 days.

Because the early British hospices focused on cancer patients, American hospital staff from the start associated hospice care with cancer and therefore more often referred such patients to hospices. Hospices themselves are more likely to accept patients with cancer because doctors can predict their life expectancy fairly accurately, and thus hospices can assume that any cancer patient they accept will die within the six-month Medicare guide-lines. In 2003, 49 percent of hospice clients had cancer (National Hospice and Palliative Care Organization, 2005). Conversely, the greatest unmet needs are found among dying patients who do not have cancer.

Costs and Financing

Hospices depend heavily on Medicare funding. Seventy-nine percent of hospice users rely on Medicare to pay the costs. Another 13 percent rely on private insurance, 5 percent on Medicaid, and the remainder on a variety of sources (National Hospice and Palliative Care Organization, 2001).

Whether hospice care saves money compared with other options remains unclear (Mor, 1987: 177–212). Direct costs appear somewhat lower for hospital-based hospices than for traditional hospital care, but indirect costs are substantial (Kidder, 1988a, 1988b; National Hospice and Palliative Care Organization, 2001). Currently, half of hospice users die in their homes, and only 7 percent die in hospices (National Hospice and Palliative Care Organization, 2005). In these circumstances, family members provide most care. They often must take time off from work or drop out of the labor market altogether. Consequently, hospice care might not reduce the costs of caring as much as it shifts the costs from hospitals and insurers to families.

Home Care

As the discussion of hospices has suggested, most individuals who experi-ence chronic or acute health problems—whether children, working-age adults, or elderly, and whether the problems are physical or mental—receive their care at home (Abel and Nelson, 1990). This is even truer now than in the recent past due to technical, demographic, and policy changes. Because of technological advances, babies born prematurely or with birth defects and persons who suffer severe trauma are increasingly likely to survive, although often with severe disabilities that require lifelong assistance. Much of this care is now given by family members in the home.

Similarly, the rise in the numbers of frail elderly, many of whom suffer both multiple physical problems and cognitive impairments, has increased the number receiving care at home. At the same time, technological advances also have made it possible for families to provide at home treatments previously available only in hospitals, ranging from chemotherapy to respiratory ventila-tion to kidney dialysis. In addition, the movement begun in the 1960s (and described in Chapters 6 and 7) to deinstitutionalize disabled and mentally ill persons, combined with the lack of community supports for such individuals once deinstitutionalized, have shifted much of the burden of care from state institutions to the home. Finally, as described earlier, policy changes now encourage hospitals to discharge patients to their homes "sicker and quicker," in essence replacing paid hospital workers with unpaid family caregivers (Glazer, 1993).

Because of the limited public or private insurance funding for home care, most who need long-term supportive care receive services only from family members and, less often, friends. The economic value of home caregiving has been estimated at $257 billion per year, much greater than the amount spent per year on paid home care or nursing-home care (National Alliance for Caregiving and AARP, 2004). Existing data suggest that home care has little impact on the costs of care or the mental or physical functioning of ill or disabled individuals but can produce small, short-term improvements in their life satisfaction (Arno, Bonuck, and Padgug, 1995; Weissert, 1991).

The Nature of Family Caregiving

A survey conducted for the nonprofit organizations National Alliance for Caregiving and AARP (2004) found that 21 percent of U.S. households include someone who is providing care for a person over age 18; it also found that the majority of these caregivers (61 percent) are women. Ethnic minorities and poorer persons also are more likely to become caregivers, probably because these groups experience higher rates of illness and disability and have less access to formal services.

Those who care for the health needs of family members typically do so out of love and often reap substantial psychological rewards. Yet, caregiving by family members should not be romanticized, nor should the financial, physical, social, or psychological costs of caregiving be underestimated (Abel, 1990; Abel and Nelson, 1990; Arras and Dubler, 1995; National Alliance for Caregiving and AARP, 2004; Reinhard and Horwitz, 1996; Tessler and Gamache, 1994).

The financial costs of caregiving are substantial. The demands of caregiving force many to shift to part-time work or even abandon paid employment. In addition, caregivers must purchase, often out of pocket, both expensive drugs and technologies and many everyday items such as diapers and bandages. In addition, caregivers typically are responsible for purchasing a variety of services and therapies from a range of companies and health care workers.

The physical costs also can be high. Caregiving often includes exhausting tasks such as lifting physically disabled or mentally incompetent individuals, some of whom either cannot help or resist being moved. The time burdens of caregiving also can become physically draining. The typical caregiver spends more than 20 hours per week on caregiving and has done so for 4.3 years; 17 percent work 40 hours or more. These hours quickly lead to exhaustion, especially for the 59 percent of caregivers who hold paid jobs, the 37 percent who have children at home, and the 31 percent who care for more than one person (National Alliance for Caregiving and AARP, 2004).

Not surprisingly, those who report giving high levels of care (helping with numerous activities of daily living for long hours) also report substantial health problems and physical strain.

Taken together, the financial and physical burdens of caregiving often leave individuals with little time, energy, or money for social relationships. Caregivers often report feeling almost totally isolated from the world outside the household (Abel, 1990; Abel and Nelson, 1990). Family relationships, too, can suffer. For example, a mother who spends hours each day caring for an ill child might feel guilty that she cannot spend more time with her other children, and those children might resent the attention given to their ill sibling. Problems are particularly acute when the person receiving care is mentally ill and throws family routines into chaos, embarrasses other family members, or physically threatens others' safety (Reinhard and Horwitz, 1996; Tessler and Gamache, 1994).

Family life also can suffer disproportionately when caregiving requires the use of high technology within the home. John D. Arras and Nancy Neveloff Dubler suggest that this

> invasion of the home by high-tech medical procedures, mechanisms, and supporting personnel exerts a cost in terms of important values associated with the notion of home. How can someone be truly "at home," truly at ease, for example, when his or her living room has been transformed into a miniature intensive care unit? . . . Rooms occupied by the paraphernalia of high-tech medicine may cease to be what they once were in the minds of their occupants; familiar and comforting family rituals, such as holiday meals, may lose their charm when centered around a mammoth Flexicare bed; and much of the privacy and intimacy of ordinary family life may be sacrificed to the institutional culture that trails in the wake of high-tech medicine. (1995: 3)

Finally, caregiving brings with it numerous psychological costs. Caregivers can easily become depressed when their efforts cannot stop or even slow the disease process. This is especially true when caregivers must routinely inflict painful treatments on their charges or when the burdens of caregiving are unceasing, as when a parent must suction the lungs of a child with cystic fibrosis hour after hour, day after day, to keep the child from dying. Moreover, as this example suggests, caregivers also often bear the enormous psychological burden of being directly responsible for another person's life. In fact, family caregivers are now expected to manage in the home—often with little training or technical support—technology considered too complex for licensed practical nurses to manage in hospitals. Finally, caregivers of persons younger than themselves face anxieties about what will happen to their charges if the caregivers die first.

Summing up the burdens of caregiving, a woman whose husband has Parkinson's disease says:

> I need some help. I am burned out. I am locked in this house. I am used to going out to work and had to retire. I didn't plan to retire so soon. We had planned our

retirement. We never did anything before because we didn't have the same vacation time. So you do all this and then bingo! . . . Two weeks ago I had a terrible pain in my ribs. But I can't run to the doctor for every little thing. How can I leave the house? I worry, what is going to happen to him, if I have to go to the hospital. . . . Medicare pays for only part of the things we need and doesn't pay for medications. That bottle of medication cost $130. . . . Sometimes he has to go to the bathroom just when I've finished eating. It is hard to get up at that instant to do it. You feel like everything [you just ate] is going to come up. You have all these things to contend with. People don't realize that unless they are in those situations themselves. . . . You have to really see it for yourself, be in it, to know what it is like. (Corbin and Strauss, 1988: 297)

Easing the Burdens of Caregiving

The problems faced by family caregivers have led to the development of new organizations, new organizational structures, and a new occupation to ease the burdens of caregiving. Two major organizations, the National Alliance for the Mentally Ill and the National Alliance for Caregiving, are now devoted to family caregiving. Both organizations work to increase assistance to family caregivers and improve access to community-based care, and the National Alliance for the Mentally Ill additionally fights to decrease the stigma of severe mental illness.

Both **respite care** (R. Montgomery, 1992) and **family leave programs** also were developed to ease the burdens of caregivers. Respite care refers to any system designed to give caregivers a break from their otherwise unrelenting responsibilities, including paid aides who provide care in the home for a few hours, day-care centers for elderly and disabled adults, and nursing homes that accept clients for brief stays. Unfortunately, only California and Pennsylvania offer formal programs for respite care. In all other states, respite care is expensive and difficult to find; only 5 percent of those included in the National Alliance for Caregiving and AARP (2004) survey had ever used respite care. Minimal data are available on the quality of these services (Kitchener and Harrington, 2004).

The concept of family leave received considerable public attention with the 1993 passage of the federal Family and Medical Leave Act. This act gives employees the right to as many as 12 weeks of unpaid leave from work yearly to care for family members. Although the law has benefited some family caregivers, its impact has been muted because only more-affluent Americans can afford to take unpaid leaves and because the law does not apply to part-time workers, temporary workers, or employees of small firms. In addition, the law is problematic because it reinforces the idea that caring for ill and disabled persons is the responsibility of the family—which, in practice, usually means women relatives—rather than the responsibility of society as a whole (Abel, 2000).

Finally, those who provide care to relatives or friends may turn for assistance to paid caregivers. Each day about 1.4 million Americans receive paid home care, most commonly in the form of help with bathing, dressing, and light housework (National Center for Health Statistics, 2005). Most paid home care is provided by **home health aides,** who typically have no formal training, or **registered nurses,** who have received at least two years of nursing training and passed national licensure requirements. Aides are overwhelmingly minorities and women, and they are highly likely to be immigrants. Few receive any job benefits, and most receive only minimum wage. Because the growth in paid home health care is so new, little more is known regarding these workers or their work.

Health Care Technologies

Since the start of medicine—and indeed, before—doctors and other healers have used technologies in their work. Two hundred years ago, doctors used knives to cut veins and "bleed" patients of their illness, and they used strips of cloth to bandage the wounds afterward. One hundred years ago, doctors used mercury compounds and electricity in attempting to cure patients of masturbation or syphilis. In modern medicine, health care technology includes everything from Band-Aids to computerized patient record systems to heart-lung machines.

The Nature of Technology

Technology refers to any human-made object used to perform a task. In addition, the term is often used to describe processes that involve such objects. For example, the term *technology* can refer both to the overall process of kidney dialysis and to the equipment used in that process.

Although we often talk about technology as if it is inherently either good or bad—"technology has made our lives easier," or "technology has depersonalized medical care"—the reality is more complex (Timmermans and Berg 2003b; Heath, Luff, and Svensson, 2003). It is true that the nature of a technology determines the *range* of ways it might be used, but whether it is harmful, helpful, or both depends on *who* uses it in *which* ways. Electricity is helpful when used by doctors to stimulate muscle healing and harmful when used by doctors who are poorly educated or are employed as torturers in dictatorships. Fetal monitors can depersonalize childbirth when nurses stare at the screens rather than pay attention to the pregnant woman. But ultrasound imaging of fetuses can *personalize* pregnancy for fathers, who literally visualize their future children as real for the first time. In addition, such technologies can create a setting in which fathers, mothers, and health care workers can discuss the emotional aspects of pregnancy and child-rearing.

Similarly, we often talk about technology as if it is either a blank slate, lacking any inherent nature, or a force outside of human control. Again, the reality is more complex. For example, there has been considerable pressure lately for doctors and hospitals to reduce medical errors by adopting computerized medical databases to standardize the collection of patient data (Timmermans and Berg, 2003a). The purpose of these databases is to eliminate human error and variability in this process. For this reason, computerized databases may prompt doctors to ask their patients a specific set of questions, in a specific sequence, with a specific set of prompts if the answers seem insufficient or inappropriate. In this way, the database program presses doctors to standardize their practices, and encourages them to focus on certain areas to the exclusion of others and to organize the data they obtain in specific ways. At the same time, doctors quickly learn how to obtain at least partial control over the database through the way they ask their questions and the answers they record to the database's questions. Similarly, although patients are pressed by the nature of the database to respond within a narrow framework, they often sidestep the questions they are asked and instead address a different set of issues.

When we study technologies sociologically, therefore, it is as important to study the cultural system that surrounds that technology and determines how it will be used, by whom, and for what purposes, as it is to study the nature of the technology itself. Yet while we explore how society and social actors shape the use of technology, we also need to explore how *technology* shapes society and social actors.

In this section we will look at how technologies develop and become adopted. We will also consider how different groups within the health care world interact with technology—and with each other.

The Social Construction of Technology

In the same way that we have talked about the social construction of illness, we can talk about the **social construction** of technology: the process through which groups decide which potential technologies should be pursued and which should be adopted. This concept in turn leads to the question of who promotes the social construction of any given technology, and who benefits from this?

As is true for the social construction of illness, the social construction of technology is a political process, reflecting the needs, desires, and relative power of various social groups. These groups can include manufacturing corporations, doctors, the government, and consumers. As a result, harmful technologies are sometimes developed and adopted, and needed technologies sometimes are not.

One fascinating example of the social construction of technology is the history of cardiopulmonary resuscitation (CPR). The purpose of CPR is to restore life to those whose hearts and lungs have stopped working. In earlier times, the very notion of such resuscitation would not have made any sense to doctors or the public. Death was considered to be in God's hands, and dead was dead. But since the rise of modern medicine, doctors have struggled to find ways to restore life to those who suddenly die.

At the same time, doctors have grown increasingly able to understand the slow trajectory of dying associated with cancer. And with the rise of the hospice movement (described earlier in this chapter), both doctors and the public have come to hold as an ideal the "good death," in which an individual comes to terms with his or her dying, has the time to make peace with family and friends, and receives appropriate terminal care to minimize physical and emotional suffering.

None of this, however, applies to the sudden—and common—deaths caused by stroke or heart disease. In his award-winning book *Sudden Death and the Myth of CPR,* sociologist Stefan Timmermans (1999) argues that CPR and associated resuscitation techniques have become part of American medical culture because they appear to offer a "good death" in these circumstances. Innumerable television dramas portray heroic doctors who save apparently dead patients through CPR, and millions of dollars have been spent teaching the general public to perform CPR and outfitting community emergency response teams and hospital emergency rooms with resuscitation equipment. Yet CPR is almost never effective except when otherwise healthy individuals drown or are struck by lightning. The typical person who receives CPR has *at best* a 1 to 3 percent chance—and probably much less—of surviving, at an estimated cost of $500,000 per survivor. Moreover, "survival" may be brief, and it may be accompanied by severe neurological damage. As a result, the emergency room doctors and emergency medical technicians Timmermans observed and interviewed overwhelmingly regarded resuscitation as futile, and so they joked, complained, or simply went through the motions when they had to use it.

Why, then, has CPR become so widely adopted? Timmermans argues that the widespread use of CPR reflects modern Americans' discomfort with death. The real benefit of CPR, according to Timmermans, is that it "takes some of the suddenness of sudden death away" (1999: 110). CPR allows families and friends to believe they have done everything possible by getting their loved ones to treatment as fast as possible. It also gives families and friends time to gather and to recognize that death may be imminent, and it gives medical personnel a sense of technical accomplishment as they fight to keep their patients' bodily organs functioning as long as possible. For these reasons, and despite all its emotional and financial costs, CPR has become a valued and expected ritual in American culture.

At the same time, adoption of CPR illustrates the economics and politics, as well as the cultural forces, that underlie the social construction of technology. CPR would not have been so widely adopted if corporations had not had a vested economic interest in promoting it. Nor is it likely that CPR would have become the norm if corporations had been required to demonstrate its effectiveness before selling it. In fact, however, there are almost no legal requirements for corporations to demonstrate the safety or effectiveness of technical devices. As a result, manufacturers of medical technologies (unlike drug manufacturers) have no reason to fund such research. Doctors thus must depend on promotional materials from manufacturers and on their own clinical experiences in deciding whether to use a technology, and patients must rely on doctors' judgments.

Conclusion

In this chapter, we examined three difficulties inherent in the ways we provide care to those who are physically or mentally ill or disabled. First, we looked at some of the inherent contradictions of trying both to provide care and to make a profit. Health care workers, from medical students to home health aides, laboring long hours under often brutal conditions to keep their employers' costs low, cannot provide the quality of care they might like. Even institutions such as hospices, for whom profit making is not a primary motive, must contend with the demands of a wider system that emphasizes cutting costs and generating profits. Meanwhile, other institutions, such as nursing homes, board and care homes, and home health agencies, have emerged specifically to make money, relegating caregiving to a secondary priority.

Second, we considered the difficulty of providing individualized care in institutional environments. Almost by definition, large institutions must provide care en masse, ignoring individual preferences and desires. Patients must follow rules, schedules, and regimens established for the sake of efficiency, regardless of the impact on patients' quality of life. This tendency to ignore the individual is further reinforced because it is far cheaper to provide regimented rather than individualized care.

Third, we explored some of the inherent difficulties of treating health care as an individual or family responsibility rather than a social responsibility. As we have seen, the burdens of caregiving can be enormous. Yet the United States offers little support to those who take on this responsibility. In contrast, other industrialized nations provide far more assistance; both Sweden and Finland, for example, allow parents of sick children to leave work for several months while still receiving most of their salaries, and they provide free or inexpensive assistance with domestic chores to elderly persons who might otherwise have to turn to family members for assistance (Swedish Institute, 1997, 1999; Zimmerman, 1993).

In sum, the data presented in this chapter regarding the virtual social abandonment of ill and disabled individuals and of their caregivers suggests the low priority this society places on caring for those who are weak or ill, especially if they also are poor. Technology is not a panacea for these problems. Nor, for that matter, is it inherently dehumanizing or otherwise problematic. Rather, technology is a tool, adopted for a combination of cultural, medical, emotional, and financial reasons, that can be used for good or ill. Only when our underlying social values and commitments change can we expect the lives of ill persons, disabled persons, or their caregivers to improve significantly.

Suggested Readings

Annas, George J. 2004. *The Rights of Patients: The Authoritative ACLU Guide to Patients' Rights.* New York: New York University Press. Written by one of America's foremost experts on health law.

Gass, Thomas E. 2004. *Nobody's Home: Candid Reflections of a Nursing Home Aide.* Ithaca, NY: ILR Press. A riveting account of life in nursing homes, describing the experiences of both residents and nursing assistants.

Timmermans, Stefan. 1999. *Sudden Death and the Myth of CPR.* Philadelphia: Temple University Press. How CPR became part of American culture and medical care—even though it almost never saves lives.

Getting Involved

National Alliance for Caregiving. 4720 Montgomery Lane, Suite 642, Bethesda, MD 20814. (301) 718-8444. www.caregiving.org. Provides information and support to family caregivers of the elderly and to health care providers working in the field. Also collects and disseminates information about the value of family caregiving and the burdens borne by caregivers.

National Citizens' Coalition for Nursing Home Reform. 1424 16th Street, N.W., Suite 202, Washington, DC 20036–2211. (202) 332-2275. www.nccnhr.org. Citizens' action group seeking reform of nursing homes and board and care homes.

Review Questions

In what ways were nineteenth-century hospitals total institutions?

What led to the development of voluntary hospitals? veterans hospitals? government hospitals? the modern hospital as we know it?

What was the original philosophy of hospices, and why and in what ways has it changed?

What is patient dumping, and why does it occur?

Who uses nursing homes?

What is the difference between nursing homes, board and care homes, and assisted living facilities?

How does the process of commodification affect nursing assistants and nursing-home residents? Why has home care grown? What are the difficulties faced by family caregivers?

What is technology? What do sociologists mean when they say that technology is inherently neither good nor bad, and neither a blank slate nor a force outside of human control?

What is the social construction of technology? What does it mean to say that this is a political process?

Why was CPR so widely adopted even though it was so ineffective?

How can society shape technology? How can technology shape society?

Internet Exercises

1. Do a search at google.com for alt.support.alzheimers, a discussion group for persons who care for individuals who have Alzheimer's disease. Once you get to the discussion group's website, read a few "threads"—queries and the answers posted to them—to identify some of the issues faced by these caregivers. Do discussion groups seem to be effective means of helping family caregivers?

2. Using the Internet, find three websites advertising nursing homes or assisted living facilities. What information would you want if you needed to place a relative in such a facility? What information do these websites leave out? How does each website encourage you to believe that its facility would be the best one for your relative?

4 Health Care Providers and Bioethics

In this final section, we shift our perspective to health care providers. Chapter 11 provides an overview of the history of medicine as a profession and describes how the social position of doctors has changed over time. In this chapter we also explore how a person becomes a doctor, including the nature of medical education and medical culture and the steps involved in building a medical career. Finally, we look at how medical education and medical culture, as well as broader social and cultural factors, affect relationships between doctors and patients.

Although doctors typically are the first persons who come to mind when we think of health care, they form only a small percentage of all health care providers. In Chapter 12, we consider some of these other providers both within and outside the mainstream health care system, including nurses, pharmacists, midwives, and acupuncturists.

The final chapter in this part, and in this book, provides a history of bioethics as well as a sociological account of how bioethics has become institutionalized and of its impact on health care and health research. We will see how issues of power underlie ethical issues and why we need a sociological understanding of bioethics.

CHAPTER **11**

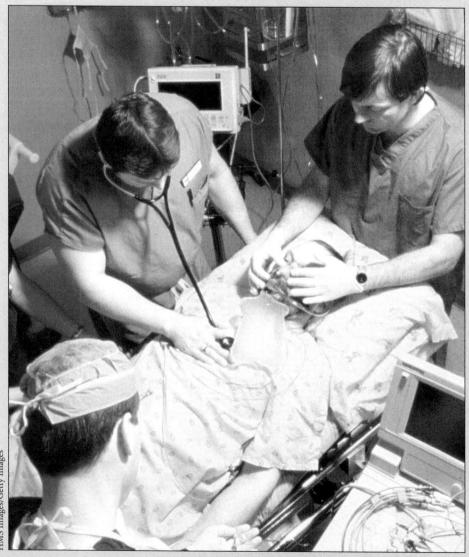

The Profession of Medicine

To become a doctor, students must spend long years studying biology, chemistry, physiology, and other subjects. In addition, students must learn the way of thinking about medicine, patients, and medical care—the worldview—that characterizes medical culture.

Michael J. Collins learned this worldview during four years as a surgical resident at the Mayo Clinic. After a particularly brutal day of surgery in which he watched a teenager die, Dr. Collins found himself emotionally traumatized, questioning the meaning of his work and the effect it had on him. Although he wished he could discuss his feelings with BJ Burke, the director of his residency program, Collins knew from experience how BJ would respond. As he wrote in his memoir,

> BJ Burke was not interested in what I thought or understood. He was interested in what I did.
>
> "If you want to learn to be sensitive and introspective," he would say, "do it on your own time."
>
> I imagined myself being called into his office. As I enter the room he is seated at his desk, reading the report in front of him. He makes certain I know I am being ignored.
>
> At length he looks at me over the top of his glasses.
>
> "Dr. Collins, what is your job?"
>
> "My job, sir?"
>
> "You have a job, don't you? You get a paycheck, don't you?"
>
> "Yes, sir."
>
> "Well, what do you do?"
>
> "I'm a second-year orthopedic resident at the Mayo Clinic."
>
> "Do you want to be a third-year resident someday, Dr. Collins?"
>
> "Yes, sir."
>
> "What is an orthopedic resident supposed to do?"
>
> Where was this going? "Following orders?" I venture.

"An orthopedic resident is supposed to practice orthopedics, Doctor. He is not supposed to go around asking patients if they have ever considered the ontological implications of their fragile, mortal state."

"I didn't exactly—"

He jumps to his feet and points his finger at me. "We fix things. Do you understand that? We don't analyze things. We don't discuss things. We don't wring our hands and cry about things. We fix them! If somebody wants to be analyzed they can see a shrink. When they come to the Department of Orthopedics at the Mayo Clinic they want only one thing: they want to be fixed.

"Now get the hell out of here and go fix things. And I better not get any more reports of touchy-wouchy, hand-holding sessions in this department." (Collins, 2005: 152–153)

Collins's story illustrates two basic elements of modern-day medical culture—emotional detachment and a belief in medical intervention. In this chapter we look at how these and other aspects of medical culture and training evolved, at the consequences for both doctors and their patients, and at the history and current status of medicine as a profession.

American Medicine in the Nineteenth Century

When confronted by disquieting illness, most modern-day Americans seek care from a doctor of medicine. Little more than a century ago, however, that would not have been the case. Instead, Americans received most of their health care from family members. If they required more complicated treatment, they could choose from an array of poorly paid and typically poorly respected health care practitioners (Starr, 1982: 31–59). These included **regular** doctors, who were the forerunners of contemporary doctors. They also included such **irregular practitioners** as patent medicine makers, who sold drugs they concocted from a wide variety of ingredients; botanic eclectics, who offered herbal remedies; bonesetters, who fixed dislocated joints and fractured bones; and midwives.

Regular doctors were also known as **allopathic doctors,** or allopaths (from the Greek for "cure by opposites"), because they sometimes treated illnesses with drugs selected to produce symptoms *opposite* to those caused by the illnesses. For example, allopaths would treat patients suffering the fevers of malaria with quinine, a drug known to reduce fevers, and treat patients with failing hearts with digitalis, a drug that stimulates the heartbeat. Their main competitors were **homeopathic doctors,** or homeopaths (from the Greek for "cure by similars"). Homeopaths treated illnesses with drugs that produced symptoms *similar* to those caused by the illnesses—treating a fever with a fever-producing drug, for example. Although in retrospect the

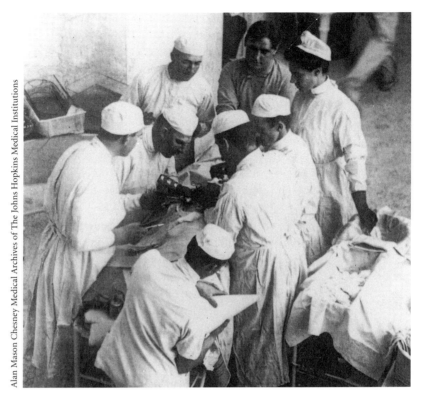

In the 1890s, even surgery at a major hospital such as Johns Hopkins required no advanced technologies.

homeopathic model might seem odd, it drew on the same logic as smallpox inoculation, the one successful inoculation available at that time: People who were inoculated with a small quantity of cowpox cells, and who therefore developed a mild form of cowpox, somehow became immune to the related but far more serious smallpox. Homeopaths therefore concluded that patients who received a small quantity of a drug that mimicked the symptoms of a given illness would become better able to resist that illness. At any rate, although homeopathy helped patients only through a **placebo** effect, it at least did not harm them.

That Americans before the twentieth century placed no greater trust in allopathic doctors than in any others who claimed knowledge of healing should not surprise us. Although by the nineteenth century, science—the careful testing of hypotheses in **controlled** experiments—had infiltrated the curricula of European medical schools, where many of the wealthiest or most dedicated Americans trained, it had gained barely a foothold in U.S. medical schools. Moreover, the United States licensed neither doctors nor medical schools (Ludmerer, 1985). Instead, and until about 1850, most doctors trained through apprenticeships lasting only a few months. After that date, most trained at any of the multitude of uncertified medical schools that had sprouted around the country, almost all of which were private, for-profit institutions, unaffiliated with colleges or universities and lacking any entrance requirements beyond the ability to pay tuition (Ludmerer, 1985). Nor were standards stricter at the few university-based medical schools. For example, in 1871, Henry Jacob Bigelow, a Harvard University professor of surgery, could protest a proposal to require written graduation examinations on the grounds that more than half of Harvard's medical students were illiterate (Ludmerer, 1985: 12). Training averaged far less than a year and depended almost entirely on lectures, so that almost no students ever examined a patient, conducted an experiment, or dissected a cadaver. Any student who regularly attended the lectures received a diploma. This situation began to change significantly only in the 1890s and only in the better university schools.

Lacking scientific research or knowledge, allopathic doctors developed their ideas about health and illness either from their clinical experiences with patients or by extrapolating from abstract, untested theories. The most popular theory of illness, from the classical Greek era until the mid-1800s, traced illness to an imbalance of bodily "humors," or fluids. Doctors had learned through experience that ill persons often recovered following episodes of fever, vomiting, or diarrhea. From this, doctors deduced—in part correctly—that fever, vomiting, and diarrhea helped the body restore itself to health. Unfortunately, lacking methods for testing their theories, doctors carried these ideas too far, often inducing life-threatening fever, vomiting, purging, and bloodletting. Consider, for example, the following description of how Boston doctors in 1833 used what was known as **heroic medicine** to treat a pregnant woman who began having convulsions a month before her delivery date:

> The doctors bled her of 8 ounces and gave her a purgative. The next day she again had convulsions, and they took 22 ounces of blood. After 90 minutes she had a headache, and the doctors took 18 more ounces of blood, gave emetics to cause vomiting, and put ice on her head and mustard plasters on her feet. Nearly four hours later she had another convulsion, and they took 12 ounces, and soon after, 6 more. By then she had lapsed into a deep coma, so the doctors doused her with cold water but could not revive her. Soon her cervix began to dilate, so the doctors gave ergot to induce labor. Shortly before delivery she convulsed again, and they applied ice and mustard plasters again and also gave a vomiting agent and calomel to purge her bowels. In six hours she delivered a stillborn child. After two days she regained consciousness and recovered. The doctors considered this a conservative treatment, even though they had removed two-fifths of her blood in a two-day period, for they had not artificially dilated her womb or used instruments to expedite delivery. (R. Wertz and D. Wertz, 1989: 69)

As this example suggests, because of the body's amazing ability to heal itself, even when doctors used heroic medicine, many of their patients survived. Thus, doctors could convince themselves they had cured their patients when in reality they either had made no difference or had endangered their patients' lives.

By the second half of the nineteenth century, most doctors, responding to the public's support for irregular practitioners and fear of heroic medicine, had abandoned the most dangerous of their techniques. Yet medical treatment remained risky. Allopathic doctors' major advantage over their competitors was their ability to conduct surgery in life-threatening situations. Unfortunately, until the development of anesthesia in the 1860s, many patients died from the inherent physical trauma of surgery. In addition, many died unnecessarily from postsurgical infections. Dr. Ignaz Semmelweis had demonstrated in the 1850s that because midwives (whose tasks included washing floors and linens) had relatively clean hands, whereas doctors routinely went without washing their hands from autopsies to obstetrical examinations and from patient to patient, more childbearing women died on medical wards than on midwifery wards. Yet it took another 30 years before hand washing became standard medical practice.

Until well into the twentieth century, then, doctors could offer their patients little beyond morphine for pain relief, quinine for malarial and other fevers, digitalis for heart problems, and, after 1910, salvarsan for syphilis—each of which presented dangers as well as benefits. According to the 1975 edition of *Cecil's Textbook of Medicine,* one of the most widely used medical textbooks, only 3 percent of the treatments described in the 1927 edition of this textbook were fully effective, whereas 60 percent were harmful, of doubtful value, or offered only symptomatic relief (Beeson, 1980). Doctors' effective pharmacopeia did not grow significantly until the development of antibiotics in the 1940s.

Beginnings of Medical Dominance

Despite the few benefits and many dangers inherent in allopathic medical care, by about 1900 doctors had eliminated most of their competitors and gained control over health care (Starr, 1982: 79–112). In this section, we will see how this change came about.

From its inception in 1847, the **American Medical Association (AMA)** had worked to restrain the practices of other health care occupations. State by state, the AMA fought to pass laws outlawing their competitors or restricting them to working only under allopathic supervision or to performing only certain techniques, such as spinal manipulation.

Most of these efforts met with little success initially, for nineteenth-century Americans considered health care an uncomplicated domestic matter, unrelated to science and not requiring complex training (Starr, 1982: 90–92). By the beginning of the twentieth century, however, as improvements in public health and in living conditions ended scourges such as cholera and typhoid, and as Americans began reaping practical dividends from scientific advances such as electric lights and streetcars, public faith in science swelled. Increasingly, Americans defined health care as a complex matter requiring expert intervention, assumed the superiority of "scientific" medicine, and turned to allopathic doctors for care (Starr, 1982: 127–142).

Like the public, homeopaths and botanic eclectics (allopathic doctors' two major groups of competitors) also had come to recognize the benefits of science and therefore to realize that a lack of scientific foundation would soon doom their fields. However, they still received considerable popular support. Moreover, because, like allopaths, most were white men, homeopaths and botanic eclectics generally held social statuses similar to those of allopaths. Thus homeopaths and botanic eclectics retained sufficient influence to pressure allopaths to accept them into medical schools and licensing programs, and their fields eventually faded away.

Other health care workers could bring far less power to their dealings with legislators and with allopathic doctors. Newly emerging occupations such as chiropractic (described in Chapter 12) lacked the long-standing history of popular support that had allowed homeopaths to push for incorporation with allopathy. Older occupations, meanwhile, such as midwives and herbalists, lacked the social status, power, and money needed to fight against doctors' lobbying. Because most of these practitioners were women or minorities, they were assumed to be incompetent by both legislators and doctors (Starr, 1982: 117, 124).

The Flexner Report and Its Aftermath

These differences between allopathic doctors and other health care practitioners increased during the early years of the twentieth century. Since the 1890s, the better medical schools had begun tightening entrance requirements, stressing higher academic standards, emphasizing research and science, and offering clinical experience. These changes placed pressures on the other medical schools to do the same. Those pressures increased following publication in 1910 of the **Flexner Report** on American medical education (Ludmerer, 1985: 166–190). The report, which was written by Abraham Flexner and commissioned by the nonprofit Carnegie Foundation at the AMA's behest, shocked the nation with its descriptions of the lax requirements and poor facilities at many medical schools. The Flexner Report increased the pressures on all medical schools to improve their programs and accelerated the process of change that was already under way. In the next few years, responding to pressure from both the public and the AMA, all U.S. jurisdictions adopted or began enforcing stringent licensing laws for medical schools (Ludmerer, 1985: 234–249). These laws hastened the closure of all proprietary and most nonprofit schools, many of which were already suffering financially from the costs of trying to meet students' growing demand for scientific training. As a result, the number of medical schools fell from 162 in 1906 to 81 in 1922 (Starr, 1982: 118, 121).

The Flexner Report, in conjunction with the changes already under way in medical education, substantially improved the quality of health care available to the American public and paved the way for later advances in health care. However, these changes in medical education also had some more problematic results. The closure of so many schools made medicine as a field even more homogeneous. Only two of the seven medical schools for African Americans survived, and only one of the seven schools for women (Ludmerer, 1985: 248; Starr, 1982: 124). In addition, because the university schools set stricter educational prerequisites than had the defunct proprietary schools, few immigrants, minorities, and poorer whites could meet their entrance requirements. Even fewer could afford the tuition required by scientifically oriented university programs. Moreover, for the next several decades many programs openly discriminated against women, African Americans, Jews, and Catholics. So, even though the technical quality of medical care increased, fewer doctors were available who would practice in minority communities and who understood the special concerns of minority or female patients. At the same time, simply because doctors were now more homogeneously white, male, and upper class, their status grew, encouraging more hierarchical relationships between doctors and patients.

Doctors and Professional Dominance

By the 1920s, doctors had become the premiere example of a **profession** (Parsons, 1951). Although definitions of a profession vary, sociologists generally define an occupation as a profession when it is considered by most to have three characteristics:

1. The autonomy to set its own educational and licensing standards and to police its members for incompetence or malfeasance;

2. Technical, specialized knowledge, unique to the occupation and learned through extended, systematic training; and

3. Public confidence that its members follow a code of ethics and are motivated more by a desire to serve than a desire to earn a profit.

For at least the first half of the twentieth century, doctors clearly met this definition of a profession. Doctors' autonomy was evidenced in the fact that they, rather than consumers or judges, were legally responsible for deciding whether to remove the license of any doctors accused of incompetence. That doctors held highly specialized knowledge, and that they spent many years acquiring that knowledge, was well known. Finally, most Americans placed great trust in the medical field, and believed that physicians placed their patients' interests first; although trust in medicine as an institution has declined precipitously in recent years, Americans still strongly trust their own physicians and rank medicine first among occupations in prestige (*Harris Poll*, 2004b).

As the leading profession in the health care world, doctors enjoyed—and to some extent still enjoy—an unusually high level of **professional dominance:** freedom from control by other occupations or groups and ability to control any other occupations working in the same economic sphere. This concept has been most fully analyzed by Eliot Freidson (1970a, 1970b, 1994). As Freidson has noted, for much of the twentieth century, most doctors worked in private practice (whether solo or group), setting their own hours, fees, and other conditions of work. Those who worked in hospitals or clinics were typically supervised by other doctors, not by nonmedical administrators. Although doctors often supervised members of other occupations, the reverse has begun taking place only in the last two decades or so. Similarly, both in the past and currently, doctors often served on boards charged with judging the education and qualifications of other health care occupations, but members of other occupations played little role in setting standards for medical education and licensing. This high level of professional dominance by doctors—otherwise known as **medical dominance**—stemmed from the public's great respect for doctors' claims to a scientific knowledge base and service orientation. This respect in turn was bolstered with active lobbying by organized medicine.

The Decline of Medical Dominance

One of the most heated debates within the sociology of health and illness is the extent to which medical dominance has declined (Freidson, 1994; Light and Levine, 1988; Starr, 1982: 379–393). Foremost among those arguing that professional dominance has declined are Marie Haug, John McKinlay, and John Stoeckle. They differ, however, in where they locate the sources of this decline, with Haug (1988) focusing on changes in public sentiment and access to medical knowledge and McKinlay and Stoeckle (1989) on changes in health care financing and organization (see Key Concepts 11.1).

Changing Patient Attitudes and Deprofessionalization

In her writings, Haug has focused on how the civil rights and feminist movements of the 1960s and 1970s increased popular emphasis on rights rather than duties and on questioning rather than obeying authorities (Haug, 1988; Starr, 1982: 379–393). At the same time, Haug argues, the general rise in educational levels and in public access to medical information has helped patients to evaluate their symptoms and treatment for themselves and to challenge their doctors' diagnoses and decisions about care. These changes, coupled with growing public awareness of how unquestioning obedience to doctors sometimes can harm patients' health, helped foster both the feminist health movement and the patients' rights movement. These movements both reflected and created more egalitarian ideas about how doctors and patients should interact.

---

*Key
Concepts 11.1*   **Divergent Views on Medical Dominance**

THE PROFESSIONAL DOMINANCE MODEL	DECLINE OF DOMINANCE MODELS
	A. Deprofessionalization
High level of prestige	Decline in public confidence and respect
Public defers to medical judgment and feel loyalty to their doctors.	Public questions medical judgment and feels little loyalty to doctors.
	B. Proletarianization
Doctors hold strong economic position.	Doctors become economically vulnerable and AMA power declines.
Doctors set own working conditions.	Doctors' working conditions set by corporate employers.
Only doctors supervise doctors.	Doctors supervised by nonmedical administrators and review boards.
Doctors supervise and control other health care occupations.	Other health care occupations gain considerable independence from medical control.
Doctors act solely or largely based on their clinical judgment.	Doctors' clinical autonomy constrained by corporate or governmental guidelines.

These new popular health movements have stimulated major changes in medical practice, ranging from the sharp decrease in use of general anesthesia during childbirth to the routine use of informed consent forms before patients receive experimental drugs. More broadly, through publications such as the many editions of the best sellers *Take Care of Yourself: The Complete Illustrated Guide to Medical Self-Care* (Fries, 2004) and *Our Bodies, Ourselves* (Boston Women's Health Book Collective, 2005), these movements have encouraged consumers to take charge of their own health, to use practitioners other than doctors, and to obtain second opinions when they do go to doctors.

The rise of the Internet has added impetus to this movement, giving consumers instant access to vast numbers of others who share their concerns and to vast quantities of medical literature, including literature on alternatives to allopathic medicine. The federal government has supported this trend; its website at http://www.healthfinder.org was established specifically to give consumers online access to publications, clearinghouses, databases, other websites, self-help groups, government agencies, and nonprofit organizations related to both allopathic and alternative medicine.

The peculiarly American propensity to file malpractice suits against doctors further suggests the public's lack of confidence in doctors (although it probably reflects even more strongly the lack of a national health care system, which can leave Americans unable to pay their medical bills if they suffer severe medical problems). A 1998 survey of obstetrician-gynecologists (the specialty with the second-highest rate of lawsuits) found that 73 percent had been sued for malpractice at least once during their careers, with an average of 2.3 lawsuits per person (American College of Obstetrician-Gynecologists, 1998). To avoid expensive lawsuits, health care institutions have worked to assert more control over the doctors who work for them. Fear of lawsuits also has encouraged doctors to change their own behaviors and to subordinate their clinical decision making in favor of **defensive medicine**—doing tests or procedures solely or primarily to reduce their risk of a malpractice suit (American College of Obstetrician-Gynecologists, 1998; Tussing and Wojtowycz, 1997). For example, one study found that, after researchers controlled for a variety of maternal, physician, clinical, and other characteristics, doctors practicing in counties with high rates of malpractice suits (in which doctors' fear of malpractice suits would likely be greater) had significantly higher than average rates of cesarean deliveries. Fear of malpractice explained 24 percent of all cesarean deliveries in the study (Tussing and Wojtowcyz, 1997).

Taken together, these changes led Haug (1988) to conclude that doctors are becoming deprofessionalized, or losing the public confidence that defines professions. This concept gains credence from national polls showing that the proportion of Americans who place a "great deal of confidence in people in charge of running medicine" dropped from 73 percent in 1966 to 42 percent in 1976 and to 29 percent in 2005 (*Harris Poll,* 2005).

The Changing Structure of Medicine and Proletarianization

In contrast, whereas McKinlay and Stoeckle (1989) agree with Haug that medical dominance has declined, they instead trace that decline to changes in health care financing. They maintain that doctors have lost substantial control over the most important professional prerogatives: deciding who may enter the profession and how, setting the conditions under which a doctor works, owning one's tools and workspace, and maintaining an individual relationship with freely chosen patients. Consequently, McKinlay and Stoeckle conclude, doctors are becoming workers (or "proletarians") rather than autonomous professionals. The authors refer to this shift as **proletarianization** and trace it to three factors: the rise of corporatization, the growth of government control, and the decline of the AMA.

The Rise of Corporatization

McKinlay and Stoeckle (1989) begin their argument by noting that before the 1960s, nonprofit or government agencies owned most hospitals and other health care institutions. With the initiation of **Medicare** and **Medicaid,** however, the potential for profit making in health care expanded tremendously, encouraging for-profit corporations to enter the field, as we saw in Chapter 8 (Starr, 1982: 428–432). During the last three decades, investor-owned corporations have purchased or developed a growing number of health care institutions. In addition, corporations increasingly have shifted from **horizontal integration** (owning multiple institutions providing the same type of service) to **vertical integration** (owning multiple institutions providing different types of services, such as both **nursing homes** and pharmaceutical companies).

This growth of corporate medicine, or **corporatization,** occurred at a time when doctors were experiencing increasing economic vulnerability (McKinlay and Stoeckle, 1989; Starr, 1982: 446–448). Since the early 1960s, the supply of doctors has grown rapidly, more than doubling between 1970 and 1998 and far surpassing the ratio in most industrialized nations (American College of Physicians, 1998). The supply of doctors now exceeds demand in the most desirable communities and specialties. For example, among doctors who completed residencies in 1996, 56 percent of those in critical care medicine and 47 percent in anesthesiology (but only 7 percent in family practice) reported difficulty finding employment (Bodenheimer, 1999). Supply is expected to continue to increase until 2020, despite recent federal legislation to reduce funding for specialty training.

Because of the current oversupply, newly graduated doctors sometimes find the competition too great to enter private or small group practice. More and more doctors now find they must accept employment with hospitals, large group practices, **managed care organizations** (MCOs), or other corporate institutions. Others, especially women with children, have more freely chosen corporate employment because they prefer its more relaxed lifestyle and shorter, more predictable hours. As of 1999, 41 percent of doctors worked as paid employees—about twice as many as did so 20 years earlier (Fraser, 2002).

As employees of salaried or group practices, whether by choice or necessity and whether in small groups or in corporate-owned hospitals, doctors' autonomy has diminished. Fearing that when left to their own devices, doctors will overuse available resources and drive up costs, administrators now make many decisions formerly made by individual doctors. According to McKinlay and Stoeckle (1989: 192), "doctors have slipped down to the position of middle management . . . , [while administrators are] organizing the necessary coordination for collaborative work, the work schedules of staff, the recruitment of patients to the practice, and the contacts with third-party purchasers, and are determining the fiscal rewards." In addition, administrators now may set such basic conditions of work as how many patients a doctor must see per hour.

Even those who do not work directly for corporations now often find that the only way they can get patients is to sign contracts with MCOs. These contracts limit doctors' autonomy both by controlling the fees they may charge for their services and by scrutinizing their clinical decision making. Many MCOs expect doctors to follow **practice protocols,** which establish norms of care for particular medical conditions under particular circumstances based on careful review of clinical research (Good, 1995; Millenson, 1997). Some MCOs forbid doctors from discussing with patients any treatments the MCO does not approve, and some require doctors to get approval before admitting patients to hospitals or administering certain treatments. Over the last few years, however, doctors' dissatisfaction with such prospective review, coupled with increasing doubts about its cost-effectiveness, has encouraged MCOs to shift to more subtle means of controlling doctors.

Moreover, both within corporate institutions and under managed care contracts, the nature of the doctor-patient relationship, and thus the power of doctors within that relationship, has changed. Doctors no longer have "their" individual patients, but now must see whatever patients their employers or MCOs assign to them. Conversely, even patients who continue to have a primary caregiver feel less loyalty to that doctor because they often see whatever doctor happens to be available when they need care. In addition, as employees, doctors feel free to move around to other practices, leaving their old patients behind and destroying any bonds of loyalty—something private doctors could not afford to do.

McKinlay and Stoeckle (1989) additionally argue that doctors' power relative to other health care occupations has declined. Many health care institutions, including MCOs, now believe they can limit costs without limiting quality by hiring cheaper, allied health personnel (such as radiation technologists or nurse practitioners) to perform specialized tasks once performed by doctors. Increasingly, pharmacists and specially trained nurses have legal authority to prescribe certain drugs. Similarly, patient management now officially belongs to the health care team, in which allied specialists often have more knowledge of specialized tasks than do doctors. As a result, McKinlay and Stoeckle argue, doctors' power to control the work of ancillary personnel has declined.

The Growth of Government Control

Government regulations also now restrict doctors' professional autonomy. Because the government pays the bills generated by Medicaid and Medicare, it has a large vested interest in controlling doctors' fees and their decisions about treatment. To do so, it has established programs such as the **diagnosis-related groups (DRG)** system and the **resource-based relative value scale (RBRVS).** The DRG system (described in Chapter 8) established preset financial limits for each diagnosis for hospital care under Medicare (and, in some states, Medicaid). Because hospitals are not reimbursed for any costs above those limits, they have a vested interest in making sure doctors stay below the limits. Consequently, hospitals may cut the wages or terminate the contracts of doctors who consistently exceed DRG limits, thus pressuring all doctors in their employ to stay within those limits (Dolenc and Dougherty, 1985). Doctors sometimes conclude that they have only two choices: to misreport a patient's diagnosis on the DRG forms so they can justify more expensive treatments, or to ignore their clinical judgment about the treatment a patient needs so they can stay within the DRG limits.

Whereas DRGs were designed to control Medicare spending on hospital care, RBRVS was designed to control spending on doctors' bills. RBRVS is a complex formula for determining appropriate compensation under Medicare for medical care, based on estimates of how much it actually costs to provide specific services in specific geographic areas. Under this system, incomes of most specialists have declined while those of generalists (other than pediatricians, who receive no Medicare funds) have increased. Although RBRVS applies legally to Medicare only, most other public and private insurance plans also have adopted RBRVS, making it, in the words of one observer, a "de facto national fee schedule" (Sigsbee, 1997).

The Decline of the AMA

Finally, McKinlay and Stoeckle (1989) argue that doctors' professional dominance has declined because the power of the AMA has declined. Although the AMA remains one of the most powerful lobbying groups in the country, its power is now counterbalanced by that of other health care organizations. Evidence for this can be found in the spending patterns of the various political action committees (PACs)—federally recognized organizations that solicit contributions from individuals, associations, and corporations and distribute this money to candidates for election who support the PACs' political agenda. The AMA still controls a larger pool of PAC lobbying money than any other health profession, contributing $2.3 million during the 2004 presidential elections (Center for Responsive Politics, 2005.) This sum is dwarfed, however, by the sums contributed (in total) by the PACs representing other health professions, pharmaceutical companies, health insurance companies, and hospitals, all of whose legislative interests sometimes compete with those of the AMA.

Similarly, whereas in the past the AMA and the doctors it represented had nearly free rein to set both admissions criteria and curricula of medical schools, this freedom has eroded substantially. Legal changes and social pressures stemming from the civil rights and feminist movements forced medical schools beginning in the 1960s to acknowledge the rights of women and minorities to enter medicine and of foreign-trained doctors to gain access to U.S. licensing. Ironically, over time medical schools came to value having a more diverse student population, but several court decisions have forced them to change admissions procedures in ways that could restrict minority enrollment. In addition, increased government, corporate, and foundation financing of medical training beginning in the 1960s has given these outside groups increased power to direct the nature of training, through choosing which educational programs to fund.

At the same time, the AMA has suffered internally from declining support among doctors. Whereas a half century ago most doctors belonged to the AMA, as of 2005, only 29 percent did (American Medical Association, 2005a). Instead, some doctors join more liberal organizations that often oppose the AMA, such as Physicians for Social Responsibility, and many join specialty organizations like the American College of Obstetricians and Gynecologists.

For all these reasons, then, McKinlay and Stoeckle (1989) argue that doctors are experiencing proletarianization. This conclusion is supported by the AMA's 1999 decision to end its long-standing opposition to unionizing doctors in order to increase doctors' bargaining power relative to MCOs (Greenhouse, 1999). About 40,000 doctors now belong to large, cross-occupational unions such as the American Federation of State, County and Municipal Employees.

The Continued Strength of Medical Dominance

Not all sociologists, however, agree that medical dominance has declined significantly. Some, such as Freidson (1984, 1994), argue that even though professional dominance has declined since its high point in the middle of the twentieth century, it remains strong. As Paul Starr notes, health care corporations depend on doctors both to generate profits and to control costs (1982: 446). As a result, these corporations, which retain a vested interest in maintaining good relationships with their physician employees, continue to give doctors considerable autonomy in day-to-day clinical matters. As noted earlier, MCOs have moved away from prospective review and rarely reject doctors' treatment recommendations (Remler et al., 1997). Similarly, although corporations increasingly hire professional managers as chief executive officers of health care institutions, they often also hire doctors as medical directors to work directly under these managers, as well as in a wide range of other administrative positions. Freidson (1985, 1986) refers to this process as the restructuring of the profession of medicine into specialties organized not by clinical territory (for example, oncologists to treat cancer, pediatricians to treat children) but by functional sector: the producers who work in clinical practice, the knowledge elite who work in research or academia, and the administrative elite. Through restructuring, Freidson argues, medicine has retained control of critical areas of professional status (such as setting licensure regulations and practice standards) and thus preserved its dominance as a field, even if the autonomy of individual physicians has eroded.

Moreover, Freidson argues, although individual doctors working in specific situations have lost some professional prerogatives, the power and dominance of doctors relative to other health care occupations have remained largely intact. Freidson notes, for example, that the rhetoric of health care "teams" hides the fact that doctors have by far the most power on these teams. By the same token, the use of medical technology by ancillary occupations tells us little about the relative power of those occupations, for medical innovations always have moved down the occupational scale over time. For example, nurses for some time have used stethoscopes and blood pressure cuffs without any increase in their power relative to doctors. Similarly, although the rise of practice protocols could decrease the autonomy of individual doctors, supporters of protocols argue that only through such self-regulation can medicine preserve public faith and, in the end, its professional autonomy (Good, 1995).

Finally, although the environment within which physicians now practice medicine has changed considerably, they retain considerable ability to manipulate and control this new environment. For example, and as mentioned earlier, many doctors now match the DRG system to their clinical decisions rather than changing their decisions to match the DRG system. Similarly, many physicians have sufficient power to ignore onerous bureaucratic directives. So, for example, hospital policies regarding how doctors should use HIV tests are only weakly correlated with doctors' behaviors (K. Montgomery, 1996). By the same token, doctors have proved surprisingly adept at maintaining their incomes because they, not consumers, largely control demand for medical services. Thus, doctors with shrinking patient pools can order more tests or treatments for their remaining patients or can expand the areas they consider suitable for medical intervention, as explained in Chapter 5's section on **medicalization.**

In addition, doctors can maintain their incomes and autonomy by performing elective procedures, in their private clinics or offices, for which patients pay out of pocket. Many doctors now heavily advertise cosmetic surgery, laser eye surgery, infertility treatment, and weight loss treatment because these procedures are both remunerative and largely free of oversight by insurance, government, or hospital bureaucrats (Sullivan, 2001). This trend partly explains why the percentage of surgeries taking place in doctors' offices rose from 5 percent in 1981 to 26 percent in 1999 (Zuger, 1999). Similarly, a small but growing number of doctors have opened "boutique" practices, in which patients pay a flat fee of several thousand dollars per year for services not covered by their insurance, including same-day appointments, heated towels, house calls, and twenty-four-hour cell phone access to their doctors (Belluck, 2002).

The continuing power of medicine as a profession is also demonstrated in the ongoing struggles, as described in Chapter 8, to place legal limits on MCOs' control over doctors. Some of the legislative proposals typically described as "patients' bills of rights" might more accurately be characterized as "doctors' bills of rights." For example, most states now legally mandate that MCOs offer women direct access to obstetrician-gynecologists rather than requiring referral from an internist or family doctor, and they allow women to use obstetrician-gynecologists as their primary care doctors. Although many women are accustomed to using obstetrician-gynecologists for primary care, logic would suggest that women would be better served by using doctors who have trained broadly in primary care rather than using doctors trained in the surgical management of one set of bodily organs. Not surprisingly, these laws received much of their political support from obstetrician-gynecologists. More generally, by portraying themselves as fighters for patients' rights against the "evils" of MCOs, doctors have burnished their public image.

For all these reasons, although Freidson's earlier model of professional dominance certainly needs modification, it remains a useful starting point for understanding the current status of medicine as a profession.

Medical Education and Medical Values

Despite the assaults on medical dominance, becoming a doctor remains an attractive option: It offers public prestige, the emotional rewards of service, and financial rewards far greater than most other professions. Although applications to medical school declined during the 1980s, they have risen considerably since then and have stayed stable for the last decade (Barzansky and Etzel, 2005). In this section, we look at how doctors-in-training learn both medical knowledge and medical values and at the consequences of this training for both doctors and patients.

The Structure of Medical Education

Becoming a doctor is not easy. Prospective doctors first must earn a bachelor's degree and then complete four years of training at a medical school. Before they can enter practice, however, and depending on their chosen specialty, they must spend another three to eight years as **residents**. Residents are doctors who are continuing their training while working in hospitals. (The term *intern,* referring to the first year of a residency, is no longer commonly used.) As a result, most do not enter practice until age 30.

For more than 80 percent of students, going to medical school means going into debt. The average debt is $100,000 for public medical schools and $135,000 for private medical schools (Jolly, 2004). These amounts are in addition to undergraduate debts.

Becoming a doctor also carries tremendous time costs. Regulations adopted following the death in 1989 of a patient treated by exhausted residents now limit surgical residents to working 100 hours per week and medical residents to working "only" 80 hours per week. Even after graduation, about one-third of doctors work more than 60 hours per week (Bureau of Labor Statistics, 2004). These time pressures, coupled with the financial pressures of training, encourage novice doctors to defer marriage, children, and other personal pursuits and to choose specialties requiring less training over those they otherwise might prefer.

Due in part to financial costs, most medical students are from the middle and upper classes. On the other hand, medicine increasingly has opened to women, who now comprise half of all first-year medical students (Barzansky and Etzel, 2005). Nonwhites have not made as much progress; all minority groups other than Asians remain underrepresented in medical schools.

Learning Medical Values

During their long years of training, doctors learn both a vast quantity of technical information and a set of **medical norms**—expectations about how doctors should act, think, and feel. As this section describes, the most important of these norms are that doctors should value emotional detachment, trust clinical experience more than scientific evidence, master uncertainty, adopt a mechanistic model of the body, trust intervention more than normal bodily processes, and prefer working with rare or acute illnesses rather than with typical or chronic illnesses.

Emotional Detachment

Undoubtedly most doctors enter the profession because they want to help others. Yet perhaps the most central medical norm is to maintain emotional detachment from patients. As illustrated by the story that opens this chapter, from Michael Collins's experiences as a resident at the Mayo Clinic, medical culture values and rewards "strength" and equates emotional involvement or expression with weakness (Hafferty, 1991).

Given doctors' daily confrontations with illness, trauma, and death, some emotional detachment is a necessary coping mechanism. Sociological research suggests, however, that doctors develop emotional detachment not only as a natural response to stress but also because their superiors teach them to (Hafferty, 1991).

Professional socialization refers to the process of learning the skills, knowledge, and values of an occupation. According to sociologist and medical school professor Frederic Hafferty (1991), who spent several years observing and interviewing medical students, this socialization typically begins even before students enter medical school. At some point during their undergraduate training, most premedical students volunteer in hospitals. Through observing the behavior of hospital doctors, students quickly learn the value placed on emotional detachment. This norm can be further reinforced during admissions interviews at medical schools. Currently enrolled students often take prospective students to see the most grotesque-looking, partially dissected human cadaver available in the school's anatomy lab. Although officially they do so to display the school's laboratory facilities, their true purpose seems to be to elicit emotional reactions from prospective students. The laughter and snickers these reactions evoke in the medical students demonstrate to prospective students that such behavior is shameful while demonstrating to the current students how "tough" they have become.

The emphasis on emotional detachment is reinforced often during medical school and residencies, as faculty and students implicitly or explicitly ridicule those who display emotions and question their ability to serve as doctors (Haas and Shaffir, 1987: 85–99; Hafferty, 1991). During daily rounds of the wards, faculty members grill residents on highly technical details of patients' diagnoses and treatments. Except in family practice residencies, however, faculty members rarely ask about even the most obviously consequential psychosocial factors. Rounds and other case presentations also teach residents to describe patients in depersonalized language. Residents learn to describe individuals as "the patient," "the ulcer," or "the appendectomy" rather than by name. As Renee Anspach (1997: 328) has described, using language like "the vagina and the cervix were noted to be clear" rather than "I noted that Mrs. Simpson's vagina and cervix were clear" reinforces the impression "that biological processes can be separated from the persons who experience them." The use of medical slang, meanwhile, which peaks during the highly stressful residency years, allows students and residents to turn their anxieties and unacceptable emotions into humor by using terms such as "crispy critters" for severe burn patients. Medical slang also enables doctors and residents to avoid emotionally distressing interactions with patients and their families by using terms that laypersons cannot understand, such as "adeno-CA" for cancer (Coombs et al., 1993).

The structure of the residency years largely prevents residents from emotionally investing in patients (Mizrahi, 1986). Long hours without sleep often make it impossible for residents to provide much beyond the minimum physical care necessary (Christakis and Feudtner, 1997). When combined with the norm of emotional detachment, such long hours can even encourage doctors to view their patients as foes. As Phillip Reilly (1987: 226) explains in his autobiographical account of medical training: "At 3 o'clock in the morning as I stood over [a comatose patient's] bedside staring at his IV, he was an enemy, part of the plot to deprive me of sleep. If he died I could sleep for another hour. If he lived, I would be up all night." According to Terry Mizrahi, who spent three years observing, interviewing, and surveying residents in internal medicine, by the end of their training, most held "attitudes towards patients ranging from apathy to antipathy" (Mizrahi, 1986: 122). These attitudes are reflected vividly in the many slang terms residents use (sometimes within earshot of patients) to describe those they dislike treating, including "trainwrecks" (seriously ill or injured patients who might not seem worth spending resources on), "scumbags" (dirty, smelly patients), and "negative wallet biopsies" (patients with neither money nor health insurance). Such terms help doctors vent frustrations regarding the difficulties of their situation and maintain needed emotional distance, but they also implicitly reinforce disparaging attitudes toward patients (Coombs et al., 1993).

Not surprisingly, given these structural factors and the resulting attitudes, the doctors Mizrahi studied sometimes appeared to care more about getting rid of patients than about providing care. The centrality of this motive to residents' lives is evidenced by the numerous slang terms for this process. For example, a resident who has "taken a hit" (received an unwanted patient on his ward) can "buff" a patient's record (making the patient seem ready to move on to another form of care) so that the resident can "turf" (transfer) the patient elsewhere (Coombs et al., 1993). Among those Mizrahi observed, the desire to get rid of patients grew as residents came to realize that many patients suffer from illnesses or social problems that medicine cannot cure. Doctors reserved their most negative attitudes for such patients, as well as for those the doctors deemed morally or socially unworthy of their time. The latter include patients whose illnesses seemed linked to self-destructive behaviors; who sought treatment for minor illnesses; who were poor, non-white, female, or old; or who suffered from common illnesses that the doctors, trained in research-oriented medical schools, found uninteresting.

Clinical Experience

In addition to teaching doctors certain attitudes toward patients, medical culture also teaches, at a more abstract level, a set of attitudes toward medical care, illness, the body, and what makes humans truly human. Ironically, given that doctors' prestige rests partly on their scientific training, medical culture values clinical experience more than scientific research and knowledge (Bosk, 2003; Ludmerer, 1985; Millenson, 1997). The structure of medical training unintentionally reinforces this notion. During the first two years of medical school (the preclinical years), students take basic science courses taught by professors who hold doctorates in fields such as biochemistry or physiology. Students spend the next two years training in hospitals and clinics under professors who are themselves doctors. This division between scientific training taught by scientists in the early years and clinical training taught by doctors in the later years teaches students that scientific training is something to be endured before the "real" work of medical training begins.

Once students begin their clinical training, they also learn to base treatment decisions primarily on their personal experiences with a given treatment rather than on scientific research (Becker et al., 1961; Ludmerer, 1985: 260–271). For example, Knafl and Burkett describe the following incidents observed during surgical rounds at a hospital they studied:

> After the residents finished presenting the case to the audience, one of the attendings [senior doctors who supervise residents] asked, "What 'bout doing a cup arthroplasty on him?" Morrison replied, "There's some *literature* to back it up but it's *my experience* that 'cups' just aren't that successful on young people." (1975: 399)

Similarly:

> The second case is presented by Dr. Lee, a 4th-year resident. He shows slides of a 13-month-old girl whose one leg is shorter than the other. The reason for presenting the case is to discuss whether or not the leg should be surgically lengthened. In presenting the case, Dr. Lee quotes from a source in favor of such a procedure. Dr. Eddy, an attending physician, interrupts with, "I know that's what he says, but that's not the way we do it here." (1975: 399)

In this way, residents have learned to value their own intuition and idiosyncratic clinical experience over scientific research. This partially explains why standard clinical procedure varies enormously from community to community and from doctor to doctor, producing high rates of medical error as well as rates of lumpectomies, prostatectomies, and back-pain surgery that are as much as thirty three times higher in some states than in others (Center for the Evaluative Clinical Sciences, 1996; Leape, 1994).

Recent events, however, suggest that scientific research may be growing as a value within medical training and the medical world in general. The practice protocols described earlier are part of a broader push for **evidence-based medicine:** the idea that medical care should be based on a thorough evaluation of the best available data from randomized, controlled clinical research. The trend toward evidence-based medicine reflects not only the concerns about cost control mentioned earlier but also the growing recognition that less than half of modern medical treatments—and only a small fraction of surgical procedures—have good scientific support (Naylor, 1995).

Almost all medical schools now explicitly incorporate evidence-based medicine into their curricula. But this does not mean that doctors now base their practices solely on scientific evidence rather than on their personal clinical experience (Timmermans and Berg, 2003). When doctors are working on a case, they rarely have time to obtain the latest research findings on the topic, let alone to evaluate that research fully. Instead, they often must settle for reading a single research article in a prestigious journal, or a single review article. In addition, because practice protocols cannot cover all the specific circumstances of each patient, doctors must rely on their clinical judgment rather than simply following practice protocols. Finally, medical training and practice remain hierarchical environments, in which doctors and medical students are expected to defer to their teachers, senior staff, or partners and are unlikely to challenge more senior doctors whose recommendations go against practice protocols. On the other hand, because junior doctors are increasingly turning to the research literature for answers, more senior doctors must do so as well to retain their reputations and status. In sum, evidence-based medicine has affected medical care, but has not supplanted clinical experience as a decision-making tool.

Mastering Uncertainty

One reason medical culture values clinical experience over scientific knowledge is that there is simply too much knowledge for students ever to learn it all. As a result, students can never be certain that they have diagnosed or treated a patient correctly. Moreover, because the answers to so many medical questions remain unknown, even a student who somehow learned all the available medical knowledge would still on occasion face uncertainty about diagnoses and treatments. From the start of medical school, then, students must learn how to cope emotionally with uncertainty and how to reduce uncertainty where possible (by, for example, focusing on memorizing the discrete facts most likely to show up on examinations) (M. Fox, 2000).

Students also must learn to question whether their difficulties in treating patients stem from a lack of available knowledge in the field or their own lack of familiarity with the available knowledge. Simultaneously, however, students' experiences in medical school classes and on the wards where they study also teach them that they must hide their sense of uncertainty if they are to be regarded as competent by their professors and patients (Atkinson, 1984; Light, 1979).

Mechanistic Model

Along with learning to master uncertainty, medical students also learn to consider the body analogous to a machine or factory and to consider illness analogous similar to a mechanical breakdown (E. Martin, 1987; Mishler, 1981; Osherson and AmaraSingham, 1981; Waitzkin, 1993). For example, medical textbooks routinely describe the biochemistry of cells as a "production line" for converting energy into different products, and they describe the female reproductive system as a hierarchically organized factory of signaling machines that "breaks down" at menopause (E. Martin, 1987). Similarly, medical writers typically describe **HIV disease** as a mechanical failure of the body's immune system (Sontag, 1988).

The mechanistic model of the body and illness leads naturally to a distrust of natural bodily processes. Doctors learn to always look for signs that the body is breaking down, and to view changes in the body as causes or consequences of such breakdowns. As a result, doctors typically view pregnancy and menopause as diseases, try to stop the effects of aging if possible, use drugs to control minor fevers (the body's natural process for fighting infection), and so on (e.g., Barker, 1998; E. Martin, 1987). Thus, for example, when Perri Klass (1987), a doctor and writer, became pregnant at age 26, her classmates were horrified that she did not have amniocentesis, a test designed to identify certain chromosomal abnormalities in fetuses. Yet for women in their twenties, who have extremely low rates of fetal abnormalities detectable by amniocentesis, the test more often causes miscarriage than detects abnormal fetuses. Klass's fellow students, however, had learned so well to distrust pregnancy and the natural body that they could not evaluate her situation objectively.

Intervention

As the example just given suggests, learning to distrust natural processes is intimately interwoven with learning to value medical intervention. During the preclinical years, doctors receive only minimal instruction in using tools such as nutrition, exercise, or biofeedback to prevent or treat illness; during the rest of their training, such tools are rarely—if ever—mentioned. Meanwhile, those medical specialties that rely most heavily on intervention historically have received the most prestige and financial rewards (although RBRVS is starting to change at least the financial balance). For example, surgeons (known in medical slang as "blades"), earn almost twice the median

---

Box 11.1   ***Making a Difference: American Medical Student Association***

The American Medical Student Association (AMSA) is an independent association (not related to the AMA) of about 30,000 medical students from schools around the country. Since 1950, it has worked to improve the quality of health care and medical education, as well as to protect the welfare of medical students and residents. Recognizing some of the traditional limitations of medical education, AMSA has for some time promoted programs designed to encourage medical students to enter primary practice in underserved areas and to develop "cultural competency": the skills needed to work effectively with persons of various ethnic groups.

With these goals in mind, AMSA (2005) has developed two model curricula. The curricula developed through this program, known as "Promoting, Reinforcing, and Improving Medical Education" (PRIME), are designed both to provide students with necessary technical and interpersonal skills and to encourage students' idealism and commitment to working with underserved populations. The cultural competency curriculum includes such topics as learning how other cultures interpret health and illness, how to use interpreters, and how to manage cross-cultural communication problems. The curriculum on primary practice in underserved areas teaches students to understand the underserved, their health care needs, the philosophy of primary care, and the importance of community public health work. The program also teaches communication skills for working with underserved populations. Finally, it covers practical issues such as the finances of working in underserved areas, ways to do so while repaying student loans, and the impact of managed care on primary care work.

Both curricula are based on experiential service learning, rather than on the lectures and demonstrations that dominate medical education. This structure, it is hoped, will make it easier for students to gain confidence and skills and more likely that they will incorporate what they learn into their personal values and career plans.

---

net income of general and family practitioners (known in medical slang as "fleas"). Similarly, medical school faculty routinely disparage general and family practitioners and discourage students from entering those fields (Block et al., 1996; Mullan, 2002). (Box 11.1 describes a program designed to change this situation.) Taken together, these forces support the **technological imperative**—the belief that technological interventions should always be used if available.

Emphasis on Acute and Rare Illnesses

As a natural corollary of valuing intervention (and a natural result of locating medical training within research-oriented universities), medical culture teaches doctors to consider **acute illness** more interesting than **chronic illness.** This is not surprising, for doctors often can perform spectacular cures for acute illnesses (such as appendicitis) but can do little for chronic illnesses (such as lupus). Similarly, medical culture teaches doctors to consider common diseases less interesting than rare ones, for the latter require complex and well-honed diagnostic skills even if no treatments are available. In sum, during the course of their training, doctors learn to value emotional detachment more than emotional involvement or expression; trust clinical experience more than scientific knowledge; adopt a mechanistic model of the body and illness; trust intervention more than natural physiological processes; master uncertainty; and prefer working with rare or acute illnesses more than working with typical or chronic illnesses.

The Consequences of Medical Values

Although each of these values serves a purpose, each also can work against the provision of high-quality health care. Emotional detachment can lead doctors to treat patients insensitively and to overlook the emotional and social sources and consequences of illness. In addition, it can cause doctors to feel disdain for patients they consider too emotional. How much emotion a person shows, however, and how that person does so, depends partly on his or her cultural socialization. In contemporary America, women and members of certain ethnic minority groups (such as Jews and Italians) are more likely than are men and nonminorities to display emotion openly (Koopman, Eisenthal, and Stoeckle, 1984). Consequently, these groups are more likely to bear the brunt of doctors' disdain.

Meanwhile, the emphasis on clinical experience, although sometimes useful, can lead doctors to adopt treatments that have not been tested through controlled clinical trials and that lack scientific validity, such as treating ulcers (which are now known to be caused by bacteria) with a bland diet and training in stress reduction (Millenson, 1997). In addition, the desire for clinical experience sometimes encourages medical students and residents to perform procedures, from drawing blood to doing surgeries, even if they cause unnecessary pain or lack sufficient training or supervision. Medical students and doctors are most likely to do so if they can define a patient as "training material" rather than as an equal human being. This is most likely to happen when patients are female, minority, poor, elderly, or otherwise significantly different both from the doctors and from the patients on whom those doctors assume they will someday practice.

Mastering uncertainty is necessary if physicians are to retain enough confidence in their clinical decisions to survive emotionally. And presenting an image of authoritative knowledge undoubtedly increases patient confidence and stimulates a placebo effect, if nothing else. At the same time, the desire for certainty—or at least an aura of certainty—also probably contributes to authoritarian relationships with patients. This is particularly problematic when proper treatment really is uncertain. For example, doctors are particularly uncomfortable with patients whose diagnoses are unclear or whose treatment is unsuccessful. Similarly, even though for years considerable evidence indicated that neither regular mammograms to screen for breast cancer among women below age 50 nor hormone replacement therapy for those who experience natural menopause were effective, many doctors—unwilling, perhaps, to give up their aura of certainty—continued to dismiss concerns about these practices and to strongly recommend them to their patients.

The emphasis on working with rare illnesses (coupled with the financial incentives of specialty practice) leads to a different set of problems. Most important, it fosters the oversupply of specialists and undersupply of **primary care doctors,** or **primary practitioners**—those doctors in family or general practice, internal medicine, and pediatrics who are typically the first doctors individuals see when they need medical care (Mullan, 2002; Stimmel, 1992). About two-thirds of U.S. doctors are specialists, although only about 20 percent of the problems patients bring to doctors require specialty care (Light, 1988: 308; U.S. Bureau of the Census, 2004: Table 149). Similarly, emphasizing acute illness leads doctors to consider patients with chronic illnesses uninteresting and makes, for example, orthopedic surgery a more appealing field than rheumatology (the study of arthritis and related disorders).

Other problems stem from medicine's mechanistic model of the body. This model leads doctors to rely on **reductionistic treatment.** This term refers to treatment in which doctors consider each bodily part separately from the whole—reducing it to one part—in much the way auto mechanics might replace an inefficient air filter without checking whether the faulty air filter was caused by problems in the car's fuel system. In contrast, sociologists (as well as a minority of doctors) argue for a more holistic image of how the body works and of how illness should be treated (Waitzkin, 1993). **Holistic treatment** refers to treatment that assumes all aspects of an individual's life and body are interconnected. For example, rather than performing wrist surgery on typists who have carpal tunnel syndrome, it might be better to begin by asking whether the problem could be cured by using a wrist rest while typing or changing the height of the typist's desk. And rather than simply excising a tumor when someone has cancer, perhaps doctors and other health care workers should also explore how their patients' social and environmental circumstances contributed to cancer growth and how psychological support might improve their odds of recovery.

Finally, emphasizing intervention can lead doctors to act when inaction might be best. An individual who has a cold, for example, will likely recover regardless of treatment. Often, however, doctors will prescribe antibiotics either because they psychologically need to intervene or because their patients pressure them to do something. Yet, antibiotics cannot cure colds but can cause unpleasant or even life-threatening health problems. Moreover, in the long run, and as described in Chapter 2, unnecessary treatment can foster the development of drug-resistant bacteria.

Probably all these values, and the problems they create, are stronger during medical training than afterward. Once doctors enter practice, economic pressures encourage them, willingly or unwillingly, to show at least somewhat more sensitivity to patients' needs. In addition, those who consistently work with the same pool of patients—a situation that, as described earlier, has become less common—can develop more meaningful relationships with them. Thus, over time, doctors may recoup some of their initial, more positive, attitudes toward patients and patient care (Mizrahi, 1986). These changes cannot, however, help the millions of Americans who lack either health insurance or the ability to pay for medical care and who therefore must rely on public clinics or hospitals for their care. These patients pay the highest costs for the medical value system.

Building a Medical Career

Two of the most important decisions any new doctor must make are choosing a specialty and a type of practice. These decisions are important not only because the nature of the work itself differs across fields and practices, but also because all medical careers are *not* created equal. Rather, some specialties and practices offer considerably more status, income, and autonomy than others do. As a result, new doctors face greater competition for some residencies and jobs than others.

A new doctor's ability to enter a prestigious medical field or type of practice depends largely on **sponsorship** (Hall, 1949). Sponsorship refers to the process through which successful professionals in a given field actively help new members to establish their careers. This process is not an egalitarian one, for established members typically choose whom to sponsor based not only on **achieved statuses,** or earned qualifications such as medical school grades, but also on **ascribed statuses,** or innate characteristics such as ethnicity and gender.

Judith Lorber's longitudinal research on the careers of men and women doctors vividly shows the impact of sponsorship. For example, one young man tells how his residency supervisors sponsored him:

> Dr. _____ made a conscious effort to interest me in gastroenterology, and he had the support of the chief of medicine. I found the two of them both excellent researchers and clinicians. They made it seem very exciting and interesting, and to some extent, they also wooed me just a little bit. Dr. _____ took me to a meeting in Boston in the fall of that year. They took me to the national GI [gastrointestinal] meeting in Philadelphia in May and I loved it. The meetings were excellent, very stimulating. I had a good time, and that's when I decided to go into gastroenterology. I also had them behind me pushing me and guiding me into my choice of fellowships. I was starting late to look for fellowships, and it would have been difficult, but I had the two of them assisting and making entrés. (Lorber, 1984: 34–35)

In contrast, the women Lorber studied lacked such sponsorship. Although they rarely experienced overt discrimination, they endured constant covert

Table 11.1   *Median Total Salary and Percentage of Residents*
Who are Female, by Specialty

SPECIALTY	MEDIAN TOTAL SALARY*	PERCENTAGE WHO ARE FEMALE**
Obstetrics/gynecology	233,000†	74
Pediatrics	153,000	68
Psychiatry	163,000	52
Family/general practice	150,000	51
Internal medicine	156,000	41
Anesthesiology	307,000	27
General surgery	255,000	25
Total	**NA**	**41**

*2002 data
**2003 data
†Insurance costs are significantly higher for obstetrics, and so net salaries in this field are relatively lower than these total salaries.
Source: Bureau of Labor Statistics (2004); American Medical Association (2005b).

discrimination. Their professors typically assumed that women would be happiest in traditionally female, low-status fields such as psychiatry, public health, and pediatrics. These professors therefore discouraged them from entering other fields and withheld the experience, recommendations, encouragement, and other forms of sponsorship needed to enter them. As a result, the women eventually found themselves in less prestigious and remunerative fields than did their male peers, despite approximately equal academic grades, research records, and desire to enter high-status fields. (Table 11.1 shows median total salary and the percentage of residents who are female for several medical specialties.) Meanwhile, women who do enter male-dominated fields typically face continual disadvantages; this problem was highlighted when in 1991 Dr. Frances Conley, the first female full professor of neurosurgery in the United States, resigned her tenured position at Stanford University in protest against years of discrimination.

Although little recent research is available on the topic, these same processes undoubtedly hinder the careers of those who differ from most doctors in ethnicity or class. Indeed, the many Catholic and Jewish nonprofit hospitals around the country were founded early in the twentieth century because most hospitals refused to hire Catholic or Jewish doctors. Over time, religious discrimination within medicine all but disappeared, and we can hope that other social barriers eventually will fall as well.

Patient-Doctor Relationships

From the beginnings of Western medicine, medical culture has stressed a paternalistic value system in which only doctors, and not patients or their families, are presumed capable of making decisions about what is best for a patient (Katz, 1984); this chapter's ethical debate on truth telling in health care (Box 11.2) gives an example of such a situation. Often, this paternalism is reinforced by patients who prefer to let their doctors make all decisions; indeed, at least part of doctors' efficacy comes simply from patients' faith in doctors' ability to heal. Paternalism is also reinforced by the structure of medical practice, in which doctors by their own (probably optimistic) estimates spend an average of only 18 minutes per patient per office visit (Mechanic, 2001b). As a result, doctors often do not have the time to inform patients fully or to assess patients' needs or desires.

Unfortunately, doctors' inclination to make decisions for patients is sometimes bolstered by doctors' racist, sexist, or classist ideas. Doctors are exposed to and sometimes adopt the same stereotypical ideas about minorities, women, and lower-class persons common among the rest of society, believing, for example, that African Americans are unintelligent, women flighty, and lower-class persons lazy. Doctors who hold such ideas sometimes make decisions for patients belonging to these groups, rather than involving the patients in the decisions, because these doctors believe it is easier and less time-consuming to do so. For example, medical residents in obstetrics and gynecology interviewed by Diana Scully (1994) made such comments as "I don't like women that think they know more than the doctor and who complain about things that they shouldn't be complaining about" and "I think the main thing is that the patient understands what I say, listens to what I say, does what I say, believes what I say." Similarly, "I don't care for the patient that gives you a fight every time you try to give them a drug. I don't care for the patient that disagrees with me" (Scully, 1994: 92).

Finally, doctors' inclination to make decisions for patients can be reinforced when cultural barriers make it difficult for doctors to gain patients' cooperation or to understand patients' beliefs or wishes. Those cultural differences are probably greatest when Western-born doctors treat immigrants from non-Western societies. In these circumstances, even the smallest gestures unintentionally can create misunderstanding and ill will. For example, in her observations of Hmong patients who had immigrated from Laos and their American doctors, Anne Fadiman found that

> when doctors conferred with a Hmong family, it was tempting to address the reassuringly Americanized teenaged girl who wore lipstick and spoke English rather than the old man who squatted silently in the corner. Yet failing to work within the traditional Hmong hierarchy, in which males ranked higher than females and old people higher than young ones, not only insulted the entire family but also yielded confused results, since the crucial questions had not been directed toward those who had the power to make the decisions. Doctors could also appear

Box 11.2   *Ethical Debate: Truth Telling in Health Care*

Jeffrey Monk, an unmarried, 26-year-old accountant, goes to see Dr. Fisher because of recurrent headaches that have made it difficult for him to concentrate at work. Jeffrey generally enjoys good physical health, although he has experienced bouts of severe depression since his mother died a few months ago.

Dr. Fisher runs a series of tests and soon discovers that Jeffrey has an inoperable brain tumor, which will probably kill him within the year. Because no treatments are available, telling Jeffrey of his diagnosis would seem to serve little purpose at this point. Jeffrey has no dependents, so he need not make a will or other financial arrangements immediately. Moreover, telling him might cause his health to deteriorate more rapidly, spark another depressive episode, or even lead him to commit suicide. Anyway, Dr. Fisher believes, few patients truly want to know they have a fatal illness. He therefore merely tells Jeffrey that the headaches are not serious and prescribes a placebo, counting on the fact that placebos significantly reduce patient symptoms in about 30 percent of cases.

Do doctors have an obligation to tell their patients the truth? Answering this question requires us to look at several significant ethical issues. The most central ethical issues in this case are autonomy versus paternalism. According to the principle of autonomy, each rational individual is assumed capable of making his or her own choices if given sufficient information, and each health care worker has the obligation to provide that information. Consequently, each individual has the right to decide what is in his or her own best interest and to act upon those decisions without

coercion from others. Counterbalancing this is the principle of personal paternalism—the idea that some individuals (in this case, doctors) have the expertise needed to decide what is in the best interest of other individuals.

Evaluating this situation requires us to weigh the benefits of disclosure against those of dissembling. Will hiding his diagnosis from Jeffrey protect him from depression or suicide, or will the anxiety caused by not knowing the meaning of his symptoms increase his emotional problems? Is suicide necessarily against Jeffrey's best interest? Is it best for a doctor to give a patient a placebo, which may offer some physical and emotional relief, or to let the patient know the truth, so the patient may make his or her own choices—from seeking unconventional treatments or a second opinion to choosing how to spend his last months? The final question, then, is can doctors know what is in their patients' best interest, and when if ever should they be given the authority to act on those judgments?

Sociological Questions

1. What social views and values about medicine, society, and the body are reflected in this debate? Whose views are these?

2. Which social groups are in conflict over this issue? Whose interests are served by the different sides of this issue?

3. Which of these groups has more power to enforce its view? What kinds of power do they have?

4. What are the intended consequences of the various policies under consideration? What are the unintended social, economic, political, and health consequences of these policies?

disrespectful if they tried to maintain friendly eye contact (which was considered invasive), touched the head of an adult without permission (grossly insulting), or beckoned with a crooked finger (appropriate only for animals). (1997: 65)

In these circumstances, doctors sometimes conclude that collaboration with patients is impossible and that paternalistic decision making is their only alternative.

Nevertheless, doctors only rarely have complete control over treatment decisions and interactions with patients. As Thomas Szasz and Marc Hollander (1956) explain, three **models of doctor-patient interactions** exist. Only in the first model, **activity-passivity,** is the doctor totally active and the patient totally passive. Emergency surgery performed on an unconscious patient would fall into this category, as would drugging a psychiatric patient against his or her will. In the second and most common model, **guidance-cooperation,** the doctor offers guidance to a cooperative but clearly submissive patient, such as one suffering from a cold. In the third model, **mutual participation,** both doctor and patient participate equally. This model occurs most often with chronic illnesses such as diabetes or multiple sclerosis, in which much of doctors' work consists of helping patients discover what works best for them.

Eliot Freidson (1970a) has looked at the power dynamics underlying these different models. Doctors' power is greatest in two situations: (1) when patients are completely incapacitated by coma, stroke, or the like; and (2) when doctors have sufficiently greater cultural authority than their patients so as to argue convincingly that they can most accurately judge patients' best interests, whether that patient is a Jehovah's Witness who refuses a blood transfusion, a pregnant woman who refuses a cesarean section, or someone labeled mentally ill who opposes hospitalization. Doctors' power also increases when they work in group practice rather than in solo practice. Because doctors in group practice obtain most of their business through referrals from colleagues or MCO contracts rather than from satisfied patients, they need not worry as much as other doctors about losing income if they assert their power and alienate patients. Finally, doctors' power is higher when interacting with patients who do not share the doctors' language, culture, and social status. In sum, doctors' power depends on their cultural authority, economic independence, cultural differences from patients, and assumed social superiority to patients. As this suggests, and given the demographic composition of contemporary medicine, doctors are most likely to adopt egalitarian interaction patterns with those they consider their equals: white, nonelderly, male, and middle- or upper-class patients (Street, 1991).

To explore *how* doctors maintain dominance during their meetings with patients, researchers have conducted detailed analyses of conversation patterns between doctors and patients (S. Fisher, 1986; Katz, 1984; Waitzkin, 1991; West, 1984). Conversations between doctors and patients typically

| Box 11.3 | *"Hi, Lucille, This Is Dr. Gold!"*
by Lucille G. Natkins |

I'm going in for a dilation and curettage (D&C) next week. But even as I worry about carcinomas and five-year survival rates, an incident from my last D&C keeps popping into my mind.

That operation occurred after I hadn't seen a gynecologist in years. On my internist's recommendation I saw a physician whom I'll call Dr. James Gold, diplomate, American Board of Obstetrics and Gynecology; fellow, American College of Surgeons; and associate attending physician at a large teaching hospital. It turned out that he was a contemporary, that he lived in my neighborhood, and that his children and mine were classmates. He'd gone to medical school with one of my friends and interned with another. No one would have worried about inviting us to the same dinner party.

One visit and several phone calls later—all conducted on a cordial "Dr. Gold" and "Mrs. Natkins" basis—surgery was scheduled and soon afterward I was wheeled into the operating room. As my vision blurred and my legs numbed, a voice cut through the anesthetic haze. "Hi, Lucille, this is Dr. Gold!" Stupor

turned to rage. "You expletive, that's not the way it goes! It goes 'Hi, Lucille, this is Jim' or 'Hi, Mrs. Natkins, this is Dr. Gold.'"

All soundless. I was out of it, zonked. The next thing I remember was a female voice saying, "Wake up, Lucille, the operation's over. Wake up, Lucille." Damn, I thought, not again.

The biopsy findings were negative. I was free to stop worrying about gynecological malignancies, but "Hi, Lucille" wouldn't leave me. There are more dignified positions in life than lying naked and horizontal, legs spread-eagle, while half a dozen strangers shove their fists into what was once (wisely) called "one's private parts." But that indignity was unavoidable. What, though, was the purpose of "Hi, Lucille, this is Dr. Gold" from someone who would have been Jim had we met socially, or "Wake up, Lucille" from someone who was ensuring my waking by slapping my face? What purpose other than to underscore my lack of dignity and helplessness?

"Hi, Lucille" was still rankling months later when my 80-year-old mother-in-law was hospitalized. Overwhelmed by crippling arthritis and

follow a pattern in which the doctor opens a topic with a question, the patient responds, and the doctor signals that the topic is closed (Mishler, 1990). The doctor can then raise the next topic or ask further questions for clarification and repeat the cycle. In either event, the doctor maintains control over the direction and length of the conversation. For example, a patient might come to a doctor complaining of various problems. The doctor will ask for further details about only some of those problems, typically ignoring how factors in patients' lives might cause health problems or how health problems might cause other problems in patients' lives. The doctor also can ask questions about problems the patient had not mentioned but the doctor expects to find, thereby defining certain problems but not others as relevant. In addition, doctors control conversations by asking

a host of other problems, she asked the nurse, whose name pin read "T. Bass," to "please get my slippers from the bedroom." "Whatever are you talking about, Bertha," snapped T. Bass, who was, perhaps, all of 30 years old. "You're in the hospital, not your house." My mother-in-law stiffened and blanched. Reality therapy with a bludgeon.

I became a first-name freak, asking friends and colleagues who addressed them by first name without expecting reciprocity and, conversely, whom they addressed by first name while expecting to be called Mr. Price or Dr. Wand. No surprises in this survey. Inferiors are called by first name: children, menial workers, the elderly, and women.

I wrote to the hospital where my mother-in-law had been a patient, noting that the hospital system that was reducing an 80-year-old woman to a child was robbing her of the will and determination she needed to ensure her recovery. The administrator replied that he could not understand my charges of abuse. I wrote to a widely syndicated medical columnist, asking why his replies to women began "Dear Amy" and to men "Dear Mr. Hall." No answer. . . .

I chose a new gynecologist. But not by using physician referrals and checking medical directories as I would have before, when I thought I was sophisticated. "Is your gynecologist a nice person?" I asked friends. "Are you treated with dignity and consideration? Called by your first name or your last?" Another survey with few surprises. Not many women answered "yes," "yes," and "last name."

But some did. (And, yes, my new gynecologist is board-certified, as nearly everyone in a metropolitan area seems to be these days.) So far, so good, but next Friday both of us will have to pass our big tests in the operating room. Will I have malignant cells on my pelvic wall? Will he resist the temptation to say "Hi, Lucille" when I'm flat on my back and going down for the count?

Health and self-respect, I've learned, are both necessities.

Source: *Journal of the American Medical Association*, May 7, 1982, 247(17): 2415. © Copyright 1982, American Medical Association. Reprinted with permission.

closed-ended rather than open-ended questions, thus making it difficult for patients to raise new topics. Doctors also can reinforce their dominance by the simple tactic of referring to the patient by first name, but expecting the patient to refer to them by their title ("Dr. Smith"), as Box 11.3 illustrates.

Other techniques also enable doctors to control interactions with patients (S. Fisher, 1986; Katz, 1984; Waitzkin, 1991; West, 1984). Doctors interrupt patients far more often than patients interrupt doctors, cutting off discussions and questions the doctors consider irrelevant or uncomfortable. They give general rather than specific answers to patients' questions, give information only when directly asked, or use euphemisms (such as "tumor" instead of "cancer") that leave patients confused about their situations. As a result, patients lack the information they need to challenge doctors' actions or make their own decisions. This in turn can create both stress and distrust when patients conclude that their doctors have withheld information.

Conclusion

Between 1850 and 1950, allopathic medicine attained and then enjoyed unprecedented autonomy and dominance, becoming the premiere example of a profession. In its battles for status with its many nineteenth-century rivals, allopathic medicine benefited from the public's growing respect for scientific knowledge and from the increase over time in the field's scientific foundations. It also benefited from the public's assumption that because allopathic doctors were disproportionately upper-class white men, they must be more competent than the minorities, women, and poorer persons who dominated competing health care fields.

Since the 1950s, however, doctors' social status has declined and their control over working conditions, relationships with patients, and finances has diminished. Yet doctors continue to have far more autonomy and dominance than do professionals in most other occupations, especially within the health care field. This continued professional dominance—and the continued internecine warfare between medicine and other health care occupations—affects all of us as consumers of health care because it sets the stage on which attempts to improve the health care system must occur.

Doctors' professional socialization, too, affects all of us as consumers. In its current form, this process is lengthy, arduous, and expensive, making it difficult if not impossible for many otherwise qualified persons to become doctors and encouraging those who do become doctors to become emotionally hardened or financially driven. To these **unintended negative consequences** of medical training must be added the problems caused by a medical culture that emphasizes emotional detachment, clinical experience, intervention, mastering uncertainty, and acute and rare illnesses rather than common and chronic illnesses.

As consumers of health care, we all benefit from the extensive training doctors receive. Those benefits, however, must be weighed against the costs we pay when our doctors also learn ways of interacting with patients and thinking about illness that can encourage overly aggressive, scientifically unjustified, or simply discourteous treatment. Only by directly confronting the nature of medical culture can we hope to change medical training and make future doctors better able to meet their patients' needs.

Currently, pressures to change medical culture and doctor-patient relationships are coming from within as well as outside the medical field. Many doctors now believe that the rise in malpractice suits largely reflects patients' disenchantment with their relationships with doctors rather than problems in the quality of care. As a result, medical journals often publish articles instructing doctors to reduce their malpractice risk by improving their relationships with patients (Annandale, 1989).

Deeply felt personal beliefs, and not just economic self-interest, have driven other doctors to work for changes in the system. Such beliefs have led to the founding of organizations such as Physicians for a National Health Plan and the American Holistic Medical Association. Similarly, the American College of Physicians, the professional organization for doctors in internal medicine, derives its strength partly from the growing number of doctors who favor its more humanistic approach to medical care.

Finally, throughout the United States, medical students and professors are working to implement innovative programs for integrating more humanistic perspectives into the medical curriculum. At Harvard Medical School, for example, students now must take a three-year course specifically designed to improve relationships with patients and to humanize medical care (Tosteson, Adelstein, and Carver, 1994). Beginning with role-playing and discussing their personal experiences of illness, students are reminded what it is like to experience illness and health care. Subsequently, students learn how to interview patients, with the emphasis on listening to patients and understanding the psychosocial circumstances in which individuals experience illness. In this way, students can learn from the beginning of their training to see health care from patients' perspectives.

Similarly, **cultural competence** is now a commonly cited goal of medical education. Cultural competence refers to the ability of health care providers to understand at least basic elements of others' cultures and thus to provide medical care in ways that better meet clients' emotional as well as physical needs. Cultural competence has been promoted as a means of increasing clients' willingness to seek medical care and satisfaction with care, and consequently as a means of improving health outcomes. Both the American Psychiatric Association and the American Academy of Family Physicians have officially endorsed including cultural competence in medical training. As of 2002, 38 percent of medical students participated in overseas programs designed to increase their understanding of other cultures, but most medical schools provided only three or fewer class sessions devoted to the topic (Champaneria and Axtell, 2004). Little data are available so far about the effectiveness of this training, but the hope is that in the long run, these programs may restructure medical culture and doctor-patient relationships.

Suggested Readings

Alvord, Lori Arviso, and Elizabeth Cohen Van Pelt. 1999. *The Scalpel and the Silver Bear.* New York: Bantam. Alvord writes of her experiences as the first female Navajo surgeon, of what she has learned from working with Navajo patients, and of what Navajo healing traditions have to offer all of us. A fascinating book.

Conley, Frances K. 1998. *Walking Out On the Boys.* New York: Farrar Straus and Giroux. Conley, the first female full professor of neurosurgery in the

United States, describes in this memoir the shocking discrimination still encountered by women in surgical training and practice.

Rothman, Ellen Lerner. 1999. *White Coat: Becoming a Doctor at Harvard Medical School.* New York: William Morrow. Rothman tells of her experiences in Harvard's revamped medical school program.

Watts, David. 2005. *Bedside Manners: One Doctor's Reflections on the Oddly Intimate Encounters Between Patient and Healer.* New York: Three Rivers Press. In this book of short essays, poet, NPR commentator, and doctor David Watts beautifully describes medicine, and the doctor-patient relationship, at its best.

Getting Involved

American Civil Liberties Union. 132 W. 43rd Street, New York, NY 10004. (212) 944-9800. www.aclu.org. Among other things, works for the civil rights of mental patients.

American Medical Students Association. 1902 Association Drive, Reston, VA 20191. (703) 620-6600. www.amsa.org. Among other things, seeks to make medical education more humanistic. Open to premedical as well as medical students.

Review Questions

What was the difference between *allopathic* and *homeopathic* doctors?

What was medical training like in 1850?

What could a doctor offer his patients in 1850? in 1900?

What does it mean to say that an occupation is a *profession?*

How did doctors achieve professional dominance? What factors have reduced doctors' professional dominance?

What are the major medical norms, how do doctors learn them, and how do they affect patient-doctor relationships?

What is cultural competence, and why is it important?

Internet Exercises

1. To find out how social class affects individuals' perceived health status, first locate the website for the University of California's Survey Documentation and Analysis (SDA) Archive. This archive contains data from several national random surveys. Enter the SDA archive, and then click on the GSS Cumulative Datafile, 1972–2002, full analysis. Find the "Select an Action" section, and then click the button for "Frequencies or Crosstabulations." Next, click on "Start." A form with several blank spaces will appear on your screen. For row variable, type *conmedic.* For column

variable, type *class*. Click on the boxes to the left of "Column Percentaging," "Statistics," and "Question Text." Then click the button to "Run the Table." Repeat, using first sex and then health as column variables. Which groups have the least confidence in the people running the institution of medicine? Which variables have the most impact?

2. Go to the website for the Center for Responsive Politics (www.opensecrets. org) and see what you can learn about how medical organizations and other health care industry groups are working to affect elections and health-related laws in the United States. You might look for information on managed care legislation, tobacco control, or gun control.

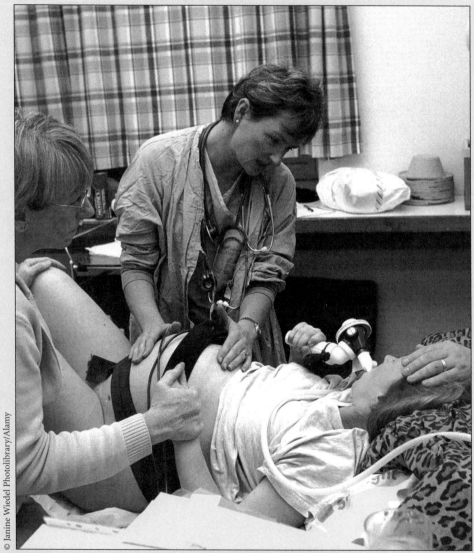

Other Mainstream and Alternative Health Care Providers

For more than a decade, Juliana van Olphen-Fehr ran an independent practice as a nurse-midwife delivering babies in women's homes. In the following story, she gives us a sense of what it is like to participate in a home birth:

Late in the evening, Mona's contractions started getting quite intense. She paced around the room while we watched. She'd sit on the toilet frequently and Dave [her husband] rubbed her back when she was on the bed. . . . We tried to encourage Dave to go take a nap but he didn't want to leave Mona for a moment. He finally fell asleep in the bed while it was our turn to rub Mona's back. The night moved into early morning. The clock ticked away. We walked and talked.

It's amazing how long it takes a baby to be born. As time passes slowly, labor gives one the opportunity to reflect on the process of birth. Each contraction comes and goes, [as] the uterus gets smaller and smaller [and] the baby is massaged down further and further into the pelvis. . . . Finally, the uterus, getting more powerful as it decreases in size, pushes the baby out of its first cradle, the pelvis, through the vagina, the passageway to life, into the outside world. The mother, feeling more and more pressure, joins the uterus in its expulsive efforts. She bears down gently and involuntarily at first but then more forcefully and purposefully as the baby approaches birth.

In its natural environment, giving birth is like a musical masterpiece, building to its crescendo when the baby enters the world. Just as a symphony pulls its audience into its powerful rhythm, so does a laboring woman pull in her onlookers. All of those present at birth must be in synch just as all of the instruments in an orchestra must be in synch. This synchronization helps the mother keep her power to create her own masterpiece. . . .

Mona's labor built up to the point where she started to feel the urge to bear down. Her cervix was completely dilated and I felt the baby's head

*low in the vagina. She squatted while she pushed during the contractions
and walked during the break between them. She found it most comfortable to lean on the banister in her hallway while she pushed. . . . Dave was
still behind her, supporting her hips. I encouraged her to push while I got
under her to monitor the baby's heartbeat.*

*Finally, the head appeared. Dave was behind Mona, sitting on the
floor, I was beneath her in the front. Together we had our hands around
the baby's head, supporting it as we coaxed her to push the baby out
slowly. A beautiful little boy was born into Dave's and my hands. I held
the baby as Dave eased Mona back onto his lap. His arms were around
her as they both welcomed the baby into their arms. My birth assistant
covered all three of them with blankets to keep the baby warm with their
body heat. We turned the light low so the baby would open his eyes. In
happy exhaustion, we sat back and through tears watched this family fall
in love with each other. (Van Olphen-Fehr, 1998: 111–113)*

Van Olphen-Fehr's story evokes for us both the joy that midwives can
find in assisting at childbirth and some of the reasons health care consumers might choose a nontraditional option like home birth. Since this
story took place, however, unaffordable insurance premiums have forced
virtually all nurse-midwives to abandon independent practice and to work
only under direct physician supervision. This situation illustrates the problems faced by nonmedical health care workers in trying to achieve professional status in a system characterized by **medical dominance.**

In this chapter, we first look at the history and current status of four
occupations now considered part of mainstream health care—nursing,
nurse-midwifery, pharmacy, and osteopathy. As we will see, nursing in general, handicapped by its historically female tradition, has achieved only
semiprofessional status, although nurse-midwives have gained a somewhat
higher status by carving out a specialized niche for themselves. Pharmacy,
on the other hand, is considered a profession, but faces continuing struggles
to retain its professional prerogatives, whereas osteopaths have attained
professional status parallel to that of medicine. We then consider the history
and status of five occupations that, to a greater or lesser extent, remain outside of mainstream health care—chiropractic, lay midwifery, *curanderismo,*
Christian Science practice, and traditional acupuncture. The history of chiropractic illustrates how, despite medical dominance, an alternative health care
occupation can secure a role for itself by limiting its services to a narrow field.
Finally, the histories of lay midwives, Christian Science practitioners, Mexican
American **curanderos,** and traditional acupuncturists show how occupations
can remain marginal to the health care system, unable in the face of medical
dominance to secure more than a small and precarious niche for themselves.
Table 12.1 compares the occupational prestige of some of these fields.

Table 12.1 **Occupational Prestige, Rated by a Random Sample of Americans, 1989**

OCCUPATION	SCORE
Doctors (MD or DO)	86 (highest score possible)
Lawyers	75
Dentists	72
Pharmacists	68
RNs	66
Legislators	61
Chiropractors	60*
LPNs	60
Dental hygienists	52
Real estate sales	49
Waiters	28
Lay midwives	23*

*From 1970 survey; data not available on 1989 survey. In general, scores are highly stable across time, so data on occupational prestige have not been collected since 1989. Note, though, that these scores were obtained prior to the resurgence of lay midwifery.

Source: National Opinion Research Center, *General Social Surveys, 1972–1991: Cumulative Codebook*, July 1991. www.norc.uchicago.edu, accessed August 2005.

Mainstream Health Care Providers

Nursing: A Semiprofession

In everyday conversations, Americans often seem to equate health care workers with doctors. The same is true for sociologists: Although many sociologists have researched doctors, very few have researched nurses. Yet nurses form the true backbone of the health care system, and hospital patients quickly learn that it is nurses who make the experience miserable or bearable and whose presence or absence often matters most. The history of nursing demonstrates the difficulties of achieving professional status for a "female" occupation.

The Rise of Nursing

Before the twentieth century, most people believed that caring came naturally to women and, therefore, that families could always call on any female relative to care for any sick family member (Reverby, 1987). Hospitals, meanwhile, relied for custodial nursing care on the involuntary labor of lower-class women who were either recovering hospital patients or inmates of public **almshouses.** These beginnings in home and hospital created the central dilemma of nursing: Nursing was considered a natural extension of women's character and duty rather than an occupation meriting either respect or rights (Reverby, 1987). Nevertheless, increasingly during the nineteenth century, unmarried and widowed women sought paid work as nurses in both homes and hospitals. Few of these, however, had any training.

The need to formalize nursing training and practice did not become obvious until the Crimean War of the 1850s, when the Englishwoman Florence Nightingale demonstrated that trained nurses could alleviate the horrors of war (Reverby, 1987). The acclaim Nightingale garnered for her war work enabled her subsequently to open new training programs and establish nursing as a respectable occupation.

Like most of her generation, Nightingale believed that men and women had inherently different characters and thus should occupy "separate spheres," playing different roles in society. To Nightingale, women's character, as well as their duty, both enabled and required them to care for others. She thus conceived of caring as nursing's central role. In addition, because her war work had convinced her of the benefits of strict discipline, she created a hierarchical structure in which nurses and nursing students would follow orders from their nursing supervisors. This structure, she hoped, would provide nurses with a power base within women's separate sphere parallel to that of doctors within their sphere. These principles became the foundation of British nursing. A few years later, when the U.S. Civil War made the benefits of professional nurses obvious to Americans, these principles were also adopted by American nursing.

By the early twentieth century, nursing schools had sprouted across the United States, as hospital administrators discovered that running a nursing school provided a ready pool of cheap labor. Within these hospital-based schools, education was secondary to patient care. A 1912 survey found that almost half of these schools had neither paid instructors nor libraries (Melosh, 1982: 41). Students worked on wards 10 to 12 hours daily, with work assignments based on hospital needs rather than on educational goals. Formal lectures or training, if any, occurred only after other work was done.

This exploitative training system stemmed directly, if unintentionally, from the Nightingale model and its emphasis on caring and duty. As historian Susan Reverby notes (1987: 75), "Since nursing theory emphasized training in discipline, order and practical skills, the ideological justification explained the abuse of student labor. And because the nursing work force was made up almost entirely of women, altruism, sacrifice, and submission were expected and encouraged."

Those women who, by the beginning of the twentieth century, sought to make nursing a profession by raising educational standards, establishing standards for licensure or registration, and improving the field's status found their hands tied by the nature of the field. According to Reverby, to raise its status, nursing reformers

Harper's Weekly, January 12, 1871. National Library of Medicine.

Nurses first won the respect of the American public during the Civil War.

had to exalt the womanly character and service ethic of nursing while insisting on the right of nurses to act in their own self-interest, yet not be "unladylike." They had to demand higher wages commensurate with their skills and position, but not appear "commercial." Denouncing the exploitation of nursing students as workers, they had to forge political alliances with hospital physicians and administrators who perpetrated this system of training. While lauding character and sacrifice, they had to measure it with educational criteria in order to formulate registration laws and set admission standards. In doing so, they attacked the background, training, and ideology of the majority of working nurses. Such a series of contradictions were impossible to reconcile. (1987: 122)

Political weaknesses also hamstrung nurses' attempts to increase their status. Like other women, few white nurses could vote until 1920, and most nonwhite nurses could not do so until considerably later. Moreover, nurses faced formidable opposition from doctors and hospitals that feared losing control over this cheap workforce. Nevertheless, by the 1920s, most states had adopted licensing laws for nursing schools and nurses. But most laws were weak and poorly enforced, and so the term *registered nurse* became truly meaningful only after World War II (Melosh, 1982: 40).

Education and the Profession of Nursing

Since World War II, the major strategy used by nursing leaders to increase nurses' autonomy and status and improve their working conditions has been to increase educational requirements for entering the field (Melosh, 1982: 67–76). Beginning in the 1960s, the American Nurses Association (ANA) promoted the development of two- and four-year college-based nursing programs and lobbied to make college education a requirement for nursing. The new college-based programs quickly proved popular, as changing social norms encouraged women to seek a college education in the hopes of improved employment opportunities.

At the same time, however, the move toward higher education challenged the qualifications of those nurses—the majority—who had not attended college, especially because the college programs did not (and still do not) accept transfer credit from noncollege training programs. The drive toward professional status, or **professionalization,** thus inadvertently limited the ANA's power by alienating most practicing nurses from the organization. As a result, only a small fraction of nurses have ever belonged to it.

The increased emphasis on educational qualifications has reinforced nursing's hierarchical structure. At the bottom of the hierarchy are nursing assistants who, as described in Chapter 10, receive minimal training. Next are the **licensed practical nurses (LPNs),** who have approximately one year of classroom and clinical training and provide mostly custodial care to patients. On the top tier are **registered nurses (RNs).**

Registered nurses themselves divide into four tiers. At the bottom of this hierarchy are diploma nurses, who receive their training through two- or three-year hospital-based diploma programs. Next are nurses who hold associate degrees in nursing from two-year community college programs, and then nurses who hold bachelor in nursing degrees from four-year colleges or universities. Finally, at the top of the RN hierarchy are **advanced practice nurses,** such as nurse practitioners and nurse-midwives, who have postgraduate training in specialized fields. All advanced practice nurses enjoy considerably more autonomy, status, and financial rewards than do other nurses, including the right to prescribe some medications in most states (Bureau of Labor Statistics, 2004; Lewin, 1993). Research published in major medical journals using randomized clinical trials—in which patients were randomly assigned to doctors or to nurse practitioners—find that care provided by nurse practitioners is as good as or better than that provided by doctors (Mundinger et al., 2000; Safriet, 1992; Sakr et al., 1999). Moreover, care provided by nurse practitioners is considerably less expensive than medical care, both because nurse practitioners are paid less and because they typically use fewer expensive tests, treatments, and medications.

Nursing's leadership has achieved considerable success in its push to increase educational qualifications. Between the 1970s and the start of the twenty-first century, the number of diploma nursing schools fell from more than 800 to less than 100; diploma nurses now comprise about 20 percent of all practicing nurses, but less than 5 percent of recent graduates (Bureau of Labor Statistics, 2004; National League for Nursing, 2004). However, because associate degree programs offer a quicker route to paid employment than do bachelors degree programs, the former enroll about 10 percent more students than the latter (National League for Nursing, 2004). But because bachelors degrees are required for most of the better-paying nursing jobs (in administration and specialized fields), many who start with associate degrees eventually seek bachelors degrees.

The greatest growth in recent years has occurred in graduate degree programs for advanced practice nurses, although only a small fraction of nurses have completed such programs. These programs first appeared during the 1960s, in response to projections of a coming shortage of doctors. They now offer a wide range of career options with considerably more autonomy and higher pay than other nursing work. Those who earn masters degrees may work in fields such as anesthesiology, nurse-midwifery, or radiology, while those who earn doctoral degrees typically seek work as researchers or college professors.

Despite the increase in nurses' education, caring has remained central to nurses' work. Using data collected during three years of observing nurse practitioners and family practice doctors, Sue Fisher (1995) concluded that nurse practitioners spend more than five times as long with each patient as do doctors, using this additional time to gain a **holistic** sense of their patients' clinical problems and social situations. Whereas, doctors typically rely on closed-ended questions, tightly control which topics are discussed during patient visits, and seek to close discussions quickly, nurse practitioners rely heavily on open-ended questions, give patients more freedom to open topics, and do not push to close discussions. In addition, whereas many doctors routinely reinforce their dominance both verbally and nonverbally (by, for example, never addressing patients by name or implying that patients cannot accurately describe their own problems), nurse practitioners downplay differences in status between themselves and patients and assume that patients can accurately assess their own situations. On the other hand, like doctors, nurse practitioners retain final authority in patient-provider interactions—opening and closing discussions, asking most of the questions and thus determining which topics will be discussed, and, in the end, defining the nature of the problem.

The Rise of Specialized Nursing

Like the move toward higher education, the rise of specialized nursing has increased the professional status of some nurses. According to sociologist Andrew Abbott (1988), occupations rarely gain full professional dominance over directly competing occupations. Instead, occupations typically achieve professional status by carving out niches for themselves where there is less competition.

Research suggests that nurses can gain increased status through doing what would otherwise be low-status work if that work affords them recognition of their specialized knowledge as well as public respect for taking on work perceived as dangerous and unpleasant (Aiken and Sloane, 1997). This happened serendipitously with the development of "dedicated" AIDS wards (devoted solely to caring for persons with AIDS) during the 1980s. A nationwide survey conducted in 1988 found that compared with nurses on other wards, nurses on dedicated AIDS wards enjoyed greater professional status, control over their work environment, and professional autonomy, as well as better relations with physicians, less burnout, and less emotional exhaustion (Aiken and Sloane, 1997). On these wards, doctors were willing to cede some autonomy and responsibility to nurses because doctors were not particularly interested in providing the low-technology, palliative care persons with AIDS most often need. In addition, on these units, nurses gained specialized knowledge as great as, if different from, that of doctors. Equally important, whereas on general wards each doctor shares only a handful of patients with each nurse, on dedicated AIDS wards doctors and nurses routinely work together on the same patients, giving doctors more opportunities to witness nurses' expertise and thus making doctors more willing to treat nurses as colleagues. Finally, nurses' willingness to do dangerous and often unpleasant work caring for stigmatized patients enhanced their public image as dedicated professionals.

The Impact of Changing Gender Roles

Changing gender roles in the broader society has the potential either to help or hinder nursing's attempts to professionalize. Over the last three decades, as women have gained entry to other fields, intelligent and motivated women increasingly have chosen to enter medicine, pharmacy, or biological research instead of nursing (*New York Times,* 1999a; D. Williams, 1988); enrollment in nursing programs dropped by almost one-quarter between 1993 and 2003 (National League for Nursing, 2004). As a result, nursing no longer attracts the type of students it once could have counted on for its future leadership. For the same reason, nursing now attracts fewer white students and middle- or upper-class students. Given existing social prejudices, these changes are likely to reduce the status of nursing even if the quality of students remains constant. Finally, because women now *can* enter medicine, the public typically assumes that no intelligent woman would instead choose to enter nursing, and so underestimates the abilities of those who do enter the field (S. Gordon, 2005).

Changing gender roles not only have encouraged women to seek careers other than nursing, but also have opened nursing to men. Men currently constitute about 6 percent of employed nurses and 10 percent of recent nursing graduates (National League for Nursing, 2004). Because nursing is so strongly identified with femininity, working as a nurse presents men with a serious conflict between their gender identity and their work identity. Christine Williams (1989) found that men typically respond to this conflict by stressing the differences between what they do and traditional nursing— de-emphasizing nurturing while emphasizing their technical skills, administrative expertise, or use of physical strength.

Despite these difficulties, working as a nurse offers men substantial benefits. Williams (1989: 95) points out that "as in other female-dominated occupations, men are over represented in the most prestigious and best paying specialties" and in administrative positions. This occurs for two reasons. First, on average male nurses have more years of education than female nurses do. Second, both male and female doctors more often respect, support, and socialize with male nurses than with female nurses, giving the men help in their careers and encouragement to enter more prestigious subfields. According to Christine Williams (1992), whereas women in nontraditional fields (such as medicine) often encounter a **glass ceiling** caused by conscious discrimination and unconscious social expectations that limit their career progress, men in nursing, as in other predominantly female fields such as social work, encounter a **glass escalator** that moves them into administrative positions unless the men actively resist. It seems, then, that entering a traditionally female field such as nursing benefits male nurses.

Whether the entry of men into nursing will improve the overall status of the field, counteract the loss of academically superior and socially prestigious women students, and raise the status of the field overall, however, remains to be seen.

Nurses and the Changing Health Care System

Since the 1970s **corporatization** and the resulting emphasis on cost control has resulted in worse working conditions and decreased job satisfaction for most hospital-based nurses. To save costs, hospitals try to release patients before their insurance coverage ends, which of necessity means patients are now released sicker and quicker than in the past. Yet to keep their staffing costs as low as possible, hospitals now hire considerably fewer RNs per patient than they used to (S. Gordon, 2005). As a result, the typical hospital ward now has fewer nurses but sicker patients than in the past.

Other changes have also worsened nurses' position. First, because RNs can perform more tasks more efficiently than LPNs, hospitals now save money by assigning to RNs many of the labor-intensive, menial tasks formerly performed by LPNs. Because RNs remain responsible for many administrative and skilled technical tasks, this shift has both deprofessionalized their daily work and dramatically increased their workload (Aiken, Sochalski, and Anderson, 1996; Brannon, 1996; S. Gordon, 2005). Second, hospitals increasingly save money by hiring nurses temporarily (without benefits) or moving full-time nurse employees from ward to ward as needed, leaving nurses with little control over their schedules, the nature of their work, and who they work with. Third, hospitals have saved costs by shifting services from inpatient wards to less-expensive outpatient clinics, where fewer RNs are needed, RN salaries are lower, and their work is less prestigious (Norrish and Rundall, 2001). Finally, nurses are increasingly pressured to work back-to-back shifts and longer hours (often unpaid). Given all these changes, it is perhaps not surprising that enrollment in nursing schools has declined almost steadily since 1990 and that the dropout rate from nursing careers is very high (S. Gordon, 2005).

The Continuing Doctor-Nurse Game

The dilemmas nurses faced in gaining acceptance as a full profession are reflected in what Leonard Stein (1967) dubbed the **doctor-nurse game.** According to Stein:

> the object of the game is as follows: The nurse is to be bold, have initiative, and be responsible for making significant recommendations, while, at the same time, she must appear passive. This must be done in such a manner so as to make her recommendations appear to be initiated by the physician. (1967: 699)

In other words, inexperienced doctors are expected to use subtle verbal cues to elicit treatment recommendations from more experienced nurses, and nurses are expected to just as subtly make their recommendations (S. Gordon, 2005). For example, an experienced surgical nurse might subtly tell an inexperienced doctor what to do by selecting which instruments to place on the table and by telling the patient step by step what the doctor is about to do. In addition, nurses often do the work of doctors—prescribing drugs, tests, or physical therapy—when doctors are unavailable, but the doctors can still reinforce the doctor-nurse game by loudly stressing that the nurses are simply following the doctors' known preferences rather than making decisions on their own. Even when patients' lives are saved by nurses' quick action, the doctors typically are given the credit by patients, administrators, and other doctors (S. Gordon, 2005).

In recent years, the doctor-nurse game has become less common in areas such as emergency rooms and intensive care units, where the need for split-second decisions makes this sort of subterfuge not only counterproductive but dangerous (Stein, Watts, and Howell, 1990). In the rest of the health care world, however, the game is alive and well (S. Gordon, 2005). Moreover, relationships between doctors and nurses remain so hierarchical that they are sometimes abusive. Almost all nurses who responded to a 2002 national survey reported either experiencing or witnessing incidents in which doctors screamed at nurses, hit or threw things at nurses, abusively criticized them, or in some other way made it difficult for them to function (Rosenstein, 2002).

As these problems suggest, the increased educational qualifications of nursing has enabled it to achieve only **semiprofessional** status, achieving some but not all of the hallmarks of a profession. Although most nurses consider themselves professionals and although nurses have more autonomy and status than in the past, they remain subordinate to doctors. In hospitals and clinics, the status difference between doctors and nurses is immediately visible (S. Gordon, 2005). Doctors rarely or never read nurses' notes on patients' charts, eat with nurses in hospital cafeterias, include nurses in discussions on hospital rounds, or invite nurses to medical mortality review meetings. And doctors expect to be referred to by their title—"Dr. Smith"—while referring to nurses by their first names or simply as "my nurse." In addition, doctors continue to determine much of nurses' working conditions and to help set educational and licensing standards for nurses. Finally, despite the growth of nursing colleges and graduate degree programs, nursing has yet to develop public confidence that it has the truly independent knowledge base that defines a profession.

Nurse-Midwifery: The Limits of Specialization

The example of nurse-midwifery, one of the oldest forms of advanced practice nursing, illustrates both the benefits and the limitations of seeking professional status for a field by carving out a specialized niche.

Throughout the nineteenth century, almost all American babies were delivered at home by lay midwives who lacked specialized training and worked within their own geographic or ethnic communities (R. Wertz and D. Wertz, 1989). By the 1920s, however, most Americans had come to believe that doctor-assisted childbirth was safer and, certainly, less painful. Yet few doctors were interested in providing care to poor or rural women. Responding to this need, in 1925 the Frontier Nursing Service opened the first school for nurse-midwives, with the aim of serving Kentucky's rural poor. **Nurse-midwives** would be registered nurses who additionally received formal, nationally accredited training in midwifery. The students who trained in Kentucky learned not only to deliver babies but also to provide all needed prenatal and postnatal care. Seven years later, the Maternity Center Association began training nurse-midwives to serve New York City's urban poor.

These two organizations remained the only sources of nurse-midwives until the 1950s, when several universities, responding to widely publicized reports of an impending shortage of doctors, opened training programs. As of 2005, more than forty colleges and universities offer accredited training programs in nurse-midwifery, mostly at the master's level. In addition, the American College of Nurse-Midwives now accredits programs to train individuals who have no nursing background as "certified midwives." These programs combine basic education in health skills and medical science with the usual graduate midwifery curriculum.

Like previous generations of nurse-midwives, current nurse-midwives are expected to work primarily for bureaucratic organizations in underserved poor and rural areas. However, whereas earlier nurse-midwives had functioned largely independent of doctors and hospitals, now nurse-midwives are expected to deliver babies solely in hospitals and to take responsibility solely for normal births, which doctors considered routine, uninteresting, and poorly paid.

From its beginnings, then, nurse-midwifery was designed to avoid threatening medical dominance. Nevertheless, during the 1970s and 1980s growing numbers of nurse-midwives began to pose a threat by opening private practices with only loose connections to the doctors who provided their backup support (Lehrman, 1992).

This threat to medical dominance, however, was short-lived, for changes in insurance coverage during the early 1990s made independent practice virtually impossible for nurse-midwives. The costs of a standard malpractice insurance policy rose from $35 in 1983 to as high as $13,500 in 1998 (R. Gordon, 1989; Rooks, 1997: 86). Insurance became even more expensive for midwives who attended home births, as well as for the doctors who worked with them. This rise in insurance costs is difficult to explain, for only about 10 percent of nurse-midwives (compared with 73 percent of obstetricians) have ever been sued for malpractice (American College of Obstetrician-Gynecologists, 1998; Lehrman, 1992). Similarly, studies consistently find that, for women at low risk of complications, care by nurse-midwives (at home or in hospitals) is at least as safe as medical care in hospitals (MacDorman and Singh, 1999; Rooks, 1997: 295–343). Unfortunately, the rise in insurance premiums has caused nurse-midwives to virtually abandon independent practice, home births, and freestanding birth centers.

On the other hand, nurse-midwives have legal authority to practice and to write prescriptions in all 50 states. Thirty-three states require private health insurers to reimburse nurse-midwives for their services, and all states reimburse midwives for serving **Medicaid** clients. However, these regulations do not apply to employers who **self-insure,** setting aside a pool of money to pay health care costs for their employees rather than offering health insurance as a benefit (American College of Nurse-Midwives, 1998); self-insurance now covers about 70 percent of insured U.S. workers.

In sum, nurse-midwives have gained considerable autonomy and public recognition, as well as an established place for themselves in the health care system, through specialized training and providing care to specific populations. Their ability to gain greater professional status and independence from medical control, however, remains restricted.

Pharmacy: The Push to Reprofessionalize

Unlike nursing, pharmacy meets the three criteria (laid out in Chapter 11) that define a profession: the autonomy to set its own educational and licensing standards and to police its members for incompetence or malfeasance; a body of specialized knowledge, learned through extended, systematic training; and public faith that its work is grounded in a code of ethics. Like medicine, however, pharmacy's history illustrates how corporatization can limit an occupation's ability to retain crucial professional prerogatives. In addition, its history shows how medical dominance limits competing occupations' ability to maintain professional status.

Gaining Education, Losing Professional Prerogatives

Pharmacists' role has changed considerably during the last half century, placing their professional status in jeopardy (Birenbaum, 1982). In the past, pharmacists needed complex skills to store, compound, and dispense the drugs that doctors prescribed. Now, however, pharmaceutical companies deliver drugs in forms suitable for dispensing, leaving pharmacists with few tasks other than counting, selling, and occasionally advising on drugs to consumers or health care providers. At the same time, whereas before about 1970 more than half of pharmacists owned their own businesses, with the associated responsibilities and rewards, now most work as employees of drugstore chains, supermarket chains, or hospitals. Incomes for pharmacists remain high, with a median income of $77,050 in 2002 (Bureau of Labor Statistics, 2004); but working conditions can be poor, especially in chain stores, where twelve-hour shifts, staffing shortages, and pressure to fill prescriptions quickly are common (Stolberg, 1999). Like doctors, then, pharmacists have experienced proletarianization: They are more economically vulnerable than in the past, have less decision-making autonomy, and no longer set their own working conditions, own their tools or workspaces, or maintain individual relationships with freely chosen clients.

As pharmacists' role has shrunk, however, their education has expanded. Virtually all of the nation's pharmacy schools have replaced their older four- and five-year degree programs with six-year programs leading to doctorates in pharmacy. These new programs place less emphasis on technical aspects of drug manufacturing and more on the complex subject of drug effects and interactions (Broadhead and Facchinetti, 1985: 427). These changes in education, combined with changes in pharmacists' role, have created an identity crisis: pharmacists consider themselves professionals, but increasingly find their professional autonomy constrained (Birenbaum, 1982; Broadhead and Facchinetti, 1985).

The Growth of Clinical Pharmacy

This identity crisis has stimulated interest among pharmacists in regaining their former level of professional status, or **reprofessionalizing.** To do so, pharmacists, beginning in the early 1970s, began touting research studies suggesting that many hospital patients become ill or die because of drug errors—summed up in the influential book *Pills, Profits, and Politics* (Silverman and Lee, 1974: 262) as "wrong drug, wrong dose, wrong route of administration, wrong patient, or failure to give the prescribed drug." Because pharmacists considered themselves more knowledgeable than doctors about drug actions, reactions, and interactions, they argued that the best way to limit drug errors was to encourage **clinical pharmacy,** in which pharmacists actively advise doctors on drug treatment, while less-skilled pharmacy technicians take over the routine tasks of storing and dispensing drugs.

The push for clinical pharmacy garnered unintended support from changes in hospital procedures (Broadhead and Facchinetti, 1985). Most hospitals now have pharmacists dispense and deliver medications to each patient daily. This system gives pharmacists regular access to patient records, including all records regarding drug treatments, health status, and progress. As a result, pharmacists can evaluate the effects—both positive and negative—of doctors' drug prescriptions and learn to predict when prescriptions are likely to cause health problems.

Other support for clinical pharmacy has come from changes in the legal system. Whereas during the 1980s, hospitals discouraged pharmacists from documenting medication errors—out of fear that such documentation might *increase* hospitals' legal liability—from the 1990s to the present, court decisions that held pharmacists legally responsible for monitoring medications have led hospitals to encourage clinical pharmacy as a means of *reducing* hospitals' legal liability. Nevertheless, pharmacists' concern about preserving cordial roles with doctors, who remain the dominant professionals in the health care arena, has led them to use caution in critiquing doctors' medication decisions. As one pharmacist described:

> I would like to talk to the physician face-to-face. You're trying to correct the mistake in a nonthreatening way. You know, "I'm not trying to put you down for making this mistake, but it's something that I want you to reconsider." It's not that I'm afraid to confront a physician, bending over backwards because he's up there and I'm down here. It's just that I want to maintain a relationship and the way you interact is important. (Broadhead and Facchinetti, 1985: 432)

The Development of Pharmaceutical Care

The growth of clinical pharmacy had little impact on pharmacists who worked outside of hospitals and increased divisions between them and hospital pharmacists. The development of "pharmaceutical care," however, has given these two groups a unified program for cementing the professional status of pharmacists (Mount, 1999). **Pharmaceutical care** refers to the idea that pharmacy's central mission should be to advise consumers (rather than doctors, as in clinical pharmacy) regarding the proper use of medications, based on knowledge gathered through **controlled** studies based on **random samples** (Hepler and Strand, 1990). Its rapid adoption by virtually all pharmacy associations reflects both insurers' concerns about cost control and pharmacists' hope that pharmacy care will improve their professional status.

The Impact of Managed Care

Ironically, **managed care** and **utilization review,** which have limited doctors' professional status, have increased the professional power and status of at least some pharmacists. In the last decade, many pharmacists working for health care businesses that use managed care (including insurance plans, hospitals, and nursing homes) have become actively involved in developing **practice protocols** for doctors to follow in prescribing drugs. Pharmacists also may participate in utilization review, monitoring doctors' use of prescription drugs. Similarly, some pharmacists now serve on committees responsible for developing **formularies**—official lists of drugs, published by insurers and other health care businesses, that are considered the most cost-effective treatments for given conditions and that doctors working with these organizations are expected to prescribe. Formularies offer pharmacists real power. For example, the previously popular antacid Tagamet lost virtually all its sales when pharmacists replaced it on formularies with Zantac, which is both safer and easier to use (*Fortune*, 1999).

The rise of managed care also has improved pharmacists' position by stimulating growth in **disease management.** Disease management (sometimes known as health management) is a form of pharmaceutical care in which pharmacists are responsible for monitoring the use of prescription drugs by certain patients (typically those with chronic conditions that require constant attention to medication). Pharmacists engaged in disease management counsel patients, monitor the impact of medications on patients, and, in some circumstances, prescribe drugs themselves. Managed care organizations have adopted disease management as a way to control costs by preventing medication errors and by shifting care from doctors to lower-paid pharmacists. Some states reimburse pharmacists under Medicaid for disease management of certain groups of patients, and more than half the states give pharmacists legal authority (in collaboration with physicians) to initiate or modify drug treatment (Garrett, 2002).

Osteopathy: A Parallel Profession

Osteopathy exemplifies a health care occupation that has achieved professional status almost equal to that of medicine. Osteopaths function as **parallel practitioners,** performing basically the same roles as **allopathic doctors** while retaining professional autonomy and at least remnants of a fundamentally different ideology about illness causation (Wardwell, 1979). The history of osteopathy demonstrates the benefits and costs of gaining professional status in the face of medical dominance.

Nineteenth-Century Roots

Osteopathy was founded by Andrew Taylor Still, a self-taught allopathic doctor (Gevitz, 1988). In 1864 three of his children died from meningitis. These deaths, coupled with his belief that the use of any drug was immoral, provoked Still to investigate alternatives to allopathic medicine. The system Still eventually developed drew on the popular contemporary concept of "magnetic healing" (Gevitz, 1988: 126–127). **Magnetic healers** theorized that an invisible magnetic fluid flowed through the body and that illness occurred when that flow was obstructed, unbalanced, inadequate, or excessive. They believed that by moving their hands along patients' spinal cords, they could correct problems in the magnetic fluid and thus cure illness. Still adopted this theory essentially intact, although he attributed health and illness to problems in the flow of blood rather than the flow of magnetic fluid.

During the next few years, Still also studied the work of local bonesetters, whose work consisted primarily of setting broken and dislocated bones and joints and secondarily of treating joint problems through extending and manipulating limbs. Still's experiences convinced him that such manipulations could cure a wide variety of illnesses.

Combining magnetic healing and bonesetting, Still concluded that disease occurs when misplaced bones, especially of the spinal column, interfere with the circulation of blood. He named his new system of spinal manipulation *osteopathy*, from the Greek words for "bone" and "sickness." After the germ theory of disease became widely accepted, Still incorporated it into his theory by arguing that spinal problems predispose individuals to infections and that correcting spinal problems can help the body fight infection. To date, no research has demonstrated clearly whether osteopathic treatment has any effect, whether positive or negative. (The same, of course, could be said for most drugs and procedures used by allopathic doctors, as we saw in Chapter 11.)

Professionalizing Osteopathy

In 1892, Still established the American School of Osteopathy and began accepting students for a four-month course of instruction. Five years later, in 1897, he helped found the American Osteopathic Association (AOA). As Gevitz describes:

> from its inception, the AOA actively worked to secure the conditions necessary for the movement to obtain professional recognition. It fought for independent boards of registration and examination to give the profession autonomy; it significantly lengthened the standard course of undergraduate training and supported ongoing research projects; and it championed a code of ethics while combating the growth of impostors and imitators. (1988: 132–133)

The AOA proved highly successful. By 1901, and despite strong opposition from doctors and medical societies, fifteen states legally recognized osteopathy (Gevitz, 1988: 132). By 1923, osteopathic colleges required as many years of education as medical colleges, and forty six of the forty eight states licensed osteopaths, although many states gave them only limited privileges and required them first to pass a basic sciences examination written and administered by allopath-controlled licensing boards.

Although threats from allopathic medicine have failed to eliminate osteopathy, changes from within raise questions about osteopathy's future as an independent field. By the 1920s, most osteopaths had concluded that to compete with allopathic doctors they would have to offer a similar range of patient services. As a result, osteopaths increasingly treated acute as well as chronic illness. Osteopathic colleges continued to teach spinal manipulation but added courses in surgery and obstetrics, often taught out of medical textbooks. By the end of the decade, in a major break with its founder, the AOA mandated that osteopathic colleges provide a course in "supplementary therapeutics," including drugs. Thus osteopathy began moving toward a merger with allopathic medicine.

Despite these changes, many allopathic doctors still disdained osteopaths. Although osteopathic education had improved, it had not kept up with the changes in allopathic education, leading many states to grant only restricted privileges to osteopaths. To combat this problem, the AOA adopted a series of reforms between 1935 and 1960, including requiring three years of college for admission to osteopathic colleges; improving the curriculum, facilities, and faculty at those colleges; and strengthening internship programs at osteopathic hospitals. Because of these changes, by 1960 osteopaths had received unrestricted privileges to practice in thirty-eight states (Gevitz, 1988: 144).

The Waning of Osteopathic Identity

Despite these reforms, osteopaths still lacked the professional autonomy and status of allopathic doctors, who outnumbered them by at least twenty to one throughout the 1900s (Gevitz, 1988: 146). This situation led osteopaths in California, the state where osteopathy was most entrenched, to strike a bargain in 1962 with their allopathic counterparts. Two thousand of the 2,300 California osteopaths agreed to dissolve their ties with the AOA, stop using their osteopathic degrees, and accept new medical degrees. The California osteopathic hospitals and colleges agreed to become allopathic institutions, and the state osteopathic organization agreed that the state would stop issuing osteopathic licenses.

Although at the time many osteopaths worried that this move would weaken osteopathy, the reverse proved true. Many allopathic and osteopathic doctors alike opposed the merger, making any further mergers unlikely. In addition, the continuing professional problems of the former California osteopaths convinced osteopaths elsewhere that merging would not end their problems. Thus, interest in pursuing a broader merger never developed. Meanwhile, both federal and state legislators and regulators interpreted American Medical Association (AMA) support for the merger to mean that osteopathic and allopathic doctors were essentially equivalent. Partly as a result, by the 1970s osteopaths had received unrestricted privileges in all fifty states and now have essentially the same relationship with insurance providers as do allopathic doctors. As of 2005, there were 54,000 osteopaths practicing in the United States—more than twice the number in practice in 1976 (American Osteopathic Association, 2005).

Osteopathy, then, no longer faces serious threats from the outside. Its existence remains threatened, however, by its success (Gevitz, 1988). Osteopaths now receive training and hospital privileges virtually identical to allopathic doctors and interact with the latter as equals. Although osteopaths occasionally use spinal manipulation, generally they use the same treatment modalities as allopaths. As a result, ties among osteopaths have waned while those to allopathic doctors have grown. At the same time, the virtual elimination of differences between allopathic and osteopathic treatment and theory has reduced osteopaths' sense of a strong separate identity.

On the other hand, the growth of the consumer health movement and the rise of interest in alternative medicine since the 1970s have given a new burst of life to osteopathy. Modern consumers are increasingly sympathetic to osteopaths' orientation toward patient care, which in general is more holistic and humanistic than that found among allopathic doctors. In addition, consumers increasingly have sought less interventionistic treatments, such as osteopathic manipulation, either instead of or in addition to allopathic treatment.

In sum, the history of osteopathy demonstrates the benefits of achieving full professional status as well as the difficulties a parallel health care profession can face in maintaining an independent identity once it no longer faces discrimination from the medical world and once the ideological justification for its separate existence wanes.

Alternative Health Care Providers

The occupations described to this point all basically share allopathic medicine's understanding of how the body works, and all enjoy significant roles within the mainstream health care system. The occupations described in the remainder of this chapter are sufficiently divorced from mainstream American medicine—neither widely used nor taught in medical schools or other medical institutions—to be considered **alternative** or **complementary therapies,** even if they sometimes are covered by health insurance.

With a few exceptions (such as chiropractic, lay midwifery, and acupuncture), little is known about the effectiveness of alternative healing techniques, which include meditation, reflexology, faith healing, herbal therapies, and colonics. Because allopathic medicine has dominated the American health care system for so long, researching alternative therapies has been all but impossible. Scientific testing requires large investments of time and money, generally available only from the government, universities, or pharmaceutical companies. Until recently, researchers who wanted to study alternative techniques faced nearly insurmountable barriers to obtaining funding, especially from pharmaceutical companies, which have no reason to fund research on herbs or techniques that they cannot patent. In addition, researchers who studied these techniques faced great difficulties in getting their results published in the prestigious medical publications that set the standards for health care practice.

In 1992, however, and in a major break with past policy, the U.S. Congress voted to establish within the National Institutes of Health (NIH) an Office for the Study of Unconventional Medical Practices (later renamed the Office of Alternative Medicine). The major impetus for this legislation came from former California Congressman Berkley Bedell, who had experimented with alternative therapies after his doctors diagnosed him with terminal cancer. His apparently successful experiences convinced him that such treatments warranted wider study and use. Bedell's success in getting this legislation

Key Concepts 12.1	*Limited and Marginal Health Care Occupations*		
		LIMITED RANGE OF CARE	
		Yes	No
Marginal social position	**Yes**	Lay midwives	Traditional healers
	No	Chiropractors	Allopathic doctors

passed reflects legislators' recognition of both the soaring costs of main-stream medical care and the growing public interest in alternative health care. In 1999, NIH budgeted $50 million for research into alternative healing, an increase from only $2 million in 1992, and upgraded the Office of Alternative Medicine into a full-fledged NIH center, the National Center for Complementary and Alternative Medicine.

Interest in alternative healing is growing not only among American consumers but also among allopathic doctors. As of 2005, almost 1,000 doctors and allied health professionals belong to the American Holistic Medical Association, and more than 95 of the nation's 125 medical schools require some kind of complementary and alternative medicine coursework (Loviglio, 2005). Even more impressive, a survey distributed to allopathic doctors in several communities in Washington State and New Mexico found that more than 60 percent had referred a patient to an alternative health care provider at least once during the preceding year (Borkan et al., 1994). However, referrals most often occurred when patients requested them, conventional treatment had failed, or physicians believed that the patients' problems were emotional rather than physical.

In the remainder of this chapter, we examine five groups of alternative health care providers. The first two, chiropractors and lay midwives, at least sometimes use the language of science to justify their work. The three remaining groups, curanderos, Christian Science healers, and traditional acupuncturists, base their practices in traditional beliefs unrelated to the Western scientific worldview.

Chiropractors: From Marginal to Limited Practitioners

Unlike osteopaths, **chiropractors** have fully retained their unique identity. (See Key Concepts 12.1.) The history of chiropractic illustrates how marginal practitioners, who treat a wide range of physical ailments and illnesses but have low social status, can become, like podiatrists, optometrists, and dentists, **limited practitioners**—confining their work to a limited range of treatments and bodily parts and thereby gaining greater social acceptance (Wardwell, 1979: 230).

Early History

The roots of chiropractic nearly mirror those of osteopathy. Chiropractic was founded in 1895 by Daniel David Palmer, who coined the term from the Greek words for "hand" and "practice." Like Still, Palmer studied magnetic healing and spinal manipulation and concluded that spinal manipulation could both prevent and cure illness. However, whereas Still argued that spinal problems foster disease by restricting blood flow, Palmer argued that spinal problems foster disease by restricting nerves.

In 1896, Palmer founded the first chiropractic school to teach his techniques of spinal manipulation. The field really began growing after his son, B. J. Palmer, took over the school in 1907. By 1916, about 7,000 chiropractors had opened practices; by 1930, that number had more than doubled, as schools opened around the country (Wardwell, 1988: 159, 174).

Although from the beginning, some allopathic doctors studied chiropractic and taught at chiropractic schools, B. J. Palmer attempted to sharply separate chiropractic and allopathic medicine. Those who shared his philosophy and used only spinal manipulation became known as "straights." Most chiropractors, however, found Palmer's theory of illness too simplistic and limiting, and so adopted a wide variety of therapeutic techniques. These "mixers" treated not only musculoskeletal problems but also other illnesses, as well as providing obstetrical and mental health care (Wardwell, 1988: 162–165).

The Fight Against Medical Dominance

The American medical establishment greeted the emergence of chiropractic with the same hostility it had demonstrated toward osteopathy. To eliminate these competitors, the AMA and its regional organizations during the 1930s and 1940s filed lawsuits—many of them successful—against more than 15,000 chiropractors for practicing medicine without a license.

To further restrict chiropractic, the AMA pressed for legislation requiring prospective chiropractors to pass statewide basic science examinations written by allopathic-controlled boards. Ironically, this requirement strengthened rather than weakened chiropractic by forcing the field to raise its previously low educational standards. (As with early allopathic and osteopathic schools, early chiropractic schools accepted essentially all who could pay tuition and offered only a few months of training.) Standards improved most dramatically during the 1940s, when the National Chiropractic Association (NCA) established accrediting standards for schools and when tuition money from veterans studying chiropractic under the federal GI Bill provided the funds schools needed to meet those standards. Since 1968, all chiropractic schools have required two years of college for admission, and most states require four years of chiropractic schooling for licensure.

Similarly, chiropractic in the end benefited from allopathic medicine's legal war against it. When **Medicare** first began in 1965, Congress bowed to pressure from the AMA and voted that Medicare would not cover services by chiropractors (or by clinical psychologists, social workers, physical therapists, and others in competition with doctors). Outraged chiropractic patients responded with a massive public letter-writing campaign, which led Congress in 1972 to pass legislation extending Medicare coverage to chiropractic services, despite the lack of scientific research available at the time on its effects. This set the stage for state legislatures to require other insurance plans to reimburse for chiropractic care, at least in certain situations (Wardwell, 1988: 179).

In 1974, the last of the fifty states passed legislation licensing chiropractors. Yet organized medicine continued to limit the ability of chiropractors to practice freely. In addition to fighting legislation designed to allow chiropractors to receive private insurance reimbursement, the AMA banned contact between chiropractors and allopaths, making it impossible for chiropractors and allopaths to refer patients to each other. In response, chiropractors and their supporters filed antitrust suits in the late 1970s against the AMA, various state medical associations, the American Hospital Association, and several other representatives of organized medicine (as well as the AOA), alleging that these organizations had restrained trade illegally. Chiropractors and their defenders eventually won or favorably settled out of court all the suits. As a result, overt opposition to chiropractic ended.

Current Status

These changes have allowed chiropractors to solidify their social position. Use of chiropractic is widespread and increasing across the country. A 2002 national random survey found that 7.4 percent of English-speaking U.S. residents had visited a chiropractor in the last year (Tindle et al., 2005). A separate survey of chiropractors' patient records found that chiropractic patients were typically between 30 and 50 years old and married, and slightly more likely to be female than male (Hurwitz et al., 1998).

For the past 30 years, the 16 U.S. schools of chiropractic have continued to graduate increasing numbers of students. Approximately 49,000 chiropractors work in the United States, most in solo practice (Bureau of Labor Statistics, 2004). Median net income for chiropractors is $81,500—considerably below the $137,000 median for general and family practitioners but for a much shorter work week, averaging about 40 hours (Bureau of Labor Statistics, 2004). These figures alone suggest chiropractic's success.

That success, however, is bounded by chiropractors' status as limited practitioners. Insurers now often pay for chiropractic services—about half of the people who use chiropractic services have full or partial coverage—but usually will do so only for treating specific conditions in specific ways (Tindle et al., 2005). State licensure laws sometimes set similar limits, as does patient demand—despite chiropractic's desires to treat a broader range of problems, most patients go to chiropractors for treatment of acute lower back pain, and only 1 percent are seen for anything other than musculoskeletal problems (Hurwitz et al., 1998).

Nevertheless, chiropractors continue to push for a wider role in health care. Many chiropractors believe spinal problems underlie all illness and that spinal manipulation can cure most health problems, from asthma to cancer (*Consumer Reports*, 1994). As a result, they believe they can serve effectively as **primary care** providers and now advertise heavily that they offer care for the whole family throughout the life course.

Current research suggests that chiropractic care may help those with acute lower back pain, but is unlikely to help others. One study in which patients with acute lower back pain were randomly assigned to receive chiropractic care, physical therapy, or simply an educational booklet on managing back pain found that both chiropractic and physical therapy were more effective in reducing symptoms than was the educational booklet. However, the improvements were slight, and chiropractic proved no more effective than physical therapy. Moreover, the three therapies did not differ significantly in number of days of reduced activity or in rate of recurrence of back pain (Cherkin et al., 1998). Other studies suggest that spinal manipulation might help some patients with neck pain, but to date none has tested whether manipulation is *more* effective than other treatments or whether its risks (including delays in seeking medical care, strokes brought on by spinal manipulation, and radiation poisoning from the full-body X-rays used by some chiropractors) outweigh any potential benefits (Shekelle, 1998). Finally, no reputable research has yet demonstrated any benefits from chiropractic for health problems other than neck and back injuries. Nor does it seem likely that future research will do so, because the basic principles of chiropractic simply do not mesh with current scientific understanding of human biology.

Lay Midwives: Limited but Still Marginal

The history of lay midwifery shows the difficulties members of an occupation face in gaining acceptance as limited practitioners when the occupation draws only from socially marginal groups—in this case, women, often from minority groups. Although until the twentieth century lay midwives delivered the majority of American babies, by 2002 lay and nurse-midwives combined delivered only 8.2 percent; of these, 95 percent were delivered by nurse-midwives and only 5 percent by lay midwives (J. Martin et al., 2003). However, these percentages, although small, have increased steadily since 1975, when the federal government began collecting statistics on midwife-assisted births. In this section we consider how these changes came about and how lay midwives have attempted to regain their lost position.

The Struggle to Control Childbirth

Until well into the nineteenth century, Americans considered childbirth solely a woman's affair (R. Wertz and D. Wertz, 1989). Almost all women gave birth at home, attended by a lay midwife or by female friends or relatives. Although a few local governments during the colonial era licensed midwives, licensure laws did not survive past U.S. independence, so anyone who wanted to call herself a midwife could practice essentially without legal restrictions. Unlike nurse-midwives, who did not exist until the twentieth century, these **lay midwives** had no formal training but rather learned their skills through experience and, sometimes, through informal apprenticeships. Typically, they served only women from their geographic or ethnic community. Doctors (all of whom were men) played almost no role in childbirth, because Americans suspected the motives of any men who worked intimately with female bodies (R. Wertz and D. Wertz, 1989: 97–98). Moreover, doctors had little to offer childbearing women beyond the ability to destroy and remove the fetus when prolonged labor threatened women with death. Midwives, meanwhile, could offer only patience, skilled hands, and a few herbal remedies.

During the late nineteenth century, Americans' willingness to have doctors attend childbirths gradually increased, as did doctors' interest in doing so. As described in Chapter 11, nineteenth-century allopathic doctors faced substantial competition not only from each other but also from many other kinds of practitioners. As a result, doctors attempted to expand into various fields, from pulling teeth to embalming the dead to assisting in childbirth (Starr, 1982: 85). Doctors considered assisting in childbirth especially crucial because they believed that families who came to a doctor for childbirth would stay with him for other services (R. Wertz and D. Wertz, 1989: 55).

As Americans' belief in science and medicine grew during the late nineteenth century, medical assistance in childbirth became more socially acceptable among the upper classes (Starr, 1982: 59). Many women supported this change because it allowed them to obtain painkillers from doctors without feeling guilty for circumventing the biblical command to bring forth children in pain (R. Wertz and D. Wertz, 1989: 110–113). In addition, because midwifery was not a respectable occupation for Victorian women, by the late nineteenth century middle- and upper-class women seeking a childbirth attendant had only two options: lower-class lay midwives or doctors of their own social class. Having a doctor attend one's childbirth thus could both reflect and increase one's social standing (Leavitt, 1986: 39; R. Wertz and D. Wertz, 1989). Ironically, however, doctors probably threatened women's health more than did midwives; although inexperienced or impatient midwives certainly could endanger women, doctors more often used surgical and manual interventions that could cause permanent injuries or deadly infections (Leavitt, 1983: 281–292, 1986: 43–58; Rooks, 1997).

Doctors' desire to obtain a monopoly on childbirth care led them, beginning in the mid-nineteenth century, to voice opposition to midwives. These attacks escalated substantially in the early twentieth century (Sullivan and Weitz, 1988: 9–14). Recent waves of immigrants had swelled the ranks of midwives and made them more visible and threatening to doctors, whose status, especially in obstetrics, remained low. Moreover, doctors now needed the business of poor women as well as wealthier women because the rise in scientific medical education had created a need for poor women patients who could serve as both research subjects and training material.

To expand their clientele, doctors attempted through speeches and publications to convince women that childbirth was inherently and unpredictably dangerous and therefore required medical assistance. In addition, doctors played on contemporary prejudices against immigrants, African Americans, and women to argue that midwives were ignorant, uneducable, and a threat to American values and that therefore midwifery should be outlawed. For example, writing in the *Southern Medical Journal,* Dr. Felix J. Underwood, the director of the Mississippi Bureau of Child Hygiene, described African American midwives as "filthy and ignorant and not far removed from the jungles of Africa, with its atmosphere of weird superstition and voodooism" (1926: 683).

Although these campaigns cost midwives many clients, they had little effect on the law. Many members of the public, and even many doctors (particularly those in public health), believed that trained midwives could provide satisfactory care, at least for poor and nonwhite women who couldn't afford doctors' services. Consequently, laws passed during this era tended to have quite lenient provisions. In the end, however, imposing lenient laws, rather than laws requiring upgraded midwifery training and skills, resulted in the deterioration of midwifery and its virtual elimination. The only exceptions were in immigrant and nonwhite communities in the rural South and Southwest, where traditional midwives continued to conduct home births until at least the 1950s (Sullivan and Weitz, 1988: 13–14).

The Resurgence of Lay Midwifery

By the second half of the twentieth century, childbirth had moved almost solely into hospital wards under medical care. Although childbearing women were grateful for the pain relief and safety that doctors promised, all too often women nonetheless found the experience painful, humiliating, and alienating. Despite the absence of scientific support for such practices, doctors routinely shaved women's pubic area before delivery, strapped them on their backs to labor and delivery tables (the most painful and difficult position for delivering a baby), isolated them from their husbands during delivery and from their infants afterwards, and gave them drugs to speed up their labors or make them unconscious—all practices that scientific research would eventually find unnecessary or dangerous (Sullivan and Weitz, 1988).

Objections to such procedures sparked the growth of the natural childbirth movement during the 1960s and 1970s and forced numerous changes in obstetric practices. Most hospitals, for example, now offer natural childbirth classes. Critics, however, argue that the real purpose of these classes is to make women patients more compliant and convince them that they have had a natural childbirth as long as they remain conscious, even if their doctors use drugs, surgery, or forceps (Sullivan and Weitz, 1988: 39).

By the late 1960s, many women had concluded that hospitals would never offer truly natural childbirth (Sullivan and Weitz, 1988: 38–39). As a result, a tiny but growing number of women chose to give birth at home. For assistance, they turned to sympathetic doctors and to female friends and relatives, some of whom were nurses. Over time, women who gained experience in this fashion might find themselves identified within their communities as lay midwives. This new generation of lay midwives who attend almost solely home births reflects the broader revolt against medicalized birth (Sullivan and Weitz, 1988: 23–59).

Working as a lay midwife means long and uncertain hours with little pay. Most midwives, however, are motivated by ideological rather than economic concerns (Sullivan and Weitz, 1988: 68–80). Although midwives recognize the need for obstetricians to manage the complications that occur in about 10 percent of births, they fear the physical and emotional dangers that arise when obstetricians employ interventionist practices, developed for the rare pathological case, during all births. Like nurse-midwives, lay midwives strongly believe in the general normalcy of pregnancy and childbirth and in the benefits of individualized, holistic maternity care in which midwife and client work as partners.

No national laws set the status of lay midwives. As of 2005, lay midwifery was definitely legal in twenty-nine states and illegal in sixteen, with their status elsewhere unclear (American College of Nurse-Midwives, 2005). In states where midwifery is illegal, midwives run the risk of prosecution for practicing medicine without a license and for child abuse, manslaughter, or homicide if a mother or baby suffers injury or death.

In states where lay midwifery is legal, midwives typically must abide by regulations restricting them to "low-risk" clients (such as women under age 35) and restricting the techniques they can use (such as forbidding them from suturing tears following deliveries). Licensed midwives typically must have a backup doctor and must transfer their clients to medical care if the doctor so orders. Thus, licensure has given midwives some degree of freedom to practice in exchange for limited subordination to medicine (Sullivan and Weitz, 1988: 97–111).

Research consistently suggests that home births conducted by experienced lay midwives working with low-risk populations are as safe as or safer than doctor-attended hospital births, even taking into account the small number of midwifery clients who develop problems needing medical attention (Lewis, 1993; Sullivan and Weitz, 1988: 112–132). For example, a recent Canadian/United States study compared 5,418 women who chose home birth with a licensed midwife with a similar group of low-risk women who chose hospital deliveries (Johnson and Daviss, 2005). In the end, both groups had similar (very low) rates of maternal and infant mortality and morbidity. However, the home-birthing women received less than half as many medical interventions. For example, only 3.7 percent of those delivered at home had cesarean deliveries, compared to 19 percent of those delivered in hospitals.

Box 12.1   **Making a Difference: Citizens for Midwifery**

Citizens for Midwifery (CFM) is a national, grassroots, consumer organization, begun by a group of mothers in 1996. The organization's primary goal is to promote the "midwifery model of care." This model is composed of two basic beliefs: (1) that pregnancy and childbearing are safe, normal processes rarely requiring medical intervention and (2) that care of pregnant women should be holistic, individualized, and delivered in an integrated fashion from the prenatal through postpartum periods. Through its website (www.cfmidwifery.org), publications, and media outreach programs, CFM offers information to consumers about the nature of midwifery, the benefits of using a midwife, and how to find and select a midwife.

CFM also works to improve the legal status of midwives (especially licensed lay midwives) and to improve access to midwives for childbearing women across the nation. Its website offers information to consumers and midwives alike about how to craft a persuasive letter to the editor or to a legislator, how to lobby effectively for legal change (including nitty-gritty details on the most effective ways to communicate by phone, in writing, or in person), and how to critically evaluate and use scientific studies on midwifery and hospital birth outcomes. As of 2005, CFM has supported midwives and consumers in their legal battles in sixteen states, three of which have since legalized licensed midwifery.

As a result, those who delivered at home avoided the lingering discomfort, pain, and loss of energy that plagues many who experience medical interventions during birth. In addition, the home births cost about one-third the price of hospital deliveries, and mothers' satisfaction with their care at home was very high.

Despite evidence such as this, medical opposition to licensed midwifery remains strong and public support weak, although insurance companies do cover midwifery services in some states. Thus lay midwives, even where licensed, cannot claim to have achieved social acceptance even as limited practitioners. Box 12.1 describes the work of Citizens for Midwifery, a grassroots organization dedicated to improving the position of midwives (especially licensed lay midwives) and promoting their use.

Curanderos

Curanderos are folk healers who function within Mexican and Mexican American communities (Perrone, Stockel, and Krueger, 1989; Roeder, 1988). In the United States, curanderos are used primarily by immigrants, as well as by some U.S.-born Mexican Americans, especially those who live in close-knit communities in the Southwest. In Denver, for example, doctors familiar with the Mexican American community estimate that between 100 and 200 curanderos work out of their homes, advertising primarily by word of mouth (*New York Times*, 1999b). Some work for free, and some charge fees ranging from

$5 to $100. A survey conducted in Denver found that 29 percent of adult Hispanic patients at a low-income clinic had visited a curandero at least once during their lives (*New York Times,* 1999b). Most did not use curanderos as a primary source of health care but instead went in addition to seeing a doctor, when medical care had failed, or when distance or poverty limited their access to medical care.

Theories and Treatments

Curanderos recognize both Western categories of disease, such as colds, and unique categories of illness, such as susto (Roeder, 1988). A common diagnosis, *susto* refers to an illness that occurs when fright "jars the soul from the body, in which case treatment consists of calling the soul back" (Roeder, 1988: 324). Curanderos also sometimes trace illness to supernatural forces such as *mal de ojo,* or the evil eye.

Curanderos treat illness in a variety of ways, including herbal remedies, massage, prayer, and rituals designed to combat supernatural forces. They believe illness reflects all aspects of an individual's life—biology, environment, social setting, religion, and supernatural forces—and thus must be treated holistically. As a result, curanderos often spend considerable time listening to their clients. The successes curanderos sometimes achieve in treating their clients' illnesses thus derive not only from their knowledge of herbs and the healing powers of their clients' faith but also from the simple healing power of a sympathetic listener.

Becoming a Curandero

Individuals become curanderos through apprenticeships, typically with family members. Successful curanderos find that their practices evolve gradually from part-time work, paid primarily in goods and services, to more or less full-time, cash businesses.

The story of Gregorita Rodriguez, a *curandera* (female curandero) living in Santa Fe, New Mexico, who specializes in massage treatments, illustrates this process:

> Gregorita traces her own career as a *curandera* back to her grandmother, Juliana Montoya, who taught Gregorita's aunt, Valentina Romero, the art of *curanderismo.* When any of Gregorita's seventeen children became ill, she took them to her Aunt Valentina for treatment. *La curandera* taught Gregorita, encouraging her by asking, "Why don't you learn? Look, touch here." Using her children's bellies as a classroom, Gregorita felt the different abdominal disorders and learned how to manipulate the intestines to relieve the ailments. Another of her patients during this learning period was her husband. Responding to his complaints, Gregorita said, "Maybe I can do something for you." Mr. Rodriguez replied, "No, no, no! You are not going to boss me!" So, off he went to see Aunt Valentina, who was elsewhere delivering a baby. Finally, Gregorita got her chance. Her husband was desperate and allowed her to learn, all the time howling about how much she

was hurting him. "Cranky," she described him, "especially when I felt a big ball in his stomach and had to work very hard. Slow, slow, I fixed him and he got better. When he went to my aunt, she said he was okay now. After that I treated my husband and one of my sisters and then her family. That's the way it started." (Perrone et al., 1989: 108–109)

After that, neighbors began to come for treatment and Gregorita's reputation grew; but she was reluctant to compete with her aunt for business. In 1950, Aunt Valentina died and Gregorita came into her own, her credibility already well established.

Because she lacks any recognized training in health care, Gregorita cannot legally charge fees or bill insurance companies as a curandera. To circumvent these legal restrictions, she has become licensed as a massage therapist and bills her clients as such. As this suggests, even a folk healer who appears to function completely outside the bounds and control of the Western scientific world cannot avoid its authority altogether.

Christian Science Practitioners

Theories and Treatments

Christian Science is a Christian sect founded in New England in about 1875. Christian Scientists believe God creates only good, while evil, sickness, suffering, and death exist only because mortals believe in them. The practitioner's job, then, is to lead the sufferer, through prayer, study, and talk, to reject the "counterfeit reality" of the "material self" and to achieve the true reality of divine perfection.

According to Margery Fox:

> Ideally, Christian Science treatment should be entirely and exclusively metaphysical. Practitioners are not even supposed to listen too attentively to patients' symptoms lest they be tempted to accept them as real; also, they idealize "undifferentiated" treatment not directed toward a specific problem. There should be no counseling of patients on a human level, no appeal to psychological processes. (1989: 107)

Reality, however, rarely matches this ideal. Practitioners spend much of their time talking with clients about the emotional and moral problems underlying clients' "counterfeit" physical problems. Healing seems to rely heavily on practitioners' persuasive verbal skills (M. Fox, 1989).

Becoming a Practitioner

As with curanderos, becoming a practitioner is a gradual process (M. Fox, 1989). Most practitioners (almost all of whom are women) begin by healing family members and friends. During weekly religious services, satisfied patients may announce successful treatment by a particular practitioner.

Over time, if a practitioner's personality and reputation seem suitable, other friends and acquaintances might turn to that practitioner for assistance. Eventually, individuals may apply to the central church office for listing in *The Christian Science Journal*. Approval comes after the practitioner submits letters of support from members of the congregation testifying to his or her effectiveness. After this, practitioners can open full-time offices. Currently, the *Journal* lists several thousand practitioners. The geographic distribution of practitioners across regions and between urban and rural communities reflects the distribution of the population as a whole. Care by practitioners is covered under Medicare, Medicaid, and many private health insurance plans (*Journal of the American Medical Association*, 1990).

Christian Scientists' opposition to medical care has precipitated a long history of legal battles in which doctors or states have sued for the right to force individuals to accept medical treatment. In general, courts have ruled that because Christian Scientists never seek medical care, doctors have no legal standing and cannot force care on adults. However, courts have ruled in favor of forcing care on children, arguing that the state has the right and duty to protect the health of children, and have found parents guilty of child abuse or involuntary manslaughter when children who received only spiritual treatment have died. (This chapter's ethical debate, Box 12.2, discusses the issues involved in the decision to refuse mainstream medical care.)

Acupuncturists

Theories and Treatments

If anything, acupuncturists' ideas regarding health and illness bear even less relationship to the ideas of Western medicine than do those of curanderos and Christian Science practitioners. Acupuncture is one of the oldest forms of healing known. Its recorded history goes back 2,000 years, with strong prehistorical evidence going back to the Bronze Age.

Like all traditional Chinese medicine, acupuncture is based on the concept of *chi* (Fulder, 1984). This concept, which has no Western equivalent, refers to the vital life force, or energy. Health occurs when chi flows freely through the body, balanced between *yin* and *yang*, the opposing forces in nature. Because any combination of problems in the mind, body, spirit, social environment, or physical environment can restrict chi, treatment must be holistic.

Following this theory, traditional Chinese healers consider both symptoms and diagnosis unimportant and focus instead on unblocking chi. Acupuncture is based on the theory that chi runs through the body to the different organs in channels known as meridians, which have no Western equivalents. To cure a problem in the colon, for example, acupuncturists apply needles to the index finger, which they believe connects to the colon via a meridian. In this way, they believe, they can stimulate an individual's chi and direct it to the parts of the body where it is needed. Acupuncturists decide on

---

Box 12.2   **Ethical Debate: Choosing Alternative Options**

John and Mary Miller, high school teachers in a medium-sized New England town, are the parents of two healthy toddlers. On the advice of their chiropractor and several of their friends, they have decided not to have their children receive the usual childhood vaccinations against measles, mumps, rubella, polio, tetanus, and other infectious diseases. John and Mary's parents, on the other hand, are horrified at their decision, for they still remember the days when many children died from infectious diseases in the United States. Although the Millers recognize the dangers these diseases can present, they argue that these diseases are now rare, and so the benefits of vaccination are outweighed by their dangers, which they believe include higher risks of autism, meningitis, and other diseases. So far, John and Mary remain committed to their decision, although they worry about the legal consequences of ignoring laws requiring childhood vaccinations, and they do sometimes wonder if they have made the right choice.

As memories of infectious disease epidemics have faded, more and more parents have decided against having their children vaccinated. In the United States, religious, philosophical, and health care concerns—as well as Internet rumors—are feeding this trend; in Great Britain, an estimated 30 percent of school-age children have not received the basic measles/mumps/rubella vaccination. Do parents have the right to refuse vaccinations or, more broadly, to refuse mainstream medical care for themselves or their children, without interference from doctors and the courts?

As in the ethical debate on truth-telling to patients (see Chapter 11), the central issues in this case are autonomy and paternalism. However, here the issue is not personal paternalism by doctors but state paternalism—the idea that the state has an obligation to protect the welfare of its citizens, even when doing so means going against citizens' wishes.

Restricting individual autonomy is a serious matter, for it implies that an individual is not competent to decide what is in his or her own best interest. As the word implies, *paternalism* suggests that an individual is more like a child or even an animal than an adult human. Requiring motorcyclists to wear helmets, for example, suggests motorcyclists are too ignorant or stupid to assess for themselves the advantages and disadvantages of helmets.

Does the need for paternalism outweigh the desire for autonomy in this case? One way to

---

treatment through taking a complete history, palpating the patient's abdomen, measuring his or her blood pressure, and reading the twelve pulses recognized by Chinese medicine.

Acupuncture is still used extensively in China, both alone and in conjunction with Western medicine, and is used increasingly in the West. To ascertain its impact, the U.S. National Institute of Health organized a Consensus Development Panel on Acupuncture in 1998. (A consensus panel is a group of experts from diverse backgrounds brought together to reach joint conclusions on a topic.) The panel's final report concluded that acupuncture definitely alleviates nausea and some types of pain and definitely does not help in stopping smoking. The report also noted that acupuncture has fewer harmful

decide is to consider in the abstract the relative value and appropriate roles of autonomy and state paternalism. We might, for example, conclude that leaving children unvaccinated is unsafe but still believe that protecting individual autonomy is more important than protecting individuals from themselves. Another way to decide is to evaluate the scientific evidence for and against vaccinations to see whether vaccinations are as safe and the risks of infection to the unvaccinated as dangerous as most doctors claim. In this case, the scientific evidence is very strong: A recent review conducted by the prestigious Institute of Medicine resoundingly supported the use of vaccinations (Stratton, Wilson, and McCormick, 2002). Still, sometimes scientists have been proven wrong in the long run.

In this situation, the ethical dilemma is complex because many people's health is at stake. Whenever an unvaccinated child becomes infected with a disease, he or she can spread the disease to other children who have not yet been vaccinated, to children who cannot be vaccinated because their immune systems are weak (due to preexisting disease or chemotherapy), and to adults whose vaccinations have worn off with time. Unvaccinated children thus place whole communities at risk; in the past few years,

several outbreaks of infectious diseases that occurred in the United States have been traceable to unvaccinated children. The issue, then, is not simply whether the Millers have the right to decide for themselves what sort of health care they want but also whether they have the right to make decisions that place both their children and others at risk. To evaluate this situation, one must also decide, first, whether parents or the state can best and most appropriately judge children's interests and, second, in what circumstances state intervention is justified.

Sociological Questions

1. What social views and values about medicine, society, and the body are reflected in this debate? Whose views are these?

2. Which social groups are in conflict over this issue? Whose interests are served by the different sides of this issue?

3. Which of these groups has more power to enforce its view? What kinds of power do they have?

4. What are the intended consequences of the various policies under consideration? What are the unintended social, economic, political, and health consequences of these policies?

side effects than modern medicine does and that many accepted Western medical practices have no greater scientific evidence of efficacy. The World Health Organization, meanwhile, considers acupuncture effective for treating about fifty disorders, including the common cold, bronchial asthma, childhood myopia, and dysentery (Wolpe, 1985: 420).

The Impact of Medical Dominance

Widespread American interest in acupuncture began during the 1970s, when the People's Republic of China first opened to U.S. travelers. Early travelers brought back near-miraculous tales of acupuncture anesthesia and treatment. Because American doctors had no scientific model that could account for acupuncture's effects, these tales threatened their position and worldview (Wolpe, 1985). As a result, various well-known doctors publicly denounced acupuncture, claiming it worked only as a placebo or only because Chinese stoicism or revolutionary zeal allowed them to ignore pain, even though acupuncture also had worked on animals and on Western travelers to China.

To remove this threat to their cultural authority, doctors endeavored to control the definition, study, and use of acupuncture (Wolpe, 1985). This proved relatively easy for, unlike chiropractic or osteopathy, acupuncture at the time had few American supporters. Consequently, in their writings and public pronouncements, doctors could strip acupuncture of its grounding in traditional Chinese medical philosophy and define it simply as the use of needles to produce anesthesia. Pressure from medical organizations led the National Institutes of Health to adopt a similar definition in funding research on acupuncture. At the same time, pressure from doctors led most states to adopt licensure laws allowing any doctors, regardless of training, to practice acupuncture but forbidding all others, no matter how well trained, from doing so except under medical supervision. Thus, for many years, most traditional acupuncturists in the United States worked illegally within Asian communities.

During the past decade, however, as acceptance of alternative healing traditions has increased, the position of acupuncturists has improved. Some insurance companies will reimburse nondoctors for acupuncture treatments, and most states now allow nondoctors to perform acupuncture, although some of these states require medical supervision or require acupuncturists to be licensed by medically dominated boards (Acupuncture Alliance, 2002). Use of acupuncture remains rare; national random surveys of English-speaking U.S. residents conducted in 1997 and 2002 found that the percentage reporting use of acupuncture held steady at 1.0 percent, with about half of these patients reporting some insurance coverage for treatment (Tindle et al., 2005). These figures suggest that acupuncture remains a marginal therapy and occupation, posing little threat to medical dominance.

Conclusion

As the discussions in this chapter have suggested, the health care arena is much broader than we usually recognize. Many alternatives to medical treatment exist far beyond those discussed herein. Most of these alternatives function not so much in opposition to mainstream health care as in parallel, with those seeking care jumping back and forth across the tracks. For example, a woman might deliver her first child with a doctor, her second with a nurse-midwife, and her third with a lay midwife; and a man who experiences chronic back pain might see a chiropractor or acupuncturist either before, after, or in addition to seeing a medical doctor.

This chapter has highlighted the factors that help health care occupations gain professional autonomy in the face of medical dominance. Timing certainly seems to play a role: Those occupations that emerged before medical dominance became cemented, such as osteopathy and chiropractic, have proved most successful. Social factors, too, consistently seem important: Health care occupations with roots in and support from higher-status social groups have a better chance of winning professional autonomy than do those with lower-status roots and supporters.

Other occupations seem to retain some autonomy—if a marginal position in the health care arena—because they pose little threat to medical dominance. Curanderos, for example, attract a small clientele of poor Mexicans and Mexican Americans who might not be able to pay for medical care or to communicate effectively with medical doctors anyway. Doctors thus have little incentive to eliminate curanderos' practices. Acupuncturists, on the other hand, have attracted not only Asians and Asian Americans but also well-educated whites—including individuals with the skills and resources to publicize the virtues of acupuncture. Consequently, doctors have had a far greater vested interest in restricting acupuncturists' practices and in co-opting acupuncture for their own purposes.

Not surprisingly, developing professional autonomy seems most difficult for those, like nurses, who work directly under medical control. In contrast, those such as Christian Scientist practitioners have considerably more leeway to develop their practices without interference from medical doctors.

Finally and ironically, strict licensing laws, even when devised by doctors opposed to a field's growth, in the end can help occupations gain professional autonomy by forcing them to increase standards and thereby enabling them to gain additional status and freedom to practice.

To date, medical doctors have succeeded in retaining their professional autonomy and dominance partly because of their greater ability to provide scientific data supporting their theories and practices—or at least to convince the public that they have such data. It remains to be seen whether, with the increased federal support for research on alternatives and despite medical control of funding and publication mechanisms, those who favor alternative health care options will be able to use this research to increase scientific credibility and public support for their practices.

Suggested Readings

Chambliss, Daniel F. 1996. *Beyond Caring: Hospitals, Nurses, and the Social Organization of Ethics.* Chicago: University of Chicago Press. A wonderful study of the social position of nurses in the health hierarchy and the consequences of that position for both nurses and patients.

Gordon, Suzanne. 2005. *Nursing Against the Odds: How Health Care Cost Cutting, Media Stereotypes, and Medical Hubris Undermine Nurses and Patient Care.* Ithaca, NY: Cornell University Press. The title tells it all. Gordon's book clearly describes the problems at the heart of modern nursing.

Root-Bernstein, Robert and Michele. 1997. *Honey, Mud, Maggots, and Other Medical Marvels.* New York: Houghton Mifflin. The authors, a physiologist and a historian, explore how modern medical researchers are investigating and, in some cases, validating "folk medicine" treatments from around the world, such as bloodletting and dirt-eating.

Vincent, Peggy. 2002. *Baby-Catcher: Chronicles of a Modern Midwife.* An engaging and fascinating memoir written by an obstetrical nurse who became a nurse-midwife specializing in home births.

Getting Involved

Midwives Alliance of North America. 4805 Lawrenceville Hwy., Suite 116–279, Lilburn, GA 30047. (888) 923-6262. www.mana.org. Promotes communication between lay midwives and nurse-midwives and the legal rights of both groups.

Review Questions

How did the early history of nursing make it difficult for nurses to increase their status or improve their working conditions?

How have nurses attempted to professionalize? Why haven't these strategies succeeded?

How have changes in the health care system affected nurses' occupational status and position?

What factors have led to the development of clinical pharmacy, pharmaceutical care, and disease management? What factors have restrained their growth, or could do so in future?

How did osteopaths attempt to professionalize? What factors enabled them to succeed? What price has osteopathy paid for its success?

To what extent and in what ways have chiropractors succeeded in improving their occupational status?

How and why did doctors gain control over childbirth?

What factors led to the growth of nurse-midwifery? of lay midwifery? What is the difference between the two?

How do individuals become traditional healers? How does medical dominance affect their work and their lives?

Internet Exercises

1. The federal government's main website for consumer health is www.healthfinder.com. Browse the site, looking for links to web pages related to fraud and quackery, accountability, and treatment errors. Do the site's organizers appear as concerned about fraud and similar problems among mainstream practitioners as among alternative practitioners? In what ways, if any,

does the site's handling of alternative medicine differ from its handling of mainstream medicine?

2. Using the Internet, find policy statements related to home birth and midwifery from a variety of organizations (such as the World Health Organization, the Midwives Alliance of North America, the American College of Obstetricians-Gynecologists, and the American College of Nurse-Midwives). How do their positions differ? What evidence do they use to justify their positions?

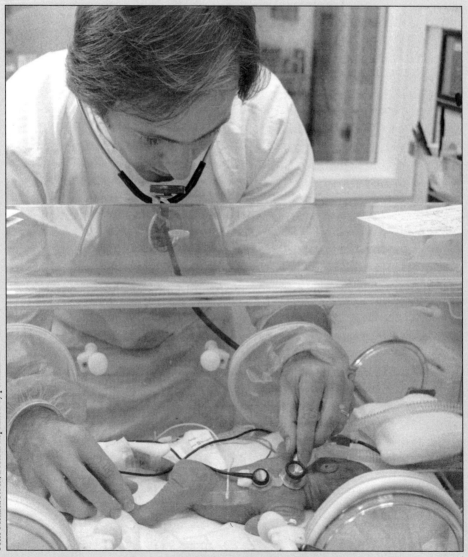

Issues in Bioethics

In January 1998 my brother-in-law, Brian, was injured in a catastrophic industrial accident that left him with second- and third-degree burns over 95 percent of his body and with strong indications that he had suffered a severe inhalation injury.

Brian's accident occurred literally in sight of a major hospital with a regional burn unit, and he was brought to the hospital within minutes. Following the accident, Brian remained in a strange limbo between life and death—unconscious although not comatose, and kept alive by aggressive medical treatment and an ever-increasing assortment of drugs and machines. Burned everywhere except his genitals and the soles of his feet, bandaged from head to toe with only his face showing, and swollen grotesquely, Brian's appearance was literally nightmarish; no one who saw him slept well afterwards. Each day brought minor crises, and each week brought a major crisis that made death seem imminent—as indeed it was, for Brian died three and a half weeks after the accident.

The severity of Brian's injuries immediately made me wonder whether it might be best to treat only his pain and let him die a natural death. Brian had never written a living will, but he had told his wife, Lisa, that he would not want to live if his quality of life was ever compromised substantially. Questions about whether treatment made sense became increasingly salient to the family as the days passed; his lungs, stomach, and kidneys failed; and bacterial, viral, and fungal infections assaulted his body.

Because Brian remained unconscious throughout his hospital stay, legally Lisa was authorized to make treatment decisions for him. The doctors acknowledged that the final decisions were up to Lisa and that they could not ethically or legally proceed without her informed consent. In practice, however, they kept decision-making authority to themselves by, among other things, defining certain decisions as purely technical matters not requiring Lisa's consent, shaping her treatment decisions through selectively providing information, ignoring her decisions when they disagreed with her opinions, cutting off her questions when they found them uncomfortable, and telling her that withholding treatment was unethical and hence out of the question. Although some nurses indicated quietly to Lisa that her concerns were valid, the hospital's pastoral counselors and social workers urged Lisa to trust the doctors' judgment.

In the end, Brian's condition began deteriorating so rapidly and completely that the doctors had no further treatments to try. Around the same time, a new resident joined the staff who took Lisa's concerns seriously. A long conversation with him greatly helped Lisa, both by allowing her to express her feelings and by helping her understand the doctors' perspective. When this resident recommended to Lisa that she give permission to withdraw the drug that kept Brian's heart beating, Lisa accepted his recommendation. Brian died that night. (Weitz, 1999)

For centuries, doctors have formally recognized that health care should be based on ethical principles. The Hippocratic oath, for example, written in about 400 B.C., instructed doctors to take only actions that would benefit their patients and to foreswear euthanasia, seducing patients, or divulging patients' secrets. As Brian and Lisa's story suggests, however, in practice health care still can fall short of meeting ethical principles. In this chapter we explore the history of **bioethics,** the study of all ethical issues involved in the biological sciences and health care, and analyze how bioethics has—and has not—affected American health care and medical research.

To some students and faculty, it might seem odd to include a chapter on bioethics in a sociology textbook. Yet the issues raised by bioethics are sociological issues, for many of the issues bioethicists ponder revolve around the impact of power differences between social groups (most importantly, between physicians and patients). Even when exploring the same issues, however, bioethicists and sociologists do so through different lenses. Robert Zussman, a sociologist who has studied bioethics extensively, succinctly summarizes the difference:

Medical ethics may be thought of as the normative study of high principles for the purpose of guiding clinical decisions. In contrast, the sociology of medical ethics may be thought of as the empirical study of clinical decisions for the purpose of understanding the social structure of medicine. Clearly then, medical ethicists and sociologists of medical ethics travel much of the same terrain, but they do so traveling in different directions. (1997: 174)

A History of Bioethics

Since its beginning in 1848, the **American Medical Association (AMA)** has required its members to subscribe to its code of ethics. The code, however, speaks more to medical etiquette—proper relations between doctors—than to medical ethics or, more broadly, bioethics. Indeed, throughout the nineteenth century and well into the twentieth century, doctors' ideas regarding bioethics remained ill-defined and their commitment to bioethics remained minimal. Although doctors undoubtedly would have identified relieving human suffering as their primary goal, both in their research and in clinical practice doctors sometimes behaved in ways that would horrify modern doctors and bioethicists. For example, Dr. J. Marion Sims, considered the father of modern obstetrics, achieved fame during the 1840s for developing a surgical procedure to correct vesico-vaginal fistulae, tears in the wall between a woman's vagina and bladder usually caused by overaggressive medical intervention during childbirth (Barker-Benfield, 1976). Women who suffered these fistulae could not control leakage of urine and often had to withdraw from social life altogether because of odor and the resulting social shame. To develop a surgical cure, Sims bought black women slaves who had fistulae and then operated on them as many as thirty times each, in an era before antibiotics and antisepsis and with only addictive drugs for anesthetics. When Sims announced his new surgical technique, the medical world and the public greeted him with acclaim. No one questioned his research ethics.

Almost a century later, Nazi doctors working in German concentration camps also used socially disvalued populations for equally barbaric—and even less justifiable—experiments. The world's response to these experiments would mark the beginnings of modern bioethics.

The Nazi Doctors and the Nuremberg Code

In 1933, the German people voted the Nazis, under Adolf Hitler's leadership, into power. At that time, Germany's medical schools and researchers were known and respected worldwide and its system of health care was considered one of the best and most comprehensive (Redlich, 1978).

Shortly after coming to power, the Nazi government passed the Law for the Prevention of Congenitally Ill Progeny (Lifton, 1986). This law required the sterilization of anyone considered likely to give birth to children with diseases that doctors considered genetic, including mental retardation, schizophrenia, manic depression, epilepsy, blindness, deafness, or alcoholism. Under this law, government-employed doctors sterilized between 200,000 and 300,000 persons. Two years later, in 1935, the government passed the Law to Protect Genetic Health, prohibiting the marriage of persons with certain diseases.

Both these laws reflected a belief in **eugenics,** the theory that the population should be "improved" through selective breeding and birth control.

The eugenics movement had many followers throughout the Western world. By 1920, twenty-five U.S. states had passed laws allowing sterilization of those believed (usually incorrectly) to carry genes for mental retardation or criminality. Several states also passed laws forbidding interracial marriage and marriage by persons with illnesses considered genetic (Lifton, 1986).

As the power of the Nazis grew in Germany, and as public response to their actions both within and outside Germany proved mild, the Nazis adopted ever-bolder eugenic actions (Lifton, 1986; Redlich, 1978). Beginning in 1939, the Nazis began systematically killing patients in state mental hospitals. Doctors played a central role in this program, selecting patients for death and supervising their poisoning with lethal drugs or carbon monoxide gas. Doctors and nurses also watched silently while many more patients starved to death. In total, between 80,000 and 100,000 adults and 5,000 children died (Lifton, 1986). Shortly after, the Nazi government began systematically killing Jews, Gypsies, and others whom they considered racially inferior. By the end of World War II, the Nazis had murdered between 5 million and 10 million people in their concentration camps.

At least 350 doctors played major roles in this genocidal policy (Lifton, 1986; Redlich, 1978). As prisoners entered the concentration camps, medical officers of the Nazi SS corps decided which to kill immediately and which to use for forced labor. When shooting those marked for death proved too expensive, doctors developed more efficient means of mass murder using carbon monoxide gassing. Medical corpsmen, supervised by doctors, conducted the murders. Those whom doctors selected for forced labor, meanwhile, usually died in a matter of weeks from starvation, overwork, or the epidemic diseases that ravaged the camps. In addition, doctors working in the concentration camps (including university professors and highly respected senior medical researchers) performed hundreds of unethical experiments on prisoners—such as studying how quickly individuals would die once exposed to freezing cold and seeing whether injecting dye into prisoners' eyes would change their eye color. Doctors also used prisoners to gain surgical experience by, for example, removing healthy ovaries or kidneys or creating wounds on which to practice surgical treatments.

Following the Nazi defeat, the Allied victors prosecuted 23 of these doctors for committing "medical crimes against humanity," eventually sentencing 7 to death and 9 to prison (Lifton, 1986). The decisions in these cases contained the basis for what is now known as the **Nuremberg Code,** a set of internationally recognized principles regarding the ethics of human experimentation (see Box 13.1). The code requires researchers to have a medically justifiable purpose, do all within their power to protect their subjects from harm, and ensure that their subjects give **informed consent,** that is, voluntarily agree to participate in the research with a full understanding of the potential risks and benefits.

Box 13.1   **Principles of the Nuremberg Code**

---

1. The voluntary consent of the human subject is absolutely essential. . . .
2. The experiment should be such as to yield fruitful results for the good of society, unprocurable by other methods or means of study, and not random and unnecessary in nature.
3. The experiment should be so designed and based on the results of animal experimentation and a knowledge of the natural history of the disease or other problem under study that the anticipated results will justify the performance of the experiment.
4. The experiment should be so conducted as to avoid all unnecessary physical and mental suffering and injury.
5. No experiment should be conducted where there is an *a priori* reason to believe that death or disabling injury will occur. . . .
6. The degree of risk to be taken should never exceed that determined by the humanitarian importance of the problem to be solved by the experiment.
7. Proper preparations should be made and adequate facilities provided to protect the experimental subjects against even remote possibilities of injury, disability, or death.
8. The experiment should be conducted only by scientifically qualified persons. The highest degree of skill and care should be required through all stages of the experiment of those who conduct or engage in the experiment.
9. During the course of the experiment, the human subject should be at liberty to bring the experiment to an end. . . .
10. During the course of the experiment, the scientist in charge must be prepared to terminate the experiment at any stage if he has probable cause to believe . . . that a continuation of the experiment is likely to result in injury, disability, or death to the experimental subject.

Source: http://www.hhs.gov/ohrp/references/nurcode.htm

The 1960s: The Rise of Bioethics

Because the trials received relatively little publicity in the United States, and because Americans typically viewed Nazi doctors as *Nazis* rather than as doctors, few drew connections between Nazi practices and American medical practices (D. Rothman, 1991). As a result, discussion of bioethics remained largely dormant in the years following the Nuremberg Trials. During the 1960s, however, as health care costs rose exponentially, ethical questions regarding access to health care became topics of popular discussion.

New technologies, too, such as the development of organ transplants and of life support systems for comatose persons, raised issues not only of equity and access but also of how to balance the benefits of new technologies against their dangers. From these issues would emerge a heightened interest in bioethics.

These issues first came to a head with the development of kidney dialysis, a technology that could keep alive persons whose kidneys had failed

(R. Fox and Swazey, 1974). Demand for dialysis far outstripped supply, forcing selection committees made up of doctors and, in some cases, laypeople to decide who would receive this life-saving treatment and who would die. Forced to choose from among the many who, on medical grounds, were equally likely to benefit from the treatment, these committees frequently based their choices on social criteria such as sex, age, apparent emotional stability, social class, and marital status. When news of these committees' work reached the public, the resulting outcry led to new federal regulations designed to allocate kidney dialysis more fairly.

Although the dialysis issue sparked public concern about medical *practice,* medical *research* still remained outside the bounds of public discussion. In 1966, however, one article changed this. Writing in the *New England Journal of Medicine,* respected medical professor Henry Beecher (1966) described twenty-two research studies, published in top journals in the recent past, that had used ethically questionable methods. In one study, for example, soldiers sick with streptococcal infections received experimental treatments instead of penicillin, causing twenty-five soldiers to develop rheumatic fever. In another, doctors working without parental consent catheterized and X-rayed the bladders of healthy newborns to see how bladders worked.

To determine the frequency of such studies, Beecher looked at 100 consecutive research studies published in a prestigious medical journal. In 12 of the 100 studies, researchers had not told subjects of the risks involved in the experiments or had not even told them they were in an experiment. Yet no journal reviewer, editor, or reader had questioned the ethics of these studies.

Beecher's article sent ripples of concern not only through the medical world but also through the general public, as news of the article spread through the mass media. This public concern translated into pressure on Congress and, in turn, pressure on the U.S. Public Health Service (PHS), the major funder of medical research. To demonstrate to Congress that they could deal with the problem on their own and to keep public concern from turning into budget cuts, the PHS in 1966 published guidelines for protecting human subjects in medical research (D. Rothman, 1991).

The responses to Beecher's article and the dialysis issue demonstrate the increased role that the mass media and the general public had begun to play in health care decision making. Meanwhile, the growth of the civil rights and women's rights movements stimulated discussion both about patients' rights generally and about birth control and abortion specifically. The patients' rights movement would also draw energy from the publication in 1969 of Dr. Elizabeth Kübler-Ross's book *On Death and Dying,* which called attention to the dehumanizing aspects of modern medical treatment of the dying.

The concept of patients' rights also found fertile ground during the 1960s because of the changing relationship between doctors and patients (D. Rothman, 1991). Before World War II, Americans typically received their health care at home or in a nearby office from general practitioners they had known for years. Doctors and their clients lived in the same neighborhoods and often shared the same ethnic and social class background. By the 1960s, however, as medical practice shifted from general to specialty care, from home and office to hospital, and from talking and direct physical interventions to impersonal technological interventions, the ties binding doctors and clients had weakened. In these circumstances, trust between doctors and clients diminished, and public demands for control over medical work grew. Similarly, medical research shifted from small-scale, rare events in which doctors typically conducted experiments first on themselves and then on their families and neighbors to large-scale business enterprises with only weak links between doctors and subjects.

By the late 1960s, writers could look at developments around the country and proclaim the birth of the bioethics movement (R. Fox, 1974; D. Rothman, 1991). Over the next few years, several important organizations devoted to bioethics were founded, including the Hastings Center for Bioethics, the Society for Health and Human Values, and the Center for Bioethics at Georgetown University, and bioethics secured at least a small place in medical education.

The 1970s: Willowbrook, Tuskegee, and Karen Quinlan

The Willowbrook Hepatitis Study

During the 1970s, three cases further stimulated popular, legal, and medical interest in bioethics. The first of these, the Willowbrook hepatitis experiments, reached public attention in 1971. Willowbrook State School, run by the state of New York, was an institution for mentally retarded children. Conditions in Willowbrook were horrendous, with children routinely left naked, hungry, and lying in urine and excrement. As a result, hepatitis, a highly contagious, debilitating, and sometimes deadly disease, ran rampant among the children and, to a lesser extent, the hospital staff.

In 1956, to document the natural history of hepatitis and to test vaccinations and treatments, two professors of pediatrics from New York University School of Medicine began purposely infecting children with the disease. In addition, to test the effectiveness of different dosages of gamma globulin, which the researchers knew offered some protection against hepatitis, they injected some children with gamma globulin but left others unvaccinated for comparison. The children's parents had consented to this research, but had received only vague descriptions of its nature and potential risks.

The researchers offered several justifications for their work. First, they argued, the benefits of the research outweighed any potential risks. Second, they had infected the children only with a relatively mild strain of the virus and therefore had decreased the odds that the children would become infected with the far less common but considerably more dangerous strain that also existed in the school. Third, the children who participated in the experiments lived in better conditions than did the others in the institution and therefore were protected against the many other infections common there. Fourth, the researchers argued that the children would probably become infected with hepatitis anyway, given the abysmal conditions in the institution. Finally, the researchers felt they should not be held accountable because the parents had given permission. Using these arguments, the researchers had obtained approval for their experiments from the state of New York, the Willowbrook State School, and New York University. Over a 15-year period, they published a series of articles based on their research, without any reviewers, editors, or readers raising ethical objections.

In 1970, however, Methodist theologian Paul Ramsey (1970) exposed the ethical flaws of these experiments in his influential book, *The Patient as Person*. Shortly thereafter, in the spring of 1971, an exchange of letters and editorials debating the ethics of these experiments appeared in the prestigious British medical journal, *The Lancet*. Ramsey and others wrote in *The Lancet* that parents had not given truly *voluntary* consent because they could get their children admitted to Willowbrook only by allowing them to participate in the hepatitis experiments. In addition, parents had not given truly *informed* consent because researchers had not told them that gamma globulin could provide long-term immunity to hepatitis. Writers to *The Lancet* also questioned why the researchers experimented on children, who could not give informed consent, rather than on the hospital staff. Finally, these writers questioned why the researchers—who, after all, were pediatricians—had chosen to take advantage of this "opportunity" to study hepatitis rather than trying to wipe out the epidemic. This debate over the Willowbrook studies was taken up by the New York media and, in the ensuing public outcry, the research ground to a halt.

The Tuskegee Syphilis Study

A year later, in 1972, the Tuskegee Syphilis Study made headlines (Jones, 1993). Begun by the federal Public Health Service (PHS) in 1932, the study, which was still under way, was intended to document the natural progression of untreated syphilis in African American men. At the time the study began, medical scientists understood the devastating course of syphilis in whites (which, in its later stages, can cause neurological damage and heart disease); but, reflecting the racist logic of the times, the scientists suspected its progression took a different and milder form in African Americans.

For this study, researchers identified 399 desperately poor and mostly illiterate African American men, all with untreated late-stage syphilis, who lived in the Tuskegee, Alabama, area. The men were neither told they had syphilis nor offered treatment. Instead, researchers informed them that they had "bad blood," a term used locally to cover a wide variety of health ailments. The researchers then told the men that if they participated in this study of bad blood, they would receive free and regular (if infrequent) health care, transportation to medical clinics, free meals on examination days, and payment of burial expenses—enormous inducements given the men's extreme poverty.

At the time the study began, treating syphilis was difficult, lengthy, and costly. The development of penicillin in the early 1940s, however, gave doctors a simple and effective treatment. Yet throughout the course of the study, researchers not only did not offer penicillin to their subjects but also kept them from receiving it elsewhere. During World War II, researchers worked with local draft boards to prevent their subjects from getting drafted into the military, where the subjects might have received treatment. When federally funded venereal disease treatment clinics opened locally, researchers enlisted the support of clinic doctors to keep research subjects from receiving treatment. Similarly, they enlisted the cooperation of the all-white County Medical Society to ensure that no local doctor gave penicillin to their subjects for any other reason.

The Tuskegee Syphilis Study, which treated African American men as less-than-human guinea pigs, was not the work of a few isolated crackpots. Rather, it was run by a respected federal agency, the PHS, with additional funding from the widely respected Milbank Fund. The study received significant cooperation from the state and county medical associations and even from doctors and nurses affiliated with the local Tuskegee Institute, a world-renowned college for African Americans. Over the years, more than a dozen articles based on the study appeared in top medical journals, without anyone ever questioning the study's ethics. Yet the study patently flouted the Nuremberg Code and, after 1966, the PHS's own research ethics guidelines. Not until 1972 did the study end, following a newspaper exposé and the resulting public outcry. By that time, at least 28 and possibly as many as 100 research subjects had died of syphilis, and an unknown number had succumbed to syphilis-related heart problems (Jones, 1993). In addition, the study indirectly caused untold additional deaths by convincing many in the African American community to distrust public health workers. That legacy has lasted to the present day, contributing to suspicions among African Americans that the federal government created **HIV** to control population growth in their community (Jones, 1993; Thomas and Quinn, 1991).

The Right to Die

Several years later, in 1975, public attention would focus on Karen Quinlan, whose case raised issues not of medical experimentation but of medical treatment. At the age of 21, after ingesting a combination of drugs at a party, Quinlan fell into a coma. Initially, her parents encouraged her doctors to make all efforts to keep her alive and return her to health. Once her parents learned that she had suffered extensive brain damage and would never regain any mental or physical functioning, they asked that she be removed from life support and allowed to die. When the doctors refused, the parents took their fight to the courts. After almost a year of legal battles, Quinlan's parents won the right to remove her from the mechanical respirator that was keeping her alive.

The Quinlan case gained enormous public attention and sympathy for the right to die and highlighted the problems involved in having too much, rather than too little, access to medical care and technology. In addition, the Quinlan case signaled the entry of lawyers and the legal system into health care decision making.

More recently, the case of Terri Schiavo raised a similar set of issues (Annas, 2005). For unknown reasons, in 1990 Schiavo fell into a "persistent vegetative state" in which, according to her doctors, she could neither feel, communicate, nor think, and from which her doctors believed she had no chance of recovering. After Schiavo had been in this condition for eight years, her husband requested that her doctors remove the feeding tube that kept her alive.

By law, Schiavo's husband, who believed she would never have wanted to be maintained in such a condition, had the legal right to make this decision on her behalf. Her doctors supported this decision, because medical norms oppose continuing futile medical interventions. Nevertheless, Schiavo's parents brought suit against her husband and doctors, arguing that she was in fact conscious and capable of recovery and that, at any rate, any life was worth continuing. After 15 years of litigation, including the unprecedented involvement of President Bush and the U.S. Congress, the federal court (supporting the decision of several lower courts) ordered Schiavo's feeding tube removed. An autopsy performed after her death a few days later confirmed that half of her brain had been destroyed, leaving her with no possibility of thought, emotion, or recovery.

In retrospect, the most striking aspect of the Schiavo case is that it raised no new medical, ethical, or legal issues. The fact that it nonetheless generated so much controversy highlights the new willingness of politicians to enter private medical decision making, the increasingly contentious atmosphere surrounding right-to-life and **right-to-die** debates, and the spread of political divisions born primarily in fights over abortion to other areas of medicine and the law.

The 1980s and 1990s: Reproductive Technology, Enhancing Human Traits, and Setting Priorities

During the last decades of the twentieth century, questions about the benefits of medical technology increased substantially. At the same time, questions increasingly were raised about inequities in access to even the most basic health care. All these questions continue to simmer in bioethical debates.

Reproductive Technology

One area that has sparked considerable debate since the late 1970s is **reproductive technology,** or medical developments that allow doctors to control the process of human conception and fetal development. Reproductive technology first came to the public's attention in 1978, with the birth of Louise Brown, the world's first "test-tube baby." Louise's mother was unable to conceive a baby because her fallopian tubes, through which eggs must descend to reach sperm and be fertilized, were blocked. Using a technique known as *in vitro fertilization,* her doctors removed an egg from her body, fertilized it with her husband's sperm in a test tube, and then implanted it in her uterus to develop. Nine months later, Louise Brown was born.

Louise Brown's birth raised questions about how far doctors should go in interfering in the normal human processes of reproduction. Subsequent cases raised even trickier questions. For example, courts have had to decide whether fetuses should be placed for adoption when the biological parents have died and whether custody of fetuses following divorce should go to the parent who wants the fetuses implanted or the one who wants them destroyed. More recently, doctors and others have debated whether couples should be allowed to hire women to carry their fetuses to term for them and whether postmenopausal women should be allowed to have a baby using another woman's egg.

More broadly, these cases have raised basic questions regarding the morality of intervening so directly in the process of human reproduction, including whether individuals are harmed or helped by having access to such technologies. Those who favor the new reproductive technologies argue that the technologies give couples greater control over their destinies. Those who oppose the new technologies, on the other hand, argue that these technologies seduce couples into spending enormous amounts of time and money in a usually futile effort to have children biologically their own, rather than finding other ways to make meaningful lives for themselves. Opponents also question whether these technologies encourage the idea that children are purchasable commodities and the idea that, for the right price, prospective parents can guarantee they will get "perfect" children (B. Rothman, 1989).

Enhancing Human Traits

The past 25 years also have witnessed growing concern about the ethics of medical interventions designed to enhance human traits. No clear definition of such enhancements exist, but the term is used to refer to techniques generally believed to improve human traits beyond a level considered normal rather than to treat conditions considered deviant or defective. This is a necessarily subjective definition, because individual judgments regarding what is normal vary greatly. Nevertheless, we would probably all acknowledge a qualitative difference between providing cosmetic surgery to a person with a severely burned face versus providing it to a professional model who desires more prominent cheekbones. Similarly, there is a qualitative difference between using psychotropic drugs to avoid schizophrenic episodes and using them to get extra energy and improve final exam grades—a process psychiatrist Peter Kramer (1993) refers to as "cosmetic psychopharmacology."

Ethical questions regarding enhancements have increased as their use has increased (Whitehouse et al., 1997). Is it ethically justifiable for individuals to improve their offspring through genetic preselection or fetal surgery, and if so, will those who do not use these technologies become a "genetic under-class"? Should health insurance cover drugs such as Viagra, which helps men achieve erections and can improve quality of life perhaps beyond the norm for a given age? Should health insurance cover cosmetic (as opposed to reconstructive) surgery, and should doctors promote surgeries (such as liposuction) whose benefits are purely cosmetic and whose potential risks include death? Should psychotropic drugs be prescribed to individuals who do not have diagnosable mental illnesses but who want to be more sociable, alert, or assertive? And is it ethical to provide potentially harmful medical care for the sake of enhancing some individuals while others still lack basic services? Finally, some have questioned whether enhancements provide unethical advantages. If Olympic athletes are forbidden from taking drugs to improve their performance, why are waitresses allowed to get breast implants to generate more tips and businesspeople allowed to take Ritalin to improve their concentration? Conversely, is it ethical to restrain the options of those who would provide or purchase such services? Questions such as these are increasingly common, as evidenced by the special supplement that the *Hastings Center Report,* an influential bioethics journal, published on this topic in January–February 1998.

Setting Priorities

For many years, policy analysts, researchers, and ethicists have raised questions about inequities in access to health care. However, whereas earlier debates on funding health care focused on deciding which *individuals* should get specific scarce resources such as kidney dialysis, beginning in the late 1980s debates focused on setting priorities to help decide which *procedures* should be funded. This debate came to the fore in 1989 with passage of legislation establishing the Oregon Health Plan (OHP), which promised to provide free care to all Oregonians who had incomes below the federal poverty level (Leichter, 1997). To extend coverage to individuals not eligible for **Medicaid,** the OHP currently provides a somewhat limited package of services. To decide each year which services to offer, OHP first prioritizes all the potential health care services it might offer. It then prospectively contracts with **managed care organizations (MCOs)** to purchase services for its members, beginning at the top of its priority list and working its way down until it reaches its budget limit. Thus, if OHP runs out of funding, some services are cut, but no individuals are dropped from the program.

The OHP legislation marked the first time that a governmental body in the United States explicitly rationed health care—deciding in advance that some procedures simply cost too much to provide to some populations. The explicit use of rationing resulted in an outcry across the country, both from those who considered it discrimination against persons with disabilities and those who believed it was unethical to ration care only for the poor. As a result, it took the state almost five years to win federal approval to pilot the program, and ethical questions continue to plague the system.

Yet rationing always has existed in the United States (Callahan, 1998). This rationing, however, is implicit rather than explicit: People do not get health care because they can't afford it, not because someone decides certain services shouldn't be offered. In the absence of some system for prioritizing services and making care accessible, health care dollars routinely are spent on services that offer little benefit or that offer great benefit only to a few, while much larger groups go without basic services. The Oregon system at least rationalizes rationing, deciding, for example, that it makes more sense to fund vaccinations for thousands of children than kidney transplants for a handful.

Current Issues

In these early years of the twenty-first century, clinicians, researchers, patients, and their families remain haunted by the ethical questions of earlier generations, such as whether there is a right to die and who should decide which medical services ethically can be offered. New technologies have added to the urgency of these and other questions.

One issue that has gained special attention in the last few years is the use of stem cells and the associated technique of cloning (Dunn, 2002). Stem cells are naturally occurring human cells that have the ability to grow into numerous types of cells. Although no successful treatments have yet been developed from stem cells, researchers hope someday to use them to replace defective cells in individuals with diseases such as diabetes and Parkinson's disease.

There are two ways to grow stem cells. First, scientists can grow stem cells in the laboratory after harvesting them from adults or from fetal blood left in a woman's blood system after giving birth. No ethical issues have been raised about this use of stem cells, which now accounts for about half of all research in this area (Kolata, 2004b). Second, scientists can grow stem cells from embryos. To do so, researchers fertilize human eggs with sperm in a laboratory to turn them into embryos. They then leave the embryos for a week or so, until each has grown into a few hundred cells, and then extract their stem cells (thus destroying the embryos). Alternatively, researchers can replace the nucleus from an unfertilized human egg with a cell nucleus taken from a donor's skin or muscle, artificially stimulate this egg (instead of fertilizing it) so it develops into an embryo, and then extract its stem cells. This second process is a form of cloning, because the embryo will be genetically identical to the donor.

To many opponents of stem cell research, the destruction of human embryos to harvest stem cells is the same as killing humans. Other critics argue that producing human cells to treat other humans is too close to selling human beings and human body parts. This is particularly worrisome because heavy political opposition to stem cell research has shifted much of this research to the for-profit sector, where it escapes most regulation. Others object specifically to the use of cloning to produce stem cells, on the grounds that it is only a matter of time before some doctors begin using cloned embryos to create cloned babies. They wonder whether in the future babies will be "farmed" and "harvested" to match parents' images of the perfect baby.

Supporters of human stem cell research argue that its potential benefits outweigh its potential problems. Most of the support for this research has come from persons who hope stem cells will provide a cure to the diseases that afflict them or their loved ones. Supporters also argue that destroying an artificially created embryo that has no potential to grow into a human being unless it is somehow implanted in a woman's uterus is not morally equivalent to destroying a human being. Finally, with regard to cloning, supporters argue that many women who want babies are already having donor eggs implanted in their uteruses and that few would choose to use cloned eggs because the chances of success are so low. (So far, no researcher has been able to keep a cloned egg alive for more than a few days, much less for a nine-month pregnancy.) For all these reasons, supporters of stem cell research argue that instead of trying to eliminate this research, we should adopt regulations to ensure that it is conducted ethically.

Institutionalizing Bioethics

Concern about bioethics has led to the development of formal mechanisms to ensure that health care and health research will be conducted ethically. In this section, we look at four of those mechanisms: hospital ethics committees, institutional review boards, professional ethics committees, and community advisory boards.

Hospital Ethics Committees

The origins of hospital ethics committees can be traced to the 1950s. Like the Seattle Kidney Center, many other hospitals used committees to select patients for kidney dialysis. Similarly, hospitals routinely used committees to decide which women merited abortions on medical grounds. At the time, the legal status of abortion was unclear, and the moral status of abortion was just starting to become a public issue (Luker, 1984). Because psychiatric problems were considered justifiable medical grounds for abortion, wealthy women easily could find doctors who would testify to committees that abortion was psychiatrically needed. Poor women, on the other hand, typically could obtain abortions only if their lives were physically endangered. In reality, therefore, these committees made their decisions more on social than on medical grounds and primarily existed to protect doctors who performed abortions from legal or social sanction (Luker, 1984).

Other hospital ethics committees arose in the aftermath of the 1982 "Baby Doe" case, in which parents of a newborn who was mentally retarded and had a defective digestive system decided that they did not want the defect corrected by surgery. The doctors complied with their decision, and the baby died six days later. After news of the case broke, the federal government implemented regulations forbidding hospitals that received federal funds from withholding medical or surgical treatment from disabled infants. The regulations also urged but did not require hospitals to establish infant care review committees to prospectively evaluate decisions regarding withholding treatment from disabled infants. Many hospitals continued to use these committees even after the Supreme Court threw out the regulations in 1986. These days, most large hospitals have ethics committees.

More recently, hospitals have come to recognize the inherent difficulties in making decisions expeditiously by committee and so have shifted from relying on ethics *committees* to relying more on ethics *consultants*—individuals trained in bioethics and hired specifically to consult with hospital personnel regarding ethical issues. Ethics committees, meanwhile, have shifted from focusing on individual cases to consulting, advising, and providing information regarding broad ethical concerns, such as how to implement requests not to resuscitate terminally ill patients and whether hospital staff have an obligation to provide care to those with **HIV disease** (Fost and Cranford, 1985).

Institutional Review Boards

Although universities and hospitals began establishing committees to review research ethics in the 1960s, such committees did not become common until the 1970s. In the aftermath of the Tuskegee scandal, Congress in 1974 created the National Commission for the Protection of Human Subjects of Biomedical and Behavioral Research. The Commission's reports laid the groundwork for current guidelines regarding research ethics. That same year, the National Research Act mandated the development of **institutional review boards (IRBs)** charged with reviewing all federally funded research projects involving human subjects. Such boards now exist at all universities and other research institutions.

In addition, *commercial* IRBs also now exist (Lemmens and Freedman, 2000). The growth of commercial IRBs reflects the movement of much drug research away from universities and to pharmaceutical companies, for-profit research organizations that contract with pharmaceutical companies, and independent doctors who contract to do research for pharmaceutical companies. Some commercial IRBs are run directly by pharmaceutical companies. Others are independent, for-profit organizations that contract with pharmaceutical companies or for-profit research organizations.

The conflict of interest involved in such IRBs is obvious. When a pharmaceutical company's employees review their company's research, these employees cannot avoid knowing that their company's success depends on getting research approved. Similarly, those who work for independent IRBs know that they are unlikely to get future contracts from pharmaceutical and research organizations unless they approve the proposed research designs.

Professional Ethics Committees

Many professional organizations now also have ethics committees that establish guidelines for professional practice. The American Fertility Society, for example, has published a statement of principles regarding the moral status of human embryos created in the laboratory, and the ethics committee of the American College of Obstetrics and Gynecology has published guidelines regarding the ethics of selectively aborting fetuses when a woman who has used fertility drugs becomes pregnant with multiple fetuses.

Community Advisory Boards

The most recent development in this area is the emergence of community advisory boards (CABs). The purpose of CABs is to bring together individuals from the community with health care providers to make difficult bioethical decisions regarding both research and treatment (Quinn, 2004). For example, when patients are unconscious or incompetent, and family members are unavailable, a CAB may be given the responsibility of representing the patient in treatment decisions.

The use of CABs to evaluate research designs is linked to the rise of genetic research. Typically, we think of genetic testing as an individual decision: Should someone whose mother died of breast cancer, or whose sister has Down syndrome, get a genetic test to ascertain their own risks of having or passing on these diseases? But genetic testing also has implications for communities. Genetic tests can lead to the stigmatizing of an entire community, can challenge ideas about who *belongs* to a community (when genetic differences are found within a population), and can challenge community ideas of their origins (as, for example, when Native American stories regarding tribal origins clash with genetic findings). For these reasons, researchers have begun involving communities in discussions of research priorities, research design, and the dissemination of research findings.

One major question raised by the use of CABs regards how hospitals, researchers, and others should decide who constitutes a community, and who should represent a community. There is, unfortunately, no easy answer to this question.

The Impact of Bioethics

The growth of the bioethics movement and the institutionalizing of bioethics in U.S. hospitals and universities have made ethical issues more visible than ever before. Articles on bioethics, virtually nonexistent before the 1960s, now appear routinely in medical journals, while in both the clinical and research worlds, ethics committees have proliferated.

These developments have led some observers to conclude that the bioethics movement has fundamentally altered the nature of medical work. According to historian David Rothman:

> By the mid-1970s, both the style and the substance of medical decision-making had changed. The authority that an individual physician had once exercised covertly was now subject to debate and review by colleagues and laypeople. Let the physician design a research protocol to deliver an experimental treatment, and in the room, by federal mandate, was an institutional review board composed of other physicians, lawyers, and community representatives to make certain that the potential benefits to the subject-patient outweighed the risks. Let the physician attempt to allocate a scarce resource, like a donor heart, and in the room were federal and state legislators and administrators to help set standards of equity and justice. Let the physician decide to withdraw or terminate life sustaining treatment from an incompetent patient, and in the room were state judges to rule, in advance, on the legality of these actions. (1991: 2)

Other observers, however, contend that the impact of the bioethics movement has been more muted (e.g., Annas, 1991). These critics argue that hospital, research, community, and professional ethics committees, like the earlier hospital abortion committees, exist primarily to offer legal protection and social support to researchers and clinicians, not to protect patients or research subjects. Further, they argue, although clinicians have become more concerned with *documenting* their allegiance to ethics guidelines, they have not become any more concerned with *following* those guidelines. Finally, sociologist Daniel F. Chambliss (1996) argues that bioethics' emphasis on helping individual health care providers make more ethical decisions simply does not apply to health care workers like nurses, who often understand clearly what they should do ethically but lack the power to do so. For example, nurses often have a much better understanding than doctors of how much a patient is suffering and thus more often believe treatment should be discontinued unless it will improve quality as well as length of life. Yet nurses rarely can act on that belief because they lack the necessary legal standing, economic independence, and social status.

The following sections evaluate the impact of bioethics on health care research, medical education, and clinical practice.

The Impact on Research

According to ethicist George Annas, the bioethics movement, as institutionalized in research ethics boards and committees, has affected medical research only slightly. In his words, the

> primary mission [of research ethics committees] is to protect the institution by providing an alternative forum to litigation or unwanted publicity.... [For this reason] its membership is almost exclusively made up of researchers (not potential subjects)

from the particular institution. These committees have changed the face of research in the U.S. by requiring investigators to justify their research on humans to a peer review group prior to recruiting subjects. But this does not mean that they have made research universally more "ethical." In at least a few spectacular instances, these committees have provided ethical and legal cover that enabled experiments to be performed that otherwise would not have been because of their potentially devastating impact on human subjects. (1991: 19)

As an example, Annas cites the case of "Baby Fae" (not her real name), who died in 1984 soon after doctors replaced her defective heart with a baboon's heart. Although all available evidence indicated that cross-species transplants could not succeed, the doctors who performed the surgery had received approval from their hospital's IRB. A subsequent review found that Baby Fae's parents had not given truly informed consent, because the doctors had not suggested seeking a human transplant, had disparaged available surgical treatments, and had unreasonably encouraged the parents to believe that a baboon transplant could succeed.

Lack of resources and conflicts of interest also limit the effectiveness of IRBs. IRB members are unpaid volunteers, who typically must review between 300 and 2,000 proposed experiments yearly and who, in many cases, have vested interests in approving research proposals so their institutions can obtain research funding (Hilts, 1999). Meanwhile, final responsibility for overseeing IRBs falls to the federal Office of Protection from Research Risks, which has only three full-time employees. These conditions make thorough review of human subjects research impossible.

Finally, even when IRBs work as designed, their authorizing statutes restrict them from addressing the broader issues of whether the benefits potentially available through research outweigh the potential for harm to society and whether the money allotted for a given research project could produce more beneficial effects if spent elsewhere (P. Williams, 1984). Yet these are often the most important questions to ask.

Nevertheless, and despite the limitations of IRBs and research ethics committees, the rise of bioethics has curbed the most egregious abuses of human subjects. According to David Rothman:

> The experiments that Henry Beecher described could not now occur; even the most ambitious or confident investigator would not today put forward such protocols. Indeed, the transformation in research practices is most dramatic in the area that was once most problematic: research on incompetent and institutionalized subjects. The young, the elderly, the mentally disabled, and the incarcerated are not fair game for the investigator. Researchers no longer get to choose the martyrs for mankind. (1991: 251)

In fact, the balance has shifted to such an extent that we now sometimes read news stories not of researchers pressuring individuals to become research subjects but, rather, of desperately ill individuals pressuring researchers to accept them as research subjects for experimental treatments.

The Impact on Medical Education

One obvious result of the bioethics movement has been the incorporation of ethics training into medical education, with courses now common at U.S. medical schools. As critics have noted, however, those courses are too often divorced from real life, aimed at teaching students ethical principles and legal norms through classroom lectures rather than at teaching students how to negotiate the everyday ethical dilemmas they face. To achieve this latter goal, the University of Pennsylvania Medical School includes in its ethics course sessions in which students discuss ethical dilemmas they have encountered during their clinical training, such as pressures placed on them to perform medical procedures on unwilling patients (Christakis and Feudtner, 1993). Discussing situations like these can help students devise strategies for responding more ethically in future.

Other observers, however, have noted that a course like this also has its limits, for it assumes that students who are already undergoing socialization to medical culture still can identify ethically problematic aspects of that culture (Hafferty and Franks, 1994, 1998). Moreover, this strategy does not challenge the ways in which ethics are discounted in the "hidden curriculum" of medical practice and culture. For example, a structure that expects students both to provide care for patients *and* to learn techniques on patients without the patients' knowledge inherently teaches students to view patients at least partly as objects rather than as subjects. From this perspective, only through "the integration of ethical principles into the everyday work of both science and medicine" can we expect new doctors to adopt more ethical approaches to care (Hafferty and Franks, 1994: 868).

The Impact on Clinical Practice

At a fundamental level, the bioethics movement challenges doctors' clinical autonomy, for it "substitutes principles and general rules for the case-by-case analysis that has long characterized medical practice . . . and attempts to reformulate medical problems as moral, rather than technical, issues" (Zussman, 1992: 10–11).

According to Annas (1991), professional ethics committees emerged to counter this challenge. Annas argues that the true purpose of these committees is not to foster more ethical behavior but to protect professional autonomy by providing clinicians with legal protection against accusations of unethical behavior. For example, published guidelines from the Ethics Committee of the American Fertility Society refer to human embryos created in the laboratory merely as "pre-embryos," even though they do not differ biologically from other embryos, and leave it up to each clinic to establish policies for their use. Similarly, published guidelines from the American College of Obstetrician-Gynecologists on whether to selectively abort fetuses when several embryos become implanted simultaneously in a woman's uterus state only that doctors and patients should make their decisions jointly. Such guidelines seem designed more to provide legal cover to clinicians than to encourage more ethical practices (Annas, 1991).

Relatively few studies have looked at the impact of bioethics on actual clinical practices. One series of studies looked at the impact of New York's 1987 law establishing formal policies for writing "do not resuscitate" orders (orders forbidding health care workers from intervening if the lungs or heart of a terminally ill patient stop functioning). These studies found that after the law's passage, doctors significantly altered how they *documented* their actions but not how they *acted* (Zussman, 1992: 162). Similarly, studies have found that hospitals sharply limit access of patients, family, and nonmedical staff to ethics consultations. As a result, consultations primarily function to provide additional institutional support to doctors confronted by families or patients they consider disruptive, such as those who challenge doctors' decisions regarding how aggressively to treat a given condition (S. Kelly et al., 1997; Orr and Moon, 1993). These findings have led researchers to conclude that the true purpose of ethics consultations is to reinforce doctors' power.

The most extensive study of the impact of bioethics on clinical practice appears in *Intensive Care: Medical Ethics and the Medical Profession* (1992), by sociologist Robert Zussman. Zussman spent more than two years observing and interviewing in the intensive care units of two hospitals. His research suggests both the impact and the limitations of the bioethics movement.

Although cases such as Karen Quinlan's and Baby Doe's might suggest that doctors often want to use aggressive treatment despite the objections of patients and families, Zussman found that on intensive care wards the reverse is usually the case. Knowing that most of their patients will die, doctors on these wards often hesitate before beginning aggressive treatment, which might only escalate costs, increase their work as well as their patients' suffering, and prolong the dying process. Patients, however—and more important, their families (for, in most cases, the patients are incapable of communicating)—often face a sudden and unexpected medical crisis. Unable to believe the situation hopeless, they demand that health care workers "do everything." In these situations, the doctors Zussman studied expressed allegiance to the principle that families have the right to make decisions regarding treatment. In practice, however, doctors found ways to assert their discretion, if no longer the authority they had in years past.

Doctors asserted their discretion in several ways. First, doctors made decisions without asking the family on the assumption that the family would agree with their decisions. Second, doctors sometimes ignored a family's stated decisions, arguing that it was cruel to force a family to make life-or-death decisions that would later cause them guilt or grief. Third, doctors might respect a family's wishes, but only after first shaping those wishes through selectively providing information. This information included defining the patient as terminally ill or not—a highly significant designation, for ethical guidelines permit health care workers to withhold or terminate treatment only for terminally ill patients. Fourth, when doctors failed to shape a family's wishes, the doctors could discount those wishes on the grounds that the family was too emotionally distraught to decide rationally.

Finally, and perhaps most important, doctors continued to assert their discretion by defining the decision to withhold or terminate treatment as a technical rather than an ethical problem. The following example from Zussman's research demonstrates this process:

> The Countryside ICU [Intensive Care Unit] staff was considering whether or not to write a Do Not Resuscitate order for Mr. Lake, a 73-year-old man who had been admitted to the unit with acute renal [kidney] failure, a gastrointestinal bleed, pneumonia, and sepsis [infection]. Ken [the medical director of the ICU] asked what they should do "if the family wanted a full court press." One of the residents started to say what he thought were the "interesting ethical issues." But Ken cut him off, arguing that the decision depended entirely on prognostics: "There are no ethical issues. . . . I'm not an ethicist. I'm a doctor." When the resident attempted to distinguish different circumstances preceding codes [decisions not to resuscitate], Ken broke in again: "A code is a code. It's a medical decision, not an ethical decision." (Zussman, 1992: 150; ellipses in original)

Once doctors succeeded in defining treatment decisions as purely technical issues, they could define the family's stated wishes as uneducated and irrelevant. Doctors could end discussion regarding treatment decisions by declaring it simply a technical fact that any treatment would be futile. Similarly, doctors might acknowledge families' general wishes regarding how aggressively treatment should proceed, but then define each specific intervention as a technical decision best left to doctors. Because most treatment decisions involve not dramatically pulling a plug but rather a series of small, minute-to-minute actions, leaving doctors in control of these "technical" matters gives doctors power far outweighing families' general statements regarding whether to pursue aggressive treatment.

Summing up his findings, Zussman writes:

> The picture I have drawn corresponds neither to an image of unbridled professional discretion nor to one of patients' rights triumphant. As many observers of contemporary medicine have argued, the discretion of physicians in clinical decisions (like the discretion of professionals in other fields) depends on their ability to make successful claims to the exclusive command of technical knowledge. Yet, while . . . physicians . . . make such claims, they do not always succeed either in convincing themselves that they are legitimate or in converting them to influence over patients and their families, for the claims of physicians are met by the counterclaims of patients and, more important, families. . . . The institutionalization of patients' rights, in law and in hospital policy, . . . empower[s] families when they

---

Box 13.2   *Making a Difference: Choosing Your Career*

By this point in the semester, some of you undoubtedly are just grateful that it is almost over. But others may now find that you are fascinated by the topic and wondering how you can somehow make a difference in this field. For those who are interested, four broad career options exist: clinical practice, administration, research and teaching, and policy work.

Many students take a course on the sociology of health, illness, and health care because they intend to become a health care practitioner of some sort. Now that you have reached the end of this semester, you probably have a better idea than when you started of the costs and benefits of entering the different health care fields. Perhaps you now recognize that you are attracted to the professional autonomy as well as the art and science of medicine, or realize that you would be more comfortable in a health care field that offers a more holistic approach to care. Perhaps you have second thoughts about

entering nursing given its struggles for professional autonomy, or find it more appealing now that you understand the intellectual challenges and financial rewards available to those who obtain masters-level training. No matter what health care occupation you might enter, you should now bring to your work a greater understanding of the underlying sources of health and illness, the culture of medicine, the experiences of persons who live with illness and of other health care consumers, and the impact of the larger health care delivery system on both consumers and providers. Working as a compassionate, ethical, and educated health care provider is an important way of making a better world, one patient at a time.

Other readers of this book may realize, when they think about their personalities, skills, and interests, that they are not really suited for the "hands-on" work of dealing directly with patients. For those who enjoy the

---

do insist on doing everything. In such a situation, physicians may continue to exercise considerable influence and enjoy considerable discretion. By no means have they been reduced to the role of technicians and nothing more. But at the same time, they must, at the very least, take the wishes of patients and families into account. (1992: 159–160)

Conclusion

In this chapter we have explored the history of the bioethics movement and its impact on health care research and practice. As we have seen, bioethics and sociology have much in common. At a very basic, if typically unacknowledged level, bioethics, like sociology, is about power. The abuses of the Nazi doctors, for example, not only illuminate the horrors possible when ethical principles are ignored but also show how social and occupational groups can obtain power over others as well as the potentially deadly consequences when this happens. Conversely, sociology, in similarly unacknowledged ways,

nitty-gritty details of the business world, there are many opportunities to work in health care administration, in everything from small non-profit agencies that provide assistance to the uninsured to major hospital chains. Your goal, as students of the sociology of health, illness, and health care, will be to find a position that allows you to help *others* deliver high-quality, equitable health care.

A third option is to enter a career in research or teaching. Such a career requires that you be primarily fascinated by the process of generating knowledge (research), evaluating research conducted by others, and figuring out how to communicate research findings to others, whether through publications or in the classroom. Research positions can be found at all levels of government (from county health departments to the federal Centers for Disease Control and Prevention), in colleges and universities, and in nonprofit organizations and "think tanks" like the Kaiser Family Foundation and the Commonwealth Fund. In some of these positions you would have the freedom to develop your own research and teaching agenda, while in others you would be assigned to a general field of study or specific research tasks. But in all cases you would have the satisfaction of generating and communicating important knowledge about health, illness, and health care.

Finally, those of you who are most interested in effecting change on a broader scale, and who have the requisite personalities and skills, should consider careers in law, government, or political advocacy. Perhaps a reader of this book will some day direct a nonprofit organization that advocates for the rights of persons with disabilities, argue a right-to-life or right-to-die case before the Supreme Court, or propose on the U.S. Senate floor a new law guaranteeing universal health coverage.

Whatever path you choose, you *can* make a difference.

is at a basic level an ethical enterprise. Underlying abstract, technical sociological discussions about the nature of society there often lurk hidden assumptions about what society *should* be like and how society should be changed. These assumptions often draw on philosophies regarding justice, autonomy, human worth, and other basic ethical issues. Yet, in the same way that bioethicists often ignore the sociological implications of their work, sociologists often ignore the ethical implications of the questions they ask, the research they conduct, and the findings their research generates. It seems, then, that bioethicists and sociologists can provide each other with broader perspectives that can only enrich our understanding of both fields—encouraging bioethicists to see not only individual cases but broader social and political issues and encouraging sociologists to see the world and their work in it as an ethical as well as a political and intellectual enterprise. These are issues that all of us should keep in mind as we seek our place in the world; Box 13.2 provides some suggestions for readers who are interested in pursuing a career related to health and health care.

Suggested Readings

Andrews, Lori B. 1999. *The Clone Age: Adventures in the New World of Reproductive Technology.* New York: Henry Holt. Andrews, a lawyer and professor respected for her work in reproductive technology, trenchantly analyzes the potential, and potential pitfalls, of the field.

Elliott, Carl. 2003. *Better Than Well: American Medicine Meets the American Dream.* New York: Norton. A thought-provoking account of the new drugs, technologies, and cultural pressures to enhance human traits, ranging from botox to speech therapy used to reduce regional accents.

Hastings Center Report. An eminently readable and always fascinating monthly journal on bioethics, published by the Hastings Center (see "Getting Involved").

Zussman, Robert. 1992. *Intensive Care: Medical Ethics and the Medical Profession.* Chicago: University of Chicago Press. An engrossing sociological analysis of the impact of modern bioethics.

Getting Involved

The Hastings Center. Rt. 9D, Garrison, NY 10524–5555. (845) 424-4040. www.thehastingscenter.org. A nonprofit organization, the center is committed to research, lobbying, and public education on bioethics. Publishes the excellent *Hastings Center Report.*

Review Questions

What is the Nuremberg Code, and how and why did it come into existence?

What factors led to the emergence of the bioethics movement in the late 1960s?

Why do researchers now consider the Tuskegee Syphilis Study and the Willowbrook experiments to have been unethical?

What are the ethical problems involved in the new reproductive technology? in enhancements? What impact has bioethics had on health care and on health research?

Internet Exercises

1. Using InfoTrac® College Edition, look for articles on cosmetic surgery from a variety of sources. (You can access InfoTrac College Edition at www.infotrac-college.com/wadsworth, if your professor ordered it when ordering this textbook.) Do these articles suggest that there are any ethical or social issues inherent in cosmetic surgery, such as whether it is morally right or wrong, or whether social forces rather than objective aesthetic concerns

press individuals to have this surgery? If yes, what ethical or social issues do they identify? If no, how do you explain why they do not recognize any ethical or social issues?

2. The ELSI program is a part of the Human Genome Project (which is itself a part of the National Institutes of Health) designed to investigate the ethical, legal, and social implications (ELSI) of human genetics research. Find the ELSI website, and learn about the types of research that have been sponsored by this program.

GLOSSARY

accommodation: Technique individuals use to smooth interactions with those they consider potential sources of trouble and to smooth interactions between those persons and others.

achieved statuses: Earned qualifications and positions, such as ranking in the top third of a class.

acquired immunodeficiency syndrome (AIDS): The end stage of HIV disease. See *HIV disease.*

active life expectancy: Number of years a person can expect to live in good health and without disabilities. Total life expectancy equals active life expectancy plus inactive life expectancy.

active voluntary euthanasia: When individuals help sick persons to kill themselves.

activity-passivity: Model of doctor-patient interaction in which doctor is active and patient is passive, such as during emergency surgery.

actuarial risk rating: A system in which insurers try to maximize their financial gain by identifying and insuring only those populations that have low health risks.

acute: Anything that had a sudden onset, such as acute illness or acute pain.

acute illness: Any illness that strikes suddenly and disappears rapidly (within a month or so). Examples include chicken pox, colds, and influenza.

ADA: See *Americans with Disabilities Act.*

addiction: The physical state in which an individual who has used a drug regularly will experience withdrawal if he or she ceases using the drug. See *withdrawal.*

advanced practice nurses: Individuals who, after becoming registered nurses, additionally receive specialized postgraduate training. Includes nurse-midwives and nurse-practitioners. See *Registered Nurses.*

age-adjusted rates: Epidemiological data that have been manipulated, using standard statistical techniques, to eliminate any effects that arise because some populations include more older or younger persons than do others. Age adjustment allows us to compare populations with different age distributions.

agency: The ability of individuals to make their own choices, free of any limitations placed on them by other people, culture, or social forces. Similar to the concept of free will.

AIDS: Acquired immunodeficiency syndrome. See *HIV disease.*

aligning actions: Actions or, more specifically, interpretations of actions designed to make behavior appear reasonable and normal in a given cultural context.

allopathic doctors: Nineteenth-century forerunners of contemporary medical doctors. Also known as "regular" doctors.

almshouse: An institution, also known as a poorhouse, in which all public wards, including orphans, criminals, the disabled, and the insane, received custodial care.

alternative therapies: Treatments rarely taught in medical schools and rarely used in hospitals.

AMA: See *American Medical Association.*

ambulatory care: Outpatient care.

American Medical Association (AMA): The main professional association for medical doctors.

Americans with Disabilities Act (ADA): Federal law, passed in 1990, that outlaws discrimination against those with disabilities in employment, public services (including transit), and public accommodations (such as restaurants, hotels, and stores). It requires that existing public transit systems and public accommodations be made accessible, along with all new public buildings and major renovations of existing buildings.

ascribed statuses: Innate characteristics such as ethnicity or gender.

assistant doctors: Chinese health care workers who receive three years of postsecondary training, similar to that of doctors, in both Western and traditional Chinese medicine.

assisted living facilities: Institutions, typically consisting of small, private apartments, that offer two levels of services: (1) basic medical and nursing services and help with basic tasks of daily living for disabled residents and (2) comfortable living situations and social activities for healthy residents.

attendings: Doctors who have completed their training and who supervise residents. These doctors can be employed by a hospital or can work in private practice. See *residents.*

avoidance: Actively working to remain ignorant in order to maintain one's emotional balance and images of the future.

balance bill: To bill patients for the difference between the amount their insurance will pay for a given procedure and the amount the doctor would normally charge for that procedure.

barefoot doctors: Former term for agricultural workers in China who received about three months of training in health care and provided basic health services to members of their agricultural production teams. Now known as *village doctors.*

bioethics: The study of all ethical issues involved in the biological sciences and health care.

blaming the victim: Process through which individuals are blamed for causing the problems from which they suffer.

Blue Cross: A group of private nonprofit companies offering insurance that reimburses individuals primarily for the costs of hospital care, not including doctors' bills. Blue Cross insurance is often offered and bought in conjunction with Blue Shield insurance. See *Blue Shield.*

Blue Shield: A group of private nonprofit companies offering insurance that reimburses individuals primarily for the costs of receiving care from doctors, especially care received in hospitals. Blue Shield insurance is often offered and bought in conjunction with Blue Cross insurance. See *Blue Cross.*

board and care homes: Residential facilities that provide assistance in daily living but not nursing or medical care.

capitation: A system in which doctors are paid a set annual fee for each patient in their practice, regardless of how many times they see their patients or what services the doctors provide for their patients.

CDC: See *Centers for Disease Control and Prevention.*

Centers for Disease Control and Prevention (CDC): Federal agency responsible for tracking the spread of diseases in the United States.

challenging: Rejecting the social norms that attach stigma to a behavior or condition, including illness or disability. See *stigma.*

chiropractors: Health care practitioners who specialize in spinal manipulation, trace illness and disability to misalignments of the spine, and believe spinal manipulation can cure a wide range of acute and chronic health problems.

chronic: Anything that continues over a long period, such as chronic illness or chronic pain.

chronic illness: Illness that develops in an individual gradually or is present from birth and that will probably continue at least for several months and possibly until the person dies. Examples include muscular dystrophy, asthma, and diabetes.

claims harassment: When insurance companies establish bureaucratic structures that make it virtually impossible for consumers to file claims or to fight the insurer if a claim is denied.

clinical pharmacy: A subfield of pharmacy in which pharmacists participate actively in decisions regarding drug treatment.

cognitive norms: Socially accepted rules regarding proper ways of thinking. For example, someone should not think that he is Napoleon or that his radio is sending him secret messages from outer space.

commercial insurance: Insurance offered by companies that function on a for-profit basis.

commodification: Process of turning people into products that can be bought or sold.

community rating: A system for calculating insurance premiums in which each individual pays a premium based on the average health risk of his or her community as a whole.

complementary therapies: Treatments rarely taught in medical schools and rarely used in hospitals.

compliance: Whether individuals do as instructed by health care workers.

conflict perspective: View that society is held together by power and coercion, with dominant groups imposing their will on subordinate groups.

control: A process through which researchers statistically eliminate the potential influence of extraneous factors. For example, because social class and race often go together, researchers who want to investigate the impact of social class have to be sure that they are not really seeing the impact of race. To study the impact of social class on mental illness, therefore, researchers would have to look separately at the relationship between social class and mental illness among whites and then at the relationship among blacks to control for any effect of race.

convergence hypothesis: The thesis that health care systems become increasingly similar over time because of similar scientific, technological, economic, and epidemiological pressures.

cooptation: A process through which an individual, organization, or movement exchanges some or all of its initial philosophy and goals for social acceptance and financial support.

co-payment: Fee paid by persons, who have certain forms of health insurance, each time they see a care provider. Fees can range from nominal sums to 20 percent of all costs.

core nations: According to world systems theory, those nations with a highly diversified, industrialized economy, providing a relatively high standard of living for most citizens.

corporatization: The growing role of investor-owned corporations in the health care field.

cost shifting: Raising prices charged some individuals for services rendered to make up for losses incurred when services are provided to other individuals who cannot or will not pay for services.

covering: Attempting to deflect attention from deviance, including illnesses or disabilities.

cultural competence: The ability of health care providers to understand at least basic elements of others' culture and, thus, to provide medical care in ways that better meet clients' emotional as well as physical needs.

curanderos: Folk healers who function within Mexican and Mexican American communities.

deductible: Dollar amount of health care expenses an individual with some forms of health insurance must pay annually before his or her insurance plan will begin covering the remaining costs of health care.

defensive medicine: Tests and procedures that doctors perform primarily to protect themselves against lawsuits rather than to protect their patients' health.

deinstitutionalize: To remove individuals, such as mentally retarded and mentally ill persons, from large institutions and return them to the community.

demedicalization: The process through which a condition or behavior becomes defined as a natural condition or process rather than an illness.

depersonalization: A sense that a person no longer is, or is not considered, fully human.

depoliticize: To define a situation in a way that hides or minimizes the political nature of that situation.

deprofessionalized: Referring to the lessening or loss of professional status of an occupational group.

developing nations: Nations characterized by a relatively low gross national product per capita. These countries typically have high rates of illiteracy, infant mortality, and other related problems, and their economies rely heavily on a few industries or products.

deviance: Behavior that violates a particular culture's norms or expectations for proper behavior, and therefore results in negative social sanctions. See *negative social sanctions.*

deviance disavowal: The process through which individuals attempt to prove that, despite their apparent deviance, they are no different from other people.

diagnosis-related groups (DRGs): System established by the federal government that sets, for all Medicaid patients and for each possible diagnosis, an average length of hospital stay and cost of inpatient treatment. Under the DRG system, hospitals are paid the established cost for each patient with a given diagnosis, regardless of actual cost of treatment.

Diagnostic and Statistical Manual (DSM): Manual published by the American Psychiatric Association and used by mental health workers to assign diagnoses to clients. Generally, this manual must be used if mental health workers want to obtain reimbursement for their services from insurance providers.

diploma nurses: Nurses who hold diplomas from hospital-based schools, rather than holding associates or bachelors degrees from colleges or universities.

disability: Restrictions or lack of ability to perform activities resulting from physical limitations or from the interplay between those limitations, social responses, and the built or social environment.

disclosing: Making one's deviance more widely known by telling others about it or making it more visible.

discrimination: Differential and unequal treatment grounded in prejudice. See *prejudice.*

disease: A biological problem within an organism.

disease management: A form of pharmaceutical care in which pharmacists are responsible for counseling certain patients (typically those with chronic conditions) on their prescription drugs, monitoring the impact of the drugs, and, sometimes, prescribing drugs. See *pharmaceutical care.*

doctor-assisted euthanasia: When doctors help patients to kill themselves.

doctor-nurse game: "Game" in which the nurse is expected to make recommendations for medical treatment in such a way that the recommendations appear to have come from the doctor.

DRG: See *diagnosis-related groups.*

DSM: See *Diagnostic and Statistical Manual.*

dysfunctional: That which threatens to undermine social stability.

emergency room abuse: Term used by hospitals to refer to patients who have neither health insurance nor money to pay for care and who therefore turn to hospital outpatient clinics and emergency rooms for treatment of chronic as well as acute health problems.

endemic: Referring to diseases that appear at a more or less stable rate over time within a given population.

enhancements: Techniques deemed to improve human traits beyond a level generally considered normal rather than to treat conditions considered deviant or defective. This distinction is artificial, but occasionally useful.

entrepreneurial system: A system based on capitalism and free enterprise.

environmental racism: The disproportionate burden of environmental pollution experienced by racial and ethnic minorities.

epidemic: Either a sudden increase in the rate of a disease or the first appearance of a new disease.

epidemiological transition: The shift from a society burdened by infectious and parasitic diseases and in which life expectancy is low to one characterized by chronic and degenerative diseases and high life expectancy.

epidemiology: The study of the distribution of disease within a population.

etiology: Cause.

eugenics: The theory that the population should be "improved" through selective breeding and birth control.

evidence-based medicine: Medical therapies whose efficacy has been confirmed by large, randomized, controlled clinical studies. See *control.*

family leave programs: Programs that allow individuals to take time off from work without risking their jobs to care for family members. Some programs offer paid leave, others only unpaid leave.

fee-for-service: The practice of paying doctors for each health care service they provide, rather than paying them a salary.

fee-for-service insurance: Insurance that reimburses patients for all or part of the costs of the health care services they have purchased. This contrasts with health maintenance organizations, in which patients pay one charge in advance in exchange for any health care they might require during a given period.

feeling norms: Socially defined expectations regarding the range, intensity, and duration of appropriate feelings and regarding how individuals should express those feelings in a given situation.

feeling work: Efforts made by individuals to avoid being labeled mentally ill by making their emotions match social expectations. Individuals can (1) change or reinterpret the situation causing them to have unacceptable feelings; (2) change their emotions physiologically, through drugs, meditation, biofeedback, or other methods; (3) change their behavior, acting as if they are feeling more appropriate emotions; or (4) reinterpret their feelings, telling themselves, for example, that they are only tired rather than worried.

feminization of aging: The fact that women comprise a larger proportion of the elderly than of younger age groups; the steady rise in the proportion of the population who are female at each successive age.

fetal rights: The growing body of legal, medical, and public opinion holding that fetuses have rights separate from and sometimes contrary to those of their mothers.

financially progressive: Describes any system in which poorer persons pay a smaller proportion of their income for a given good or service than do wealthier persons.

Flexner Report: The report on the status of American medical education produced in 1910 by Abraham Flexner for the Carnegie Foundation. This report identified serious deficiencies in medical education and helped to produce substantial improvements in that system.

for-profit hospitals: Hospitals run with the primary goal of producing a profit each year for shareholders.

formulary: Official list of drugs that doctors in a managed care organization can prescribe without special authorization. See *managed care.*

functionalism: View of society as a harmonious whole held together by socialization, mutual consent, and mutual interests.

gatekeeper: A primary care doctor, in managed care plans such as HMOs, who serves as the intermediary between patients and specialists. Patients must get a referral from their gatekeeper doctor before they see a specialist, or they must pay out of pocket to see the specialist. See *primary care doctors* and *health maintenance organizations.*

gender: The social categories of masculine and feminine, and the social expectations of masculinity and femininity.

genetic paradigm: A way of looking at the world that emphasizes genetic causes.

glass ceiling: The invisible but real forces of discrimination and prejudice that keep members of a group from rising above a certain level in an organization or profession.

glass escalator: The invisible but real social forces that give members of a group an extra assist in rising in an organization or profession.

globalization: The process through which ideas, resources, and persons increasingly operate within a worldwide rather than local framework. For example, the globalization of tourism means that U.S. tourists now consider Africa a plausible destination.

government hospitals: Hospitals established by state and federal governments to provide services to those groups that would not otherwise receive care.

Great Confinement: The shift, from the 1830s on in both Europe and the United States, toward confining mentally ill persons in large public institutions instead of in almshouses, small private "madhouses," or at home with relatives.

group rates: Insurance rates set by an insurance company for all members of a large group, regardless of their individual health status or risk factors. Typically much lower than individual rates.

guidance-cooperation: Model of doctor-patient interaction in which the doctor guides and the patient cooperates, such as when a patient follows a doctor's advice regarding treating an injury.

health belief model: Model predicting that individuals will follow medical advice when they (1) believe they are susceptible to a particular health problem, (2) believe the health problem they risk is a serious one, (3) believe compliance will significantly reduce their risk, and (4) do not perceive any significant barriers to compliance.

health lifestyle theory: A theory that attempts to predict why groups adopt patterns of healthy or unhealthy behavior by showing how demographic circumstances and cultural memberships combine with socialization and experiences to produce both life chances and life choices. These life chances and choices in turn lead to habitual dispositions toward healthy or unhealthy behaviors, which then lead to actual behaviors.

health maintenance organizations (HMOs): Organizations that provide health care based on prepaid group insurance. Patients pay a fixed yearly fee in exchange for a full range of health care services, including hospital care as well as doctor's services.

health social movements: Informal networks of individuals who band together to collectively challenge health policy, politics, beliefs, or practices.

heroic medicine: System of treatment used by allopathic doctors before about 1860 that emphasized curing illnesses by purging the body through bloodletting, causing extreme vomiting, or using repeated laxatives and diuretics. See *allopathic doctors.*

HIV (human immunodeficiency virus): The virus that causes HIV disease.

HIV disease: A disease in humans caused by HIV infection that gradually breaks down the body's immune system and typically leads to death about twelve years after infection.

HMO: See *health maintenance organization.*

holistic treatment: Treatment based on the premise that all aspects of an individual's life and body are interconnected—that, for example, to treat an individual with cancer effectively, health care workers must look at all organs of the body, not only the one that currently has a tumor, as well as at the individual's psychological and social functioning.

home health aides: Workers, typically untrained, who provide essentially custodial care within individuals' homes.

homeopathic doctors: Popular nineteenth-century health care workers who treated illnesses with extremely dilute solutions of drugs that, at full strength, produced similar symptoms to a given illness.

horizontal integration: Situation in which a large corporation owns several institutions that provide the same type of service, such as several nursing homes.

hospices: Institutions designed to meet the needs of the dying.

human immunodeficiency virus: See *HIV.*

illness: The social experience of having a disease.

illness behavior: The process of responding to symptoms and deciding whether to seek diagnosis and treatment.

immersion, illness as: Situation in which illness becomes so demanding that a person must structure his or her life around it.

incidence: Number of new cases of an illness or health problem occurring within a given population during a given time period (for example, the number of children born with Down syndrome in the United States during 2007).

inactive life expectancy: Number of years a person can expect to live in poor health and with disabilities. Total life expectancy equals active life expectancy plus inactive life expectancy.

individualism: A set of cultural beliefs and practices that encourages the autonomy, equality, and dignity of individuals and that downplays the importance of connections to social groups.

industrialized nations: Nations characterized by a relatively high gross national product per capita. These countries typically have diversified economies and low rates of illiteracy, infant mortality, and other related problems.

informed consent: Voluntary agreement to participate in medical research or to receive a medical procedure or treatment, with a full understanding of the potential risks and benefits.

inpatient: Hospital patient who is formally admitted and kept overnight.

institutional review boards (IRBs): Federally mandated committees charged with reviewing the ethics of research projects involving human subjects. No research can be conducted using federal funds unless it first receives IRB approval.

internal colonialism: The treatment of minority groups within a country in ways that resemble the treatment of native peoples by foreign colonizers.

interruption, illness as: Situation in which illness is experienced as only a small and temporary part of one's life.

intersex: Individuals having both male and female biological characteristics, such as a penis and a uterus, or a vagina and an unusually large clitoris.

intrusion, illness as: Situation in which illness demands time, accommodation, and attention and forces one to live from day to day.

IRB: See *institutional review boards.*

irregular practitioners: Nineteenth-century health care practitioners other than allopathic doctors, such as homeopaths, midwives, botanic doctors, bonesetters, and patent medicine makers.

lay midwives: Midwives who do not have formal training in midwifery and who typically learn through experience or apprenticeship.

licensed practical nurses (LPNs): Individuals, not registered nurses, who assist nurses primarily with the custodial care of patients. LPNs usually have completed approximately one year of classroom and clinical training.

life events: Any changes that force readjustments in individuals' lives, including marriage or divorce, starting or leaving school, and gaining or losing a job.

life expectancy: The average number of years that individuals of a given group born in a given year are expected to live.

lifeworld: The everyday needs of people and ways in which they interact and live their lives.

limited practitioners: Occupational groups, such as chiropractors and optometrists, that confine their work to a limited range of treatments and certain parts of the body.

long-term care insurance: Insurance designed specifically to pay the costs of nursing-home care, board and care homes, in-home nursing care, and other long-term health-related needs.

LPN: See *licensed practical nurses.*

magic bullets: Drugs that prevent or cure illness by attacking one specific etiological factor.

magnetic healers: Nineteenth-century health workers who believed that an invisible magnetic fluid flowed through the body and that illness occurred when that flow was obstructed, unbalanced, inadequate, or excessive. Their treatments consisted of moving their hands along patients' spinal cords to "free" blocked magnetic fluid.

managed care: A system that controls health care spending by monitoring closely how health care providers treat patients and where and when patients receive their health care.

managed care organizations (MCOs): Health insurance providers, such as health maintenance organizations, that operate under the principles of managed care.

manufacturers of illness: Those groups, such as alcohol and tobacco manufacturers, that promote illness-causing behaviors and social conditions.

marginal practitioners: Occupational groups such as faith healers that have low social status.

master status: A status viewed by others as so important that it overwhelms all other information about that individual. For example, if we know someone as the local scoutmaster, know he is a Republican and likes to play chess, and then learn he is gay, we might start thinking about him and interacting with him solely on the basis of his sexual orientation, essentially forgetting or ignoring the other information we have about him.

MCOs: See *managed care organizations.*

Medicaid: Joint federal-state health insurance program that pays the costs of health care for people with incomes below a certain (very low) amount. Most Medicaid recipients are poor mothers and their children. Medicaid can cover the costs of both preventive and therapeutic medical care and both inpatient and outpatient hospital care, but details of coverage vary considerably from state to state, with some providing considerably more services than others.

medical dominance: Professional dominance by doctors. See *professional dominance.*

medical model of illness: The way in which doctors conceptualize illness. This model consists of five doctrines: that disease is deviation from normal, specific and universal, caused by unique biological forces, analogous to the breakdown of a machine, and defined and treated medically through a neutral scientific process.

medical model of mental illness: A model of mental illness assuming that (1) objectively measurable conditions define mental illness, (2) mental illness stems largely or solely from something within individual psychology or biology, (3) mental illness will worsen if left untreated but might improve or disappear if treated promptly by a medical authority, and (4) treating someone who might be healthy is safer than not treating someone who might be ill.

medical norms: Expectations doctors hold regarding how they should act, think, and feel.

medicalization: Process through which a condition or behavior becomes defined as a medical problem requiring a medical solution, or through which the definition of an illness is broadened to cover a wider population.

medically indigent: Persons who earn too much to receive government-provided health care but too little to pay medical bills or purchase health insurance.

Medicare: Federal insurance, based on the Social Security system, that offers hospital insurance and medical insurance to those over age 65 and to permanently disabled persons.

medigap policies: Insurance policies available for purchase by persons who receive Medicare to pay for prescription drugs and other medical services not available through Medicare. See *Medicare.*

miasma: According to pre-twentieth-century doctors, disease-causing air "corrupted" by foul odors or fumes.

microcredit: Programs that offer loans of $100 or less to residents of developing nations.

minority group: Any group that, because of its physical or cultural characteristics, is considered inferior and subjected to differential and unequal treatment.

model programs: Programs for treating persons with serious mental illnesses that aim to avoid stigma, dehumanization, and hierarchical patient/staff relationships and that offer a range of social and economic services as well as psychiatric care.

models of doctor-patient interactions: See definitions for three models: *activity-passivity, guidance-cooperation,* and *mutual participation.*

moral status: A status that identifies in society's eyes whether a person is good or bad, worthy or unworthy.

moral treatment: A nineteenth-century practice aimed at curing persons with mental illness by treating them with kindness and giving them opportunities for both work and play.

morbidity: Symptoms, illnesses, injuries, or impairments.

mortality: Deaths.

mortification: A process, occurring in total institutions, through which a person's prior self-image is partially or totally destroyed and replaced by a personality suited for life in the institution. See *total institutions.*

mutual participation: Model of doctor-patient interaction in which doctor and patient are equal participants, with both assumed to have useful information regarding how to deal with a particular health problem. For example, doctors and patients might work together to establish the balance between diet and insulin for a person with diabetes.

national health insurance: A system in which all citizens of a country receive their health coverage from a single governmental insurance plan.

National Health Service (NHS): A system in which the government directly pays all costs of health care for its citizens.

naturalistic theories of illness: Theories tracing illness to heat, cold, wind, damp, or other natural events that upset the body's equilibrium.

negative social sanctions: Punishments meted out to those considered deviant by society. Negative social sanctions can range from ridicule and isolation to imprisonment and execution.

neonatal infant mortality: Deaths of infants during the first twenty-seven days after birth.

network model HMOs: HMOs that contract with multiple group medical practices to provide services to HMO patients. See *health maintenance organizations.*

new social movements: Groups of individuals who reject modern society's emphasis on science and rationality, value human interaction, and hope to create a more humane society primarily by living their lives in ways that reflect their ideals rather than through organized political activity.

NHS: See *National Health Service.*

normalize: Make something seem like the normal course of events. In the context of medical error, this refers to emphasizing how medical errors can happen to anyone. In the context of mental illness, this refers to explaining to oneself and others how unusual behavior is not really a sign of mental illness.

norms: Social expectations for appropriate behavior.

Nuremberg Code: A set of internationally recognized principles regarding the ethics of human experimentation that emerged during the post–World War II Nuremberg trials for medical crimes against humanity. The code stipulates that researchers must have a medically justifiable purpose, do all within their power to protect their subjects from harm, and ensure that their subjects give voluntary, informed consent.

nurse-midwives: Registered nurses who receive additional formal, nationally accredited training in midwifery.

nursing assistants: Individuals, often untrained, who provide basic custodial care for patients, most often in nursing homes and hospitals. See *nursing homes.*

nursing homes: Facilities that primarily provide nursing and custodial care to many individuals over a long period of time. Skilled nursing homes also provide some medical care.

outpatient: Hospital patient who is neither formally admitted nor kept overnight.

pandemic: A worldwide epidemic. See *epidemic.*

parallel practitioners: Those occupational groups, such as osteopaths, that perform basically the same roles as allopathic doctors while retaining occupational autonomy. See *allopathic doctors.*

pass: To hide one's deviance (such as illnesses or disabilities) from others.

passive euthanasia: When health care workers allow patients to die through inaction.

patient dumping: When voluntary or for-profit hospitals surreptitiously transfer to public hospitals those patients who cannot pay for their care.

performance norms: Socially accepted rules for how a person should perform his or her roles. For example, we expect mothers to keep their children clean and paid workers to arrive on time each day.

peripheral nations: According to world systems theory, those nations where modernization and industrialization have developed slowly, if at all, and the standard of living is generally low.

personalistic theories of illness: Theories holding that illness occurs when a god, witch, spirit, or other supernatural power deservedly or maliciously lashes out at an individual.

pharmaceutical care: The idea that pharmacists' central mission should be to advise consumers regarding the proper use of medications, based on pharmacists' knowledge of randomized controlled studies.

physician extenders: Health care providers who have less education than physicians but who can, at lower costs, take over some of the tasks traditionally done by physicians.

placebo: Anything offered as a cure that has no known biological effect. Approximately 30 percent of the time, placebos will produce cures through their psychological effects.

positive social sanctions: Rewards of any sort, from good grades to public esteem.

postneonatal infant mortality: Deaths of infants between the twenty-eighth day after birth and eleven months after birth.

practice protocols: Guidelines that establish norms of care for particular medical conditions under particular circumstances based on careful review of clinical research.

preferred provider organizations (PPOs): Health insurance plans in which doctors agree to charge lower, preset fees in exchange for the additional business, and consumers agree to obtain care from these doctors in exchange for lower premiums and deductibles.

prejudice: Unwarranted suspicion or dislike of individuals because they belong to a particular group.

prevalence: Total number of cases of an illness or health problem within a given population at a particular point in time (for example, the number of persons living in the United States who have Down syndrome). This includes both those first diagnosed that year and those diagnosed in previous years but still alive.

primary care: Health care provided by physicians (such as family care doctors) and others who are trained to offer treatment and prevention services when individuals first seek health care and, ideally, as part of an ongoing provider-patient relationship.

primary care doctors: Those doctors in family or general practice, internal medicine, and pediatrics who are typically the first doctors individuals see when they need medical care.

primary practice; primary practitioners: See *primary care doctors.*

primary prevention: Strategies designed to keep people from becoming ill, including vaccinating, using seat belts, encouraging exercise, and pasteurizing milk.

profession: An occupation that (1) has the autonomy to set its own educational and licensing standards and to police its members for incompetence or malfeasance; (2) has its own technical, specialized knowledge, learned through extended, systematic training; and (3) has the public's confidence that it follows a code of ethics and works more from a sense of service than a desire for profit.

professional dominance: A profession's freedom from control by other occupations or groups and ability to control other occupations working in the same sphere. Only priests, for example, can decide whether someone can become a priest, and priests control the training and work responsibilities of lay religious workers in their churches.

professional socialization: The process of learning the skills, knowledge, and values of an occupation.

professionalization: Process through which an occupation achieves professional status.

proletarianization: Process through which the status of members of an occupation declines from professionals to workers.

prospective reimbursement: For doctors, a system in which they are paid in advance a set fee per patient regardless of how many times they see that patient or what procedures they perform. For hospitals, a system in which the government pays hospitals a set amount for each Medicare or Medicaid patient based on the average cost of treating someone with that patient's diagnosis.

random samples: Samples selected in such a way that each member of a population has an equal chance of being selected. When a sample is randomly selected, we can be fairly certain that the selected individuals will represent the population as a whole well.

rates: Proportions of populations that experience certain circumstances.

rate spiral: A situation in which insurers raise prices, encouraging relatively healthy persons to risk going without coverage while relatively ill persons continue to purchase insurance out of necessity. As a result, the costs of providing coverage increase for insurers, leading them to raise prices still further and leading even more relatively healthy persons to drop their coverage.

RBRVS: See *resource-based relative value scale.*

Red Cross health workers: See *street doctors.*

reductionistic treatment: Treatment based on the assumption that each part can be treated separately from the whole, in the same way that an air filter can be replaced in a car without worrying whether the problem with the air filter has caused or stemmed from problems in the car's electrical system.

registered nurses (RNs): Individuals who have received at least two years of nursing training and passed national licensure requirements. In everyday conversation, the word *nurse* generally means "registered nurse."

regressive financing system: Any system in which low-income individuals pay a higher proportion of their income than high-income individuals pay.

regular doctors: Nineteenth-century forerunners of contemporary medical doctors. Also known as allopathic doctors. See *allopathic doctors.*

reliability: The likelihood that different people using the same measure will reach the same conclusions.

remedicalization: The process through which mental illness is increasingly regarded by doctors and others as rooted in biology and amenable only to biological treatments.

reprofessionalizing: Regaining former professional status.

reproductive technology: Medical developments that offer control over human conception and fetal development.

residents: Individuals who have graduated medical school and received their MD degrees, but who are now engaging in further on-the-job training needed before they can enter independent practice.

resource-based relative value scale (RBRVS): A complex formula designed to curb the costs of Medicare by limiting reimbursement to doctors to the estimated actual costs of services in a particular geographic area.

respite care: Any system designed to give family caregivers a break from their responsibilities.

restructuring of medicine: Concept that medicine as a profession has maintained its dominance by reorganizing specialties not by clinical territory but by functional sector.

retrospective reimbursement: A system in which insured individuals first receive care from health care providers and pay their bills, and then their insurance provider reimburses them for all or part of these costs.

right to die: The right to make decisions concerning one's own death.

RNs: See *registered nurses.*

role strain: Problems individuals experience within their major social roles as workers, parents, students, and so on.

secondary prevention: Strategies designed to reduce the prevalence of disease through early detection and prompt intervention, such as screening for diabetes and pap smears.

self-fulfilling prophecy: A situation in which individuals become what they are expected to be. For example, when it is assumed that no girls can throw a ball properly, girls might never be taught to do so, might never think it worth trying on their own, and thus would not be able to do so.

self-insuring: Putting aside a pool of money from which to pay all health care expenses for one's employees rather than contracting with an insurance firm.

semiperipheral nations: According to world systems theory, those nations whose economy, standard of living, and relative power place them at a level midway between the core and peripheral nations.

semiprofessional: Referring to those occupations that have achieved some but not all of the hallmarks of a profession.

sex: The biological categories of male and female, to which we are assigned based on our chromosomal structure, genitalia, hormones, and so on. Generally, individuals are considered male if they have XY sex chromosomes, and female if they have XX sex chromosomes.

sick role: The set of four social expectations in Western society regarding how society should view sick people and how sick people should behave. First, the sick person is considered to have a legitimate reason for not fulfilling his or her normal social role. Second, sickness is considered beyond individual control, something for which the individual is not held responsible. Third, the sick person must recognize that sickness is undesirable and work to get well. Fourth, the sick person should seek and follow medical advice.

single-payer system: A health care system in which a single government health insurance organization covers all residents of a nation.

snowballing: A process through which the perceived effect on life of each problematic behavior and emotion increases as the total number of problems increases. This process increases the odds that individuals will define the person experiencing the problems—whether self or other—as mentally ill.

social class: The combination of an individual's education, income, and occupational status or prestige; some researchers use only one of these indicators to measure social class, whereas others combine two or more indicators.

social construction: Ideas created by a social group, as opposed to something that is objectively or naturally given.

social control: Means used by a social group to ensure that individuals conform to the norms of that group. Social control can be formal (such as execution or commitment to a mental hospital) or informal (such as ridicule or shunning). See *norms*.

social control agents: Those individuals or groups of individuals who have the authority to enforce social norms, including parents, teachers, religious leaders, and doctors. See *norms*.

social drift theory: A theory holding that lower-class persons have higher rates of illness because middle-class persons who become ill drift over time into the lower class.

social epidemiology: The study of the distribution of disease within a population according to social factors (such as social class, use of alcohol, or unemployment) rather than biological factors (such as blood pressure or genetics).

social insurance: See *sickness funds*.

Social Security: Federally funded program that, since 1935, has provided financial assistance to any mentally or physically disabled adults who have held a job for a specified period of time, as well as to the elderly, the blind, and disabled children.

social stress theory: A theory holding that lower-class persons have higher rates of mental illness due to the stresses of lower-class life.

sociological model of disability: A model that defines disabilities as restrictions or lack of ability to perform activities resulting largely or solely either from social responses to bodies that fail to meet social expectations or from assumptions about the body reflected in the social or physical environment.

sociological perspective: A perspective regarding human life and society that focuses on identifying social patterns and grappling with social problems rather than on analyzing individual behavior and finding solutions for personal troubles.

sociology in medicine: An approach to the sociological study of health, illness, and health care that focuses on research questions of interest to doctors.

sociology of medicine: An approach that emphasizes using the area of health, illness, and health care to answer research questions of interest to sociologists in general. This approach often requires researchers to raise questions that could challenge medical views of the world and existing power relationships within the health care world.

sponsorship: Process through which successful professionals in a given field actively help new members establish their careers.

stakeholder mobilization: Organized political opposition or support by groups with vested interest in the outcome.

stereotypes: Oversimplistic assumptions regarding the nature of group members, such as assuming that black people are unintelligent.

stigma: Any personal attribute that would be deeply discrediting should it become known.

street doctors: Chinese health care workers with little formal training who work in urban outpatient clinics under the supervision of a doctor. Sometimes known as "Red Cross health workers," these workers offer primary and basic emergency care as well as health education, immunization, and assistance with birth control.

stress: Situations that make individuals feel anxious and unsure how to respond, the emotions resulting from exposure to such situations, or the bodily changes occurring in response to these situations and feelings.

stressor: Any specific source of stress, such as buying a new house or losing a job.

structure: The social forces around us, including cultural pressures, economic standing, gender expectations, presence of absence of resources (time, money, prestige), and so on. When used as an opposite to *agency*, refers to the concept that individual choices are limited by all these social forces.

superego: According to Freud, that portion of the personality that represents internalized ideas about right and wrong. Similar to the idea of a conscience.

symbolic interactionism: A theoretical perspective arguing that identity develops as part of an ongoing process of social interaction. Through this process, individuals learn to see themselves through the eyes of others, adopt the values of their community, and measure their self-worth against those values.

technological imperative: Belief that technology is always good, so any existing technological interventions should be used.

technology: Any human-made object used to perform a task, or a process using such objects. For example, the term *technology* can refer both to the overall process of kidney dialysis and to the specific pieces of equipment used in that process.

tertiary prevention: Strategies designed to minimize deterioration and complications among those who already have a disease.

third-party reimbursement: Payment from insurance programs for services rendered.

total institutions: Institutions in which all aspects of life are controlled by a central authority and in which large numbers of like-situated persons are dealt with en masse. Examples include mental hospitals, prisons, and the military.

tuberculosis: An infectious, airborne disease caused by the bacillus *Mycobacterium tuberculosis,* which attacks and destroys lung tissue. Can be fatal if not treated.

unintended negative consequences: Unplanned, harmful effects of actions that had been expected to produce only benefits.

universal coverage: Health care systems that provide access to health care for all legal residents of a nation.

utilization review: A system in which insurance companies require doctors to get approval before ordering certain tests, performing surgery, hospitalizing a patient, or keeping a patient hospitalized more than a given number of days.

validity: The likelihood that a given measure accurately reflects reality and measures what researchers believe it measures.

vertical integration: Situation in which a large corporation owns several institutions that provide different types of service within the health care field, such as both nursing homes and drug manufacturing companies.

veterans hospitals: Hospitals established by the U.S. federal government to serve the health needs of those who have served in the Armed Forces.

vigilance: Seeking knowledge so that one can feel able to respond appropriately and therefore feel more in control.

village doctors: Chinese agricultural workers who receive a few months of training in health care and provide basic health services to members of their agricultural production team. More rigorously trained than the "barefoot doctors" they replaced. See *barefoot doctors.*

voluntary hospitals: Hospitals that are financially based in voluntarism, or charity, rather than a profit motive. Same as nonprofit institutions.

WHO: See *World Health Organization.*

withdrawal: The predictable combination of distressing physical symptoms experienced by a person who stops using a drug to which he or she is addicted.

World Health Organization (WHO): United Nations organization charged with documenting health problems and improving world health.

world systems theory: Theory stating that the wealthier nations of the world have achieved and maintained their present economic position by exploiting the resources of the poorer nations.

Abbott, Andrew. 1988. *The System of Professions*. Chicago: University of Chicago Press.

Abel, Emily K. 1986. "The hospice movement: Institutionalizing innovation." *International Journal of Health Services* 16: 71–85.

———. 1990. "Family care of the frail elderly." Pp. 65–91 in *Circles of Care: Work and Identity in Women's Lives*, edited by Emily K. Abel and Margaret K. Nelson. Albany: State University of New York Press.

———. 2000. *Hearts of Wisdom: American Women Caring for Kin, 1850–1940*. Cambridge, MA: Harvard University Press.

Abel, Emily K., and Margaret K. Nelson. 1990. "Circles of care: An introductory essay." Pp. 4–34 in *Circles of Care: Work and Identity in Women's Lives*, edited by Emily K. Abel and Margaret K. Nelson. Albany: State University of New York Press.

Abelson, Reed, with Patricia Leigh Brown. 2002. "Alternative medicine is finding its niche in nation's hospitals." *New York Times* April 13: B1.

Abraham, Laurie Kaye. 1993. *Mama Might Be Better Off Dead: The Failure of Health Care in Urban America*. Chicago: University of Chicago Press.

Abramson, John. 2004. *Overdosed America: The Broken Promise of American Medicine*. New York: Harper Collins.

Acupuncture Alliance. April 2002. Personal communication.

Aday, Lu Ann. 2001. *At Risk in America: The Health and Health Care Needs of Vulnerable Populations in the United States*. 2nd ed. San Francisco: Jossey-Bass.

Aiken, Linda H., and Douglas M. Sloane. 1997. "Effects of specialization and client differentiation on the status of nurses: The case of AIDS." *Journal of Health and Social Behavior* 38: 203–222.

Aiken, Linda H., Julie Sochalski, and Gerard F. Anderson. 1996. "Downsizing the hospital nursing workforce." *Health Affairs* 15: 88–92.

Albrecht, Gary L. 1992. *The Disability Business: Rehabilitation in America*. Newbury Park, CA: Sage.

Allan Guttmacher Institute. June 2005. "An overview of abortion in the United States." http://www.agiusa.org/media/presskits/2005/06/28/abortionoverview .html, accessed November 2005.

Altman, Lawrence K. 1994. "Infectious diseases on the rebound in the U.S., a report says." *New York Times* May 10: B7.

Amaro, Hortensia. 1999. "An expensive policy: The impact of inadequate funding for substance abuse treatment." *American Journal of Public Health* 89: 657–659.

American Bar Association, Commission on Mental and Physical Disability Law. 1998. "Study finds employers win most ADA Title I judicial and administrative complaints." *Mental and Physical Disability Law Reporter* 22: 403–407.

American College of Nurse-Midwives. 1998. *Basic Facts about Certified Nurse-Midwives.* Washington, DC: author.

———. 2005. *State Laws Governing Direct Entry Midwifery.* Washington, DC: author.

American College of Obstetrician-Gynecologists. 1998. "Malpractice costs for ob-gyns continue to rise; Risks affecting physicians earlier in their career." Press release. March 30. Washington, DC: author.

American College of Physicians. 1998. "The physician workforce and financing of graduate medical education." *Annals of Internal Medicine* 128: 142–148.

American Hospital Association. 1998. *Hospital Statistics.* Chicago: author.

———. October 2002. Personal communication.

American Medical Association. July 2005a. Personal communication.

———. 2005b. "Women Physicians Congress: Table 4, Women residents by specialty, 2003." http://www.ama-assn.org/ama/pub/category/12915.html, accessed July 2005.

American Medical Student Association. 2005. "Promoting, Reinforcing and Improving Medical Education." http://www.amsa.org/programs/prime.cfm, accessed August 2005.

American Osteopathic Association. 2005. "Frequently Asked Questions." http://www.osteopathic.org/index.cfm?PageID=faq_cons#wheremany, accessed June 2005.

American Pain Society. 2000. *Pain Assessment and Treatment in the Managed Care Environment.* Glenview, IL: author.

American Pain Society and American Academy of Pain Medicine. 1996. *Consensus Statement on the Use of Opioids for the Treatment of Chronic Pain.* Glenview, IL: author.

American Society of Plastic and Reconstructive Surgeons. 1989 (July 1). *Comments on the Proposed Classification of Inflatable Breast Prosthesis and Silicone Gel Filled Breast Prosthesis.* Arlington Heights, IL: author.

Anderson, Robert N. 2001. "Deaths: Leading causes for 1999." *National Vital Statistics Report* 49(11).

Andrulis, Dennis P., Katherine L. Acuff, Kevin B. Weiss, and Ron J. Anderson. 1996. "Public hospitals and health care reform: Choices and challenges." *American Journal of Public Health* 86: 162–165.

Angell, Marcia. 2004. *The Truth About the Drug Companies: How They Deceive Us and What to Do About It.* New York: Random House.

Annandale, Ellen C. 1989. "The malpractice crisis and the doctor-patient relationship." *Sociology of Health and Illness* 11: 1–23.

Annas, George J. 1991. "Ethics committees: From ethical comfort to ethical cover." *Hastings Center Report* 21(May–June): 18–21.

———. 2005. "'Culture of life' politics at the bedside: The case of Terri Schiavo." *New England Journal of Medicine* 352: 1710–1715.

Anspach, Renee R. 1997. "The language of case presentation." Pp. 320–338 in *The Sociology of Health and Illness: Critical Perspectives,* 5th ed., edited by Peter Conrad and Rochelle Kern. New York: St. Martin's Press.

Apovian, Caroline M. 2004. "Sugar-sweetened soft drinks, obesity, and type 2 diabetes." *Journal of the American Medical Association* 291: 978–979.

Armstrong, Elizabeth M. 1998. "Diagnosing moral disorder: The discovery and evolution of fetal alcohol syndrome." *Social Science and Medicine* 47: 2025–2042.

———. 2000. "Lessons in control: Prenatal education in the hospital." *Social Problems* 47: 583–605.

Armstrong, Pat, and Hugh Armstrong. 1996. *Wasting Away: The Undermining of Canadian Health Care.* New York: Oxford University Press.

———. 1998. *Universal Health Care: What the United States Can Learn from the Canadian Experience.* New York: New Press.

Arno, Peter S., Karen Bonuck, and Robert Padgug. 1995. "The economic impact of high-tech home care." Pp. 220–234 in *Bringing the Hospital Home: Ethical and Social Implications of High-tech Home Care,* edited by John D. Arras. Baltimore: Johns Hopkins University Press.

Aronowitz, Robert A. 1998. *Making Sense of Illness: Science, Society, and Disease.* New York: Cambridge University Press.

Arras, John D., and Nancy Neveloff Dubler. 1995. "Ethical and social implications of high-tech home care." Pp. 1–34 in *Bringing the Hospital Home: Ethical and Social Implications of High-tech Home Care,* edited by John D. Arras. Baltimore: Johns Hopkins University Press.

Association of Asian Pacific Community Health Organizations. 1997. *Taking Action: Improving Access to Health Care for Asian Pacific Islanders.* Oakland, CA: author.

Astin, John A. 1998. "Why patients use alternative medicine: Results of a national study." *Journal of the American Medical Association* 279: 1548–1553.

Atkinson, Paul. 1984. "Training for certainty." *Social Science and Medicine* 19: 949–956.

Azevedo, Kathryn, and Hilda Ochoa Bogue. 2001. "Health and occupational risks of Latinos living in rural America." Pp. 359–380 in *Health Issues in the Latino Community,* edited by Marilyn Aguirre-Molina, Carlos W. Molina, and Ruth Enid Zambrana. San Francisco: Jossey-Bass.

Banister, Judith. 1999. *Son Preference in Asia: Report of a Symposium.* www.census. gov/ipc/www/ebspr96a.html, accessed March 2002.

Baquet, C. R., J. W. Horm, T. Gibbs, and P. Greenwald. 1991. "Socioeconomic factors and cancer incidence among blacks and whites." *Journal of the National Cancer Institute* 83: 551–557.

Barker, K. K. 1998. "A ship upon a stormy sea: The medicalization of pregnancy." *Social Science and Medicine* 47: 1067–1076.

Barker, Kristin. 2005. *The Fibromyalgia Story: Biomedical Authority and Women's Worlds of Pain.* Philadelphia: Temple University Press.

Barker-Benfield, Graham J. 1976. *The Horrors of the Half-Known Life: Male Attitudes Toward Women and Sexuality in Nineteenth Century America.* New York: Harper.

Barrow, Susan M., Daniel B. Herman, Pilar Cordova, and Elmer L. Struening. 1999. "Mortality among homeless shelter residents in New York City." *American Journal of Public Health* 89: 529–534.

Barstow, Anne Llewellyn. 1994. *Witchcraze: A New History of the European Witch Hunts.* San Francisco: Pandora.

Barzansky, Barbara, and Sylvia I. Etzel. 2005. "Educational programs in U.S. medical schools, 2004–2005." *Journal of the American Medical Association* 294: 1068–1074.

Basch, Paul F. 1999. *Textbook of International Health.* 2nd ed. New York: Oxford University Press.

Becker, Howard S., Blanche Geer, Everett C. Hughes, and Anselm Strauss. 1961. *Boys in White: Student Culture in Medical School.* Chicago: University of Chicago Press.

Becker, Marshall H. (ed.). 1974. *The Health Belief Model and Personal Health Behavior.* San Francisco: Society for Public Health Education.

———. 1993. "A medical sociologist looks at health promotion." *Journal of Health and Social Behavior* 34: 1–6.

Beecher, Henry K. 1966. "Ethics and clinical research." *New England Journal of Medicine* 274: 1354–1360.

Beeson, Paul B. 1980. "Changes in medical therapy during the past half century." *Medicine* 59: 79–99.

Belluck, Pam. 2002. "Doctors' new practices offer deluxe service for deluxe fee." *New York Times* January 15: A1.

Billings, Paul R., Mel A. Kohn, Margaret de Cuevas, Jonathan Beckwith, Joseph S. Alper, and Marvin R. Natowicz. 1992. "Discrimination as a consequence of genetic testing." *American Journal of Human Genetics* 50: 476–482.

Birenbaum, Arnold. 1982. "Reprofessionalization of pharmacy." *Social Science and Medicine* 16: 871–878.

Blackless, Melanie, Anthony Charuvastra, Amanda Derryck, Anne Fausto-Sterling, Karl Lauzanne, and Ellen Lee. 2000. "How sexually dimorphic are we? Review and synthesis." *American Journal of Human Biology* 12: 151–166.

Blair, Stephen N., and Timothy S. Church. 2004. "The fitness, obesity, and health equation: Is physical activity the common denominator?" *Journal of the American Medical Association* 292: 1232–1234.

Blalock, Kaija, and Sidney M. Wolfe. 2001. *Questionable Hospitals.* Washington, DC: Public Citizen.

Bland, Karina. 1999. "Tucson facility offers some women a new beginning." *Arizona Republic* September 19: A1+.

Blauner, Robert. 1972. *Racial Oppression in America.* New York: Harper & Row.

Block, Susan D., Nancy Clark-Chiarelli, Antoinette S. Peters, and Judith D. Singer. 1996. "Academia's chilly climate for primary care." *Journal of the American Medical Association* 276: 677–682.

Blue Cross and Blue Shield Association. 2005. "Healthcare claims top $206 billion for BCBS companies in 2004." http://onlinepressroom.net/bcbsa/, accessed August 2005.

Bodenheimer, Thomas. 1999. "The American health care system—physicians and the changing medical marketplace." *New England Journal of Medicine* 340(7): 584–588.

———. 2000. "Uneasy alliance: Clinical investigators and the pharmaceutical industry." *New England Journal of Medicine* 342: 1539–1543.

Boonstra, Heather, and Adam Sonfield. 2000. "Rights without access: Revisiting public funding of abortion for the poor." *Guttmacher Report on Public Policy* April: 8–11.

Borkan, Jeffrey, Jon O. Neher, Ofra Anson, and Bret Smoker. 1994. "Referrals for alternative therapies." *Journal of Family Practice* 39: 545–550.

Bosk, Charles L. 2003. *Forgive and Remember: Managing Medical Failure.* 2nd ed. Chicago: University of Chicago Press.

Boston Women's Health Book Collective. 2005. *Our Bodies, Ourselves: A New Edition for a New Era.* New York: Touchstone.

Bradshaw, David H., Yoshio Nakamura, and C. Richard Chapman. 2005. "National Institutes of Health Grant Awards for pain, nausea, and dyspnea research: An assessment of funding patterns in 2003." *Journal of Pain* 6: 277–293.

Brandt, Allan M., and Paul Rozin. 1997. *Morality and Health.* New York: Routledge.

Brannon, Robert L. 1996. "Restructuring hospital nursing: Reversing the trend toward a professional workforce." *International Journal of Health Services* 26: 643–654.

Brickner, Philip W., Linda Keen Scharer, Barbara A. Conanan, Marianne Savarese, and Brian C. Scanlan. 1990. *Under the Safety Net: The Health and Social Welfare of the Homeless in the United States.* New York: Norton.

Broadhead, Robert S., and Neil J. Facchinetti. 1985. "Drug iatrogenesis and clinical pharmacy: The mutual fate of a social problem and a professional movement." *Social Problems* 32: 425–436.

Brodie, Mollyann. 2001. *Understanding the Effects of Direct-to-Consumer Prescription Drug Advertising.* Washington, DC: Henry J. Kaiser Family Foundation.

Brooks, Nancy A., and Ronald R. Matson. 1987. "Managing multiple sclerosis." *Research in the Sociology of Health Care* 6: 73–106.

Brothwell, Don R. 1993. "Yaws." Pp. 1096–1100 in *Cambridge World History of Human Disease,* edited by Kenneth F. Kiple. New York: Cambridge University Press.

Brown, Phil. 1985. *The Transfer of Care: Psychiatric Deinstitutionalization and Its Aftermath.* Boston: Routledge & Kegan Paul.

———. 1990. "The name game: Toward a sociology of diagnosis." *Journal of Mind and Behavior* 11: 385–406.

Brown, Phil, Stephen Zavestoski, Sabrina McCormick, Brian Myer, Rachel Morello-Frosch, and Rebecca Gasior Altman. 2004. "Embodied health movements: New approaches to social movements in health." *Sociology of Health and Illness* 26: 50–80.

Brown, Theodore M., and Elizabeth Fee. 2001. "Preemptive biopreparedness: Can we learn anything from history?" *American Journal of Public Health* 91: 721–725.

Brumberg, Joan Jacobs. 1997. *The Body Project: An Intimate History of American Girls.* New York: Random House.

Brundin, Jennifer. 1993. "How the U.S. press covers the Canadian health care system." *International Journal of Health Services* 23: 275–277.

Bullard, Robert D. 1983. "Solid waste sites and the black Houston community." *Sociological Inquiry* 53: 273–288.

———. 1993. *Confronting Environmental Racism: Voices from the Grassroots.* Boston: South End.

Bullard, Robert D., Rueben C. Warren, and Glenn S. Johnson. 2001. Pp. 471–488 in *Health Issues in the Black Community,* edited by Ronald L. Braithwaite and Sandra E. Taylor, 2nd ed. San Francisco: Jossey-Bass.

Bunker, John P., Howard S. Frazier, and Frederick Mosteller. 1994. "Improving health: Measuring effects of medical care." *Milbank Quarterly* 72: 225–258.

Bureau of Labor Statistics, U.S. Department of Labor. 2004. *Occupational Outlook Handbook, 2004–2005.* Washington, DC: U.S. Government Printing Office.

Burns, Thomas J., Andrew I. Batavia, and Gerben DeJong. 1993. "The health insurance work disincentive for persons with disabilities." *Research in the Sociology of Health Care* 11: 57–68.

Bury, Michael. 1982. "Chronic illness as biographical disruption." *Sociology of Health and Illness* 4: 167–182.

Butler, Sandra, and Barbara Rosenblum. 1991. *Cancer in Two Voices.* San Francisco: Spinsters.

Butterfield, Fox. 1999. "Prisons brim with mentally ill, study finds." *New York Times* July 12: A10.

Caldwell, John C. 1993. "Health transition: The cultural, social, and behavioral determinants of health in the Third World." *Social Science and Medicine* 36: 125–135.

Caldwell, John C., Pat Caldwell, and Israel O. Orubuloye. 1992. "The family and sexual networking in sub-Saharan Africa: Historical regional differences and present day implications." *Population Studies* 46: 385–392.

Caldwell, John C., Pat Caldwell, and Pat Quiggin. 1989. "The social context of AIDS in sub-Saharan Africa." *Population Development Review* 15: 185–234.

Caldwell, John C., Israel O. Orubuloye, and Pat Caldwell. 1991. "The destabilization of the traditional Yoruba sexual system." *Population Development Review* 17: 229–262.

Callahan, Daniel. 1998. *False Hopes: Why America's Quest for Perfect Health Is a Recipe for Failure.* New York: Simon & Schuster.

Camacho, David E. (ed.). 1998. *Environmental Injustices, Political Struggles: Race, Class, and the Environment.* Durham, NC: Duke University Press.

Campos, Paul. 2004. *The Obesity Myth: Why America's Obsession with Weight Is Hazardous to Your Health.* New York: Gotham Books.

Cancian, Francesca M., and Stacey J. Oliker. 2000. *Caring and Gender.* Thousand Oaks, CA: Pine Forge.

Casper, Monica J. 1998. *The Making of the Unborn Patient: A Social Anatomy of Fetal Surgery.* New Brunswick, NJ: Rutgers University Press.

Castle, Nicholas G., and Vincent Mor. 1998. "Physical restraints in nursing homes: A review of the literature since the Nursing Home Reform Act of 1987." *Medical Care Research and Review* 55: 139–170.

Center for Responsive Politics. 2005. "2004 Election Overview: Top Industries." http://opensecrets.org/overview/industries.asp?cycle=2004, accessed June 2005.

Center for the Evaluative Clinical Sciences, Dartmouth Medical School. 1996. *The Dartmouth Atlas of Health Care.* Chicago: American Hospital Association.

Centers for Disease Control and Prevention. 1998. *National Diabetes Fact Sheet: National Estimates and General Information on Diabetes in the United States.* Rev. ed. Atlanta: author.

———. 2004. *HIV/AIDS Surveillance Report* 15(1), Table 17.

———. 2005. "Overweight and Obesity: Home." http://www.cdc.gov/nccdphp/dnpa/obesity, accessed May 2005.

Chambliss, Daniel F. 1996. *Beyond Caring: Hospitals, Nurses, and the Social Organization of Ethics.* Chicago: University of Chicago Press.

Champaneria, Manish C., and Sara Axtell. 2004. "Cultural competence training in U.S. medical schools." *Journal of the American Medical Association* 291:2142.

Chapin, Rosemary, and Debra Dobbs-Kepper. 2001. "Aging in place in assisted living: Philosophy versus policy." *Gerontologist* 41: 43–50.

Charmaz, Kathy. 1991. *Good Days, Bad Days: The Self in Chronic Illness and Time.* New Brunswick, NJ: Rutgers University Press.

Chase-Dunn, Christopher. 1989. *Global Formation: Structure of the World Economy.* London: Basil Blackwell.

Chasnoff, Ira J., Harvey J. Landress, and Mark E. Barrett. 1990. "The prevalence of illicit drug use during pregnancy and discrepancies in mandatory reporting in Pinellas County, Florida." *New England Journal of Medicine* 322: 1202–1206.

Chavis, Benjamin F. 1993. "Foreword." Pp. 3–5 in *Confronting Environmental Racism: Voices from the Grassroots,* edited by Robert D. Bullard. Boston: South End.

Chen, Meei-Shia. 2001. "The great reversal: Transformation of health care in the People's Republic of China." Pp. 456–482 in *The Blackwell Companion to Medical Sociology,* edited by William C. Cockerham. Malden, MA: Blackwell.

Cherkin, Daniel C., Richard A. Deyo, Michele Battié, Janet Street, and William Barlow. 1998. "A comparison of physical therapy, chiropractic manipulation, and provision of an educational booklet for the treatment of patients with low back pain." *New England Journal of Medicine* 339: 1021–1029.

Cherry, Mark J. 2005. *Kidney for Sale by Owner: Human Organs, Transplantation, and the Market.* Washington DC: Georgetown University Press.

Christakis, Dmitri A., and Christopher Feudtner. 1993. "Ethics in a short white coat: The ethical dilemmas that medical students confront." *Academic Medicine* 68: 249–254.

———. 1997. "Temporary matters: The ethical consequences of transient social relationships in medical training." *Journal of American Medical Association* 278: 739–743.

Claiborne, William. 1999. "Indian health chief crusades for his people: In family tradition, Trujillo seeks better care for natives." *Washington Post* March 29: A17+.

Cockerham, William. 2005. "Health lifestyle theory and the convergence of agency and structure." *Journal of Health and Social Behavior* 46: 51–67.

Cohen, Hillel W., Robert M. Gould, and Victor W. Sidel. 2001. "Good intentions and the road to bioterrorism preparedness." *American Journal of Public Health* 91: 716–717.

Collins, Michael J. 2005. *Hot Lights, Cold Steel: Life, Death and Sleepless Nights in a Surgeon's First Years.* New York: St. Martin's Press.

Conrad, Peter. 1985. "The meaning of medications: Another look at compliance." *Social Science and Medicine* 20: 29–37.

———. 1987. "The experience of illness: Recent and new directions." *Research in the Sociology of Health Care* 6: 1–32.

———. 1997. "Public eye and private genes: Historical frames, news construction, and social problems." *Social Problems* 44: 139–154.

———. 2005. "The shifting engines of medicalization. *Journal of Health and Social Behavior* 46: 3–14.

Conrad, Peter, and Deborah Potter. 2004. "Human growth hormone and the temptations of biomedical enhancement." *Sociology of Health and Illness* 26: 184–215.

Conrad, Peter, and Joseph W. Schneider. 1992. *Deviance and Medicalization: From Badness to Sickness.* Philadelphia: Temple University Press.

Consumer Reports. 1992. "Health care in crisis: The search for solutions." 57: 579–592.

———. 1994. "Chiropractors." 59: 383–390.

———. 1996. "Drug advertising: Is this good medicine?" 61(6): 62–63.

———. 2001. "Is assisted living the right choice?" January: 26–31.

Contexts. 2004. "Assisted suicide." *Contexts* 3(3): 58.

Coombs, Robert H., Sangeeta Chopra, Debra R. Schenk, and Elaine Yutan. 1993. "Medical slang and its functions." *Social Science and Medicine* 36: 987–998.

Corbin, Juliet M., and Anselm Strauss. 1987. "Accompaniments of chronic illness: Changes in body, self, biography, and biographical time." *Research in the Sociology of Health Care* 6: 249–282.

———. 1988. *Unending Work and Care: Managing Chronic Illness at Home.* San Francisco: Jossey-Bass.

Costello, Anthony, and Harshpal S. Sachdev. 1998. "Protecting breast feeding from breast milk substitutes." *British Medical Journal* 316: 1103–1104.

Costello, Timothy W., and Joseph T. Costello. 1992. *Abnormal Psychology.* New York: Harper Perennial.

Council on Ethical and Judicial Affairs, American Medical Association. 1990. "Black-white disparities in health care." *Journal of the American Medical Association* 263: 2344–2346.

———. 1991. "Gender disparities in clinical decision making." *Journal of the American Medical Association* 266: 559–562.

Council on Scientific Affairs, American Medical Association. 1992. "Violence against women." *Journal of the American Medical Association* 267: 3184–3189.

Cousineau, Michael R. 1997. "Health status of and access to health services by residents of urban encampments in Los Angeles." *Journal of Health Care for the Poor and Underserved* 8: 70–82.

Crawford, Robert. 1979. "Individual responsibility and health politics." Pp. 247–268 in *Health Care in America: Essays in Social History,* edited by Susan Reverby and David Rosner. Philadelphia: Temple University Press.

Crisler, Greg. 2003. *Fat Land: How Americans Became the Fattest People in the World.* New York: Houghton Mifflin.

Crosby, Alfred J. 1986. *Ecological Imperialism: The Biological Expansion of Europe, 900–1900.* New York: Cambridge University Press.

Crossette, Barbara. 2001. "Wars enlist young legions, report says." *New York Times* June 14: A14.

Cutler, David, and Grant Miller. 2005. "The role of public health improvements in health advances: The twentieth-century United States." *Demography* 42: 1–22.

Daniell, William Edward, Charles David Treser, and Donald Clark Wetter. 2001. "Hospital Preparedness for Victims of Chemical or Biological Terrorism." *American Journal of Public Health* 91: 710–715.

Daniels, Cynthia R. 1993. *At Women's Expense: State Power and the Politics of Fetal Rights.* Cambridge, MA: Harvard University Press.

D'Arcy, Carl, and Joan Brockman. 1976. "Changing public recognition of psychiatric symptoms? Blackfoot revisited." *Journal of Health and Social Behavior* 17: 302–310.

Davis, Fred. 1961. "Deviance disavowal: Management of strained interaction by the visibly handicapped." *Social Problems* 9: 120–132.

DeCew, Judith Wagner. 1994. "Drug testing: Balancing privacy and public safety." *Hastings Center Report* 24(2): 17–23.

DeFrances, Carol J., and Margaret J. Hall. 2004. "2002 National Health Discharge Survey." *Advance Data from Vital and Health Statistics.* No. 342.

DeJong, Gerben, Andrew I. Batavia, and Robert Griss. 1989. "America's neglected health minority: Working-age persons with disabilities." *Milbank Quarterly* 67(suppl. 2): 311–351.

De Lew, Nancy, George Greenberg, and Kraig Kinchen. 1992. "A layman's guide to the U.S. health care system." *Health Care Financing Review* 14: 151–165.

DeNavas-Walt, Carmen, Bernadette D. Proctor, and Robert J. Mills. 2004. *Current Population Reports,* P60-226, "Income, Poverty, and Health Insurance Coverage in the United States: 2003." Washington, DC: U.S. Government Printing Office.

Dettwyler, Katherine A. 1995. "Beauty and the breast." Pp. 167–213 in *Breastfeeding: Biocultural Perspectives*, edited by Patricia Stuart-Macadam and Katherine A. Dettwyler. New York: Aldine De Gruyter.

Dey, Achintya N. 1997. "Characteristics of elderly nursing home residents: Data from the 1995 National Nursing Home Survey." *Advance Data* 289.

Diamond, Timothy. 1992. *Making Gray Gold: Narratives of Nursing Home Care.* Chicago: University of Chicago Press.

Diller, Lawrence H. 1998. *Running on Ritalin: A Physician Reflects on Children, Society, and Performance in a Pill.* New York: Bantam.

DiMaggio, Paul J., and Walter W. Powell. 1983. "The iron cage revisited: Institutional isomorphism and collective rationalizing in organizational fields." *American Sociological Review* 48: 147–160.

Dixon, Mim, and Yvette D. Roubideaux. 2001. *Promises to Keep: Public Health Policy for American Indians and Alaska Natives in the 21st Century.* Washington, DC: American Public Health Association.

Dobash, Russell P., and Rebecca Emerson Dobash. 1998. *Rethinking Violence Against Women.* Newbury Park, CA: Sage.

Dolenc, Danielle A., and Charles J. Dougherty. 1985. "DRGs: The counterrevolution in financing health care." *Hastings Center Report* 15(3): 19–29.

Donelan, Karen, Robert J. Blendon, John Benson, Robert Leitman, and Humphrey Taylor. 1996. "All payer, single payer, managed care, no payer: Patients' perspectives in three nations." *Health Affairs* 15: 254–265.

Draper, Debra A., Robert E. Hurley, Cara S. Lesser, and Bradley C. Strunk. 2002. "The changing face of managed care." *Health Affairs* 21: 11–23.

Dreze, Jean, and Amartya Sen. 1989. *Hunger and Public Action.* Oxford, UK: Clarendon.

Dubos, Rene. 1961. *Mirage of Health.* New York: Anchor.

Dunn, Kyla. 2002. "Cloning Trevor." *Atlantic Monthly* 289(6): 31–53.

Durán, Luis. 2002. Mexican Institute of Social Security, personal communication, January.

Durán-Arenas, Luis, Malaquias Lopéz-Cervantes, Octavio Gómez-Dantes, and Sandra Sosa-Rubí. 2002. "The unbearable homogeneity of reform: The Mexican health care system reform." Pp. 241–266 in *Health Care Reform Around the World*, edited by Andrew C. Twaddle. Westport, CT: Auburn House.

Eaton, William W., and Carles Muntaner. 1999. "Social stratification and mental disorder." In *A Handbook for the Study of Mental Health: Social Contexts, Theories, and Systems,* edited by Allan V. Horwitz and Teresa L. Scheid. Cambridge, UK: Cambridge University Press.

Eckholm, Erik. 2002. "Desire for sons drives use of prenatal scans in China." *New York Times* June 21: A3.

Economic Report of the President. 1993. Washington, DC: U.S. Government Printing Office.

Eisenberg, David M., Roger B. Davis, Susan L. Ettner, Scott Appel, Sonja Wilkey, Maria Van Rompay, and Ronald C. Kessler. 1998. "Trends in alternative medicine use in the United States, 1990–1997: Results of a follow-up national survey." *Journal of the American Medical Association* 280: 1569–1575.

Elliott, Carl. 2004. "Pharma goes to the laundry: Public relations and the subject of medical education." *Hastings Center Review* 34: 18–23.

Emanuel, Ezekiel J., Elisabeth R. Daniels, Diane L. Fairclough, and Brian R. Clarridge. 1998. "The practice of euthanasia and physician-assisted suicide in the United States: Adherence to proposed safeguards and the effects on physicians." *Journal of the American Medical Association* 280: 507–513.

Engelhardt, H. Tristram Jr. 1986. *Foundations of Bioethics.* New York: Oxford University Press.

Ensel, Walter M., and Nan Lin. 1991. "The life stress paradigm and psychological distress." *Journal of Health and Social Behavior* 32: 321–341.

Epstein, Steven. 1996. *Impure Science: AIDS, Activism, and the Politics of Knowledge.* Berkeley: University of California Press.

Equal Employment Opportunity Commission. 1999. *Americans with Disabilities Act of 1990 (ADA) Charges, FY 1992–FY 1998.* www.eeoc.gov/stats/ada.html

Fadiman, Anne. 1997. *The Spirit Catches You and You Fall Down: A Hmong Child, Her American Doctors, and the Collision of Two Cultures.* New York: Farrar, Straus and Giroux.

Farmer, Paul. 1999. *Infections and Inequalities: The Modern Plagues.* Berkeley: University of California Press.

Feagin, Joe R., and Melvin P. Sikes. 1994. *Living with Racism: The Black Middle-Class Experience.* Boston: Beacon.

Federal Interagency Forum on Child and Family Statistics. 1999. *America's Children 1999.* Washington, DC: U.S. Government Printing Office.

Federal Trade Commission. 1999. *Self-Regulation in the Alcohol Industry.* Washington, DC: U.S. Government Printing Office.

Feinstein, Jonathan S. 1993. "The relationships between socioeconomic status and health: A review of the literature." *Milbank Quarterly* 71: 279–322.

Feshbach, Morris. 1999. "Dead souls." *Atlantic Monthly* 283(1): 26–27.

Feshbach, Morris, and Alfred Friendly. 1992. *Ecocide in the USSR: Health and Nature Under Siege.* New York: Basic.

Figert, Anne E. 1996. *Women and the Ownership of PMS: The Structuring of a Psychiatric Disorder.* New York: Aldine De Gruyter.

Fine, Michelle, and Adrienne Asch. 1988a. "Disability beyond stigma: Social interaction, discrimination, and activism." *Journal of Social Issues* 44: 3–21.

———. 1988b. "Introduction: Beyond pedestals." Pp. 1–38 in *Women with Disabilities: Essays in Psychology, Culture, and Politics,* edited by Michelle Fine and Adrienne Asch. Philadelphia: Temple University Press.

Finn Paradis, Lenora, and Scott B. Cummings. 1986. "The evolution of hospice in America toward organizational homogeneity." *Journal of Health and Social Behavior* 27: 370–386.

Fisher, Barbara, Mel Hovell, C. Richard Hofstetter, and Richard Hough. 1995. "Risks associated with long-term homelessness among women: Battery, rape, and HIV infection." *International Journal of Health Services* 25: 351–369.

Fisher, Elliott S., David E. Wennberg, Therese A. Stukel, Daniel J. Gottlieb, F. L. Lucas, and Etoile L. Pinder. 2003. "Implications of regional variation in Medicare spending." *Annals of Internal Medicine* 138: 273–287.

Fisher, Ian. 1998. "Families provide medical care, tubes and all." *New York Times* June 7: 1+.

Fisher, Sue. 1986. *In the Patient's Best Interest: Women and the Politics of Medical Decisions.* New Brunswick, NJ: Rutgers University Press.

———. 1995. *Nursing Wounds: Nurse Practitioners, Doctors, Women Patients and the Negotiation of Meaning*. New Brunswick, NJ: Rutgers University Press.

Flegal, Katherine M., Barry I. Graubard, David F. Williamson, and Mitchell H. Gail. 2005. "Excess Deaths Associated with Underweight, Overweight, and Obesity." *Journal of the American Medical Association* 293:1861–1867.

Foner, Nancy. 1994. *The Caregiving Dilemma: Work in an American Nursing Home*. Berkeley: University of California Press.

Fortune Magazine. 1999. "Why drug companies fear and loathe pharmacists: The *real* power in health care." March 29: 46–47.

Fost, Norman, and Ronald E. Cranford. 1985. "Hospital ethics committees: Procedural aspects." *Journal of the American Medical Association* 253: 2687–2692.

Foster, George. 1976. "Disease etiologies in non-western medical systems." *American Anthropologist* 78: 773–782.

Fox, Margery. 1989. "The socioreligous role of the Christian Science practitioner." Pp. 98–114 in *Women as Healers: Cross-Cultural Perspectives*, edited by Carol Shepherd McClain. New Brunswick, NJ: Rutgers University Press.

Fox, Renee C. 1974. "Ethical and existential developments in contemporaneous American medicine: Their implications for culture and society." *Milbank Memorial Fund Quarterly* 52: 445–483.

———. 1977. "The medicalization and demedicalization of American society." *Daedalus* 106: 9–22.

———. 2000. "Medical uncertainty revisited." Pp. 409–425 in *The Handbook of Social Studies in Health and Medicine*, edited by Gary L. Albrecht, Ray Fitzpatrick, and Susan C. Scrimshaw. Thousand Oaks, CA: Sage.

Fox, Renee C., and Judith Swazey. 1974. *The Courage to Fail*. Chicago: University of Chicago Press.

Frank, Deborah A., Marilyn Augustyn, Wanda Grant Knight, Tripler Pell, and Barry Zuckerman. 2001. "Growth, development, and behavior in early childhood following prenatal cocaine exposure: A systematic review." *Journal of the American Medical Association* 285: 1613–1625.

Frank, Richard G., and Thomas G. McGuire. 1998. "The economics of behavioral health carve-outs." Pp. 41–50 in *Managed Behavioral Health Care: Current Realities and Future Potential*, edited by David Mechanic. San Francisco: Jossey-Bass.

Franks, Peter, Carolyn M. Clancy, and Martha R. Gold. 1993. "Health insurance and mortality." *Journal of the American Medical Association* 270: 737–741.

Fraser, Ross. October 2002. American Medical Association. Personal communication.

Freidson, Eliot. 1970a. *Profession of Medicine: A Study of the Sociology of Applied Knowledge*. New York: Dodd, Mead.

———. 1970b. *Professional Dominance: The Social Structure of Medical Care*. New York: Atherton.

———. 1975. *Doctoring Together: A Study of Professional Social Control*. New York: Elsevier.

———. 1984. "The changing nature of professional control." *Annual Review of Sociology* 10: 1–20.

———. 1985. "The reorganization of the medical profession." *Medical Care Review* 42: 11–35.

———. 1986. "The medical profession in transition." Pp. 63–79 in *Applications of Social Science to Clinical Medicine and Health Policy*, edited by Linda Aiken and David Mechanic. New Brunswick, NJ: Rutgers University Press.

———. 1994. *Professionalism Reborn*. Chicago: University of Chicago Press.

French, S. A., M. Story, D. Neumark-Sztainer, J. A. Fulkerson, and P. Hannan. 2001. "Fast food restaurant use among adolescents: Associations with nutrient intake, food choices and behavioral and psychosocial variables." *International Journal of Obesity* 25: 1823–1833.

Fries, James F. 2004. *Take Care of Yourself: The Complete Illustrated Guide to Medical Self-Care*. New York: Da Capo.

Freud, Sigmund. 1925 [1971]. *The Standard Edition of the Complete Psychological Works of Sigmund Freud*. Volume 19. London: Hogarth.

Fulder, Stephen. 1984. *Handbook of Complementary Medicine*. London: Hodder and Stoughton.

Garrett, Daniel. March 2002. American Pharmaceutical Association Foundation. Personal communication.

Garrett, Laurie. 1994. *The Coming Plague: Newly Emerging Diseases in a World Out of Balance*. New York: Farrar, Straus & Giroux.

———. 2000. *Betrayal of Trust: The Collapse of Global Public Health*. New York: Hyperion.

Gaston, Robert S., Ian Ayres, Laura G. Dooley, and Arnold G. Dietheim. 1993. "Racial equity in renal transplantation: The disparate impact of HLA-based allocation." *Journal of the American Medical Association* 270: 1352–1356.

Geiger, H. Jack, and Robert M. Cook-Deegan. 1993. "The role of physicians in conflicts and humanitarian crises: Case studies from the field missions of Physicians for Human Rights, 1988 to 1993." *Journal of the American Medical Association* 270: 616–620.

General Social Survey. 2002. www.icpsr.umich.edu/GSS, accessed April 2002.

Gerber, J. 1990. "Enforced self-regulation in the infant formula industry." *Social Justice* 17: 98–112.

Gerberding, Julie L., and James S. Marks. 2004. "Making America fit and trim." *American Journal of Public Health* 94: 1478–1479.

Gething, Lindsay. 1992. "Judgments by health professionals of personal characteristics of people with a visible physical disability." *Social Science and Medicine* 34: 809–815.

Gevitz, Norman. 1988. "Osteopathic medicine: From deviance to difference." Pp. 124–156 in *Other Healers: Unorthodox Medicine in America*, edited by Norman Gevitz. Baltimore: Johns Hopkins University Press.

Gibbs, W. Wayt. 2005. "Obesity: An overblown epidemic?" *Scientific American* 292(6): 70–77.

Gilman, Sander L. 1999. *Making the Body Beautiful: A Cultural History of Aesthetic Surgery*. Princeton, NJ: Princeton University Press.

Givel, Michael, and Stanton A. Glantz. 2004. "The 'global settlement' with the tobacco industry: 6 years later." *American Journal of Public Health* 94: 218–229.

Glasser, Ronald J. 2005. "A war of disabilities: Iraq's hidden costs are coming home." *Harper's Magazine* 311(1862): 59–62.

Glazer, Nona Y. 1993. *Women's Paid and Unpaid Labor: The Work Transfer in Health Care and Retailing*. Philadelphia: Temple University Press.

Goffman, Erving. 1961. *Asylums*. Garden City, NY: Doubleday.

———. 1963. *Stigma: Notes on the Management of Spoiled Identity.* Englewood Cliffs, NJ: Prentice Hall.

Goleman, Daniel. 1995. "Making room on the couch for culture." *New York Times* December 5: C1+.

Gonzalez, Daniel. 2005. "Guatemalans tops at filling area roofing jobs." *Arizona Republic* May 31: 1+.

Good, Mary-Jo DelVecchio. 1995. *American Medicine: The Quest for Competence.* Berkeley: University of California Press.

Gordon, Rena. 1989. "The effects of malpractice insurance on certified nurse-midwives." *Journal of Nurse-Midwifery* 35: 99–106.

Gordon, Suzanne. 2005. *Nursing Against the Odds: How Health Care Cost Cutting, Media Stereotypes, and Medical Hubris Undermine Nurses and Patient Care.* Ithaca, NY: Cornell University Press.

Gostin, Lawrence O., Chai Feldblum, and David W. Webber. 1999. "Disability discrimination in America: HIV/AIDS and other health conditions." *Journal of the American Medical Association* 281: 745–752.

Gostin, Lawrence O., Zita Lazzarini, T. Stephen Jones, and Kathleen Flaherty. 1997. "Prevention of HIV/AIDS and other blood-borne diseases among injection drug users: A national survey on the regulation of syringes and needles." *Journal of the American Medical Association* 277: 53–62.

Gottfried, Robert S. 1983. *The Black Death.* New York: Free Press.

Goyal, Madhav, Ravindra L. Mehta, Lawrence J. Schneiderman, and Ashwini R. Sehgal. 2002. "Economic and health consequences of selling a kidney in India." *Journal of the American Medical Association* 288: 1589–1593.

Gray, Bradford. 1991. *The Profit Motive and Patient Care.* Cambridge, MA: Harvard University Press.

Green, Carolyn J., Arminée Kazanjian, and Diane Helmer. 2004. "Informing, advising, or persuading? An assessment of bone mineral density testing information from consumer health websites." *International Journal of Technology Assessment in Health Care* 20:156–166.

Greenberg, Brigitte. 1999. "Study questions use of growth hormones." *Arizona Republic* March 14: A27.

Greenberg, Michael R. 1987. "Health and risk in urban-industrial society." Pp. 3–24 in *Public Health and the Environment: The United States Experience,* edited by Michael R. Greenberg. New York: Guilford.

Greenfield, S., W. Rogers, M. Mangotich, M. F. Carney, and A. R. Tarlov. 1995. "Outcomes of patients with hypertension and non-insulin dependent diabetes mellitus treated by different systems and specialties." *Journal of the American Medical Association* 274: 1436–1444.

Greenhouse, Steven. 1999. "AMA's delegates decide to create union of doctors." *New York Times* June 24: A1+.

———. 2001. "Fear and poverty sicken many migrant workers in U.S." *New York Times* May 13: A14.

Grob, Gerald N. 1997. "Deinstitutionalization: The illusion of policy." *Journal of Policy History* 9: 48–73.

Gwyther, Marni E., and Melinda Jenkins. 1998. "Migrant farmworker children: Health status, barriers to care, and nursing innovations in health care delivery." *Journal of Pediatric Health Care* 12: 60–66.

Haan, M., G. Kaplan, and T. Camacho. 1987. "Poverty and health: Prospective evidence from the Alameda County study." *American Journal of Epidemiology* 125: 989–998.

Haas, Jack, and William Shaffir. 1987. *Becoming Doctors: The Adoption of a Cloak of Competence.* Greenwich, CT: JAI.

Habermas, Jürgen. 1981. "New Social Movements." *Telos* 49: 33–37.

Hafferty, Frederic W. 1991. *Into the Valley: Death and the Socialization of Medical Students.* New Haven, CT: Yale University Press.

———. 1998. "Beyond curriculum reform: Confronting medicine's hidden curriculum." *Academic Medicine* 73: 403–407.

Hafferty, Frederic W., and Ronald Franks. 1994. "The hidden curriculum, ethics teaching, and the structure of medical education." *Academic Medicine* 69: 861–871.

Hahn, Harlan. 1985. "Toward a politics of disability definitions, disciplines, and policies." *Social Science Journal* 22 (October): 87–105.

Hall, Oswald. 1949. "Types of medical careers." *American Journal of Sociology* 55: 243–253.

Hall, Wayne. 1986. "Social class and survival on the S.S. *Titanic.*" *Social Science and Medicine* 22: 687–690.

Halpern, S. A. 1990. "Medicalization as a professional process: Postwar trends in pediatrics." *Journal of Health and Social Behavior* 31: 28–42.

Hamilton, Brady E., Joyce A. Martin, and Paul D. Sutton. 2004. "Births: Preliminary Data for 2003." *National Vital Statistics Reports* 53(9).

Hammond, Ross. 1998. *Addicted to Profit: Big Tobacco's Expanding Global Reach.* Washington, DC: Essential Action.

Hancock, Lyn Nell. 2002. *Hands to Work: The Stories of Three Families Racing the Welfare Clock.* New York: William Morrow.

Harlan, Sharon L., and Pamela M. Robert. 1998. "The social construction of disability in organizations: Why employers resist reasonable accommodation." *Work and Occupations* 25: 397–435.

Harrington, Charlene, Steffie Woolhandler, Joseph Mullan, Helen Carrillo, and David U. Himmelstein. 2001. "Does investor ownership of nursing homes compromise the quality of care?" *American Journal of Public Health* 91: 1452–1455.

Harris, Gardiner. 2004. "2 cancer drugs, no comparative data." *New York Times* February 26: C1+.

Harris Poll. 2004a. "Email, news, and weather top the list of how people use the Internet." Number 98.

———. 2004b. "Doctors, scientists top list of most prestigious occupations." Number 65.

———. 2005. "Overall confidence in leaders of major institutions declines slightly." Number 21.

Haug, Marie. 1988. "A re-examination of the hypothesis of physician deprofessionalization." *Milbank Quarterly* 66(supplement 2): 48–56.

Hayward, Mark D., and Melonie Heron. 1999. "Racial inequality in active life among adult Americans." *Demography* 36: 77–91.

Health Care Financing Administration. 2000. *Medicare 2000: 35 Years of Improving Americans' Health and Security.* Washington, DC: author.

HealthGrades. 2004. *Patient Safety in American Hospitals.* Lakewood, CO: author.

Heath, Christian, Paul Luff, and Marcus Sanchez Svensson. 2003. "Technology and medical practice." *Sociology of Health and Illness* 25: 75–96.

Helzer, John L., Lee N. Robin, Mitchell Taibleson, Robert A. Woodruff, Theodore Reich, and Eric D. Wish. 1977. "Reliability of psychiatric diagnosis: A methodological review." *Archives of General Psychiatry* 34: 129–133.

Hendlin, Herbert, Chris Rutenfrans, and Zbigniew Zylicz. 1997. "Physician-assisted suicide and euthanasia in the Netherlands: Lessons from the Dutch." *Journal of the American Medical Association* 277: 1720–1722.

Henshaw, Stanley K., and Lawrence B. Finer. 2003. "The accessibility of abortion services in the United States, 2001." *Perspectives on Sexual and Reproductive Health* 35: 16–24.

Henshaw, Stanley K., Susheela Singh, and Taylor Haas. 1999. "The incidence of abortion worldwide." *International Family Planning Perspectives* 25(supplement): S30–38.

Hepler Charles D., and Linda M. Strand. 1990. "Opportunities and responsibilities in pharmaceutical care." *American Journal of Hospital Pharmacy* 47: 533–543.

Higgins, Paul C. 1992. *Making Disability: Exploring the Social Transformation of Human Variation.* Springfield, IL: Charles C. Thomas.

Hilts, Philip J. 1996. *Smokescreen: The Truth Behind the Tobacco Industry Cover-up.* Reading, MA: Addison-Wesley.

———. 1999. "In tests on people, who watches the watchers?" *New York Times* May 25: D1+.

Himmelstein, David U., and Steffie Woolhandler. 1994. *The National Health Program Book: A Source Guide for Advocates.* Monroe, ME: Common Courage.

———. 2003. "National Health Insurance or Incremental Reform: Aim high or at our feet?" *American Journal of Public Health* 93: 102–105.

Himmelstein, David U., Steffie Woolhandler, Ida Hellander, and Sidney M. Wolfe. 1999. "Quality of care in investor-owned versus not-for-profit HMOs." *Journal of the American Medical Association* 282: 159–163.

Hintz, Raymond L., Kenneth M. Attie, Joyce Baptista, and Alex Roche. 1999. "Effect of growth hormone treatment on adult height of children with idiopathic short stature." *New England Journal of Medicine* 340: 502–507.

Hochschild, Arlie. 1983. *The Managed Heart: The Commercialization of Human Feelings.* Berkeley: University of California Press.

Hoffman, Diane E., and Anita J. Tarzian. 2001. "The girl who cried pain: A bias against women in the treatment of pain." *Journal of Law, Medicine, and Ethics* 29: 13–27.

Holmberg, Lars, Anna Bill Axelson, Fred Helgesen, Jaakko O. Salo, Per Folmerz, Michael Haggman, Swen Andersson, Anders Spangberg, Christer Busch, Steg Nordling, Juni Palmgren, Hans Adami, Jan Erik Johansson, Bo Johan Norlen. 2002. "A randomized trial comparing radical prostatectomy with watchful waiting in early prostate cancer." *New England Journal of Medicine* 347(11): 781–789.

Horn, Joshua S. 1969. *"Away with all Pests . . .": An English Surgeon in People's China.* London: Paul Hamlyn.

Horwitz, Allan V. 1982. *Social Control of Mental Illness.* New York: Academic.

———. 1999. Personal communication.

———. 2002. *Creating Mental Illness.* Chicago: University of Chicago Press.

Horwitz, Allan V., and Jeffrey S. Mullis. 1998. "Individualism and its discontents: The response to the seriously mentally ill in late twentieth century America." *Sociological Focus* 31: 119–133.

Houghton, John Theodore, L. Meria Filho, B. A. Callander, and N. Harris (eds.). 1996. *Climate Change 1995: The Science of Climate Change.* New York: Cambridge University Press.

House, James S. 2002. "Understanding social factors and inequalities in health: 20th century progress and 21st century prospects." *Journal of Health and Social Behavior* 43: 125–142.

Hunt, Charles W. 1989. "Migrant labor and sexually transmitted disease: AIDS in Africa." *Journal of Health and Social Behavior* 30: 353–373.

———. 1996. "Social vs. biological: Theories on the transmission of AIDS in Africa." *Social Science and Medicine* 42: 1283–1296.

Hurwitz, Eric L., Ian D. Coulter, Alan H. Adams, Barbara J. Genovese, and Paul G. Shekelle. 1998. "Use of chiropractic services from 1985 through 1991 in the United States and Canada." *American Journal of Public Health* 88: 771–776.

Iglehart, John K. 1999. "The American health care system: Expenditures." *New England Journal of Medicine* 340: 70–76.

Institute of Medicine. 2002. *Care Without Coverage: Too Little Too Late.* Washington, DC: National Academy Press.

International Baby Food Action Network. 2005. http://www.ibfan.org/english/codew00.html, accessed July 2005.

Jacobson, Matthew Frye. 1998. *Whiteness of a Different Color: European Immigrants and the Alchemy of Race.* Cambridge, MA: Harvard University Press.

James, W. Philip T., Michael Nelson, Ann Ralph, and Suzi Leather. 1997. "Socioeconomic determinants of health. The contribution of nutrition to inequalities in health." *British Medical Journal* 314(7093): 1545–1549.

Janis, Irving L., and Leon Mann. 1977. *Decision Making: A Psychological Analysis of Conflict, Choice, and Commitment.* New York: Free Press.

Janoff, Barry. 1999. "Centers of attention." *Progressive Grocer* 78(4): 75–80.

Johnson, Kenneth C., and Betty-Anne Daviss. 2005. "Outcomes of planned home births with certified professional midwives: Large prospective study in North America." *British Medical Journal* 330: 1416.

Johnson, Kirk. 2004. "Harm to fetuses becomes issue in Utah and elsewhere." *New York Times* March 27: A9.

Jolly, Paul. 2004. *Medical School Tuition and Young Physician Indebtedness.* Washington, DC: Association of American Medical Colleges.

Jones, James. 1993. *Bad Blood: The Tuskegee Syphilis Experiment.* Rev. ed. New York: Free Press.

Journal of the American Medical Association. 1990. "Christian Scientists claim healing efficacy equal if not superior to that of medicine." 264: 1379–1381.

Kahn, Joseph P. 1989. "Radner's humor good medicine, even though it couldn't save her." *Arizona Daily Star* (May 30): 1C–2C.

Kaiser Commission on Medicaid and the Uninsured. 2004. *The Uninsured: A Primer.* Washington, DC: Kaiser Family Foundation.

Katz, Jay. 1984. *The Silent World of Doctor and Patient.* New York: Free Press.

Kaufman, Martin. 1971. *Homeopathy in America: The Rise and Fall of a Medical Heresy.* Baltimore: Johns Hopkins University Press.

Kaye, H. Stephen, Mitchell P. LaPlante, Dawn Carlson, and Barbara L. Wenger. 1996. "Trends in disability rates in the United States, 1970–1994." *Disability Statistics Abstract* (17). Washington, DC: U.S. Department of Education, National Institute on Disability and Rehabilitation Research.

Kaysen, Susan. 1993. *Girl, Interrupted.* New York: Random House.

Kellerman, Arthur L., Frederick P. Rivara, Norman B. Rushforth, Joyce G. Banton, Donald T. Reay, Jerry T. Francisco, Ana B. Locci, Janice Prodzinski, Bela B. Hackman, and Grant Somes. 1993. "Gun ownership as a risk factor for homicide in the home." *New England Journal of Medicine* 329(15): 1084–1091.

Kellogg, J. H. 1880. *Plain Facts for Young and Old.* Burlington, IA: Segner and Condit.

Kelly, John. 2005. *The Great Mortality: An Intimate History of the Black Death, the Most Devastating Plague of All Time.* New York: HarperCollins.

Kelly, Susan E., Patricia A. Marshall, Lee M. Sanders, Thomas A. Raffin, and Barbara A. Koenig. 1997. "Understanding the practice of ethics consultation: Results of an ethnographic multi-site study." *Journal of Clinical Ethics* 8: 136–149.

Kessler, Ronald C., Patricia Berglund, Olga Demler, Robert Jim, and Ellen E. Walters. 2005a. "Lifetime prevalence and age-of-onset distributions of *DSM IV* disorders in the National Comorbidity Survey Replication." *Archives of General Psychiatry* 62: 593–602.

Kessler, Ronald C., Roger B. Davis, David F. Foster, Maria I. Van Rompay, Ellen E. Walters, Sonja A. Wilkey, Ted J. Kaptchuk, and David M. Eisenberg. 2001. "Long-term trends in the use of complementary and alternative medical therapies in the United States." *Annals of Internal Medicine* 135(4): 262–268.

Kessler, Ronald C., Olga Demler, Richard G. Frank, Mark Olfson, Harold Alan Pincus, Ellen E. Walters, Philip Wang, Kenneth B. Wells, and Alan M. Zashavsky. 2005b. "Prevalence and treatment of mental disorders, 1990–2003." *New England Journal of Medicine* 352: 2515–23.

Kessler, Ronald C., James S. House, and J. Blake Turner. 1987. "Unemployment and health in a community sample." *Journal of Health and Social Behavior* 28: 51–59.

Kessler, Ronald C., Katherine A. McGonagle, Shanyang Zhao, Christopher B. Nelson, Michael Hughes, Suzann Eshleman, Hans-Ulrich Wittchen, and Kenneth S. Kendler. 1994. "Lifetime and 12-month prevalence of *DSM-III-R* psychiatric disorders in the United States. Results from the National Comorbidity Survey." *Archives of General Psychiatry* 51: 8–19.

Kessler, Ronald C., and Harold W. Neighbors. 1986. "A new perspective on the relationships among race, social class, and psychological distress." *Journal of Health and Social Behavior* 27: 107–115.

Kessler, Suzanne J. 1998. *Lessons from the Intersexed.* New Brunswick, NJ: Rutgers University Press.

Kidder, David. 1988a. "The impact of hospices on the health-care costs of terminal cancer patients." Pp. 48–68 in *The Hospice Experiment,* edited by Vincent Mor, David S. Greer, and Robert Kastenbaum. Baltimore: Johns Hopkins University Press.

———. 1988b. "Hospice services and cost savings in the last weeks of life." Pp. 69–87 in *The Hospice Experiment,* edited by Vincent Mor, David S. Greer, and Robert Kastenbaum. Baltimore: Johns Hopkins University Press.

Kiesler, Charles A., and Amy E. Sibulkin. 1987. *Mental Hospitalization: Myths and Facts About a National Crisis.* Newbury Park, CA: Sage.

Kiple, Kenneth F. 1993. *Cambridge World History of Human Disease.* New York: Cambridge University Press.

Kirk, Stuart A. 1992. *The Selling of DSM: The Rhetoric of Science in Psychiatry.* New York: Aldine de Gruyter.

Kirp, David L. 1989. *Learning by Heart: AIDS and Schoolchildren in America's Communities.* New Brunswick, NJ: Rutgers University Press.

Kitchener, Martin, and Charlene Harrington. 2004. "The U.S. long-term care field: A dialectic analysis of institution dynamics." *Journal of Health and Social Behavior* 45: 87–101.

Klass, Perri. 1987. *A Not Entirely Benign Procedure: Four Years as a Medical Student.* New York: Putnam's.

Kleinman, Lawrence C., Howard Freeman, Judy Perlman, and Lillian Gelberg. 1996. "Homing in on the homeless: Assessing the physical health of homeless adults in Los Angeles County using an original method to obtain physical examination data in a survey." *Health Services Research* 31: 533–549.

Kligman, David. 1999. "Institutions: Mental or penal?" *Arizona Republic* May 9: A22.

Klima, Edward S., and Ursula Bellugi. 1979. *The Signs of Language.* Cambridge, MA: Harvard University Press.

Knafl, Kathleen, and Gary Burkett. 1975. "Professional socialization in a medical specialty: Acquiring medical judgment." *Social Science and Medicine* 9: 397–404.

Kochanek, Kenneth D., Sherry L. Murphy, Robert N. Anderson, and Chester Scott. 2005. "Deaths: Final data for 2002." *National Vital Statistics Report* 53(5).

Koerner, Brendan I. 2003. "Dr. No Free Lunch." *Mother Jones* 28(2): 24.

Kohn, Linda T., Janet M. Corrigan, and Molla S. Donaldson. 1999. *To Err is Human: Building a Safer Health System.* Washington, DC: National Academy Press.

Kolata, Gina. 1994. "Pharmacists paid to suggest drugs." *New York Times* July 29: A1.

———. 2004a. "Diet and lose weight? Scientists say 'Prove it!'" *New York Times* Jan. 4:D1+.

———. 2004b. "Stem cells: Promise, in search of results." *New York Times* August 24: D1+.

———. 2005. "PSA test no longer gives clear answers." *New York Times* June 20: E1+.

Kolata, Gina, and Kurt Eichenwald. 1999. "For the uninsured, drug trials are health care." *New York Times* June 22: A1+.

Kolder, Veronika E. B., Janet Gallagher, and Michael T. Parsons. 1987. "Court-ordered obstetrical interventions." *New England Journal of Medicine* 316: 1192–1196.

Kolko, Gabriel. 1999. "Ravaging the poor: The International Monetary Fund indicted by its own data." *International Journal of Health Services* 29: 51–57.

Koopman, Cheryl, Sherman Eisenthal, and John D. Stoeckle. 1984. "Ethnicity in the reported pain, emotional distress, and requests of medical outpatients." *Social Science and Medicine* 18(6): 487–490.

Koren, Gideon, and Naomi Klein. 1991. "Bias against negative studies in newspaper reports of medical research." *Journal of the American Medical Association* 266: 1824–1826.

Koren, Gideon, Heather Shear, Karen Graham, and Tom Einarson. 1989. "Bias against the null hypothesis: The reproductive hazards of cocaine." *Lancet* 2(8677): 1440–1443.

Kramer, Peter. 1993. *Listening to Prozac.* New York: Viking.

Kravitz, Richard L., Ronald M. Epstein, Mitchell D. Feldman, Carol E. Franz, Rahman Azari, Michael S. Wilkes, Ladson Hinton, and Peter Franks. 2005. "Influence of patients' requests for direct-to-consumer advertised antidepressants: A randomized controlled trial." *Journal of the American Medical Association* 293: 1995–2002.

Kübler-Ross, Elizabeth. 1969. *On Death and Dying*. New York: Macmillan.

Kunitz, Stephen J. 1996. "The history and politics of U.S. health care policy for American Indians and Alaskan natives." *American Journal of Public Health* 86: 1464–1473.

Kuo, JoAnn, and Kathryn Porter. 1998. "Health status of Asian Americans: 1992–1994." *Advance Data from Vital and Health Statistics* No. 298.

Kurz, Demie. 1987. "Emergency department responses to battered women: Resistance to medicalization." *Social Problems* 34: 69–81.

Kutner, Nancy G. 1987. "Social worlds and identity in end-stage renal disease (ESRD)." *Research in the Sociology of Health Care* 6: 33–71.

Lacey, Marc. 2005. "Beyond the bullets and blades." *New York Times* March 20: WK1+.

Laing, Ronald D. 1967. *The Politics of Experience*. New York: Ballantine.

Lancet. 1990. "Marketing of breast milk substitutes." 335(8698): 11511–11552.

Lane, Harlan. 1992. *The Mask of Benevolence: Disabling the Deaf Community*. New York: Knopf.

Lappé, Frances Moore, Joseph Collins, and Peter Rosset. 1998. *World Hunger: Twelve Myths*. 2nd ed. New York: Grove.

Lassey, Marie L., William R. Lassey, and Martin J. Jinks. 1997. *Health Care Systems Around the World: Characteristics, Issues, Reforms*. Upper Saddle River, NJ: Prentice Hall.

LaVeist, Thomas A. 1993. "Segregation, poverty, and empowerment: Health consequences for African Americans." *Milbank Quarterly* 71: 41–64.

Lawn, Joy E., Simon Cousens, and Jelka Zupan. 2005. "4 million neonatal deaths: When? Where? Why?" *Lancet* 365: 891–900.

Lawrence, Ruth A. 1997. *A Review of the Medical Benefits and Contraindications to Breastfeeding in the United States (Maternal and Child Health Technical Information Bulletin)*. Arlington, VA: National Center for Education in Maternal and Child Health.

Leape, Lucian L. 1992. "Unnecessary surgery." *Annual Review of Public Health* 13: 363–383.

———. 1994. "Error in medicine." *Journal of the American Medical Association* 272: 1851–1857.

Leape, Lucian L., and Donald M. Berwick. 2005. "Five years after *To Err is Human:* What have we learned?" *Journal of the American Medical Association* 293: 2384–2390.

Leavitt, Judith Walzer. 1983. "Science enters the birthing room: Obstetrics in America since the eighteenth century." *Journal of American History* 70: 281–304.

———. 1986. *Brought to Bed: Childbearing in America, 1750–1950*. New York: Oxford University Press.

Leavitt, Judith Walzer, and Ronald L. Numbers. 1985. *Sickness and Health in America*. Madison: University of Wisconsin Press.

Lehrman, Ella-Joy. 1992. "Findings of the 1990 annual ACNM Membership Survey." *Journal of Nurse-Midwifery* 37: 33–47.

Leichter, Howard M. 1997. "Rationing of health care in Oregon: Making the implicit explicit." Pp. 138–162 in *Health Policy Reform in America: Innovations from the States,* edited by Howard M. Leichter. Armonk, NY: Sharpe.

Lemmens, Trudo. 2004. "Piercing the veil of corporate secrecy about clinical trials." *Hastings Center Review* 34: 14–18.

Lemmens, Trudo, and Benjamin Freedman. 2000. "Ethics review for sale? Conflict of interest and commercial research review boards." *Milbank Quarterly* 78: 547–584.

Lewin, Tamar. 1993. "Nursing invades turf once ruled by doctors." *New York Times* November 22: A1+.

Lewis, Caroline T. 1993. "Midwife-attended births." Pp. 247–250 in *Encyclopedia of Childbearing: Critical Perspectives,* edited by Barbara Katz Rothman. Phoenix, AZ: Oryx.

Libov, Charlotte. 1999. "Beat your risk factors." *Ladies' Home Journal* 116(4): 86–90.

Lifton, Robert J. 1986. *The Nazi Doctors: Medical Killing and the Psychology of Genocide.* New York: Basic.

Light, Donald W. 1979. "Uncertainty and control in professional training." *Journal of Health and Social Behavior* 20: 310–322.

———. 1988. "Toward a new sociology of medical education." *Journal of Health and Social Behavior* 29: 307–322.

———. 1992. "The practice and ethics of risk-rated health insurance." *Journal of the American Medical Association* 267: 2503–2508.

Light, Donald W., and Sol Levine. 1988. "Changing character of the medical profession: A theoretical overview." *Milbank Quarterly* 66(supplement 2): 10–32.

Link, Bruce G. 1987. "Understanding labeling effects in the area of mental disorders: An assessment of the effects of expectations of rejection." *American Sociological Review* 52: 96–112.

Link, Bruce G., Francis T. Cullen, James Frank, and John F. Wozniak. 1987. "The social rejection of former mental patients: Understanding why labels matter." *American Journal of Sociology* 92: 1461–1500.

Link, Bruce G., Francis T. Cullen, Elmer Struening, Patrick E. Shrout, and Bruce P. Dohrenwend. 1989. "A modified labeling theory approach to mental disorders: An empirical assessment." *American Sociological Review* 54: 400–423.

Link, Bruce G., Bruce P. Dohrenwend, and Andrew E. Skodol. 1986. "Socioeconomic status and schizophrenia: Noisome occupational characteristics as a risk factor." *American Sociological Review* 51: 242–258.

Link, Bruce G., and Jo C. Phelan. 1995. "Social conditions as fundamental causes of illness." *Journal of Health and Social Behavior* Extra issue: 80–94.

Link, Bruce G., Jo C. Phelan, Michaeline Bresnahan, Ann Stueve, and Bernice A. Pescosolido. 1999. "Public conceptions of mental illness: Labels, causes, dangerousness, and social distance." *American Journal of Public Health* 89: 1328–1333.

Link, Bruce G., Elmer L. Struening, Michael Rahav, Jo C. Phelan, and Larry Nuttbrock. 1997. "On stigma and its consequences: Evidence from a longitudinal study of men with dual diagnoses of mental illness and substance abuse." *Journal of Health and Social Behavior* 38: 177–190.

Lips, Hilary M. 1993. *Sex & Gender: An Introduction.* Mountain View, CA: Mayfield.

Liska, Ken. 1997. *Drugs and the Human Body.* Upper Saddle River, NJ: Prentice Hall.

Little, Ruth E. 1998. "Public health in central and eastern Europe and the role of environmental pollution." *Annual Review of Public Health* 19: 153–172.

Litwin, Mark S., David J. Pasta, Marcia L. Stoddard, James M. Henning, and Peter R. Carroll. 1998. "Epidemiological trends and financial outcomes in radical prostatectomy among Medicare beneficiaries, 1991 to 1993." *Journal of Urology* 160: 445–448.

Loe, Meika. 2004. *The Rise of Viagra: How the Little Blue Pill Changed Sex in America.* New York: New York University Press.

Lonsdale, Susan. 1990. *Women and Disability.* Basingstoke, UK: Macmillan.

Lorber, Judith. 1984. *Women Physicians: Careers, Status, and Power.* New York: Tavistock.

Loring, Marti, and Brian Powell. 1988. "Gender, race, and *DSM-III*: A study of the objectivity of psychiatric diagnostic behavior." *Journal of Health and Social Behavior* 29: 1–22.

Loviglio, Joann. 2005. "Alternative goes mainstream." *Washington Times* June 10, http://washtimes.com/culture/20050609-114805-1956r.htm, accessed July 2005.

Ludmerer, Kenneth M. 1985. *Learning to Heal: The Development of American Medical Education.* New York: Basic.

Luhrmann, T. M. 2000. *Of Two Minds: The Growing Disorder in American Psychiatry.* New York: Knopf.

Lundberg, George D. 2001. *Severed Trust: Why American Medicine Hasn't Been Fixed.* New York: Basic.

Luker, Kristin. 1984. *Abortion and the Politics of Motherhood.* Berkeley: University of California Press.

Lu-Yao, Grace L., Dale McCloran, John Wasson, and John E. Wennberg. 1993. "An assessment of radical prostatectomy: Time trends, geographic variation, and outcomes." *Journal of the American Medical Association* 269: 2633–2636.

Lynch, Michael. 1983. "Accommodation practices: Vernacular treatments of madness." *Social Problems* 31: 152–164.

MacDorman, Marian F., and Gopal K. Singh. 1999. "Midwifery care, social and medical risk factors and birth outcomes in the USA." *Journal of Epidemiology and Community Health.* May: 310–317.

Mairs, Nancy. 1986. *Plaintext.* Tucson: University of Arizona Press.

Manderscheid, Ronald W., and Mary Anne Sonnenschein (eds.). 1992. *Mental Health, United States, 1992.* Washington, DC: U.S. Government Printing Office.

Mann, Charles C. 1993. "The prostate-cancer dilemma." *Atlantic* November: 102+.

Markoff, John. 2002. "Technology's toxic trash is sent to poor nations." *New York Times* February 25: C1+.

Marmot, Michael G. 2002. "The influence of income on health." *Health Affairs* 21: 31–46.

———. 2004. *The Status Syndrome: How Your Social Standing Directly Affects Your Health and Life Expectancy.* London: Bloomsbury.

Marmot, Michael G., and Martin J. Shipley. 1996. "Do socioeconomic differences in mortality persist after retirement?" *British Medical Journal* 313: 1177–1180.

Martin, Emily. 1987. *The Woman in the Body.* Boston: Beacon.

Martin, Joyce A., Brady E. Hamilton, Paul D. Sutton, Stephanie J. Ventura, Fay Menacker, and Martha L. Munson. 2003. "Births: Final Data for 2002." *National Vital Statistics Report* 52(10).

Martin, Philip. 2002. *Guest Workers: New Solution, New Problem?* Washington, DC: Pew Hispanic Center.

McCarthy, John D., and Mayer N. Zald. 1973. *The Trend of Social Movements in America: Professionalization and Resource Mobilization.* Morristown, NJ: General Learning.

McCord, Colin, and Harold P. Freeman. 1990. "Excess mortality in Harlem." *New England Journal of Medicine* 322: 173–177.

McKeown, Thomas. 1979. *The Role of Medicine: Dream, Mirage, or Nemesis?* Princeton, NJ: Princeton University Press.

McKinlay, John B. 1994. "A case for refocussing upstream: The political economy of illness." Pp. 509–530 in *The Sociology of Health and Illness,* edited by Peter Conrad and Rachelle Kern. New York: St. Martin's Press.

McKinlay, John B., and Sonja J. McKinlay. 1977. "The questionable effect of medical measures on the decline of mortality in the United States in the twentieth century." *Milbank Memorial Fund Quarterly* 55: 405–428.

McKinlay, John B., and John D. Stoeckle. 1989. "Corporatization and the social transformation of doctoring." *International Journal of Health Services* 18: 191–205.

McNeil, Donald G. 2002. "With folk medicine on rise, health group is monitoring." *New York Times* May 17: A9.

Mechanic, David. 1989. *Mental Health and Social Policy.* 3rd ed. Englewood Cliffs, NJ: Prentice Hall.

———. 1995. "Sociological dimensions of illness behavior." *Social Problems* 41: 1207–1216.

———. 1997. "Managed mental health care." *Society* 35(1): 44–52.

———. 1999. *Mental Health and Social Policy: The Emergence of Managed Care.* 4th ed. Boston: Allyn & Bacon.

———. 2001a. "The managed care backlash: Perceptions and rhetoric in health care policy and the potential for health care reform." *Milbank Quarterly* 79: 35–51.

———. 2001b. "Are patients' office visits with physicians getting shorter?" *New England Journal of Medicine* 344(3): 198–203.

———. 2004. "The rise and fall of managed care." *Journal of Health and Social Behavior* 45: 76–86.

Mechanic, David, and David A. Rochefort. 1990. "Deinstitutionalization: An appraisal of reform." *Annual Review of Sociology* 16: 301–327.

———. 1996. "Comparative medical systems." *Annual Review of Sociology* 22: 239–270.

Melosh, Barbara. 1982. *"The Physician's Hand": Work Culture and Conflict in American Nursing.* Philadelphia: Temple University Press.

Melucci, Alberto. 1995. "The process of collective identity." Pp. 41–63 in *Social Movements and Culture,* edited by Hank Johnston and Bert Klandermans. Minneapolis: University of Minnesota Press.

Messer, Ellen. 1997. "Intra-household allocation of food and health care: Current findings and understandings—introduction." *Social Science and Medicine* 44: 1675–1684.

Meyer, Pamela A., Timothy Pivetz, Timothy A. Dignam, David M. Homa, Jaime Schoonover, and Debra Brody. 2003. "Surveillance for elevated blood lead levels among children—United States, 1997–2001." *Morbidity and Mortality Weekly Report* 52(No. SS-10): 1–21.

Meyerowitz, Beth E., Janice G. Williams, and Jocelyne Gessner. 1987. "Perceptions of controllability and attitudes toward cancer and cancer patients." *Journal of Applied Social Psychology* 17: 471–492.

Millenson, Michael L. 1997. *Demanding Medical Excellence: Doctors and Accountability in the Information Age.* Chicago: University of Chicago Press.

Miller, Judith, Stephen Engelberg, and William Broad. 2001. *Germs: Biological Weapons and America's Secret War.* New York: Simon & Schuster.

Miller, Robert H., and Harold T. Luft. 1997. "Does managed care lead to better or worse quality of care?" *Health Affairs* 16(5): 7–25.

Millman, Marcia. 1976. *The Unkindest Cut: Life in the Backrooms of Medicine.* New York: Morrow.

Mills, C. Wright. 1959. *The Sociological Imagination.* New York: Grove.

Mills, Robert J. 2002. "Health insurance coverage: 2001." *Current Population Reports* P60–220.

Mirowsky, John, and Catherine E. Ross. 1989. "Psychiatric diagnosis as reified measurement." *Journal of Health and Social Behavior* 30: 11–25.

Mishler, Elliot G. 1981. "Viewpoint: Critical perspectives on the biomedical model." Pp. 1–23 in *Social Contexts of Health, Illness, and Patient Care,* edited by Elliot G. Mishler. Cambridge, UK: Cambridge University Press.

———. 1990. "The struggle between the voice of medicine and the voice of the lifeworld." Pp. 295–307 in *The Sociology of Health and Illness: Critical Perspectives,* edited by Peter Conrad and Rachelle Kern. New York: St. Martin's Press.

Mizrahi, Terry. 1986. *Getting Rid of Patients: Contradictions in the Socialization of Physicians.* New Brunswick, NJ: Rutgers University Press.

Mokdad, Ali H., James S. Marks, Donna F. Stroup, and Julie L. Gerberding. 2004. "Actual causes of death in the United States, 2000." *Journal of the American Medical Association* 291:1238–1245.

Montgomery, Kathleen. 1996. "Responses by professional organizations to multiple and ambiguous institutional environments: The case of AIDS." *Organizational Studies* 17: 649–671.

Montgomery, Rhonda J. V. 1992. "Examining respite: Its promise and limits." Pp. 75–96 in *In-Home Care for Older People: Health and Supportive Services,* edited by Marcia G. Ory and Alfred P. Duncker. Newbury Park, CA: Sage.

Mor, Vincent. 1987. *Hospice Care Systems: Structure, Process, Costs, and Outcome.* New York: Springer.

Morbidity and Mortality Weekly Report. 1998. "Tobacco use among U.S. racial/ethnic minority groups—African Americans, American Indians and Alaska Natives, Asian Americans and Pacific Islanders, Hispanics." 47(RR-18): 1–16.

Morgan, Patricia A. 1988. "Power, politics, and public health: The political power of the alcohol beverage industry." *Journal of Public Health Policy* 9: 177–197.

Mosher, James F. 1995. "The merchants, not the customers: Resisting the alcohol and tobacco industries' strategy to blame young people for illegal alcohol and tobacco sales." *Journal of Public Health Policy* 16: 412–432.

Mount, Jeanine. Personal communication. August 1999.

Mullan, Fitzhugh. 2002. *Big Doctoring in America: Profiles in Primary Care.* Berkeley: University of California Press.

Mundinger, Mary O., Robert L. Kane, Elizabeth R. Lenz, Annette M. Totten, Wei-Yann Tsai, Paul D. Cleary, William T. Friedewald, Albert L. Siu, Michael L. Shelanski. 2000. "Primary care outcomes in patients treated by nurse practitioners or physicians: A randomized trial." *Journal of the American Medical Association* 283: 59–68.

Murdock, George P. 1980. *Theories of Illness: A World Survey.* Pittsburgh, PA: University of Pittsburgh Press.

Murray, Christopher J. L., and Alan D. Lopez. 1996. *The Global Burden of Disease.* Cambridge, MA: Harvard University Press.

Murtaugh, Christopher M., Peter Kemper, Brenda C. Spillman, and Barbara Lepidus Carlson. 1997. "The amount, distribution, and timing of lifetime nursing home use." *Medical Care* 35: 204–218.

Nader, Ralph. 1965. *Unsafe at Any Speed.* New York: Knightsbridge.

National Alliance for Caregiving and AARP. 2004. Caregiving in the U.S. Bethesda, MD: National Alliance for Caregiving.

National Center for Health Statistics. 2001. *Health, United States, 2001.* Hyattsville, MD: U.S. Public Health Service.

———. 2004. *Health, United States, 2004.* Hyattsville, MD: U.S. Public Health Service.

———. 2005. "Home health care patients: Data from the 2000 National Home and Hospice Care Survey," http://www.cdc.gov/nchs/pressroom/04facts/patients.htm, accessed July 2005.

National Coalition on Health Care. 2005. *Building a Better Health Care System: Specifications for Reform.* Washington, DC: author.

National Genome Research Institute. 2005. *Genetic Discrimination Fact Sheet: Genetic Discrimination in Health Insurance.* Bethesda, MD: National Institutes of Health.

National Hospice and Palliative Care Organization. 2001. *NHPCO Facts and Figures.* Alexandria, VA: author.

———. 2005. *Hospice Facts and Figures.* Alexandria, VA: author.

National Institute for Health Care Management Foundation. 2002. *Prescription Drug Expenditures in 2001: Another Year of Escalating Costs.* Washington DC: author.

National Institute of Health Consensus Development Panel on Acupuncture. 1998. "Acupuncture." *Journal of the American Medical Association* 280: 1518–1524.

National League for Nursing. 2004. *Nursing Data Review Academic Year 2003.* New York: author.

National Low Income Housing Coalition. 2001. *Out of Reach 2001: America's Growing Wage-Rent Disparity.* Washington, DC: author.

National Opinion Research Center, General Social Surveys. 1972–1991: *Cumulative Codebook,* July 1991. www.norc.uchicago.edu, accessed August 2005.

National Organization on Disability. 2001. *N.O.D./Harris Survey of Americans with Disabilities.* Washington, DC: author.

Natowicz, Marvin R., Jane K. Alper, and Joseph S. Alper. 1992. "Genetic discrimination and the law." *American Journal of Human Genetics* 50: 465–475.

Navarro, Mireya. 2004. "When gender isn't a given." *New York Times* September 19: I1+.

Navarro, Vincente. 1990. "Race or class versus race and class: Mortality differentials in the United States." *Lancet* 336(8725): 1238–1240.

Naylor, C. David. 1995. "Grey zones of clinical practice: Some limits to evidence-based medicine." *Lancet* 345: 840–842.

Neisser, Arden. 1983. *The Other Side of Silence: Sign Language and the Deaf Community in America.* New York: Knopf.

Nelkin, Dorothy, and Laurence Tancredi. 1989. *Dangerous Diagnostics: The Social Power of Biological Information.* New York: Basic.

Nelson, Alan R., Brian D. Smedley, and Adrienne Y. Stith. 2002. *Unequal Treatment: Confronting Racial and Ethnic Disparities in Health Care.* Washington, DC: Institute of Medicine, National Academy Press.

Nersesian, William S. 1988. "Infant mortality in socially vulnerable populations." *Annual Review of Public Health* 9: 361–377.

Nestlé, Marion. 2002. *Food Politics: How the Food Industry Influences Nutrition and Health.* Berkeley: University of California Press.

Neubauer, Deane. 1997. "Hawaii: The health state revisited." Pp. 163–188 in *Health Policy Reform in America: Innovations from the States,* edited by Howard M. Leichter. Armonk, NY: M. E. Sharpe.

New York Times. 1999a. "Registered nurses in short supply at hospitals nationwide." March 23: A14.

———. 1999b. "Third of Hispanic Americans do without health coverage." April 9: A1.

Nichols, Judy. 2002. "Indian health care: Separate, unequal." *Arizona Republic* April 14: A1.

Nichter, Mark, and Elizabeth Cartwright. 1991. "Saving the children for the tobacco industry." *Medical Anthropology Quarterly* 5: 236–256.

Nord, Mark, Margaret Andrews, and Steven Carlson. 2004. "Household Food Security in the United States, 2003." *Food Assistance and Nutrition Research Report* No. 42, Economic Research Service, United States Department of Agriculture.

Norrish, Barbara R., and Thomas G. Rundall. 2001. "Hospital restructuring and the work of registered nurses." *Milbank Quarterly* 79: 55–79.

North, Fiona M., S. Leonard Syme, Amanda Feeney, Martin Shipley, and Michael Marmot. 1996. "Psychosocial work environment and sickness absence among British civil servants: The Whitehall II Study." *American Journal of Public Health* 86: 332–340.

Novello, Antonia C., Mark Rosenberg, Linda Saltzman, and John Shosky. 1992. "A medical response to domestic violence." *Journal of the American Medical Association* 267: 3132.

Omran, Abdel R. 1971. "The epidemiological transition." *Milbank Memorial Fund Quarterly* 49: 509–538.

Organization for Economic Cooperation and Development. 2004. *OECD Health Data 2004.* Paris, France: author.

Orlander, Jay D., and Graeme Fincke. 2003. "Morbidity and mortality conferences: A survey of academic internal medicine departments." *Journal of General Internal Medicine* 18: 656–661.

Orr, Robert D., and Eliot Moon. 1993. "Effectiveness of an ethics consultation service." *Journal of Family Practice* 36: 49–53.

Osherson, Samuel, and Lorna Amara Singham. 1981. "The machine metaphor in medicine." Pp. 218–249 in *Social Contexts of Health, Illness, and Patient Care,* edited by Elliot G. Mishler. Cambridge, UK: Cambridge University Press.

Otten, Mac W., Steven M. Teutsch, David F. Williamson, and James F. Marks. 1990. "The effect of known risk factors on the excess mortality of black adults in the United States." *Journal of the American Medical Association* 263: 845–850.

Pan American Health Organization. 2005. "Mexico." http://www.paho.org/English/DD/AIS/cp_484.htm, accessed July 2005.

Parsons, Talcott. 1951. *The Social System.* New York: Free Press.

Paul, Peter V. 1998. *Literacy and Deafness: The Development of Reading, Writing, and Literate Thought.* Boston: Allyn & Bacon.

Payer, Lynn. 1996. *Medicine and Culture.* Rev. ed. New York: Holt.

Peabody, John W. 1996. "Economic reform and health sector policy: Lessons from structural adjustment programs." *Social Science and Medicine* 43: 823–835.

Pear, Robert. 2002a. "HMO's for 200,000 pulling out of Medicare." *New York Times* September 9: A25.

———. 2002b. "Many on Medicaid lack drugs, study says." *New York Times* April 9: A22.

———. 2002c. "9 in ten nursing homes lack adequate staff, study finds." *New York Times* February 18: A1+.

Pear, Robert, and Robin Toner. 2002. "Grim choices face states in making cuts in Medicaid." January 14: A1+.

Pearlin, Leonard I. 1989. "The sociological study of stress." *Journal of Health and Social Behavior* 30: 241–256.

Pearlin, Leonard I., and Carol S. Aneshensel. 1986. "Coping and social supports: Their functions and applications." Pp. 417–437 in *Applications of Social Science to Clinical Medicine and Health Policy,* edited by Linda H. Aiken and David Mechanic. New Brunswick, NJ: Rutgers University Press.

Pérez-Peña, Richard. 2004. "Study says 50% of children enter shelters with asthma." *New York Times* March 2: A23.

Perrone, Bobette, H. Henrietta Stockel, and Victoria Krueger. 1989. *Medicine Women, Curanderas, and Women Doctors.* Norman: University of Oklahoma Press.

Pescosolido, Bernice A. 1992. "Beyond rational choice: The social dynamics of how people seek help." *American Journal of Sociology* 97: 1096–1138.

Pescosolido, Bernice A., Carol Brooks Gardner, and Keri M. Lubell. 1998. "How people get into mental health services: Stories of choice, coercion and 'muddling through' from 'first-timers.'" *Social Science and Medicine* 46: 275–286.

Phelan, Jo C., Bruce G. Link, Ana Diez-Roux, Ichiro Kawachi, and Bruce Levin. 2004. "'Fundamental causes' of social inequality in mortality: A test of the theory." *Journal of Health and Social Behavior* 45: 265–85.

Phillips, Marilynn J. 1985. "'Try harder': The experience of disability and the dilemma of normalization." *Social Science and Medicine* 22: 45–57.

———. 1990. "Damaged goods: Oral narratives of the experience of disability in American culture." *Social Science and Medicine* 30: 849–857.

Pierluissi, Edgar, Melissa A. Fischer, Andre R. Campbell, and C. Seth Landefeld. 2003. "Discussion of medical errors in morbidity and mortality conferences." *Journal of the American Medical Association* 290: 2838–2842.

Plotnick, Robert D. 1992. "The effects of attitudes on teenage premarital pregnancy and its resolution." *American Sociological Review* 57: 800–811.

Polakovic, Gary. 2002. "Asia's wind-born pollution a hazardous export to U.S." *Los Angeles Times* April 26: A1+.

Polednak, Anthony P. 1996. "Trends in U.S. urban black infant mortality, by degree of residential segregation." *American Journal of Public Health* 86: 723–726.

Pollack, Andrew. 2005. "Marketing a disease, and also a drug to treat it." *New York Times* May 9: C1+.

PollingReport.com. 2005. "Abortion and birth control." http://www.pollingreport.com/abortion.htm, accessed May 2005.

Pollitt, Katha. 1990. "A new assault on feminism." *Nation* 250: 409–418.

Population Reference Bureau. 2004. *World Population Data Sheet.* Washington, DC: author.

———. 2005. PRB Datafinder. www.prb.org/datafind, accessed June 2005.

Preves, Sharon E. 2003. *Intersex and Identity: The Contested Self.* New Brunswick, NJ: Rutgers University Press.

Quadagno, Jill. 2005. *One Nation Uninsured: Why The U.S. Has No National Health Insurance.* New York: Oxford University Press.

Quinn, Sandra Crouse. 2004. "Protecting human subjects: The role of community action boards." *American Journal of Public Health* 94: 918–922.

Raisler, Jeanne, Cheryl Alexander, and Patricia O'Campo. 1999. "Breast-feeding and infant illness: A dose–response relationship?" *American Journal of Public Health* 89: 25–30.

Ramsey, Paul. 1970. *The Patient as Person.* New Haven, CT: Yale University Press.

Rawls, John. 1971. *A Theory of Justice.* Cambridge, MA: Harvard University Press.

Reading, Richard. 1997. "Social disadvantage and infection in childhood." *Sociology of Health and Illness* 19: 395–414.

Redlich, Fredrick C. 1978. "Medical ethics under National Socialism." Pp. 1015–1019 in *Encyclopedia of Bioethics,* edited by Warren T. Reich. New York: Free Press.

Register, Cheri. 1987. *Living with Chronic Illness: Days of Patience and Passion.* New York: Free Press.

Reilly, Phillip. 1987. *To Do No Harm: A Journey Through Medical School.* Dover, MA: Auburn House.

Reinhard, Susan, and Allan V. Horwitz. 1996. "Caregiver burden: Differentiating the content and consequences of family caregiving." *Journal of Marriage and the Family* 57: 741–750.

Reinhardt, Uwe E., Peter S. Hussey, and Gerard F. Anderson. 2004. "U.S. health care spending in an international context." *Health Affairs* 23: 10–25.

Remler, Dahlia K., Karen Donelan, Robert J. Blendon, George D. Lundberg, Lucian L. Leape, David R. Calkins, Katherine Binns, and Joseph P. Newhouse. 1997. "What do managed care plans do to affect care? Results from a survey of physicians." *Inquiry* 34: 196–204.

Reverby, Susan. 1987. *Ordered to Care: The Dilemma of American Nursing.* New York: Cambridge University Press.

Richards, Peter. 1977. *The Medieval Leper and His Northern Heirs.* Totowa, NJ: Rowman & Littlefield.

Rieker, Patricia R., and Chloe E. Bird. 2000. "Sociological explanations of gender differences in mental and physical health." Pp. 98–113 in *Handbook of Medical Sociology,* edited by Chloe E. Bird, Peter Conrad, and Allan Fremont. New York: Prentice Hall.

Riley, Gerald F., Arnold L. Potosky, Carrie N. Klabunde, Joan L. Warren, and Rachel Ballard-Barbash. 1999. "Stage at diagnosis and treatment patterns among older women with breast cancer." *Journal of the American Medical Association* 281: 720–726.

Risse, Guenter B. 1988. "Epidemics and history: Ecological perspectives and social responses." Pp. 33–66 in *AIDS: The Burdens of History,* edited by Elizabeth Fee and Daniel M. Fox. Berkeley: University of California Press.

Roberts, Les, Riyadh Lafta, Richard Garfield, Jamal Khudhairi, and Gilbert Burnham. 2004. "Mortality before and after the 2003 invasion of Iraq: Cluster sample survey." *Lancet* 364: 1857–1864.

Roeder, Beatrice A. 1988. *Chicano Folk Medicine from Los Angeles, California.* Vol. 34 in Folklore and Mythology Series. Los Angeles: University of California Press.

Rogers, Richard G., Robert A. Hummer, Charles B. Nam, and Kimberley Peters. 1996. "Demographic, socioeconomic, and behavioral factors affecting ethnic mortality by cause." *Social Forces* 74: 1419–1438.

Rogler, Lloyd. 1991. "Acculturation and mental health status among Hispanics." *American Psychologist* 46: 585–597.

Rohter, Larry. 2004. "Tracking the sale of a kidney on a path of poverty and hope." *New York Times* May 23: A1+.

Rooks, Judith. 1997. *Midwifery and Childbirth in America.* Philadelphia: Temple University Press.

Roos, Leslie L., Elliot S. Fisher, Ruth Brazauskas, Sandra M. Sharp, and Evelyn Shapiro. 1992. "Health and surgical outcomes in Canada and the United States." *Health Affairs* 11: 56–72.

Rosenau, Pauline Vaillancourt. 1997. "Migration for medical care and pharmaceuticals: A research note on the NAFTA countries." *Social Science Quarterly* 78: 578–590.

Rosenberg, Charles E. 1987. *The Care of Strangers: The Rise of America's Hospital System.* New York: Basic.

Rosenfeld, Dana, and Christopher A. Faircloth (eds.). 2005. *Medicalized Masculinities.* Philadelphia: Temple University Press.

Rosenfield, Sarah. 1997. "Labeling mental illness: The effects of received services and perceived stigma on life satisfaction." *American Sociological Review* 62: 660–672.

Rosenhan, David L. 1973. "On being sane in insane places." *Science* 179: 250–258.

Rosenstein, Alan H. 2002. "Nurse-physician relationships: Impact on nurse satisfaction and retention." *American Journal of Nursing* 102: 26–34.

Rosenstock, Irwin M. 1966. "Why people use health services." *Milbank Memorial Fund Quarterly* 44: 94–127.

Rothman, Barbara Katz. 1986. *The Tentative Pregnancy: Prenatal Diagnosis and the Future of Motherhood.* New York: Penguin.

———. 1989. *Recreating Motherhood: Ideology and Technology in a Patriarchal Society.* New York: Norton.

Rothman, David J. 1971. *The Discovery of the Asylum.* Boston: Little, Brown.

———. 1991. *Strangers at the Bedside: A History of How Law and Bioethics Transformed Medical Decision-Making.* New York: Basic.

———. 1997. *Beginnings Count: The Technological Imperative in American Health Care.* New York: Oxford University Press.

Rothman, Sheila M., and David J. Rothman. 2003. *The Pursuit of Perfection.* New York: Pantheon.

Ryan, William. 1976. *Blaming the Victim.* Rev. ed. New York: Pantheon.

Sade, Robert M. 1971. "Medical care as a right: A refutation." *New England Journal of Medicine* 285: 1288–1292.

Safran, Stephen P. 1998. "The first century of disability portrayal in film: An analysis of the literature." *Journal of Special Education* 31: 467–479.

Safriet, Barbara J. 1992. "Health care dollars and regulatory sense: The role of advanced practice nursing." *Yale Journal on Regulation* 9: 417–487.

Sakr, M., J. Angus, J. Perrin, C. Nixon, J. Nicholl, and J. Wardrope. 1999. "Care of minor injuries by emergency NPs or junior MDs: A randomized controlled trial." *Lancet* 354: 1321–1326.

Sandhaus, Sonia. 1998. "Migrant health: A harvest of poverty." *American Journal of Nursing* 98: 52–53.

Sandweiss, Stephen. 1998. "The social construction of environmental justice." Pp. 31–58 in *Environmental Injustices, Political Struggles: Race, Class, and the Environment,* edited by David E. Camacho. Durham, NC: Duke University Press.

Schechter, Susan. 1982. *Women and Male Violence.* Boston: South End.

Scheff, Thomas J. 1984. *Being Mentally Ill: A Sociological Theory.* Rev. ed. Chicago: Aldine.

Scheid, Teresa L. 2001. "Rethinking professional prerogative: Managed mental health care providers." Pp. 153–171 in *Rethinking the Sociology of Mental Health,* edited by Joan Busfield. Oxford, UK: Blackwell.

Schemo, Diana Jean. 2002. "Education suffers in Africa as AIDS ravages teachers." *New York Times* May 8: A10.

Schmidt, Charles W. 2000. "Bordering on environmental disaster." *Environmental Health Perspectives* 108: A308–313.

Schneider, Andrew and David McCumber. 2004. *An Air That Kills: How the Asbestos Poisoning of Libby, Montana, Uncovered a National Scandal.* New York: Putnam's Sons.

Schneider, Joseph W., and Peter Conrad. 1983. *Having Epilepsy: The Experience and Control of Illness.* Philadelphia: Temple University Press.

Schneirov, Matthew, and Jonathan David Geczik. 1996. "A diagnosis for our times: Alternative health's submerged networks and the transformation of identities." *Sociological Quarterly* 37: 627–644.

Schoen, Cathy, Michelle M. Doty, Sara R. Collins, and Alyssa L. Holmgren. 2005. "Insured but not protected: How many adults are underinsured? *Health Affairs* (forthcoming). Web exclusive, prior to publication: http://content.healthafairs .org/cgi/content/abstract/hlthaff.w5.289

Schoendorf, Kenneth C., Carol J. R. Hogue, Joel C. Kleinman, and Diane Rowley. 1992. "Mortality among infants of black as compared with white college-educated parents." *New England Journal of Medicine* 326: 1522–1526.

Schooler, Caroline, Ellen Feighery, and June A. Flora. 1996. "Seventh graders' self-reported exposure to cigarette marketing and its relationship to their smoking behavior." *American Journal of Public Health* 86: 1216–1221.

Scott, Wilbur J. 1990. "PTSD in *DSM-III:* A case in the politics of diagnosis and disease." *Social Problems* 37: 294–310.

Scull, Andrew. 1977. *Decarceration, Community Treatment and the Deviant: A Radical View.* Englewood Cliffs, NJ: Prentice Hall.

———. 1989. *Social Order/Mental Disorder: Anglo-American Psychiatry in Historical Perspective.* Berkeley: University of California Press.

Scully, Diana. 1994. *Men Who Control Women's Health: The Miseducation of Obstetrician-Gynecologists.* New York: Teachers College Press.

Sen, Amartya. 1999. *Development as Freedom.* New York: Knopf.

Sered, Susan Starr, and Rushika Fernandopulle. 2005. *Uninsured in America: Life and Death in the Land of Opportunity.* Berkeley, CA: University of California Press.

Shapiro, Dan. 2004. "Drug companies get too close for med school's comfort." *New York Times* January 20: D7+.

Shapiro, Joseph P. 1993. *No Pity: People with Disabilities Forging a New Civil Rights Movement.* New York: Random House.

Shaw, Jonathan E., Maximilian de Courten, Edward J. Boyko, and Paul Z. Zimmet. 1999. "Impact of new diagnostic criteria for diabetes on different populations." *Diabetes Care* 22: 762–766.

Shekelle, Paul G. 1998. "What role for chiropractic in health care?" *New England Journal of Medicine* 339: 1074–1075.

Shilts, Randy. 1987. *And the Band Played On.* New York: St. Martin's Press.

Shortell, Stephen M., Teresa M. Waters, Kenneth W. B. Clarke, and Peter P. Budetti. 1998. "Physicians as double agents: Maintaining trust in an era of multiple accountabilities." *Journal of the American Medical Association* 280: 1102–1108.

Siegel, Bernie. 1990. *Love, Medicine and Miracles.* Rev. ed. New York: Harper & Row.

Siegman, Aron W., and Theodore M. Dembroski (eds.). 1989. *In Search of Coronary Prone Behavior: Beyond Type A.* Hillsdale, NJ: Erlbaum.

Siegrist, Johannes. 1996. "Adverse health effects of high-effort/low-reward conditions." *Journal of Occupational Health Psychology* 1: 27–41.

———. 2001. "Work stress and health." Pp. 114–125 in *The Blackwell Companion to Medical Sociology,* edited by William Cockerham. Oxford, UK: Blackwell.

Sigsbee, Bruce. 1997. "Medicare's resource-based relative value scale, a de facto national fee schedule: Its implications and uses for neurologists." *Neurology* 49: 315–320.

Sills, Trish. 2002. NDCHealth, personal communication, February.

Silverman, Milton, and Philip R. Lee. 1974. *Pills, Profits, and Politics.* Berkeley: University of California Press.

Simmons, Janie, Paul Farmer, and Brooke G. Schoepf. 1996. "A global perspective." Pp. 39–90 in *Women, Poverty, and AIDS: Sex, Drugs, and Structural Violence,* edited by Paul Farmer, Margaret Connors, and Janie Simmons. Monroe, ME: Common Courage.

Singer, Lynn T., Robert Arendt, Sonia Minnes, Kathleen Farkas, Ann Salvator, H. Lester Kirchner, and Robert Kliegman. 2002. "Cognitive and Motor Outcomes of Cocaine-Exposed Infants." *Journal of the American Medical Association* 287: 1952–1960.

Skocpol, Theda. 1996. *Boomerang: Clinton's Health Security Effort and the Turn Against Government in U.S. Politics.* New York: Norton.

Skolnick, Andrew. 1995. "Along U.S. southern border, pollution, poverty, ignorance, and greed threaten nation's health." *Journal of the American Medical Association* 273: 1478–1482.

Smyke, Patricia. 1991. *Women & Health.* London: Zed.

Sontag, Susan. 1978. *Illness as Metaphor.* New York: Farrar, Strauss, and Giroux.

———. 1988. *AIDS and Its Metaphors.* New York: Farrar, Strauss, and Giroux.

Specter, Michael. 1997. "Deep in the Russian soul, a lethal darkness." *New York Times* June 8: A1+.

Spitzer, Robert L., Janet B. W. Williams, and Andrew E. Skodol. 1980. "*DSM-III:* The major achievements and an overview." *American Journal of Psychiatry* 137: 151–164.

Starr, Paul. 1982. *The Social Transformation of American Medicine.* New York: Basic.

———. 1994. *The Logic of Health Care Reform: Why and How the President's Plan Will Work.* New York: Penguin.

Stein, Leonard. 1967. "The doctor-nurse game." *Archives of General Psychiatry* 16: 699–703.

Stein, Leonard, David T. Watts, and Timothy Howell. 1990. "The doctor-nurse game revisited." *New England Journal of Medicine* 322: 546–549.

Steinberg, Jacques. 1999. "Expanded school drug tests face a challenge." *New York Times* August 18: A14.

Steingart, Richard M. 1991. "Sex differences in the management of coronary artery disease." *New England Journal of Medicine* 325: 226–230.

Stevens, Rosemary. 1989. *In Sickness and in Wealth: American Hospitals in the Twentieth Century.* New York: Basic.

Stevens, Simon. 2004. "Reform strategies for the English NHS." *Health Affairs* 23: 37–44.

Stewart, David C., and Thomas J. Sullivan. 1982. "Illness behavior and the sick role in chronic disease: The case of multiple sclerosis." *Social Science and Medicine* 16: 1397–1404.

Stewart, Walter F., Judith A. Ricci, Elsbeth Chee, David Morganstein, and Richard Lipton. 2003. "Lost productive time and cost due to common pain conditions in the U.S. workforce." *Journal of the American Medical Association* 290: 2443–2454.

Stimmel, Barry. 1992. "The crisis in primary care and the role of medical schools: Defining the issues." *Journal of the American Medical Association* 268: 2060–2065.

Stine, Gerald J. 2005. *AIDS Update 2005.* San Francisco: Benjamin Cummings.

Stolberg, Sheryl Gay. 1999. "The boom in medications brings rise in fatal risks." *New York Times* June 3: A1+.

Stratton, Kathleen, Christopher B. Wilson, and Marie C. McCormick (eds.). 2002. *Immunization Safety Review: Multiple Immunizations and Immune Dysfunction.* Washington, DC: National Academy Press.

Straus, Robert. 1957. "The nature and status of medical sociology." *American Sociological Review* 22: 200–204.

Street, Richard. 1991. "Information-giving in medical consultations: The influence of patients' communicative styles and personal characteristics." *Social Science and Medicine* 32: 541–548.

Stretesky, Paul, and Michael J. Hogan. 1998. "Environmental justice: An analysis of Superfund sites in Florida." *Social Problems* 45: 268–287.

Strote, Mary Ellen. 2002. "Ten Easy Ways to Boost Your Immunity." *Shape,* http://findarticles.com/p/articles/mi_m0846/is_7_21/ai_82823102/pg_2, accessed November 2005.

Sullivan, Deborah A. 2001. *Cosmetic Surgery: The Cutting Edge of Commercial Medicine in America.* Brunswick, NJ: Rutgers University Press.

Sullivan, Deborah A., and Rose Weitz. 1988. *Labor Pains: Modern Midwives and Home Birth.* New Haven, CT: Yale University Press.

Sutton, John R. 1991. "The political economy of madness: The expansion of the asylum in progressive America." *American Sociological Review* 56: 665–678.

Swedish Institute. 1997. *Social Insurance in Sweden.* Stockholm: author (http://www.si.se).

———. 1999. *The Care of the Elderly in Sweden.* Stockholm: author (http://www.si.se).

Swiss, Shana, and Joan E. Giller. 1993. "Rape as a crime of war: A medical perspective." *Journal of the American Medical Association* 270: 612–615.

Szasz, Thomas. 1970. *The Manufacture of Madness.* New York: Dell.

———. 1974. *The Myth of Mental Illness.* Rev. ed. New York: Harper & Row.

Szasz, Thomas S., and Mark H. Hollander. 1956. "A contribution to the philosophy of medicine." *Archives of Internal Medicine* 97: 585–592.

Taylor, Anna. 1998. "Violations of the international code of marketing of breast milk substitutes: Prevalence in four countries." *British Medical Journal* 316: 1117–1122.

Tesh, Sylvia. 1988. *Hidden Arguments: Political Ideology and Disease Prevention Policy.* New Brunswick, NJ: Rutgers University Press.

Tessler, Richard, and Gail Gamache. 1994. "Continuity of care, residence, and family burden in Ohio." *Milbank Quarterly* 72: 149–169.

Thoits, Peggy A. 1985. "Self-labeling processes in mental illness: The role of emotional deviance." *American Journal of Sociology* 91: 221–249.

Thomas, Stephen B., and Sandra C. Quinn. 1991. "The Tuskegee Syphilis Study, 1932–1972: Implications for HIV education and AIDS risk education programs in the black community." *American Journal of Public Health* 81: 1498–1504.

Tiefer, Leonore. 1994. "The medicalization of impotence: Normalizing phallocentrism." *Gender & Society* 8: 363–377.

Timmermans, Stefan. 1999. *Sudden Death and the Myth of CPR.* Philadelphia: Temple University Press.

Timmermans, Stefan, and Marc Berg. 2003a. *The Gold Standard: The Challenge of Evidence-Based Medicine and Standardization in Health Care.* Philadelphia: Temple University Press.

———. 2003b. "The practice of medical technology." *Sociology of Health and Illness* 25: 97–114.

Tindle, Hilary A., Roger B. Davis, Russell S. Phillips, and David M. Eisenberg. 2005. "Trends in use of complementary and alternative medicine by U.S. adults: 1997–2002." *Alternative Therapies* 11: 42–49.

Tjaden, Patricia, and Nancy Thoennes. 1998 (November). "Prevalence, incidence, and consequences of violence against women: Findings from the national violence against women survey." *National Institute of Justice Research in Brief.*

Toole, Michael J., and Ronald J. Waldman. 1993. "Refugees and displaced persons: War, hunger, and public health." *Journal of the American Medical Association* 270: 600–605.

Tosteson, Daniel C., S. James Adelstein, and Susan T. Carver (eds.). 1994. *New Pathways to Medical Education: Learning to Learn at Harvard Medical School.* Cambridge, MA: Harvard University Press.

Turner, R. Jay, and William R. Avison. 2003. "Status variations in stress exposure: Implications for the interpretation of research on race, socioeconomic status, and gender." *Journal of Health and Social Behavior* 44: 488–505.

Turshen, Meredith. 1989. *The Politics of Public Health.* New Brunswick, NJ: Rutgers University Press.

Tussing, A. Dale, and Martha A. Wojtowycz. 1997. "Malpractice, defensive medicine, and obstetric behavior." *Medical Care* 35: 172–191.

UNAIDS/WHO. 2002. *Report on the Global AIDS/HIV Epidemic: July 2002.* Geneva, Switzerland: World Health Organization.

———. 2004. *AIDS Epidemic Update.* Geneva, Switzerland: World Health Organization.

Underwood, Felix J. 1926. "Development of midwifery in Mississippi." *Southern Medical Journal* 19: 683–685.

U.K. Department of Health. 2005. "Primary Care Trusts." http://www.info.doh.gov.uk/nhsfactsheets.nsf/vwHelp/Primary%20care%20trusts?OpenDocument, accessed July 2005.

U.S. Bureau of the Census. 1975. *Historical Statistics of the United States, Colonial Times to 1970.* Washington, DC: U.S. Government Printing Office.

———. 2004. *Statistical Abstract of the United States, 2004–2005.* Washington, DC: U.S. Government Printing Office.

U.S. Congressional Budget Office. 1995. *The Effect of Managed Care and Managed Competition.* Washington, DC: U.S. Government Printing Office.

U.S. Department of Agriculture. 1999. Personal communication. August.

U.S. Department of Health and Human Services. 1990. *Indian Health Conditions.* Washington, DC: U.S. Government Printing Office.

U.S. Environmental Protection Agency. 1992. *Environmental Equity: Reducing Risk for All Communities.* Washington, DC. U.S. Government Printing Office.

———. 1996. *Environmental Health Threats to Children.* (EPA 175-F-96–001). Washington, DC: U.S. Government Printing Office.

U.S. General Accounting Office. 1991. *Canadian Health Insurance: Lessons for the United States.* Washington, DC: U.S. Government Printing Office.

Valenstein, Elliot S. 1986. *Great and Desperate Cures.* New York: Basic.

Van Olphen-Fehr, Juliana. 1998. *Diary of a Midwife: The Power of Positive Childbearing.* Westport, CT: Bergin & Garvey.

Vastag, Brian. 2001. "Pay attention: Ritalin acts much like cocaine." *Journal of the American Medical Association* 286: 905–906.

Vega, William A., Bohdan Kolody, Sergio Aguilar-Gaxiola, Ethel Alderete, Ralph Catalano, and Jorge Caraveo-Anduaga. 1998. "Lifetime Prevalence of DSM-III-R Psychiatric Disorders Among Urban and Rural Mexican Americans in California." *Archives of General Psychiatry* 55: 771–778.

Vitek, C. R., and M. Wharton. 1998. "Diphtheria in the former Soviet Union: Reemergence of a pandemic disease." *Emerging Infectious Diseases* 4: 539–550.

Vuckovic, Nancy, and Mark Nichter. 1997. "Changing patterns of pharmaceutical practice in the United States." *Social Science and Medicine* 44: 1285–1302.

Waddell, Charles. 1982. "The process of neutralisation and the uncertainties of cystic fibrosis." *Sociology of Health and Illness* 4: 210–220.

Waitzkin, Howard. 1981. "The social origins of illness: A neglected history." *International Journal of Health Services* 11: 77–103.

———. 1991. *The Politics of Medical Encounters.* New Haven, CT: Yale University Press.

———. 1993. *The Second Sickness: Contradictions of Capitalist Health Care.* Rev. ed. New York: Free Press.

Wallerstein, Immanuel M. 1974. *The Modern World-System.* New York: Academic.

Wardwell, Walter I. 1979. "Limited and marginal practitioners." Pp. 230–250 in *Handbook of Medical Sociology,* edited by Howard E. Freeman, Sol Levine, and Leo G. Reeder. Englewood Cliffs, NJ: Prentice Hall.

———. 1988. "Chiropractors: Evolution to acceptance." Pp. 157–191 in *Other Healers: Unorthodox Medicine in America,* edited by Norman Gevitz. Baltimore: Johns Hopkins University Press.

Weil, Andrew, and Winifred Rosen. 1998. *From Chocolate to Morphine.* Rev. ed. New York: Houghton Mifflin.

Weinreb, Linda, Robert Goldberg, Ellen Bassuk, and Jennifer Perloff. 1998. "Determinants of health and service use patterns in homeless and low-income housed children." *Pediatrics* 102: 554–562.

Weissert, William G. 1991. "A new policy agenda for home care." *Health Affairs* 10: 67–77.

Weitz, Rose. 1989. "Uncertainty in the lives of persons with AIDS." *Journal of Health and Social Behavior* 30: 270–281.

———. 1991. *Life with AIDS.* New Brunswick, NJ: Rutgers University Press.

———. 1999. "Watching Brian die: The rhetoric and reality of informed consent." *Health: An Interdisciplinary Journal for the Social Study of Health, Illness and Medicine* 3: 209–227.

Werner, Anne and Kirsti Malterud. 2003. "'It is hard work behaving as a credible patient': Encounters between women with chronic pain and their doctors." *Social Science & Medicine* 57: 1409–1419.

Werth, Barry. 1991. "How short is too short? Marketing human growth hormone." *New York Times Magazine* June 16: 14+.

Wertz, Dorothy C., and John C. Fletcher. 1998. "Ethical and social issues in prenatal sex selection: A survey of geneticists in 37 nations." *Social Science and Medicine* 46: 255–273.

Wertz, Richard, and Dorothy Wertz. 1989. *Lying-In.* New Haven, CT: Yale University Press.

West, Candace. 1984. *Routine Complications: Troubles with Talk between Doctors and Patients.* Bloomington: Indiana University Press.

White, Larry C. 1988. *Merchants of Death: The American Tobacco Industry.* New York: Morrow.

Whitehouse, Peter J., Eric Juengst, Maxwell Mehlman, and Thomas H. Murray. 1997. "Enhancing cognition in the intellectually intact." *Hastings Center Report* 27(3): 14–22.

Whitt, Hugh P., and Richard L. Meile. 1985. "Alignment, magnification, and snowballing: Processes in the definition of 'symptoms of mental illness.'" *Social Forces* 63: 682–697.

WHO/UNICEF Joint Monitoring Programme for Water Supply and Sanitation. 2004. *Meeting the MDG Drinking Water and Sanitation Target: A Midterm Assessment of Progress.* Geneva, Switzerland: World Health Organization.

Wilkinson, Richard G. 1996. *Unhealthy Societies: The Afflictions of Inequality.* London: Routledge.

———. 2005. *The Impact of Inequality.* London: New Press.

Williams, Christine. 1989. *Gender Differences at Work.* Berkeley: University of California.

———. 1992. "The glass escalator: Hidden advantages for men in the 'female' professions." *Social Problems* 39: 253–267.

Williams, David R. 1998. "African-American health: The role of the social environment." *Journal of Urban Health: Bulletin of the New York Academy of Medicine* 75(2): 300–321.

Williams, David R., and Chiquita Collins. 1995. "U. S. socioeconomic and racial differences in health: Patterns and explanations." *Annual Review of Sociology* 21: 349–386.

Williams, David R., and Pamela Braboy Jackson. 2005. "Social sources of racial disparities in health." *Health Affairs* 24: 325–335.

Williams, David R., Yan Yu, James S. Jackson, and Norman B. Anderson. 1997. "Racial differences in physical and mental health: Socio-economic status, stress and discrimination." *Journal of Health Psychology* 2: 335–351.

Williams, Peter C. 1984. "Success in spite of failure: Why IRBs falter in reviewing risks and benefits." *IRB: A Review of Human Subjects Research* 6 (May–June): 1–4.

Wines, Michael. 1999. "Russians drown sorrows, and selves." *New York Times* June 28: A10.

Wingood, Gina M., and Ralph J. DiClemente. 1997. "The effects of an abusive primary partner on the condom use and sexual negotiation practices of African-American women." *American Journal of Public Health* 87: 1016–1018.

Wintemute, Garen J. 1999. "Future of firearm violence prevention: Building on success." *Journal of the American Medical Association* 282: 475–478.

Wirth, Louis. 1985. "The problem of minority groups." Pp. 309–315 in *Theories of Society: Foundations of Modern Sociological Theory,* edited by T. Parsons, E. Shils, K. D. Naegele, and J. R. Pitts. New York: Free Press.

Wise, Jacqui. 1998. "Companies still breaking milk marketing code." *British Medical Journal* 316: 1111.

Wolpe, Paul Root. 1985. "Acupuncture and the American physician." *Social Problems* 32: 409–424.

Woodward, Christel A., and Catherine A. Charles. 2002. "The changing faces of health care in Canada." Pp. 78–96 in *Health Care Reform Around the World,* edited by Andrew C. Twaddle. Westport, CT: Auburn House.

World Bank. 1998. *Assessing Aid: What Works, What Doesn't, and Why.* New York: Oxford University Press.

World Health Organization. 1980. *International Classification of Impairments, Disabilities, and Handicaps.* Geneva, Switzerland: author.

———. 1985. "Appropriate technology for birth." *Lancet* 2(8452): 436–437.

———. 1993. *Infant and Young Child Nutrition.* Geneva, Switzerland: author.

———. 1997. *Fact Sheet No. 153: Female Genital Mutilation.* Geneva, Switzerland: author.

———. 1998a. *Fact Sheet No. 154: Tobacco Epidemic: Health Dimensions.* Geneva, Switzerland: author.

———. 1998b. *World Health Report, 1998.* Geneva, Switzerland: author.

———. 2000a. *Fact Sheet No. 241: Female Genital Mutilation.* Geneva, Switzerland: author.

———. 2000b. *The World Health Report 2000—Health Systems: Improving Performance.* Geneva, Switzerland: author.

———. 2001. *The Optimal Duration of Exclusive Breastfeeding.* April 2: Press note #7.

———. 2005a. "Summary of probable SARS cases with onset of illness from 1 November 2002 to 31 July 2003." www.who.int/csr/sars/country/table2004 _04_21/en/index.html, accessed July 2005.

———. 2005b. *World Health Report, 2005.* Geneva: World Health Organization.

World Health Organization, Child and Adolescent Health. 2005. http://www.who.int/child-adolescent-health/OVERVIEW/CHILD_HEALTH/map_00-03_world.jpg, accessed May 2005.

World Health Organization, Reproductive Health and Research Department. 2004. *Unsafe Abortion: Global and Regional Estimates of the Incidence of Unsafe Abortion and Associated Mortality in 2000.* 4th ed. Geneva, Switzerland: author.

Wren, Christopher S. 1999. "Bid for alcohol in antidrug ads hits resistance." *New York Times* May 31: A1+.

Yago, Glenn. 1984. *The Decline of Transit: Urban Transportation in Germany and U.S. Cities, 1900–1970.* New York: Cambridge University Press.

Yardley, Jim. 2005. "A deadly fever, once defeated, lurks in a Chinese lake." *New York Times* February 22: A2.

Zimmerman, Mary K. 1993. "Caregiving in the welfare state: Mothers' informal health care work in Finland." *Research in the Sociology of Health Care* 10: 193–211.

Ziporyn, Terra D. 1992. *Nameless Diseases.* New Brunswick, NJ: Rutgers University Press.

Zola, Irving K. 1972. "Medicine as an institution of social control." *Sociological Review* 20: 487–504.

———. 1985. "Depictions of disability—Metaphor, message and medium in the media: A research and political agenda." *Social Science and Medicine* 22: 5–17.

Zuger, Abigail. 1999. "Surgeons leaving the O.R. for the office." *New York Times* May 18: D1+.

Zussman, Robert. 1992. *Intensive Care: Medical Ethics and the Medical Profession.* Chicago: University of Chicago Press.

———. 1997. "Sociological perspectives on medical ethics and decision-making." *Annual Review of Sociology* 23: 171–189.

INDEX

Page numbers in *italic* indicate figures or tables. Page numbers in **bold** indicate glossary terms.